	To convert from	to	Multiply by
Moment	dyne-centimeter	newton-meter (N-m)	$1.000\ 000\ (10^{-7})$*
	pound-foot	newton-meter (N-m)	$1.355\ 818$
Power	Btu/hour	watt (W)	$2.930\ 711\ (10^{-1})$
	erg/second	watt (W)	$1.000\ 000\ (10^{-7})$*
	foot-pound/second	watt (W)	$1.355\ 818$
	horsepower (550 ft–lbs)	watt (W)	$7.456\ 999\ (10^{2})$
Pressure and Stress	atmosphere	pascal (Pa)	$1.013\ 25\ (10^{5})$
	bar	pascal (Pa)	$1.000\ 000\ (10^{5})$*
	centimeter of mercury (0°C)	pascal (Pa)	$1.333\ 22\ (10^{3})$
	dyne/centimeter^2	pascal (Pa)	$1.000\ 000\ (10^{-1})$*
	pound/foot^2	pascal (Pa)	$4.788\ 026\ (10^{1})$
	pound/inch^2 (psi)	pascal (Pa)	$6.894\ 757\ (10^{3})$
Speed	foot/second	meter/second (m/s)	$3.048\ 000\ (10^{-1})$*
	inch/second	meter/second (m/s)	$2.540\ 000\ (10^{-2})$*
	kilometer/hour	meter/second (m/s)	$2.777\ 778\ (10^{-1})$
	knot	metef/second (m/s)	$5.144\ 444\ (10^{-1})$
	mile/hour	meter/second (m/s)	$4.470\ 400\ (10^{-1})$*
Volume	foot^3	meter^3 (m^3)	$2.831\ 685\ (10^{-2})$
	gallon (U.S. liquid)	meter^3 (m^3)	$3.785\ 412\ (10^{-3})$
	inch^3	meter^3 (m^3)	$1.638\ 706\ (10^{-5})$
	liter	meter^3 (m^3)	$1.000\ 000\ (10^{-3})$*
	quart (U.S. liquid)	meter^3 (m^3)	$9.463\ 529\ (10^{-4})$

*Denotes an exact quantity.

DYNAMICS

JERRY H. GINSBERG

JOSEPH GENIN

School of Mechanical Engineering
Purdue University

JOHN WILEY & SONS
New York   Santa Barbara   London
Sydney   Toronto

DYNAMICS

This book was printed and bound by Kingsport Press.
It was set in Times Roman by Graphic Arts Composition, Inc.
The copyeditor was Lynne Lackenback.
Kenneth R. Ekkens supervised production.

Text and cover design by Edward A. Butler.

Library of Congress Cataloging in Publication Data:

Ginsberg, Jerry H. 1944-
 Dynamics.

 Includes index.
  1. Dynamics. I. Genin, Joseph, 1932-   joint
author. II. Title.

TA352.G56   620.1'04   76-40409
ISBN 0-471-29606-6

Printed in the United States of America

10 9 8 7 6 5 4 3 2 1

There are two fundamental goals to be attained in the basic engineering mechanics courses of statics and dynamics. The first is to develop an understanding of the physical laws governing the response of engineering systems to forces. The second is to enhance the reasoning powers required in engineering; that is, to develop the ability to solve problems logically, using the concept of mathematical models for physical systems.

To meet these goals this book, like its companion volume on statics, is written in modular form. The modules are broader in extent than the conventional grouping by chapters, and are largely self-contained to minimize the amount of cross-referencing necessary to develop the material. We also chose this approach because we wish the book to communicate directly to the student, with a minimum of amplification and clarification required of the instructor.

To achieve this direct communication it was necessary to develop a consistent approach to problem solving that could be applied to a broad class of problems. This approach is founded on the recognition that there are two distinct aspects to the learning experience in mechanics. Clearly, one must first develop an understanding of the fundamental principles. Only then can a group of these principles be applied selectively to the solution of a broad class of problems.

In this book the comprehension phase is addressed in a conventional manner. After each principle is derived, it is keynoted by remarks regarding common systems that illustrate its implications, as well as critical comments regarding its applicability. This is then followed by one or more solved examples. The examples directly illustrate how to employ the derived principle and, equally important, enhance the understanding of the physical meaning of the derived principles and of the system reponses that result from these principles. In addition to the discussions following the derivations, care is taken in the solved problems to discuss the qualitative aspects of the numerical value of the solution. Hopefully, this will also give the student better insight into the ways engineers think.

The application phase is addressed after a broad body of principles and techniques has been developed. This begins with the presentation of a set of sequential steps detailing the multiple operations necessary for the solution of general problems. These steps are merely a logical sequence to follow. (Certainly, they are not the only possible sequence.) With these steps, where appropriate, we indicate places where common errors are made.

By presenting a series of logical steps that systematize the approach to problem solving, our purpose is to enhance the student's senses of logic and deductive reasoning. Thus, the steps are not intended for memorization. Rather, as a student gains proficiency in problem solving many of the steps will be performed intuitively.

The steps are then employed in a series of four or more solved problems, called illustrative problems, that emphasize the synthesis of the material developed with the appropriate mathematical tools. These illustrative problems are cross-referenced directly to the steps for problem solving, to allow the student to isolate a particular aspect of the solution which may prove troublesome. Homework problems are presented after both examples and illustrative problems; in general, the homework problems following the latter are broader in scope.

In the solution process, emphasis is placed on the logical application of the appropriate principle(s); each solution being implemented with the aid of only that level of mathematics appropriate to solve the problem at hand.

To assist in the development of the methodology, and also to remove a past source of confusion for students, as few alternative methods are presented as are adequate for treating the same system. Furthermore where alternative principles in kinematics or kinetics are given, care is taken to explain their relative assets and, equally important, their relative liabilities.

Physical understanding of the subject matter is enhanced by the organization of topics. After the introductory material, the text is divided into three main groupings that discuss in turn the kinematics and kinetics of particles, rigid bodies in planar motion, and rigid bodies in three-dimensional motion. The kinetics study of each class of systems involves treatments of the equations of motion, followed by those of the principles of work and energy, and of linear and angular impulse and momentum. Since the primary tool in kinetics, because of its wide applicability, is the concept of equations of motion, the steps for problem solving in kinetics are generally presented at that juncture. The later material in the module is then developed consistently with the basic steps. In contrast, the problem-solving steps in kinematics usually appear near the end of the module.

Three special topics, on vibrations, orbital motion, and particle flow were selected to conclude the book. If desired, these application topics may be treated at the end of the appropriate modules as indicated in the Instructor's Manual. Alternatively, they may be completely omitted without sacrifice to the basic level of understanding of dynamics. These three topics were selected because they are not only valuable for motivating students and in developing a high level of technical skill, but also because they are consistent with a first course in engineering dynamics.

It is our belief that the presentation of a systematic approach to the subject of dynamics is one of the major contributions of the book. In particular, the systematization of the methodology for treating the kinematics and kinetics of rigid bodies in three-dimensional motion is entirely new. Our classroom experience with this material at Purdue University has shown that many students who would have otherwise been overwhelmed by the topic are able to comprehend and apply it. A direct result of their comprehension is that they are able to place particle and planar kinetics in proper perspective. Indeed, we hope that the overall presentation in this book will enable the student to gain an appreciation of the orderly philosophy of dynamics and of the viewpoint of dynamicists.

Another feature of the book, beyond the systemization it provides, is the flexibility it affords the instructor. It was written to give the greatest opportunity for adjustments and accommodations in instruction, so as to better serve the range of individual experience. The book is as suitable for courses using innovative instructional approaches, such as self-paced and self-taught courses, as for those using the more conventional methods.

It should be noted that this text does not address the organizational problems involved in self-paced and self-taught courses. However, it is possible for instructors wishing to utilize this approach to employ the book as a framework that can be supplemented by material addressing the problem of the interaction between the student and the instructor. The self-contained aspect of the modules makes the book especially valuable for courses of this kind.

A final technical matter for consideration here is the question of physical units. We strongly believe that the SI system of metric units is the best for engineering. Hopefully, by the time this is read, the SI system will have been adopted universally. However, recognizing the transitory nature of this problem, and noting the large number of physical systems that have already been built according to the U.S.-British system of units, the numerical problems have been divided approximately into a two to one proportion between SI and British units. This ratio allows for a sufficient number of problems for courses using only one system of units.

We gratefully acknowledge the encouragement of our colleagues at Purdue University, whose helpful comments proved invaluable. We also thank the faculty of the École Nationale Supérieure d'Electricité et de Mécanique at Nancy, France; particularly Professor Michel Lucius. By providing an academic environment for Professor Ginsberg during his sabbatical leave, they greatly expedited the completion of this work. We are indebted to Rona Ginsberg for her excellent editorial comments, as well as for typing the manuscript, and to Dr. Joseph W. Klahs for his technical review prior to submission of the manuscript. Finally, we note the knowledgeable assistance of the staff of John Wiley, especially Thurman R. Poston, in all phases of the production process.

JERRY H. GINSBERG
JOSEPH GENIN

CONTENTS

MODULE I
FUNDAMENTAL
CONCEPTS

On a daily basis we encounter physical systems, such as automobiles, elevators, and other types of machinery, in which objects are made to move by the application of forces. To use such objects, it is not necessary to have a detailed understanding of the exact manner in which the systems behave, although in some cases we do have an intuitive understanding of what is taking place. However, as engineers designing a system to perform a specific task, we need considerably more than a qualitative feel for the different ways in which systems respond to forces. In this text on *dynamics* we shall study the quantitative relationships among the responses of many types of physical systems and the forces that act on those systems. The primary new feature of the field of dynamics that distinguishes it from statics is that here systems are always in motion.

A. BASIC DEFINITIONS

The investigation of the dynamic response of a physical system can generally be broken into two parts, one treating the subject of *kinematics,* and the other the subject of *kinetics.* A kinematical study is one that deals with methods for describing the motion of a system without regard to the role of the forces causing the motion. A kinetics study is one where the relationship of the forces to the kinematical variables representing the motion of the system is considered.

A key phrase used above is *physical system.* By it we mean a collection of basic atomic elements that combine to form bodies. In the field of mechanics, for the convenience of problem solving, we tend to categorize these collections of elements into three broad groupings: particles, rigid bodies, and deformable bodies.

PARTICLE   A body whose dimensions are negligible is said to be a particle. It follows, then, that a particle occupies only a single point in space.

RIGID BODY   A body occupying more than one point in space is said to be rigid if all of the constituent elements of matter within the body are always at fixed distances from each other.

DEFORMABLE BODY   A body is said to be deformable if its constituent elements of matter experience changes in their distances from each other that are significant to the problem being investigated.

Clearly there is a certain amount of ambiguity in these definitions. For instance, is the group of molecules that compose a gas a system of particles or a deformable body? Another basic question we could ask is whether any body can correctly be considered to be rigid, because all real materials deform when forces are exerted on them.

There are no absolute answers to these questions, for the *model* of the system we form depends on what knowledge we wish to gain about the response of the system. This leads us to the next topic, which is the modeling process.

B. THE MODELING PROCESS

Our general approach in studying the response of a physical system will be first to construct a conceptual model of the system. That is, we shall consciously consider its components to be either particles, rigid bodies, or deformable bodies. We shall then employ our knowledge of kinematics to describe the motion of the chosen model, and thus of the physical system it represents. The response of the system will then be determined by relating the results of the kinematical study to the force concepts of kinetics. The general approach to the subject of dynamics can be summarized in the conceptual diagram given as Figure 1.

Figure 1 Dynamics study flow chart.

The dotted paths in Figure 1 illustrate a key feature of the modeling process. That is, to be sure of the validity of the model initially chosen to represent the system, it is necessary that the model display any relevant phenomena that are observed experimentally. If not, the model must be improved. Hence, modeling is an educated trial-and-error procedure. That is, the modeling process is, in part, an art based on prior experience and physical intuition. On the other hand, the modeling process is also a science based on a knowledge of the purpose of the physical system and of the available analytical methods to study the phenomena exhibited by the system.

As an example of the foregoing, consider the case of the flight of a golf ball. For a first attempt, one would tend to model the ball as a particle occupying a single point in space. As will be seen in our studies, there are some situations where such an approach results in a successful analysis; that is, the theoretical and experimental results are in good agreement. On the other hand, we shall also see that considering the golf ball to be a particle leaves one unable to explain why the ball hooks, slices, and otherwise deviates from a straight course when the wind is calm. The latter phenomena stem from the aerodynamic forces produced by the spin of the golf ball, so an accurate study of its flight requires that the ball be modeled as a rigid body, as opposed to a simple particle. Carrying these thoughts one step further, if the problem was one of determining how a golf club imparts energy to a ball, we would quickly find that the correct model would be one where the golf ball (and club) must be modeled as deformable bodies. From the foregoing example we see that the model chosen to represent the physical system is a crucial element in the solution process.

The study of the dynamics of deformable bodies is not presented in this book, because the scope of the mathematical tools necessary for this subject far exceeds that required to study the motion of particles and rigid bodies. This would be especially so if our goal was to attain the same level of comprehension and versatility in each subject area. Therefore, only those physical systems that can successfully be modeled as particles and rigid bodies are considered in this first course in dynamics.

C. FOUNDATIONS OF NEWTONIAN MECHANICS

We wish to be able to determine completely the motion of all points in the physical system we are studying. To achieve this goal we must first quantify the concept of motion.

As you probably recall from your previous courses in elementary physics, the three fundamental kinematical quantities used to describe the motion of a point are position, velocity, and acceleration. These basic terms are defined with respect to a frame of reference that describes the three-dimensional space within which the motion occurs. This frame of reference can be visualized as a set of rectangular Cartesian axes.

The laws of motion stated by Sir Isaac Newton (1666–1727) in his historic work, *Principia* (1687), are the foundation of the study of dynamics. To employ these laws we must postulate the existence of an "absolute" (fixed) reference frame. This requirement is in conflict with the concepts of the theory of relativity, which tells us that nothing is fixed in space. Fortunately, it can be shown that the laws of the theory of relativity reduce to the simpler laws of Newtonian mechanics in the special case where bodies are moving much slower than the speed of light. This is certainly the case for most systems of engineering significance. Hence, we shall assume that there is an absolute reference frame, which we shall denote as XYZ. (Throughout the text capital letters will indicate that the coordinate axes are fixed in space.) For most problems we can use the earth to define the XYZ frame. However, to set things in proper perspective, in Module VI, Section E, we shall study some effects of the motion of the earth.

Let us now return to the basic kinematical quantities. The location of a point P within an arbitrarily chosen XYZ reference frame cannot be

Figure 2

specified until we know both the length and orientation of the line from the origin O to point P. For example, let us locate point P in Figure 2. To describe the *position* of point P with respect to point O we employ the position vector $\bar{r}_{P/O}$ (which should be read as r of P with respect to O).

You will note that Figure 2 indicates that as point P moves in space its position vector changes, so that $\bar{r}_{P/O}$ is a function of time. The vector $\Delta\bar{r}_{P/O}$ represents the change in the vector $\bar{r}_{P/O}$ in the time interval from t to $t + \Delta t$. According to the rules for adding vectors diagrammatically, from Figure 2 we may write

$$\bar{r}_{P/O}(t) + \Delta\bar{r}_{P/O} = \bar{r}_{P/O}(t + \Delta t)$$

so that

$$\Delta\bar{r}_{P/O} = \bar{r}_{P/O}(t + \Delta t) - \bar{r}_{P/O}(t)$$

When we divide $\Delta\bar{r}_{P/O}$ by Δt, we obtain a quantity that resembles the time derivative of a scalar function $f(t)$. Recall from basic calculus that for a scalar function

$$\frac{df}{dt} \equiv \lim_{\Delta t \to 0} \frac{\Delta f}{\Delta t} = \lim_{\Delta t \to 0} \frac{f(t + \Delta t) - f(t)}{\Delta t}$$

A similar definition is used for differentiation of a vector function, where the time rate of change of the position vector $\bar{r}_{P/O}$ is called the *velocity* $\bar{v}_P$. This is the second of the three basic kinematical quantities. Thus

$$\boxed{\begin{aligned} \bar{v}_P &\equiv \frac{d}{dt}\bar{r}_{P/O} \equiv \lim_{\Delta t \to 0} \frac{\Delta\bar{r}_{P/O}}{\Delta t} \\ &= \lim_{\Delta t \to 0} \frac{\bar{r}_{P/O}(t + \Delta t) - \bar{r}_{P/O}(t)}{\Delta t} \end{aligned}} \qquad (1)$$

In denoting the velocity as $\bar{v}_P$ we eliminated any indication that this quantity is defined with respect to a fixed point O. This is done to facilitate notation, and equally important, because we shall see later that the velocity of a point with respect to a fixed reference frame is independent of the location of the origin O of that frame.

Recalling that a vector describes a quantity having magnitude and direction, let us examine the velocity vector $\bar{v}_P$. The magnitude of this vector is defined as the *speed* v_P: $v_P = |\bar{v}_P|$. The speed expresses a rate of motion (e.g., 50 km/hr) and is a scalar quantity. When we are requested to determine how fast a point is moving, it is synonymous to being asked to find the speed. The direction of $\bar{v}_P$ tells us in which direction the point P is traveling. We shall have more to say about speed and velocity in the next module.

Now that we have the mathematical definition for velocity, equation (1), we may follow a similar approach in defining *acceleration,* which is the third of the basic kinematical quantities. The acceleration $\bar{a}_P$ is the time rate of change of the velocity, thus we have

$$\bar{a}_P \equiv \frac{d}{dt}\bar{v}_P \equiv \frac{d^2}{dt^2}\bar{r}_{P/O} \tag{2}$$

The kinematical quantities of position and velocity are pieces of information that our eyes and brain are attuned to sense. Were it not for Newton's laws of motion, it would not be obvious that we should also be interested in acceleration. Because these laws are the basis for our future studies we restate them here, in modern language.

FIRST LAW   A particle will remain at rest or move with constant speed along a straight line, unless it is acted upon by a resultant force.

SECOND LAW   When a resultant force is exerted on a particle, the acceleration of that particle is parallel to the direction of the force and the magnitude of the acceleration is proportional to the magnitude of the force.

THIRD LAW   Each force exerted upon a body is the result of an interaction with another body, the forces of action and reaction being equal in magnitude, opposite in direction, and collinear.

In our studies we shall find that the first law is included in the second; it merely emphasizes that the state of motion of a particle can be changed only by the action of forces. The third law helps to define what forces are acting on a body.

The first and third laws are the only ones needed to solve problems in statics. The added feature in dynamics is the application of the second law. This law may be reworded to state that the force $\bar{F}$ acting on a particle is proportional to the acceleration $\bar{a}_P$ of the particle. As you probably know, the constant of proportionality in this relation is the mass m of the particle. Thus, we have the familiar equation

$$\bar{F} = m\bar{a}_P \tag{3}$$

Before we can use this equation to solve kinetics problems, we must be sure that the units chosen to describe each term are mutually consistent. In the next section we consider this topic.

D. SYSTEMS OF UNITS

The quantities appearing in Newton's second law, $\bar{F} = m\bar{a}$, involve measurements of length, time, force, and mass, which we denote for brevity as $L, T, F,$ and $M,$ respectively. The units chosen for the measurement of these four dimensional quantities cannot be defined independently. They must obey the *law of dimensional homogeneity,* which requires that the physical units of all terms in an equation be identical. Stated in colloquial language, "apples cannot be equated to oranges."

In terms of the four basic measurements, the dimensions of acceleration are L/T^2. Dimensional homogeneity of equation (3) requires that

$$F = M\frac{L}{T^2}$$

In other words, we are free to choose the units of three of the four quantities, $L, T, F,$ and $M.$ The fourth must be derived. Historically, the units by which length and time are measured are well defined, and, through the efforts of international standards committees, universally accepted.

This leaves us with the choice of defining either a standard unit of mass or of force, and then deriving the units for the undefined quantity. When the unit of mass is defined according to some standard, the system is called *absolute,* whereas a system in which the unit of force is defined is called *gravitational.* This terminology originates from the relationship between the weight W and the acceleration g of a particle that is falling freely in a vacuum at the surface of the earth. According to Newton's second law this relationship is

$$W = mg \tag{4}$$

In using equation (4) we must be aware that the values of W and g depend on the location of the particle with respect to the surface of the earth, and equally important on the choice of measurement system. Presently, the two systems that are in wide usage are the metric absolute system and the British gravitational system. In terms of the units of these systems, for most problems we shall use the value $g = 9.806$ m/s^2 (metric) or $g = 32.17$ ft/s^2 (British) as average values. These values will be looked at in considerable detail in Example 5, Module VI.

In Table 1 we summarize the commonly used systems of units and indicate how the derived units are obtained. The basic units in each system are defined with respect to standard bodies or with respect to physical phenomena. For example, a kilogram is the mass of a particular bar of platinum alloy. On the other hand, a meter is defined as 1,650,763.73 wavelengths in a vacuum of the orange-red line of krypton-86. The derived quantities in the table are the starred values.

Table 1 System of Units

UNIT/SYSTEM	LENGTH	TIME	MASS	FORCE
Metric absolute	Meter (m)	Second (s)	Kilogram (kg)	Newton (N)* = kg-m/s²
Metric gravitational	Meter (m)	Second (s)	Metric slug* = kg-s²/m	Kilogram (kg)
British absolute	Foot (ft)	Second (s)	Pound (lb)	Poundal (pdl)* = lb-ft/s²
British gravitational	Foot (ft)	Second (s)	Slug* = lb-s²/ft	Pound (lb)

The choice for systems of units presented in Table 1 is further complicated by the fact that certain multiples and submultiples of the basic units in each system are given their own names, such as the mile, which is 5280 ft, or the dyne, which is one hundred-thousandth of a newton. To simplify this situation the SI (Standard International) system is now being adopted worldwide.

Essentially, the SI system is the metric absolute system with only one unit being used to describe each type of quantity. The SI units for length, time, mass, and force are meters, seconds, kilograms, and newtons, respectively. Then, if one desires, decimal multiples and submultiples of the basic units may be indicated by the appropriate prefixes. Thus, in the SI system, length dimensions of 0.2 meters and 200 millimeters are identical and correct. On the other hand, $2(10^{-3})$ newtons, or equivalently 2 millinewtons, is the correct description for a force whose magnitude is 200 dynes. Lists of the SI units and of the preferred prefixes may be found in Appendix A.

Many of the examples and problems in this text will be presented in SI units. Unfortunately, this system is not yet universally used in the United States, where many engineers, as well as nontechnical people, still use the British gravitational system. We shall therefore present some examples and problems that are posed in both systems of units. In general, we shall solve problems in the same system of units as that used for the given information, and we shall not convert between different systems of units. Obviously, the units of any physical quantity can be changed with the aid of conversion factors. These conversion factors are numbers that are physically unity, but not numerically. For example, 25.40 mm = 1 in. A brief set of factors is given in Appendix A.

Conversion factors can be used consecutively. For instance, to change from a speed in miles per hour to a speed in meters per second, we can compute

$$1 \frac{\text{mile}}{\text{hr}} \left(\frac{1 \text{ hr}}{3600 \text{ s}} \right) \left(\frac{5280 \text{ ft}}{1 \text{ mile}} \right) \left(\frac{12 \text{ in.}}{1 \text{ ft}} \right) \left(\frac{25.40 \text{ mm}}{\text{in.}} \right) \left(\frac{1 \text{ m}}{1000 \text{ mm}} \right) = 0.4470 \text{ m/s}$$

Incidentally, a conversion factor that is contained in the preceding equation is that 1 mile/hr = (88/60) ft/s. We shall use this factor frequently in solving problems having data in the British system of units. Remember, the units of all quantities in an equation must be consistent in order to satisfy the law of dimensional homogeneity.

As practice in obtaining a consistent set of units, and also to gain an intuitive understanding of the magnitude of British units relative to SI units, we present the following example.

EXAMPLE 1

Using the facts that 1 in. equals 25.40 mm and that a body weighing 1 lb on the surface of the earth has a mass of 0.4536 kg, determine the factors for converting the British gravitational units of pounds force, slugs, mass, and density (slugs/ft^3) into their SI equivalents.

Solution

The conversion factors between pounds and newtons is determined by using $W = mg$. Let us compute the weight in newtons of a 0.4536-kg body. Thus, not knowing where the body was weighed, we use the average value $g = 9.806$ m/s^2 to obtain

$$W = (0.4536 \text{ kg})(9.806 \text{ m/s}^2) = 4.448 \text{ N}$$

In other words,

$$1 \text{ lb} = 4.448 \text{ N}$$

Then, because 1 lb-s^2/ft is 1 slug, we have

$$1 \text{ slug} = \left(1\frac{\text{lb-s}^2}{\text{ft}}\right)\left(\frac{4.448 \text{ N}}{1 \text{ lb}}\right)\left(\frac{1 \text{ ft}}{12 \text{ in.}}\right)\left(\frac{1 \text{ in.}}{25.40 \text{ mm}}\right)\left(\frac{1000 \text{ mm}}{1 \text{ m}}\right)$$

$$= 14.593 \frac{\text{N-s}^2}{\text{m}} = 14.593 \text{ kg}$$

To convert the units of density, we use

$$1\frac{\text{slug}}{\text{ft}^3} = \left(1\frac{\text{slug}}{\text{ft}^3}\right)\left(\frac{14.593 \text{ kg}}{\text{slug}}\right)\left(\frac{1 \text{ ft}}{12 \text{ in.}}\right)^3\left[\left(\frac{1 \text{ in.}}{25.40 \text{ mm}}\right)\left(\frac{1000 \text{ mm}}{1 \text{ m}}\right)\right]^3$$

$$= 515.3 \text{ kg/m}^3$$

The differences between the values obtained here and those in the table in Appendix A are attributable to the fact that we were only using four significant figures for our calculations, which is the maximum allowed by our choice for the value of g.

HOMEWORK PROBLEMS

In the following problems, t is time in seconds, x is distance, v is speed, a is the magnitude of acceleration, and m is mass.

I.1 The work done by a force is Fx and the power developed by this force is Fv. In the SI system the basic unit of work and energy is joules (J), a joule being the work done by a 1-N force in moving a body a distance of 1 m. The basic SI unit for power is a watt (W), a watt being the power developed by a 1-N force when it moves a body at a speed of 1 m/s. Determine what (a) 1 J is in terms of meters, kilograms, and seconds, (b) 1 W is in terms of meters, kilograms, and seconds, and (c) 1 W is in terms of joules and seconds.

I.2 The power developed by a force is Fv. In the SI system the basic unit of power is a watt (W), a watt being the power developed by a 1-N force when it moves a body at a speed of 1 m/s. In the British gravitational system the basic unit of power is a horsepower, a horsepower being the power developed by a 550-lb force when it moves a body at a speed of 1 ft/s. Determine (a) how many kilowatts there are in a horsepower, and (b) how many horsepower there are in a kilowatt.

I.3 Derive conversion factors for changing the following British gravitational units to their SI equivalents. (a) area: square inch; (b) volume: cubic foot; (c) force: ounce; (d) pressure: pounds per square inch; (e) pressure: pounds per square foot; (f) speed: foot per second; (g) speed: inches per hour; (h) acceleration: miles per hour squared.

I.4 In each of the following formulas c_1, c_2, etc., denote constants and θ is an angle in radians. Determine the units of these constants if the formula is to be dimensionally correct. (a) $a = c_1 v^2/x$, (b) $\frac{1}{2}mv^2 = \frac{1}{2}c_1 x^2$, (c) $x = c_1 x_0 + c_2 v_0 + c_3 t^2$, (d) θ (degrees) $= c_1 \theta$ (radians), (e) $d\theta/dt = c_1 + c_2 t$, (f) $t = c_1\sqrt{x}$, (g) $mc_1{}^2\, d^2\theta/dt^2 = Fx$.

I.5 At a certain large university known for its athletic prowess, the following new set of physical units are in use. The basic length unit is the *touchdown,* which is 100 yd, the basic time unit is the *dash,* which is 9 s, and the basic force unit is the *golf ball,* which is 1.620 oz. Determine the conversion factors between these units and their SI equivalents.

I.6 At the university in Problem I.5 the unit of mass is the *basket,* one basket being one golf ball-dash²/touchdown. Determine the number of kilograms in one basket.

I.7 A well-known fluid mechanics equation for the pressure drop along a pipe of length l and diameter D is

$$P_1 - P_2 = f\frac{l}{D}\; \rho\, \frac{v^2}{2}$$

where P denotes pressure, ρ is the density of the fluid, v is its velocity, and f is the resistance coefficient, which is given by the expression

$$\sqrt{f} = 2.30 \log_{10}\left(\frac{Dv}{\nu}\sqrt{f}\right) - 0.91$$

Determine the dimensions of ν, the kinematic viscosity.

I.8   A restaurant buys 1000 lb of coffee at a standard locality and also a spring scale calibrated at that locality. Determine what the coffee weighs on the spring scale in a locality where (a) $g = 32.11$ ft/s², (b) $g = 9.791$ m/s².

E.   VECTORS

In establishing the definitions for the kinematical variables of position, velocity, and acceleration, we saw that a basic tool in the study of dynamics is the description of vector quantities. As was true in the subject of statics, we shall frequently need to perform the algebraic operations of addition, subtraction, and multiplications. Equally important, in view of the relationship of the kinematical variables, we shall also be required to differentiate vectors. In what follows we shall briefly review some fundamental operations and results of vector algebra and then study some basic theorems of vector calculus.

1   Vector Algebra

A vector is any quantity that is fully specified when its magnitude and direction are known. Graphically, to depict any vector $\bar{A}$ in a diagram we draw an arrow, as we did in Figure 2 to illustrate the position vector $\bar{r}_{P/O}$. The length of this arrow is proportional to the magnitude $|\bar{A}|$ of the vector, and the orientation of the arrow shows its direction.

Multiplication by a Scalar   If q is a scalar number, then the product $q\bar{A}$ is defined to have the property that

$$|q\bar{A}| = |q||\bar{A}|$$

The resulting vector $q\bar{A}$ is parallel to $\bar{A}$, in the same sense as $\bar{A}$ if q is positive, and in the opposite sense if q is negative.

The concept of a unit vector is an important application of the multiplication by a scalar. Suppose that we multiply $\bar{A}$ by $(1/|\bar{A}|)$; the result of this operation is a vector $\bar{e}_A$, parallel to $\bar{A}$. Because the magnitude of $\bar{e}_A$ is the dimensionless number one, we call $\bar{e}_A$ a unit vector. Thus

$$\bar{e}_A = \frac{\bar{A}}{|\bar{A}|} \tag{5}$$

As you may recall from statics, this equation proves to be very useful for representing vector quantities.

Figure 3

Addition of Vectors   The basic rule for adding any two vectors $\bar{A}$ and $\bar{B}$ is the parallelogram law, according to which the tails of the two vectors are made to coincide. The resultant, $\bar{R} = \bar{A} + \bar{B} = \bar{B} + \bar{A}$, is then obtained from the parallelogram formed by these vectors, as shown in Figure 3. This figure also illustrates an equivalence rule for adding vectors, specifically, that of forming a triangle by placing the tail of one vector at the head of the other and obtaining the resultant as the third side of the triangle.

The difference of two vectors can be obtained as a corollary of the addition rule. This is so because

$$\bar{A} - \bar{B} = \bar{A} + (-\bar{B})$$

where, according to the law for multiplication by a scalar, $-\bar{B} \equiv (-1)\bar{B}$ is a vector having the same magnitude as $\bar{B}$, but opposite direction.

Figure 4

Dot Product of Vectors   The product $\bar{A} \cdot \bar{B}$ is defined to be the product of the magnitude of one vector with the component of the other vector in the direction of the first. From Figure 4 we see that this definition can be written mathematically as

$$\bar{A} \cdot \bar{B} = \bar{B} \cdot \bar{A} \equiv |\bar{A}||\bar{B}| \cos \theta$$

where θ is the angle between the vectors $\bar{A}$ and $\bar{B}$. Notice that the dot product will be negative if $\theta > 90°$, which means that the projection of $\bar{A}$ onto $\bar{B}$ is opposite the sense of $\bar{B}$, or vice versa. Also, from this definition, $\bar{A} \cdot \bar{A} = |\bar{A}|^2$.

Cross Product of Vectors   The definition of the cross product $\bar{A} \times \bar{B}$ is based on the fact that two intersecting vectors $\bar{A}$ and $\bar{B}$ define a flat plane. The cross product is defined to be a vector normal (perpendicular) to the plane of the two vectors. To specify the sense of this normal we use the "right-hand" rule, in accordance with which we curl the fingers of our right hand from the first vector to the second and define the sense of the result to be in the direction of the thumb, as shown in Figure 5.

Figure 5

The definition of the magnitude of the cross product is

$$|\bar{A} \times \bar{B}| \equiv |\bar{A}| |\bar{B}| \sin \theta$$

From Figure 5 it can be seen that the definition involves the component of one vector normal to the other vector.

An important property of the cross product, illustrated in Figure 5, is that

$$\bar{B} \times \bar{A} = -\bar{A} \times \bar{B}$$

This is so because according to the right-hand rule, both products have the same magnitude but opposite direction.

Theorems of Vector Algebra The definition of the basic operations leads to rules for combining these operations that resemble those of scalar algebra. Thus

$$q(\bar{A} + \bar{B}) = q\bar{A} + q\bar{B} \; ; \; q(\bar{A} \cdot \bar{B}) = (q\bar{A}) \cdot \bar{B} = \bar{A} \cdot (q\bar{B})$$

$$q(\bar{A} \times \bar{B}) = (q\bar{A}) \times \bar{B} = \bar{A} \times (q\bar{B})$$

$$\bar{C} \cdot (\bar{A} + \bar{B}) = (\bar{C} \cdot \bar{A}) + (\bar{C} \cdot \bar{B})$$

$$\bar{C} \times (\bar{A} + \bar{B}) = \bar{C} \times \bar{A} + \bar{C} \times \bar{B}$$

Rectangular Components So far we have described vectors diagrammatically by using arrows, an approach that has two serious shortcomings. First, three-dimensional vectors are difficult to represent in a diagram. Second, even for two-dimensional vectors, the addition of vectors according to the parallelogram law requires extensive use of the trigonometry of scalene triangles. To circumvent these difficulties, throughout the text we shall employ the alternative approach of representing vectors by their components with respect to a set of orthogonal (mutually perpendicular) directions.

To describe the components of a vector $\bar{A}$ with respect to a rectangular Cartesian coordinate system XYZ, we place the origin of the coordinate axes at the tail of the vector. We then drop perpendiculars (indicated by dashed lines in Figure 6) from the tip of the vector onto the three coordinate axes. From the intersections of these perpendiculars with the axes, we form the rectangular parallelipiped depicted in the figure. The lengths A_X, A_Y, and A_Z are the components of the vector $\bar{A}$ with respect to the XYZ coordinate system.

In terms of the angles α, β, γ between the vector $\bar{A}$ and the respective coordinate axes, we have

Figure 6

$$A_X = |\bar{A}| \cos \alpha \qquad A_Y = |\bar{A}| \cos \beta \qquad A_Z = |\bar{A}| \cos \gamma \qquad (6)$$

The angles α, β, γ are called the *direction angles;* the cosines of the direction angles are the *direction cosines.* Because $\bar{A}$ is a diagonal for the parallelipiped, it follows that

$$|\bar{A}| = \sqrt{A_X{}^2 + A_Y{}^2 + A_Z{}^2} \tag{7}$$

Substituting equations (6) into equation (7) yields

$$\cos^2 \alpha + \cos^2 \beta + \cos^2 \gamma = 1$$

which shows that only two of the direction angles are independent quantities.

From the foregoing we see that the components of a vector completely describe the properties of a vector, provided that we know which coordinate system is associated with the components. To specify the coordinate system we use the unit vectors $\bar{I}$, $\bar{J}$, and $\bar{K}$, which are oriented in the directions of the positive X, Y, and Z axes, respectively. Then, for example, equation (5) shows that $A_X\bar{I}$ is a vector parallel to the X axis whose magnitude is the component A_X. Figure 7, which illustrates the three component vectors, also shows that the vector sum of these components is $\bar{A}$. Hence

$$\boxed{\bar{A} = A_X\bar{I} + A_Y\bar{J} + A_Z\bar{K}} \tag{8}$$

This is the vector component representation of $\bar{A}$.

An important result of representing vectors by their components is that the equality of vectors takes on a simpler description. Two vectors can be equal only when their components are equal. In other words,

$$\bar{A} = \bar{B} \quad \text{means} \quad A_X = B_X, \quad A_Y = B_Y, \quad A_Z = B_Z \tag{9}$$

Equation (9) is particularly useful for our studies, as it shows that a vector equation can be subdivided into the three scalar equations obtained by equating corresponding components from each side of the equation.

In general, the various vector operations are more easily performed in terms of components than from their basic definitions. To obtain the unit vector $\bar{e}_A$, we substitute equations (7) and (8) into equation (5), with the result that

$$\bar{e}_A = \frac{\bar{A}}{|\bar{A}|} = \frac{A_X\bar{I} + A_Y\bar{J} + A_Z\bar{K}}{\sqrt{A_X{}^2 + A_Y{}^2 + A_Z{}^2}} \tag{10a}$$

For an alternative form of $\bar{e}_A$, we can employ equation (6), with the result that

$$\bar{e}_A = \cos \alpha \bar{I} + \cos \beta \bar{J} + \cos \gamma \bar{K} \tag{10b}$$

Figure 7

The computation of the sum or difference of two vectors is also straightforward, for

$$\bar{A} \pm \bar{B} = (A_X \pm B_X)\bar{I} + (A_Y \pm B_Y)\bar{J} + (A_Z \pm B_Z)\bar{K} \qquad (11)$$

By induction, equation (11) can be extended to treat more than two vectors.

The scalar and vector products are obtained by first considering the various products of the unit vectors, using the definitions of these operations. For example, for the scalar product

$$\bar{I} \cdot \bar{I} = |\bar{I}||\bar{I}| \cos 0 = 1$$

Hence, it follows that

$$\bar{I} \cdot \bar{I} = \bar{J} \cdot \bar{J} = \bar{K} \cdot \bar{K} = 1$$

$$\bar{I} \cdot \bar{J} = \bar{J} \cdot \bar{I} = \bar{J} \cdot \bar{K} = \bar{K} \cdot \bar{J} = \bar{K} \cdot \bar{I}$$

$$= \bar{I} \cdot \bar{K} = 0 \qquad (12)$$

whereas, for the vector product,

$$|\bar{I} \times \bar{I}| = |\bar{I}||\bar{I}| \sin 0 = 0$$

Thus,

$$\bar{I} \times \bar{I} = \bar{J} \times \bar{J} = \bar{K} \times \bar{K} = \bar{0}$$

$$\bar{I} \times \bar{J} = -\bar{J} \times \bar{I} = \bar{K}$$

$$\bar{J} \times \bar{K} = -\bar{K} \times \bar{J} = \bar{I}$$

$$\bar{K} \times \bar{I} = -\bar{I} \times \bar{K} = \bar{J} \qquad (13)$$

As an aid in obtaining the resulting direction and sign for the nonzero cross products in equation (13), we can consider the cross product of two unit vectors to be positive if the operation coincides with the clockwise order shown in Figure 8 and negative if the operation coincides with the counterclockwise order depicted there.

The products of two vectors then follows from the rules of vector algebra. Thus the scalar product is

$$\bar{A} \cdot \bar{B} = (A_X\bar{I} + A_Y\bar{J} + A_Z\bar{K}) \cdot \bar{B}$$

$$= A_X[\bar{I} \cdot (B_X\bar{I} + B_Y\bar{J} + B_Z\bar{K})]$$

$$+ A_Y[\bar{J} \cdot (B_X\bar{I} + B_Y\bar{J} + B_Z\bar{K})]$$

$$+ A_Z[\bar{K} \cdot (B_X\bar{I} + B_Y\bar{J} + B_Z\bar{K})]$$

$$= A_XB_X + A_YB_Y + A_ZB_Z \qquad (14)$$

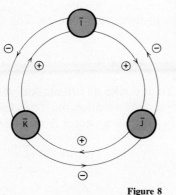

Figure 8

A similar procedure for the vector product gives

$$\bar{A} \times \bar{B} = (A_X\bar{I} + A_Y\bar{J} + A_Z\bar{K}) \times \bar{B}$$

$$= A_X[\bar{I} \times (B_X\bar{I} + B_Y\bar{J} + B_Z\bar{K}]$$

$$+ A_Y[\bar{J} \times (B_X\bar{I} + B_Y\bar{J} + B_Z\bar{K})]$$

$$+ A_Z[\bar{K} \times (B_X\bar{I} + B_Y\bar{J} + B_Z\bar{K})]$$

$$= A_X(B_Y\bar{K} - B_Z\bar{J}) + A_Y(-B_X\bar{K} + B_Z\bar{I})$$

$$+ A_Z(B_X\bar{J} - B_Y\bar{I})$$

$$= (A_YB_Z - A_ZB_Y)\bar{I} + (A_ZB_X - A_XB_Z)\bar{J}$$

$$+ (A_XB_Y - A_YB_X)\bar{K} \tag{15a}$$

An alternative way of computing the cross product, which involves the evaluation of a determinant, is a convenient device for expediting calculations. This method gives

$$\bar{A} \times \bar{B} = \begin{vmatrix} \bar{I} & \bar{J} & \bar{K} \\ A_X & A_Y & A_Z \\ B_X & B_Y & B_Z \end{vmatrix} \tag{15b}$$

To illustrate the various vector operations, let us now solve a problem that reviews the basic operations of statics.

EXAMPLE 2

The crate shown in the sketch has a mass of 50 kg and its center of mass coincides with the geometric centroid of the volume. A force $\bar{F}$ of magnitude 650 N parallel to the diagonal line BC pushes the crate at point C. Determine (a) the force-couple system acting at point A that is equivalent to the combined effect of gravity and $\bar{F}$; (b) the moment of these forces about the edge AE.

Solution

We begin by writing the forces in terms of their components. The weight force mg is downward, and we have

$$\bar{W} = -50(9.806)\bar{K} = -490.3\bar{K} \text{ N}$$

The force $\bar{F}$ has a magnitude of 650 N. To describe its orientation we use a unit vector $\bar{e}_F$, which according to the given information must be along the line from point C to point B. Because $\bar{r}_{B/C}$ is also a vector from point C to point B, equation (5) gives

$$\bar{F} = 650\bar{e}_F = 650 \frac{\bar{r}_{B/C}}{|\bar{r}_{B/C}|}$$

The components of the vector from point C to point B are 1.2 m in the negative X direction, 0.4 m in the positive Y direction, and 0.3 m in the negative Z direction, so that

$$\bar{r}_{B/C} = -1.2\bar{I} + 0.4\bar{J} - 0.3\bar{K}$$

$$\bar{F} = 650\left[\frac{-1.2\bar{I} + 0.4\bar{J} - 0.3\bar{K}}{\sqrt{(1.2)^2 + (0.4)^2 + (0.3)^2}}\right]$$

$$= 500(-1.2\bar{I} + 0.4\bar{J} - 0.3\bar{K})$$

$$= -600\bar{I} + 200\bar{J} - 150\bar{K} \text{ N}$$

Recalling a theorem from statics, an equivalent force system acting at any point consists of a force $\bar{R}$, which is the sum of the forces in the system, and a couple $\bar{M}$, which is the moment of the force system about the chosen point. Hence

$$\bar{R} = \bar{F} + \bar{W} = -600\bar{I} + 200\bar{J} - 150\bar{K} - 490.3\bar{K}$$

$$= -600\bar{I} + 200\bar{J} - 640.3\bar{K} \text{ N}$$

The moment of a force about a point is the cross product of the position vector from that point to any point on the line of action of the force with the force. In that we want a force-couple system at point A, we compute moments about this point. Denoting the center of mass as G, we therefore have

$$\bar{M}_A = \bar{r}_{C/A} \times \bar{F} + \bar{r}_{G/A} \times \bar{W}$$

It is given that point G is at the center of the crate, so its coordinates are (0.6, 0.2, 0.15) m. Hence

$$\bar{M}_A = (1.2\bar{I} + 0.3\bar{K}) \times (-600\bar{I} + 200\bar{J} - 150\bar{K})$$

$$+ (0.6\bar{I} + 0.2\bar{J} + 0.15\bar{K}) \times (-490.3\bar{K})$$

$$= 1.2(200)\bar{K} + 1.2(-150)(-\bar{J}) + 0.3(-600)\bar{J}$$

$$+ 0.3(200)(-\bar{I}) + 0.6(-490.3)(-\bar{J}) + 0.2(-490.3)\bar{I}$$

$$= -158.1\bar{I} + 294.2\bar{J} + 240\bar{K} \text{ N-m}$$

To solve the last portion of the problem, recall that the moment about an axis is the component parallel to that axis of the moment about any point on the axis. In this problem point A is on the axis defined by edge AE, so we need the component of $\bar{M}_A$ parallel to this edge. Such a component is obtained by calculating the scalar product of $\bar{M}$ and a *unit* vector parallel to line AE, which in this case is $\bar{I}$, so

$$\bar{M}_{AE} = (\bar{M}_A \cdot \bar{I})\bar{I} = -158.1\bar{I} \text{ N-m}$$

HOMEWORK PROBLEMS

I.9 A force $\bar{F}$ of magnitude 200 N is applied at point A. Determine the moment of $\bar{F}$ about point O.

I.10 A force $\bar{R}$ of magnitude 300 N is applied at point B. Determine the moment of $\bar{R}$ about point C.

Problems I.9 and I.10

Prob. I.11

I.11 Each of the cables supporting the 500-kg flat plate has a tension of 5000 N. The weight force acts at the centroid of the plate. Determine the force-couple system acting at point D that is equivalent to the forces exerted by gravity and the supporting cables for the plate.

I.12 Three cables are attached to the vertical post AB, as shown, in order to support it against the 1-kN force $\bar{F}$. Determine the tension force in each of the cables if the tension in cable EB is 2 kN and the resultant moment about point A is zero.

I.13 Determine the three vertex angles θ_A, θ_B, and θ_C for the triangle ABC shown. Also determine the unit vector normal to the plane of the triangle that is oriented outward from the origin.

I.14 A 1000-N force parallel to the XY plane is applied at point D along edge AB of the triangle, as shown. Resolve this force into two force vectors, one perpendicular to the plane of triangle ABC, the other parallel to that plane.

I.15 For the two forces acting on the disk, determine the equivalent force-couple system acting about (a) point O, (b) point C.

I.16 A cable is wrapped around pulleys A and B on a block that slides on the horizontal guide bar. The tension in the cable is 50 lb. Determine (a)

Prob. I.12

Probs. I.13 and I.14

100 N

$30°$

$30°$

$45°$

200 mm

O

200 N

Prob. I.15

the components of the resultant force exerted by the cable on the pulley block assembly in directions parallel and perpendicular to the guide bar, (b) the component of the resultant cable force parallel to the line connecting the centers of the two pulleys, and (c) the resultant moment of the cable forces about the center of pulley A.

50 lb

9 in.

$R = 3$ in.

B

$R = 3$ in.

2 in.

2 in.

$30°$

50 lb

Prob. I.16

2   Vector Calculus

In dynamics, derivatives with respect to time are of primary importance, so we shall use t as the independent variable in the following discussion. Letting $\bar{A}(t)$ be any vector whose value depends on t, the derivative is

$$\frac{d\bar{A}}{dt} = \lim_{\Delta t \to 0} \frac{\bar{A}(t + \Delta t) - \bar{A}(t)}{\Delta t} \tag{16}$$

The following example is intended to give you a basic appreciation of the significance of equation (16), and, equally important, to emphasize the meaning of the basic concept of velocity.

EXAMPLE 3

A radar station tracks an airplane by measuring the distance R to the airplane and the angle of elevation θ. At a certain instant an airplane is sighted at $R = 4100$ m, $\theta = 33.7°$. After an interval of 0.75 s the airplane is at $R = 4240$ m and $\theta = 29.3°$. What is the approximate speed of the airplane and the approximate angle α at which it is diving or climbing?

R

α

θ

v_P

Solution

Both v_P and α can be found once the velocity is known. Because the functions defining R and θ are unknown, simple differentiation cannot be performed. Noting that the given values of R and θ can be used to locate $\bar{r}_{P/O}(t)$ and $\bar{r}_{P/O}(t + \Delta t)$, we can form an approximation to the derivative for the velocity. Thus

$$\bar{v}_P \simeq \frac{\bar{r}_{P/O}(t + \Delta t) - \bar{r}_{P/O}(t)}{\Delta t}$$

where, in this problem, $\Delta t = 0.75$ s.

For general values of R and θ the vector $\bar{r}_{P/O}$ going from the radar station to the airplane is shown in the diagram to be

$$\bar{r}_{P/O} = R(\cos \theta \bar{I} + \sin \theta \bar{J})$$

Thus

$$\bar{r}_{P/O}(t) = 4100 \, (\cos 33.7° \bar{I} + \sin 33.7° \bar{J})$$

$$= 3411 \bar{I} + 2275 \bar{J} \text{ m}$$

$$\bar{r}_{P/O}(t + \Delta t) = 4240 \, (\cos 29.3° \bar{I} + \sin 29.3° \bar{J})$$

$$= 3698 \bar{I} + 2075 \bar{J} \text{ m}$$

$$\bar{v}_P = \frac{1}{0.75} \, [(3698 - 3411)\bar{I} + (2075 - 2275)\bar{J}]$$

$$= 383 \bar{I} - 267 \bar{J}$$

Then from the following sketch we find the magnitude and direction of $\bar{v}_P$ to be

$$v_P = |\bar{v}_P| = \sqrt{(383)^2 + (267)^2}$$

$$= 466 \text{ m/s}$$

$$\alpha = \tan^{-1} \frac{267}{383} = 34.9°$$

Thus the speed of the airplane is 467 m/s and it is diving at an angle of 34.9° from the horizontal.

The similarity of the definitions of the derivatives of scalar and vector variables has the important consequence that the basic theorems of vector calculus are directly analogous to those for scalar calculus. These theorems are presented below; the corresponding proofs follow the same steps as those for scalar variables and are therefore not presented here. Let $\bar{A}$ and $\bar{B}$ be two vector functions and let c be a scalar function. Then

$$\frac{d}{dt}(\bar{A} \pm \bar{B}) = \frac{d\bar{A}}{dt} \pm \frac{d\bar{B}}{dt}$$

$$\frac{d}{dt}(c\bar{A}) = \frac{dc}{dt}\bar{A} + c\frac{d\bar{A}}{dt}$$

$$\frac{d}{dt}(\bar{A} \cdot \bar{B}) = \frac{d\bar{A}}{dt} \cdot \bar{B} + \bar{A} \cdot \frac{d\bar{B}}{dt}$$

$$\frac{d}{dt}(\bar{A} \times \bar{B}) = \frac{d\bar{A}}{dt} \times \bar{B} + \bar{A} \times \frac{d\bar{B}}{dt} \tag{17}$$

Note that in evaluating the derivative of a cross product, the order of multiplication must be maintained, because the cross product is not commutative ($\bar{A} \times \bar{B} \neq \bar{B} \times \bar{A}$).

The first and second identities in equations (17) will be of particular use to us, as they provide the basis for differentiating the component representation of a vector. Also, we shall frequently apply the *chain rule* for differentiation, because some vector $\bar{A}$ may be a function of a parameter α, whose value in turn depends on t. In such a situation,

$$\frac{d\bar{A}}{dt} = \frac{d\alpha}{dt}\frac{d\bar{A}}{d\alpha} \tag{18}$$

Next, let us consider the process of integration, which also has the same basis for vectors as for scalars. In dynamics, we are generally concerned with definite integrals, hence

$$\bar{A} = \frac{d\bar{B}}{dt} \quad \text{means} \quad \bar{B}(t) = \bar{B}(t_0) + \int_{t_0}^{t} \bar{A}(t)\, dt \tag{19}$$

Equations (17)–(19) will be employed extensively in the next module.

To simplify writing derivatives, we shall hereafter denote a derivative with respect to time by a dot over the quantity. For example, the basic definitions, equations (1) and (2), may now be rewritten as

$$\bar{v}_P \equiv \dot{\bar{r}}_{P/O} \qquad \bar{a}_P = \dot{\bar{v}}_P = \ddot{\bar{r}}_{P/O}$$

As can be seen, two dots over a quantity denotes the second derivative.

In closing this section we call attention to an important fact that follows from the foregoing development. Consider the representation of a vector in terms of its unit vector, as given by equation (5). Then

$$\frac{d\bar{A}}{dt} = \frac{d}{dt}(|\bar{A}|\bar{e}_A) = \left[\frac{d}{dt}|\bar{A}|\right]\bar{e}_A + |\bar{A}|\frac{d}{dt}\bar{e}_A$$

from which we see that a vector is constant only if both its magnitude is constant ($d|\bar{A}|/dt = 0$) and its orientation is constant ($d\bar{e}_A/dt = \bar{0}$).

F. PROBLEM SOLVING

The goal of a course in dynamics is the development of an understanding of the laws governing the motion of physical systems, with a primary emphasis on the application of this knowledge to solving problems. In general, engineering is the process of applying established physical laws to the solution of practical problems. Therefore, a common methodology has evolved for solving problems, although in each area the physical laws are different. The pertinent steps are detailed below.

1   Read (or formulate) the problem. If necessary draw a sketch of the system.
2   Show any known information in the sketch, or else list it. Also, list the information that is to be determined from the solution.
3   Choose a model to represent the system. For this model, write the equations corresponding to all physical principles relating the given and desired information.
4   Substitute the given information into these equations.
5   Count the number of equations and number of unknowns. An excess of unknowns usually means that either some given information was not used or a principle governing the model was not considered. In contrast, an excess of equations usually means that improper assumptions were made, or a principle employed in step 3 was incorrectly formulated, or possibly, that such a principle is not truly applicable to the problem.
6   Solve the problem.

These steps are very general. At crucial points in the text we shall present detailed steps, followed by a series of *Illustrative Problems*. The purpose of these problems is to demonstrate the applications of the procedural steps to a broad class of problems. This is in contrast to *Examples*, such as the three we have already studied, whose purpose is to demonstrate specific principles or techniques.

To conclude this module, let us consider the question that invariably occurs regarding numerical calculations, specifically, accuracy and precision. As you may have already noticed, all quantities are considered to have four significant figures. This approach is motivated by the extensive use of electronic calculators, which can easily handle numbers of such precision. We therefore have introduced $g = 9.806$ m/s² $= 32.17$ ft/s². However, as you will see, the fourth significant figure in g varies with the location of the physical system on the earth's surface. Hence, without any specification of this location, we generally have only three-place accuracy in a solution, because calculations are only as accurate as the least accurate quantity. Furthermore, three-place accuracy is more than sufficient for most engineering problems. Hence it makes sense to compute to four places and round off the final figures for the result to three places.

HOMEWORK PROBLEMS

I.17 The radar station in Example 3 tracks another aircraft. At one sighting the aircraft is at $R = 3500$ m, $\theta = 55.2°$. If this aircraft is traveling at 1836 km/hr at a climb angle of 30°, determine what the values of R and θ will be when the aircraft is sighted 0.50 s later.

I.18 At a third sighting of the aircraft in Example 3, 0.75 s after the second sighting, the aircraft is at $R = 4450$ m, $\theta = 25.6°$. Determine an approximation to the acceleration of the aircraft. Would you say that the airplane is under control?

Pin

$Y = X^2/200$

X

Prob. I.19

I.19 A pin in a machine follows a parabolic groove defined by $Y = X^2/200$, where X and Y are measured in millimeters. When first observed the pin is at $X = 255.2$ mm, and 2 ms later it is at $X = 239.4$ mm. Determine the approximate velocity of the pin.

I.20 An automobile is observed while traversing a level circular turn of 500-m radius. An observer at the center O of the curve initially sees the vehicle at a position where $\theta = 30°$ east of south. One second later, the car is at the position where $\theta = 33°$ east of south. What are the speed and the direction in which the vehicle is moving, approximately?

Probs. I.20 and I.21

I.21 Two seconds after the first sighting, a third observation of the automobile in Problem I.20 places it at $\theta = 36°$ east of south. Determine an approximation to the acceleration of the vehicle.

I.22 To track a UFO whose flight path does not pass over the station, a radar station measures the azimuth angle ϕ, as well as the angle of elevation θ and the distance R, as shown. Two sightings of the UFO 0.40 s apart place the object first at $R = 22,050$ ft, $\theta = 46.1°$, $\phi = 35.7°$, and then at $R = 21,130$ ft, $\theta = 47.3°$, $\phi = 32.2°$. Determine its speed and describe the direction in which it is moving.

Prob. I.22

MODULE II
KINEMATICS
OF A
PARTICLE

Because a particle occupies only a single point in space, it represents the simplest model we can construct for a physical system. Thus, the study of the kinematics of a particle reduces to an investigation of methods for relating the position, velocity, and acceleration of a point.

A. BASIC KINEMATICAL PROPERTIES

One of the principal properties of the motion of a particle is its *path*, which is the line (curved or straight) the particle follows as it travels through space. To depict the path of a particle we trace the point occupied by the particle at each instant of time, thereby forming the locus of these points. A major topic for consideration in this module is the relationship between the particle path and the kinematical quantities of position, velocity, and acceleration, whose definitions we shall repeat here from Module I.

Referring to Figure 1, assume that at an instant t the particle is at a point P on its path. To serve as a frame of reference, we choose a fixed coordinate system XYZ whose origin is point O. Then the vector $\bar{r}_{P/O}$ shown in the figure describes the position of point P with respect to point O.

Having defined the position vector $\bar{r}_{P/O}$, we next define the velocity vector $\bar{v}_P$, which is the absolute velocity of point P with respect to the fixed reference frame. By definition, velocity is the time rate of change of position, so that

Figure 1

$$\bar{v}_P \equiv \frac{d}{dt}\bar{r}_{P/O} \equiv \dot{\bar{r}}_{P/O} \qquad (1)$$

There is no need to introduce the fixed origin O into the notation for velocity because, as we shall show later, the velocity with respect to the fixed frame of reference is independent of the location of the origin.

Now that we have a mathematical definition for velocity, a similar approach may be followed in defining acceleration. Acceleration is the time rate of change of the velocity, so that

$$\bar{a}_P \equiv \dot{\bar{v}}_P \equiv \ddot{\bar{r}}_{P/O} \qquad (2)$$

The three vectors $\bar{r}_{P/O}$, $\bar{v}_P$, and $\bar{a}_P$, can be described by their components with respect to the XYZ coordinate system shown in Figure 1. This coordinate system can also be used to write the equations describing the particle's path. These and other uses of a coordinate system will be discussed in the following sections. For now we merely call your attention to the fact that XYZ is a fixed rectangular Cartesian coordinate system with right-hand orientation. We shall use capital letters, such as X, Y, Z and $\bar{I}$, $\bar{J}$, $\bar{K}$, throughout the text to indicate that the axes and unit vectors of the reference frame are fixed in space.

In summary, it is important that you understand clearly the basic definitions of position, velocity, and acceleration. The next section will utilize these definitions in demonstrating different methods for describing the basic kinematical vectors. Indeed, the definitions will be of major significance throughout our study of dynamics.

B. KINEMATICAL RELATIONSHIPS

There are several methods for describing the basic kinematical properties of a particle. A fundamental task in learning to set up problems for solution is to develop the ability to identify which method is best. Here, we shall first derive the basic equations for the several methods and then see how they are used.

1 Comparison of Fixed Coordinate Systems and Path Variables

Referring to Figure 1, let us consider two ways in which we can locate the point P at a particular instant in time. One possible way of specifying the position is to give the rectangular Cartesian coordinates of the particle as functions of time, that is, to give $X(t)$, $Y(t)$, and $Z(t)$. Clearly, using these functions at any instant t we can find the corresponding values of X, Y, and Z. A plot of the values of X, Y, and Z corresponding to each instant will yield a trace of the path followed by the particle.

Now consider an alternative approach. Suppose that we are given the particle's path, and the position of the particle along its path is defined by the arc length that the particle has traveled from its initial position. The arc length is called a *path variable*. An example of such a description is to be told that a particle follows a circular path in the XY plane and that at $t = t_0$ it has traveled one-quarter of the circumference of the circle from its initial position in a clockwise direction. As we can see in Figure 2, this locates the position of the particle, provided that we know its initial position.

The fundamental difference between the Cartesian coordinate description and path variables is that in Cartesian coordinates the specification of the position at any instant is dependent on the choice of coordinate axes, and, as we shall see, is independent of knowledge of the path. The same is true for the velocity and acceleration vectors. Hence in the Cartesian coordinate approach the choice of coordinate system will change the description of the basic variables. On the other hand, when using a path variable approach all quantities, including position, are defined with respect to the path. In this method the description is essentially independent of the choice of coordinate system; the main purpose of the coordinate

Figure 2

system is to describe the path. Because the path variables are intrinsically tied to the properties of the path, not to the fixed coordinate system, this description is sometimes referred to as *intrinsic coordinates*.

In Cartesian coordinates, and later for cylindrical coordinates, as well as many other coordinate systems that we shall not consider here, all of the variables are defined independent of knowledge of the path. They are sometimes referred to as *extrinsic coordinates*.

All mathematical descriptions of the kinematical properties of a physical system relate to one or the other of the types of coordinate systems just described. Hence, we must study them in detail. In summary, then, we wish to study both

$$\left\{\begin{array}{c} \text{Fixed} \\ \text{coordinate} \\ \text{systems} \end{array}\right\} \leftrightarrow \left\{\begin{array}{c} \text{Extrinsic} \\ \text{coordinates} \end{array}\right\} \leftrightarrow \left\{\begin{array}{c} \text{Path independent} \\ \text{and coordinate} \\ \text{axes dependent} \end{array}\right\}$$

and

$$\left\{\begin{array}{c} \text{Path} \\ \text{variables} \end{array}\right\} \leftrightarrow \left\{\begin{array}{c} \text{Intrinsic} \\ \text{coordinates} \end{array}\right\} \leftrightarrow \left\{\begin{array}{c} \text{Path dependent} \\ \text{and coordinate} \\ \text{axes indepedent} \end{array}\right\}$$

2 Velocity in Path Variables

We begin our study by determining the kinematical relationship for velocity in terms of path variables.

To find the velocity we must take the time derivative of the position vector. In path variables the path is given and the position vector is defined in terms of the arc length $s(t)$ that the particle has traveled. Restating equation (1), we have

$$\bar{v}_P = \dot{\bar{r}}_{P/O} = \frac{d\bar{r}_{P/O}}{dt}$$

Note that we were given s as an explicit function of t. This suggests using the chain rule for differentiation. Hence

$$\bar{v}_P = \frac{d\bar{r}_{P/O}}{dt} = \frac{d\bar{r}_{P/O}}{ds}\frac{ds}{dt} = \frac{d\bar{r}_{P/O}}{ds}\,\dot{s} \tag{3}$$

In order for equation (3) to be useful we must define $d\bar{r}_{P/O}/ds$. Proceeding toward this objective, in Figure 3 the vector $\Delta\bar{r}_P$ is the change in $\bar{r}_{P/O}$ when the particle moves an arc length Δs from position P. From Figure 3 we may write

$$\Delta\bar{r}_{P/O} = \bar{r}_{P/O}(s + \Delta s) - \bar{r}_{P/O}(s)$$

Figure 3

Geometrically, $\Delta \bar{r}_{P/O}$ is a chord for the path connecting points P and P'.

Now consider when happens as Δs becomes smaller and smaller. Because the path is a continuous line, point P' approaches P, and the chord and the tangent to the path become indistinguishable. In other words, in the limit the cord length $|\Delta \bar{r}_{P/O}|$ and arc length Δs are identical. Thus if we define $\bar{e}_t$ to be a *unit vector tangent to the path at point P* in the direction of increasing s, we have

$$\frac{d\bar{r}_{P/O}}{ds} = \lim_{\Delta s \to 0} \frac{\Delta \bar{r}_{P/O}}{\Delta s} = \lim_{\Delta s \to 0} \frac{\Delta s \bar{e}_t}{\Delta s} = \bar{e}_t$$

Returning to eqation (3), we now may write

$$\bar{v}_P = \frac{d\bar{r}_{P/O}}{ds} \dot{s} = \dot{s} \bar{e}_t \qquad (4)$$

Equation (4) yields a valuable piece of information. Noting that the path we chose in Figure 3 was quite arbitrary, it must be true that the velocity is always tangent to the path.

Now let us compute the speed, which is the absolute value of the velocity:

$$v_P = |\bar{v}_P| = \sqrt{\bar{v}_P \cdot \bar{v}_P} = \dot{s}\sqrt{\bar{e}_t \cdot \bar{e}_t} = \dot{s}$$

Thus we can conclude that

$$\boxed{\begin{aligned} \bar{v}_P &= v_P \bar{e}_t \\ v_P &= \dot{s} \end{aligned}} \qquad (5)$$

In summary, equations (5) say that the velocity (a vector) is tangent to the path and that the speed (a scalar) is the time rate of change of the distance traveled along the path. Also note that because $\bar{e}_t$ is tangent to the path, the direction of $\bar{e}_t$ will change as the location of point P changes, with one exception. Which?

EXAMPLE 1
A locomotive follows a circular track that lies in a horizontal plane. The track has a radius of 2000 ft. If the distance the train has traveled from the start of the curve is given by $s = 40t + t^2$ (s is feet, and t is seconds), what is the velocity of the train when $t = 10$ seconds?

Solution
We are given $s(t)$. Then, knowing that $\bar{v}_P = \dot{s} \bar{e}_t$ we compute

$$\dot{s} = 40 + 2t$$

For $t = 10$ seconds, this yields $\dot{s} = 60$ ft/s, or

$$\bar{v}_P = 60\bar{e}_t \text{ ft/s}$$

The unit vector $\bar{e}_t$ is illustrated in the accompanying sketch.

As an additional exercise, let us express $\bar{v}_P$ in terms of its X and Y components. From the diagram,

$$\theta = \frac{s}{R} \quad \text{and} \quad \bar{e}_t = \cos\theta\bar{I} + \sin\theta\bar{J}$$

When $t = 10$ seconds, $s = 40(10) + 10^2 = 500$ ft, so that

$$\theta = \frac{500}{2000} \text{ rad} = 0.25 \text{ rad} \left(\frac{180°}{\pi \text{ rad}}\right) = 14.32°$$

Thus

$$\bar{v}_P = 60 (\cos 14.32° \, \bar{I} + \sin 14.32°\bar{J})$$
$$= 58.13\bar{I} + 14.84\bar{J} \text{ (ft/s)}$$

3 Acceleration in Path Variables

We can now obtain a relationship for acceleration by differentiating the velocity (using the laws for vector differentiations).

$$\bar{a}_P \equiv \frac{d}{dt}(v_P\bar{e}_t) = \dot{v}_P\bar{e}_t + v_P\dot{\bar{e}}_t \tag{6}$$

In equation (6) the only term with which we are unfamiliar is $\dot{\bar{e}}_t$. Remember that $\bar{e}_t$ is tangent to the path at point P. Thus $\bar{e}_t$ is a function of the arc length s, and the chain rule gives

$$\frac{d\bar{e}_t}{dt} = \frac{d\bar{e}_t}{ds}\frac{ds}{dt} = \dot{s}\frac{d\bar{e}_t}{ds} = v_P\frac{d\bar{e}_t}{ds}$$

Placing this into equation (6), we have

$$\bar{a}_P = \dot{v}_P\bar{e}_t + v_P{}^2\frac{d\bar{e}_t}{ds} \tag{7}$$

We now have as an unknown $d\bar{e}_t/ds$. To deal with it consider the unit vectors $\bar{e}_t(s)$ and $\bar{e}_t(s + \Delta s)$, illustrated in Figure 4.

Although we are dealing with an arbitrary smooth curve in space, without loss of generality we may consider the arc length in question to have a center of curvature O, which is defined as the center of the circle that most closely fits the path at position s. Further, for ease of visualization, consider the curve to be contained in the plane of the page. [We shall

Figure 4

discuss the matter of center of curvature for a three-dimensional curve in Section II.B.5].

Let us denote the angle between points $\bar{e}_t(s)$ and $\bar{e}_t(s + \Delta s)$ in Figure 4 as $\Delta\theta$. We shall now slide the unit vectors to a position where their tails coincide. Hence we may form the triangle shown in Figure 5. The angle between the unit vectors is $\Delta\theta$. In Figure 5 the parameter $\Delta\theta$ appears explicitly, whereas Δs does not. Because of this, let us again use the chain rule to evaluate $d\bar{e}_t/ds$.

Figure 5

$$\frac{d\bar{e}_t}{ds} = \frac{d\bar{e}_t}{d\theta}\frac{d\theta}{ds} \tag{8}$$

Because

$$\left|\bar{e}_t(s)\right| = \left|\bar{e}_t(s + \Delta s)\right| = 1$$

the triangle in Figure 5 is an isosceles triangle. Thus we may write

$$\left|\Delta\bar{e}_t\right| = 2\left|\bar{e}_t\right| \sin\frac{\Delta\theta}{2} = 2\sin\frac{\Delta\theta}{2} \tag{9}$$

In equation (8) we require $d\bar{e}_t/d\theta$. Hence we divide both sides of equation (9) by $\Delta\theta$ to get

$$\frac{\left|\Delta\bar{e}_t\right|}{\Delta\theta} = \frac{\sin\Delta\theta/2}{\Delta\theta/2}$$

Now, in the limit as $\Delta\theta \to 0$, we get

$$\lim_{\Delta\theta\to0}\frac{\left|\Delta\bar{e}_t\right|}{\Delta\theta} = \frac{d\left|\bar{e}_t\right|}{d\theta} = \lim_{\Delta\theta\to0}\frac{\sin\Delta\theta/2}{\Delta\theta/2} \equiv 1 \tag{10}$$

Do you remember this limit from your elementary calculus course?

From Figure 5 we see graphically that as $\Delta\theta \to 0$, the vector $\Delta\bar{e}_t/\Delta\theta$ approaches the perpendicular to $\bar{e}_t(s)$. From Figure 4 we see that this perpendicular is directed toward the center of curvature. Finally, from equation (10) we see that the perpendicular has magnitude unity.

When we put these facts together we find that $d\bar{e}_t/d\theta$ is a unit vector perpendicular (normal) to $\bar{e}_t$ and directed toward the center of curvature. We denote this unit vector as $\bar{e}_n$, where the letter n is for normal; hence

$$\frac{d\bar{e}_t}{d\theta} = \bar{e}_n$$

The plane defined by $\bar{e}_t$ and $\bar{e}_n$ is called the *osculating plane*. The center of curvature, of course, also lies in the osculating plane. For a planar path the osculating plane is the plane of the path.

Let us once again return to the central theme of determining the acceleration. Equation (8) now reads:

$$\frac{d\bar{e}_t}{ds} = \frac{d\theta}{ds}\,\bar{e}_n \tag{11}$$

Inspecting equation (11), we must now express $d\theta/ds$ in a more usable form. Referring to Figure 4, you will note that as Δs becomes infinitesmal, it may be considered to form the arc of a circle of radius ρ, which is called the radius of curvature. Thus

$$\rho = \frac{ds}{d\theta}$$

Do you recognize this formula from basic calculus? From this formula equation (11) then becomes

$$\frac{d\bar{e}_t}{ds} = \frac{1}{\rho}\bar{e}_n$$

Hence equation (7) may be written in the standard form

$$\boxed{\bar{a}_P = \dot{v}_P\bar{e}_t + \frac{v_P{}^2}{\rho}\,\bar{e}_n} \tag{12}$$

Before investigating some examples involving acceleration in path variables, we should note that there are basically two classes of problems that occur in dynamics. In the simpler class of problems the given information describes the position, and the velocity and acceleration are desired. As we have seen in the derivations of the basic formulas, the velocity and acceleration are obtained through the process of differentiation.

A more complicated problem occurs when information about the acceleration is given and velocity and position are the desired quantities. A typical situation, involving equation (12), is to be given the tangential acceleration $\dot{v}_P$ in equation form. An equation that contains a derivative of one of the unknown parameters is called a *differential equation*. For example, $\dot{v}_P = t^2$ is a differential equation involving v_P and t in which v_P is called the dependent variable and t the independent variable. We shall study the topic of differential equations as it applies to dynamics problems in the next module on the equations of motion in particle motion. In this module we restrict ourselves to cases where we find velocity and/or acceleration using the differentiation process only.

A   B   C

EXAMPLE 2
Shown are the path of a particle and representations of the velocity and acceleration of the particle at that instant in time. Indicate if each of the conditions shown is possible or not.

Solution

Figure A: We know that the velocity *must* be tangent to the path (equation (4)). Therefore this is not a possible situation. Also, we know that the acceleration *must* point toward the inside of the curve, for its two components are directed tangent to the path and normal to the path towards the center of curvature. Hence, this too is not possible.

Figure B: These are possible situations, from the arguments for Figure A.

Figure C: The velocity is tangent to the path. The path is a straight line, hence ρ is infinite and the only component of $\bar{a}_P$ is the one that is tangent to the path. Hence this situation is possible. What does the condition that the direction of $\bar{a}$ is opposite to the direction of $\bar{v}$ mean?

EXAMPLE 3

The diagram shown is a portion of a roller coaster track. It consists of a circle in a vertical plane of radius 150 m. Find the magnitude of the total acceleration of the roller coaster if, at the time shown, the car is moving at 50 km/hr and is gaining speed at the rate of 0.4g (1g = 9.806 m/s²).

Solution

The given information is clearly in the form of path variables (path, speed, and rate of increase of speed). Therefore select the unit vectors at P as shown. In general we know that

$$\bar{a}_P = \dot{v}_P \bar{e}_t + \frac{v_P^2}{\rho} \bar{e}_n$$

The given information is

$$v_P = 50{,}000 \text{ m/hr} \left(\frac{1}{3600}\frac{s}{hr}\right) = 13.889 \text{ m/s}$$

$$\dot{v}_P = 0.4(9.806) = 3.922 \text{ m/s}^2$$

Hence we compute

$$\bar{a}_P = 3.922\bar{e}_t + \frac{(13.889)^2}{150}\bar{e}_n$$

$$= 3.922\bar{e}_t + 1.286\bar{e}_n \text{ m/s}^2$$

Then the magnitude of $\bar{a}_P$ is

$$|\bar{a}_P| = \sqrt{(3.922)^2 + (1.286)^2}$$

$$= 4.13 \text{ m/s}^2$$

HOMEWORK PROBLEMS

II.1 A bicyclist whose speed is 18 miles/hr wishes to turn left. What is the minimum radius turn for which the acceleration will not exceed 5 ft/s²?

Prob. II.3

Prob. II.4

Prob. II.5

II.2 An automobile on a 40-m-radius curve has a speed of 60 km/hr. What rate of increase of this speed will result in a 50% increase in the magnitude of the acceleration of the vehicle?

II.3 At a particular instant an airplane is diving at a 30° angle with a speed of 800 km/hr. Its instantaneous acceleration is $3.0g$ upward. Determine (a) the rate of increase of the speed, (b) the unit vector normal to the path of the airplane, and (c) the radius of curvature of the path.

II.4 A freight train heading east enters a 1500-m-radius curve to the north. The arc length traveled along the curve is given by $s = 80t - t^2$ (s is in meters and t is in seconds). What is the speed, the heading in degrees with respect to the easterly direction, and the acceleration of the train after it has traveled 1200 m?

II.5 The motion of an automobile along a 800-ft-radius track is defined by $s = 6t^2 - 0.2t^3$, where t is measured in seconds and s is in feet. Determine the distance traveled, the speed, and the acceleration of the vehicle (a) when $t = 5$ seconds, (b) when $t = 15$ seconds.

II.6 Solve Problem II.5 for the instant when (a) the speed is maximum, (b) the tangential acceleration is maximum.

II.7 The constant speed of the particle spinning in a circular horizontal path at the end of a cable of length l is known to depend on the height h illustrated in the diagram according to $v^2 = v_0^2 + 2gh$. For what value of v_0 will the acceleration have a magnitude of $2v_0^2/l$ when the height is $0.8l$?

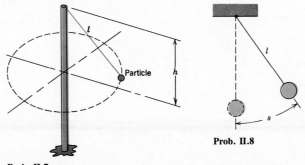

Prob. II.7

Prob. II.8

II.8 The arc length s for the pendulum in a vertical plane shown in the diagram is given by $s = (l/10) \sin\sqrt{g/l}\, t$. Determine the acceleration of the pendulum, and show the result in a sketch, (a) for $s = 0$, (b) for $s = l/10$, (c) for $s = l/20$.

4 Further Remarks on Path Variable Kinematics

Equation (12) shows that the total acceleration is the vector sum of two components that lie in the osculating plane: The tangent component is the rate of increase of the speed, $a_t = \dot{v}_P$, whereas in the normal direction the centripetal component, $a_n = v_P^2/\rho$, is directed toward the center of curvature. Symbolically the acceleration vector is often written in the form

$$\bar{a} = a_t\bar{e}_t + a_n\bar{e}_n \tag{13}$$

From this expression we can formulate the magnitude of the total acceleration as

$$|\bar{a}| = \sqrt{\bar{a} \cdot \bar{a}} = \sqrt{a_t^2 + a_n^2} = \sqrt{\dot{v}^2 + v^4/\rho^2}$$

Let us briefly turn to two special cases that frequently occur. In many problems the speed will be given as a function of the distance traveled by the particle. In other words, $v_P(s)$ will be given. Then to determine the rate of increase of the speed we can use the chain rule.

$$\dot{v}_P = \frac{dv_P}{ds}\frac{ds}{dt} = \dot{s}\,\frac{dv_P}{ds} = v_P\,\frac{dv_P}{ds} \tag{14}$$

Another special case occurs when the curved path is circular. As can be seen in Figure 6a, $\rho \equiv R$ and $\bar{e}_n$ always points toward the center of the circle. For an arbitrary path, Figure 6b, the value of ρ changes as the location of the particle changes and, in general, $\bar{e}_n$ is no longer directed toward a single point.

Figure 6a   Figure 6b

The process of functionally determining $\bar{e}_t$, $\bar{e}_n$, and ρ for a general three-dimensional path is a mathematical topic treated in the next section. For the case of a two-dimensional path, without loss of generality we can describe the path by selecting XYZ so as to make the particle move in the XY plane. Then the functional equation $Y = f(X)$ defines the path.

For the planar path shown in Figure 7, formulas developed in elementary calculus give

$$s = \int_{X_0}^{X_1} \left[1 + \left(\frac{dY}{dX}\right)^2\right]^{1/2} dX \tag{15}$$

Figure 7

$$\theta = \tan^{-1} \frac{dY}{ds} \tag{16}$$

$$\frac{1}{\rho} = \frac{\left| d^2 Y/dX^2 \right|}{[1 + (dY/dX)^2]^{3/2}} \tag{17}$$

We should note, however, that equation (15) will usually lead to a complicated integral.

Once the value of θ is found at point s, we can, if we so desire, transform the expressions for velocity and acceleration by expressing the unit vectors $\bar{e}_t$ and $\bar{e}_n$ in terms of their components with respect to the Cartesian unit vectors $\bar{I}$ and $\bar{J}$.

As a concluding remark, we again call attention to the fact that the basic formulas developed for path variables depend on knowing either $s(t)$ and/or v_P at any position along the path. It follows that if we have knowledge about the path a particle follows and information about the distance traveled along the path (or speed), it will prove advantageous to use path variables. Without such information, as we shall see, there are better ways to proceed.

Let us now look at an example that utilizes most of the equations governing path variables for planar motion.

EXAMPLE 4

A ski jump is a curve in the vertical plane defined by $Y = 0.005X^2$, where X and Y are the horizontal and vertical coordinates of a point relative to the bottom of the ski jump measured in meters. In one trial a skier began her run at $X = 150$ m and her speed was observed to vary according to $v_P = 30 \sin(\pi s/2s_0)$ m/s, where s is the distance she has traveled and s_0 is the arc length from the start to the bottom. Determine the acceleration of the skier when she is at a horizontal distance of 50 m from the bottom of the ski jump.

Solution

The given information about the speed is expressed in terms of the parameter s, thus suggesting the path variable approach.

$$\bar{a}_P = \dot{v}_P \bar{e}_t + \frac{v_P^2}{\rho} \bar{e}_n$$

Let us begin with a sketch of the ski jump that also shows the path variable unit vectors.

We must determine the value of s_0 and also the value of s corresponding to the position where $X = 50$ m. From equation (15) we have

$$s_0 = \int_0^{150} \left[1 + \left(\frac{dY}{dX} \right)^2 \right]^{1/2} dX$$

$$= \int_0^{150} (1 + 0.0001X^2)^{1/2} \, dX$$

$$= 0.005\{X(100^2 + X^2)^{1/2} + 100^2 \ln[X + (100^2 + X^2)^{1/2}]\} \, \Big|_0^{150}$$

$$= 194.95 \text{ m}$$

(The general solution of the integral was obtained from a table of integrals.) As indicated earlier, for most paths whose shapes are common curves the evaluation of the arc length is not a trivial step. Following a similar step for the value of s at point P yields

$$s = 194.95 - \int_0^{50} \left[1 + \left(\frac{dY}{dX} \right)^2 \right]^{1/2} dX = 142.94 \text{ m}$$

The corresponding value of v_P may now be determined by direct substitution.

$$v_P = 30 \sin \left(\frac{\pi \times 142.93}{2 \times 194.95} \right) = 30 \sin(1.1517 \text{ rad})$$

$$= 30 \sin(65.98°) = 27.40 \text{ m/s}$$

Because we are given $v_P(s)$, we use the chain rule to find $\dot{v}_P(s)$.

$$\dot{v}_P(s) = v_P \frac{dv_P}{ds} = 30 \sin\left(\frac{\pi s}{2s_0}\right) \times 30 \left(\frac{\pi}{2s_0}\right) \cos\left(\frac{\pi s}{2s_0}\right)$$

Thus when $s = 142.93$ m,

$$\dot{v}_P(s) = \left(\frac{450\pi}{194.95}\right) \sin(65.98°) \cos(65.98°)$$

$$= 2.696 \text{ m/s}^2$$

Finally, we use equation (19) to find $1/\rho$.

$$\frac{1}{\rho} = \frac{|d^2Y/dX^2|}{[1 + (dY/dX)^2]^{3/2}} = \frac{0.01}{[1 + (0.01X)^2]^{3/2}}$$

$$= \frac{0.01}{[1 + 0.25]^{3/2}} = 0.00716 \frac{1}{\text{m}}$$

Substituting these values into the acceleration equation gives

$$\bar{a}_P = \dot{v}_P \bar{e}_t + \frac{v_P^2}{\rho} \bar{e}_n$$

$$= 2.70\bar{e}_t + 5.38\bar{e}_n \text{ m/s}^2$$

The solution is now complete because we have the components of acceleration with respect to a set of unit vectors that are defined in the diagram.

In the next step, which is only necessary if the problem requires it, we show how the answer may be converted to Cartesian components. This can be accomplished by expressing $\bar{e}_t$ and $\bar{e}_n$ in their Cartesian components. The diagram of the unit vectors shows that

$$\theta = \tan^{-1} \frac{dY}{dX} = \tan^{-1}(0.01X) = \tan^{-1}(0.5) = 26.57°$$

$$\bar{e}_t = -\cos\theta \bar{I} - \sin\theta \bar{J} = -0.894\bar{I} - 0.447\bar{J}$$

$$\bar{e}_n = -\sin\theta \bar{I} + \cos\theta \bar{J} = -0.447\bar{I} + 0.894\bar{J}$$

Then substituting these results into the equation for $\bar{a}_P$ gives

$$\bar{a}_P = 2.70(-0.894\bar{I} - 0.447\bar{J}) + 5.38(-0.447\bar{I} + 0.894\bar{J})$$

$$= -4.82\bar{I} + 3.60\bar{J} \text{ m/s}^2$$

Prob. II.9

HOMEWORK PROBLEMS

II.9   A toy slot car follows a 4-ft-radius track. The speed of the toy is related to the distance traveled, s, according to $v = 2\sqrt{s}$ (s is in feet and v is in feet per second). Determine the acceleration (a) when $s = 1$ ft., (b) when $s = 6.25$ ft. Draw a sketch of the results.

II.10   The computerized controller of a high-speed passenger train slows the train according to the given graph. A train is on a 10-km curve with an initial speed $v_0 = 360$ km/hr when the brakes are applied. Determine the magnitude of the acceleration when $s = 1.5$ km.

II.11   For Problem II.10 determine the position (a) where $\dot{v}$ has its maximum magnitude, (b) where the acceleration has its maximum magnitude.

Prob. II.10

II.12   An automobile moves with a constant speed of 90 km/hr along a highway whose vertical elevation Y in meters is given by $Y = 2\cos(\pi X/100)$, where X is the horizontal distance traveled, also in meters.

Determine the acceleration of the automobile and sketch the vector along the curve when $X = 75$ m.

Prob. II.12

II.13 The brakes of the automobile in Problem II.12 are suddenly applied when $X = 75$ in, causing the speed to decrease instantaneously at the rate of $0.5g$. Determine and sketch the resulting acceleration in this position.

II.14 Arm AB rotates at a rate such that the pin follows the parabolic groove $Y = 4 - X^2$ (X and Y are in feet) at a constant speed of 3 ft/s. Determine the horizontal and vertical components of the velocity and acceleration of the pin when $X = 2$ ft.

Prob. II.14

5 Path Variables in Three Dimensions

From equation (12) we see that the acceleration of a particle in space consists of two scalar components perpendicular to each other. In two dimensions it is easy to visualize that the plane that contains the vector $\bar{a}_P$ is the plane of motion. What about a particle traversing a three-dimensional curvilinear path? Let us see how the $\bar{e}_t\bar{e}_n$ plane can be described for such a general case.

A general three-dimensional path is shown in Figure 8. As Δs becomes infinitesimal, but nonzero, it can be proved that $\bar{e}_t(s)$ and $\bar{e}_t(s + \Delta s)$ lie in a common plane, but are not collinear. This plane is the *osculating plane*. To get some idea of the meaning of the osculating plane, we

Figure 8

suggest the following. Bend a wire clothes hanger into an arbitrary shape and pick out some point on the hanger. Put the hanger into contact with a flat table (as shown in Figure 9) and rotate the wire until you find the orientation for which a small region of the wire in the vicinity of the point seems to be fully in contact with the table. You have found the osculating

Figure 9

plane for the chosen point on the wire. Note that the center of curvature for the infinitesmal arc length ds containing the chosen point also lies in the osculating plane. As mentioned previously, if the wire is bent into a planar curve, the osculating plane for such a curve is the plane of the wire. In the special case where the wire forms a straight line, *any* plane that contains the line is the osculating plane.

In summary, at an arbitrary point on a particle's path, the unit vector $\bar{e}_n$ is perpendicular with the tangent vector $\bar{e}_t$, lies in the osculating plane, and is directed toward the center of curvature.

When we think of the Cartesian coordinates XYZ, we associate the unit vectors $\bar{I}$, $\bar{J}$, $\bar{K}$ to the system. Similarly, for the path-dependent variables $\bar{e}_t$ and $\bar{e}_n$ we can define a unit vector perpendicular to the osculating plane (the plane of $\bar{e}_t$ and $\bar{e}_n$) in order to complete the triad. That direction is called the *binormal* and is defined as

$$\bar{e}_b \equiv \bar{e}_t \times \bar{e}_n$$

(Clearly, this is directly analogous to $\bar{I} \times \bar{J} = \bar{K}$.) Thus the vector triad is $\bar{e}_t$, $\bar{e}_n$, $\bar{e}_b$.

The importance of $\bar{e}_b$ for particle motion is that when one considers Newton's second law, $\bar{F} = m\bar{a}$, from equation (12) we see that the component of $\bar{a}$ in the direction of $\bar{e}_b$ is zero. This means that we have an equation of statics to call upon in problem solving, as will be seen in Illustrative Example 3 in the next module.

In general, the determination of the path variable quantities for three dimensional motion involves more mathematical manipulations than that required for path variables in two dimensional motion. Let us look at an example to see what is involved.

EXAMPLE 5
A common way of defining a three-dimensional path is in parametric form. A helix is defined in parametric form by

$$X = A \cos \alpha \qquad Y = A \sin \alpha \qquad Z = C\alpha$$

where α is the parameter. Determine the tangent, normal, and binormal unit vectors and the radius of curvature for an arbitrary value α.

Solution
The basic equations for the desired quantities are

$$\bar{e}_t = \frac{d\bar{r}_{P/O}}{ds} \qquad \bar{e}_n = \rho \frac{d\bar{e}_t}{ds} \qquad \bar{e}_b = \bar{e}_t \times \bar{e}_n$$

The point P has coordinates (X, Y, Z) so

$$\bar{r}_{P/O} = X\bar{I} + Y\bar{J} + Z\bar{K}$$

$$= A \cos \alpha \bar{I} + A \sin \alpha \bar{J} + C\alpha \bar{K}$$

Because $\bar{r}_{P/O}$ is a function of α, we will use the chain rule with respect to α:

$$\bar{e}_t = \frac{d\bar{r}_{P/O}}{ds} = \frac{d\bar{r}_{P/O}}{d\alpha}\frac{d\alpha}{ds}$$

$$= \frac{d\alpha}{ds}[-A \sin \alpha \bar{I} + A \cos \alpha \bar{J} + C\alpha \bar{K}]$$

We do not know $(d\alpha/ds)$. To determine this quantity we use the fact that $\bar{e}_t$ must be a unit vector, so

$$|\bar{e}_t| = 1 = \frac{d\alpha}{ds}[(-A \sin \alpha)^2 + (A \cos \alpha)^2 + C^2]^{1/2}$$

$$= \frac{d\alpha}{ds}[A^2 + C^2]^{1/2}$$

Hence

$$\frac{d\alpha}{ds} = (A^2 + C^2)^{-1/2}$$

so that

$$\bar{e}_t = (A^2 + C^2)^{-1/2}(-A \sin \alpha \bar{I} + A \cos \alpha \bar{J} + C\alpha \bar{K})$$

Now, using the chain rule again,

$$\bar{e}_n = \rho \frac{d\bar{e}_t}{ds} = \rho \frac{d\bar{e}_t}{d\alpha}\frac{d\alpha}{ds}$$

$$= \rho(A^2 + C^2)^{-1/2}(-A \cos \alpha \bar{I} - A \sin \alpha \bar{J})(A^2 + C^2)^{-1/2}$$

$$= -\rho(A^2 + C^2)^{-1}(A \cos \alpha \bar{I} + A \sin \alpha \bar{J})$$

However, $\bar{e}_n$ is also a unit vector.

$$|\bar{e}_n| = 1 = \rho(A^2 + C^2)^{-1}[(A \cos \alpha)^2 + (A \sin \alpha)^2]^{1/2}$$

$$= \rho(A^2 + C^2)^{-1}A$$

This enables us to find the radius of curvature and $\bar{e}_n$.

$$\rho = \frac{A^2 + C^2}{A}$$

and

$$\bar{e}_n = -\frac{1}{A}(A \cos \alpha \bar{I} + A \sin \alpha \bar{J})$$

Finally, we find $\bar{e}_b$.

$$\bar{e}_b = \bar{e}_t \times \bar{e}_n$$

$$= -\frac{1}{A}(A^2 + C^2)^{-1/2}(-A \sin \alpha \bar{I} + A \cos \alpha \bar{J} + C\alpha \bar{K})$$
$$\times (A \cos \alpha \bar{I} + A \sin \alpha \bar{J})$$

$$= -(A^2 + C^2)^{-1/2}[-C \sin \alpha \bar{I} + C \cos \alpha \bar{J}$$
$$-A(\sin^2 \alpha + \cos^2 \alpha)\bar{K}]$$

$$= (A^2 + C^2)^{-1/2}(C \sin \alpha \bar{K} - C \cos \alpha \bar{J} + A\alpha \bar{K})$$

◿

HOMEWORK PROBLEMS

II.15   A conical helix is defined in parametric form by

$$X = 40\alpha \cos \pi\alpha \qquad Y = 40\alpha \sin \pi\alpha \qquad Z = 30\alpha$$

where X, Y, and Z have units of millimeters. Determine the tangent, normal, and binormal unit vectors when $Z = 18$ mm.

II.16   In the position where $\alpha = 3.25$ a particle following the conical helix of Problem II.15 has a speed of 50 m/s and is gaining speed at 10 m/s². Determine its corresponding position, velocity and acceleration.

6   Fixed Rectangular Cartesian Coordinates

The underlying principle for the Cartesian coordinate description of the kinematical relationships is to express the components of vector quantities in a fixed XYZ coordinate system. The unit vectors $\bar{I}, \bar{J}, \bar{K}$ describe the positive X, Y, and Z directions. Note again that the use of capital letters for $\bar{I}, \bar{J}$, and $\bar{K}$ indicate that they correspond to the fixed XYZ system.

Figure 10a shows a point P whose coordinates are (X, Y, Z). The components of the position vector $\bar{r}_{P/O}$ parallel to the $\bar{I}, \bar{J}$, and $\bar{K}$ unit vectors are

$$\bar{r}_{p/O} = X\bar{I} + Y\bar{J} + Z\bar{K} \tag{18}$$

In Figure 10b we selected a Cartesian coordinate system having the same origin O but different orientation from the one shown in Figure 10a. Note that the values of X, Y, and Z are different, as are, of course, the $\bar{I}, \bar{J}$, and $\bar{K}$ directions. However, the resulting vector $\bar{r}_{P/O}$ represents the same physical position in space.

Figure 10a   **Figure 10b**

When the particle moves, the value of the X, Y, and Z components change with time. The functions $X(t)$, $Y(t)$, and $Z(t)$ define the path of the particle in parametric form, with t being the parameter. For instance, $X = Y = 0$, $X = -1/2t^2$ defines a path that is the Z axis.

The velocity is found by differentiating the position vector with respect to time. Because XYZ is a fixed system, $\bar{I}$, $\bar{J}$, and $\bar{K}$ have constant magnitude and direction, so that $\dot{\bar{I}} = \dot{\bar{J}} = \dot{\bar{K}} = \bar{0}$. Then

$$\bar{v}_P \equiv \dot{\bar{r}}_{P/O} = \dot{X}\bar{I} + X\dot{\bar{I}} + \dot{Y}\bar{J} + Y\dot{\bar{J}} + \dot{Z}\bar{K} + Z\dot{\bar{K}}$$

$$\boxed{\bar{v}_P \equiv \dot{X}\bar{I} + \dot{Y}\bar{J} + \dot{Z}\bar{K}} \qquad (19)$$

The speed is the magnitude of the velocity of $\bar{v}_P$.

$$v_P = |\bar{v}_P| = \sqrt{\bar{v}_P \cdot \bar{v}_P}$$
$$= \sqrt{(\dot{X})^2 + (\dot{Y})^2 + (\dot{Z})^2}$$

Differentiation of the velocity expression yields the acceleration formula:

$$\bar{a}_P = \ddot{X}\bar{I} + \dot{X}\dot{\bar{I}} + \ddot{Y}\bar{J} + \dot{Y}\dot{\bar{J}} + \ddot{Z}\bar{K} + \dot{Z}\dot{\bar{K}}$$

$$\boxed{\bar{a}_P = \ddot{X}\bar{I} + \ddot{Y}\bar{J} + \ddot{Z}\bar{K}} \qquad (20)$$

The Cartesian coordinate formulation is one alternative to path variables for determining velocity and acceleration. Depending on the way a problem is posed, one method or the other will yield the solution in a simpler and more straightforward manner.

Basically, we use Cartesian coordinates in problems where the important variables are most easily expressed as functions of the Cartesian coordinates, as in the case where quantities are given with respect to fixed directions, e.g., horizontal and vertical. For any given motion a solution by Cartesian coordinates will, of course, give the same vector resultants for velocity or acceleration as a solution by path variables.

EXAMPLE 6
A ball is thrown horizontally off the top of a building and follows a parabolic path in the vertical plane defined by $Z = 0$, $Y = -cX^2$, where Y is its elevation measured from the top of the building and X is the horizontal distance traveled by the ball. We know that the horizontal component of the velocity is constant. Find the acceleration of the particle as a function of its position.

Solution
In this problem the path is given. We are also given information about the velocity component in the (fixed) horizontal direction. These data strongly indicate that Cartesian coordinates will give an easier solution than path variables.

We begin the solution by drawing a sketch of the path. Because $Z \equiv 0$, the acceleration is

$$\bar{a}_P = \ddot{X}\bar{I} + \ddot{Y}\bar{J}$$

The component of velocity in the X direction, V_X, was given to the constant, so that $\dot{X} = V_X$, $\ddot{X} = 0$. We must now determine $\ddot{Y}$. Given that $Y = -cX^2$, the chain rule yields

$$\dot{Y} = \frac{dY}{dt} = \frac{dY}{dX}\frac{dX}{dt} = (-2cX)\dot{X} = -2cV_X X$$

Then differentiating again gives

$$\ddot{Y} = -2c\dot{V}_X X + 2cV_X \dot{X} = -2cV_X^2$$

Substituting the results for $\ddot{X}$ and $\ddot{Y}$ into the basic acceleration equation gives

$$\bar{a}_P = -2cV_X^2\bar{J}$$

Alternative Approach   The foregoing illustrates how the derived formulas may be utilized to solve problems. As an alternative, one can always fall back on the basic definitions to solve *any* problem as is illustrated below.

For any particle motion problem in Cartesian coordinates we can write

$$\bar{r}_{P/O} = X\bar{I} + Y\bar{J} + Z\bar{K}$$

In this case, from the given information we write

$$\bar{r}_{P/O} = X\bar{I} - cX^2\bar{J} + 0\bar{K}$$

Note that we eliminate Y rather than X because: (a) we know something about X, and (b) the mathematical representation is cleaner. Differentiating, we find the velocity.

$$\bar{v}_P = \dot{X}\bar{I} - 2cX\dot{X}\bar{J}$$

Differentiating again, we find the acceleration.

$$\bar{a}_P = \ddot{X}\bar{I} - (2c\dot{X}^2 + 2cX\ddot{X})\bar{J}$$

It is given that $\dot{X} = V_X = $ constant, therefore

$$\bar{a}_P = -2cV_X^2\bar{J}$$

EXAMPLE 7
The coordinates of a moving particle are given parametrically by $X = 7t$, $Y = 3 + t^2$, and $Z = t^3/3$ m, where t is in seconds. Find the normal and tangential components of velocity and acceleration, the radius of curvature, and the Cartesian components of the unit normal and tangential vectors when $t = 3$ s.

Solution
We were given that

$$\bar{r}_{P/O} = 7t\bar{I} + (3 + t^2)\bar{J} + \frac{t^3}{3}\bar{K}$$

Differentiating this gives

$$\bar{v}_P \equiv \dot{\bar{r}}_{P/O} = 7\bar{I} + 2t\bar{J} + t^2\bar{K} \text{ m/s}$$

from which

$$v_P = |\bar{v}_P| = \sqrt{(7)^2 + (2t)^2 + (t^2)^2} \text{ m/s}$$

This information allows us to determine $\bar{e}_t$ using the facts that the velocity is parallel to $\bar{e}_t$ and that the speed is the $\bar{e}_t$ component of $\bar{v}_P$; that is,

$$v_P = v_P\bar{e}_t$$

so that

$$\bar{e}_t = \frac{\bar{v}_P}{v_P} = \frac{7\bar{I} + 2t\bar{J} + t^2\bar{K}}{\sqrt{(7)^2 + (2t)^2 + (t^2)^2}}$$

For $t = 3$ s, we find that

$$v_P = \sqrt{49 + 36 + 81} = 12.9 \text{ m/s}$$

and

$$\bar{e}_t = 0.543\bar{I} + 0.465\,\bar{J} + 0.698\,\bar{K}$$

What do the three components of $\bar{e}_t$ represent?

To find the path variable components of acceleration, we first find $\bar{a}_P$ in Cartesian coordinates.

$$\bar{a}_P \equiv \dot{\bar{v}}_P = 2\bar{J} + 2t\bar{K}$$

$$= 2\bar{J} + 6\bar{K} \qquad \text{when } t = 3 \text{ s}$$

But we also know that

$$\bar{a}_P = \dot{v}_P\bar{e}_t + \frac{v_P{}^2}{\rho}\bar{e}_n = a_t\bar{e}_t + a_n\bar{e}_n$$

Recall that the dot product of a vector and a unit vector $\bar{e}$ gives the component the vector in the direction of $\bar{e}$. Hence to find a_t we compute

$$\bar{a}_P \cdot \bar{e}_t = a_t = \dot{v}_P$$

so that

$$a_t = \frac{7\bar{I} + 2t\bar{J} + t^2\bar{K}}{\sqrt{49 + 4t^2 + t^4}} \cdot (2\bar{J} + 2t\bar{K})$$

$$= \frac{4t + 2t^3}{\sqrt{49 + 4t^2 + t^4}}$$

At $t = 3$ s, this yields

$$a_t = 5.12 \text{ m/s}^2$$

To find a_n, we match the magnitude of $\bar{a}_P$ as given by path variables and by Cartesian coordinates.

$$\left|\bar{a}_P\right| = \sqrt{a_t{}^2 + a_n{}^2} = \sqrt{(\ddot{X})^2 + (\ddot{Y})^2}$$

At $t = 3$ s, this gives

$$\sqrt{(5.12)^2 + (a_n)^2} = \sqrt{(2)^2 + (6)^2}$$

Then solving for a_n yields

$$a_n = \sqrt{4 + 36 - (5.12)^2} = 3.71 \text{ m/s}^2$$

Finally, we know that $a_n = v_P{}^2/\rho$, so that

$$\rho = \frac{v_P{}^2}{a_n} = \frac{(12.88)^2}{3.71} = 44.7 \text{ m} \qquad \text{when } t = 3 \text{ s} \qquad \triangle$$

Once we have determined a_n, we can find $\bar{e}_n$ by equating the expressions for $\bar{a}_P$ is path variables and Cartesian coordinates. When $t = 3$ s,

$$a_t \bar{e}_t + a_n \bar{e}_n = 2\bar{J} + 6\bar{K}$$

$$(5.12)\frac{1}{12.9}(7\bar{I} + 6\bar{J} + 9\bar{K}) + 3.71\bar{e}_n = 2\bar{J} + 6\bar{K}$$

Thus

$$\bar{e}_n = \frac{1}{3.71}\left[-\frac{7 \times 5.12}{12.88}\bar{I} + \left(2 - \frac{6 \times 5.12}{12.88}\right)\bar{J} \right.$$

$$\left. + \left(6 - \frac{9 \times 5.12}{12.88}\right)\bar{K} \right]$$

$$= -0.75\bar{I} - 0.10\bar{J} + 0.65\bar{K} \qquad \triangle$$

Question: How can we define the osculating plane?

___

HOMEWORK PROBLEMS

II.17 The planar motion of a particle is defined by $X = 4t - 3$, $Y = \sqrt{24t - 12}$, where X and Y are expressed in meters and t is in seconds. Determine the velocity and acceleration of the particle (a) when $t = 1$ s, (b) when $t = 3$ s. Show the results in a sketch of the path.

II.18 The coordinates of a particle in planar motion are defined by $X = e^{t/2}$, $Y = 16e^{-t/2}$, where X and Y are measured in feet and t is in seconds. Show that the path is a hyperbola and determine the velocity and acceleration when $t = 2$ s. Show the solution in a sketch.

II.19 An electron in a cathode-ray tube follows the path

$$Y = 0.7X, Z = \sqrt{X^2 - 2500}$$

where X, Y, and Z are measured in millimeters. The value of X varies with time t according to $X = 1.5(10^6)t^4$, where t is in seconds. Determine the velocity and acceleration of the electron when $X = 150$ mm.

II.20 The roof of a building has the shape of a hyperbolic paraboloid $Z =$

Prob. II.20

Prob. II.22

$XY/10$, where the length unit is meters. A ball rolling on the roof is observed to have the motion $X = 10 - 20t$, $Y = 10 - 20t^2$, where t is in seconds. Determine the velocity and acceleration of the ball when it reaches the edge where $X = 0$.

II.21 Determine the tangent and normal vectors and the radius of curvature of the path of the ball in Problem II.20 at the instant when the ball reaches the edge where $X = 0$. Also determine the rate of increase of the speed in this position.

II.22 The (X, Y) coordinates of a sliding pin connecting the two grooved members of the machine shown in the diagram simultaneously satisfy the relations $X^2 + Y^2 = 2500$, $X = 40 \sin 5\pi t$, where X and Y have units of millimeters and t is in seconds. Determine the velocity and acceleration of the pin when $t = 0$.

II.23 Solve Problem II.22 for the position where $t = \frac{1}{30}$ second.

II.24 For the pin of Problem II.22, determine the tangent and normal components of acceleration and the radius of curvature of the path when $t = \frac{1}{30}$ second.

7 Cylindrical Coordinates

For Cartesian coordinates we selected three independent quantities, namely the X, Y, and Z coordinates of a particle, to uniquely determine its position in the chosen reference frame. Then, in terms of the unit vectors of the coordinate system, we obtained position, velocity, and acceleration formulas for the particle. In this section we follow a similar procedure using a different set of quantities to define the position of a particle.

Suppose that we select an XYZ coordinate system, Figure 11, and wish to define the position of a particle at point P in this reference system by another method. We may begin by dropping a perpendicular point P

Figure 11

onto the XY plane, and calling this projected point P'. For notation, we shall denote the length of the line OP' by R, the angle between the line OP' and the X axis by ϕ, and the length of the line PP' by h. If we know the values of R, ϕ, and h, we obviously know where point P is. These quantities R, ϕ, and h are by definition *extrinsic coordinates,* because they locate the point P, even when the path is unknown.

If we wish to, we can write a unique coordinate transformation to convert between (X, Y, Z) and (R, ϕ, h) if the range of values of ϕ is limited to 2π, for example, $0 \le \phi < 2\pi$. Specifically, these transformations are

$$Z = h \qquad X = R \cos \phi \qquad Y = R \sin \phi$$

so that

$$R = \sqrt{X^2 + Y^2} \quad \text{and} \quad \phi = \tan^{-1} \frac{Y}{X}$$

The terminology we shall use is to call R the radial distance, ϕ the transverse angle, and h the axial distance. The values (R, ϕ, h) are called *cylindrical coordinates.* This name arises because if we hold R constant and vary ϕ, the line PP' follows the surface of a right circular cylinder. Note that for the special case of plane motion in the XY plane, h is zero and the remaining coordinate system (R, ϕ) is called *polar coordinates.*

Having defined the three coordinates, we now require unit vectors to describe their directions. Let $\bar{e}_R$ be the vector that points in the direction in which point P would move if the value of R were increased. Similarly, let a unit vector $\bar{e}_\phi$ point in the direction of increasing ϕ (perpendicular to $\bar{e}_R$), and let $\bar{e}_h$ point in the direction of increasing h. Thus, as we view Figure 11, $\bar{e}_R$ is perpendicular to the Z axis directed outward from point P, $\bar{e}_h$ is parallel to the Z axis, so that $\bar{e}_h \equiv \bar{K}$. Then, because $\bar{e}_R$, $\bar{e}_\phi$, and $\bar{e}_h$ are mutually perpendicular unit vectors, in accordance with the right-hand rule they are related by $\bar{e}_R \times \bar{e}_\phi = \bar{e}_h$. Note that with these definitions only $\bar{e}_h$ is a constant vector; the orientation of $\bar{e}_R$ and $\bar{e}_\phi$ will change as the angle ϕ changes.

Once the set of three independent unit vectors $\bar{e}_R$, $\bar{e}_\phi$, and $\bar{e}_h$ has been defined, it follows that we can express any vector in terms of their radial, transverse, and axial components. Using these unit vectors we shall now determine expressions for the position, velocity, and acceleration of a point.

In Figure 11 we see that the radial component of $\bar{r}_{P/O}$ is R, and the axial component of $\bar{r}_{P/O}$ is h. Note that $\bar{r}_{P/O}$ lies in the plane formed by $\bar{e}_R$ and $\bar{e}_h$. Hence there is no transverse component of position.

$$\boxed{\bar{r}_{P/O} = R\bar{e}_R + h\bar{e}_h} \tag{21}$$

Previously we showed that the values of R, ϕ, and h were required to define the position. Then why doesn't the value of ϕ appear in this expression for $\bar{r}_{P/O}$? The answer is that the equation for $\bar{r}_{P/O}$ really does depend on the value of ϕ, but not explicitly, because one must know ϕ in order to define (locate) the direction of $\bar{e}_R$.

Now let us find the velocity by differentiating $\bar{r}_{P/O}$ in equation (21).

$$\bar{v}_P = \dot{\bar{r}}_{P/O} = \dot{R}\bar{e}_R + R\dot{\bar{e}}_R + \dot{h}\bar{e}_h + h\dot{\bar{e}}_h \tag{22}$$

As in the case of path variables, we must determine the time derivatives of the nonconstant unit vectors. The unit vector $\bar{e}_h$ has constant direction, so $\dot{\bar{e}}_h = 0$. The unit vectors $\bar{e}_R$ and $\bar{e}_\phi$ depend on ϕ only. Using the chain rule with respect to ϕ yields

$$\dot{\bar{e}}_R \equiv \frac{d\bar{e}_R}{dt} = \frac{d\bar{e}_R}{d\phi}\frac{d\phi}{dt} = \dot{\phi}\,\frac{d\bar{e}_R}{d\phi}$$

$$\dot{\bar{e}}_\phi = \dot{\phi}\frac{d\bar{e}_\phi}{d\phi}$$

Figure 12a

Let us look down the Z axis in the negative direction and observe how $\bar{e}_R$ and $\bar{e}_\phi$ change when ϕ is increased by $\Delta\phi$, as shown in Figure 12a. The argument is exactly the same as was developed for unit vectors in path variables. First we form the difference vectors

$$\Delta\bar{e}_R = \bar{e}_R(\phi + \Delta\phi) - \bar{e}_R(\phi)$$

and

$$\Delta\bar{e}_\phi = \bar{e}_\phi(\phi + \Delta\phi) - \bar{e}_\phi(\phi)$$

Figure 12b

as shown in Figures 12b and 12c.

Because $\bar{e}_R(\phi)$ and $\bar{e}_R(\phi + \Delta\phi)$ are unit vectors, we can write

$$\left|\Delta\bar{e}_R\right| = 2\left|\bar{e}_R\right|\sin\frac{\Delta\phi}{2} = 2\sin\frac{\Delta\phi}{2}$$

Figure 12c

Dividing both sides by $\Delta\phi$ and taking the limit as $\Delta\phi \to 0$,

$$\frac{\left|d\bar{e}_R\right|}{d\phi} = \lim_{\Delta\phi\to 0}\frac{(\Delta\bar{e}_R)}{\Delta\phi} = \lim_{\Delta\phi\to 0}\frac{\sin(\Delta\phi/2)}{(\Delta\phi/2)} = 1$$

Therefore $d\bar{e}_R/d\phi$ is a unit vector. From Figure 12b we see that as $\Delta\phi$ goes to zero, $\Delta\bar{e}_R$ approaches the perpendicular to $\bar{e}_R(\phi)$. From Figure 12a we see that the perpendicular in question is in the $\bar{e}_\phi$ direction. Thus,

$$\frac{d\bar{e}_R}{dt} = \frac{d\bar{e}_R}{d\phi}\frac{d\phi}{dt} = \dot{\phi}\bar{e}_\phi \tag{23}$$

By similar reasoning we can show that

$$\frac{d\bar{e}_\phi}{dt} = \frac{d\bar{e}_\phi}{d\phi}\frac{d\phi}{dt} = -\dot{\phi}\bar{e}_R \tag{24}$$

We shall have use for equation (24) shortly. Placing equation (23) into equation (22) yields the standard form of the velocity equation,

$$\boxed{\bar{v}_P = \dot{R}\bar{e}_R + R\dot{\phi}\bar{e}_\phi + \dot{h}\bar{e}_h} \tag{25}$$

Equation (25) is the velocity equation in cylindrical coordinates. The radial component $v_R = \dot{R}$ occurs when the particle moves in the radial direction. The axial component $v_h = \dot{h}$ occurs when the particle moves in the axial direction. The transverse component $v_\phi = R\dot{\phi}$ occurs when ϕ is not a constant. We can see that, for a given value of $\dot{\phi}$, the particle will move faster for larger values of R. The quantity $\dot{\phi}$ is termed the *angular speed*.

We find the acceleration formula by differentiating the velocity (25).

$$\bar{a}_P = \dot{\bar{v}}_P = \ddot{R}\bar{e}_R + \dot{R}\dot{\bar{e}}_R + \dot{R}\dot{\phi}\bar{e}_\phi + R\ddot{\phi}\bar{e}_\phi + R\dot{\phi}\dot{\bar{e}}_\phi + \ddot{h}\bar{e}_h + \dot{h}\dot{\bar{e}}_h \tag{26}$$

Placing equations (23) and (24) into equation (26) and grouping like components gives

$$\boxed{\bar{a}_P = (\ddot{R} - R\dot{\phi}^2)\bar{e}_R + (R\ddot{\phi} + 2\dot{R}\dot{\phi})\bar{e}_\phi + \ddot{h}\bar{e}_h} \tag{27}$$

Let us look at the acceleration equation (27) by considering motion along some simple paths. If a particle moves along a radial line, then only R will change with time so that equation (27) reduces to $\bar{a}_P = \ddot{R}\bar{e}_R$. If a particle moves along a line parallel to the Z axis, only h changes with time and equation (27) reduces to $\bar{a}_P = \ddot{h}\bar{e}_h$. If a particle moves on a circular path of constant radius R parallel to the XY plane, then equation (27) reduces to

$$\bar{a}_P = -R\dot{\phi}^2\bar{e}_R + R\ddot{\phi}\bar{e}_\phi \tag{28}$$

The first term on the right-hand side of equation (28) is the centripetal component of acceleration normal to the path. The second term is the tangential component. For a general motion, equation (27) contains an additional effect. The radial acceleration is a combination arising from a varying radial distance and a centripetal-like motion. The transverse acceleration is a combination of the effect of angular acceleration and an as yet unexplained term $2\dot{R}\dot{\phi}$ which is an interaction brought about by a combination of radial and angular motion. This term is called the *Coriolis acceleration* effect, so named after G. Coriolis (1792–1843), who first disclosed its existence. In a later module we shall deal with this term at length. For now, recognize $2\dot{R}\dot{\phi}$ as the Coriolis component of acceleration, and take solace in the fact that like the other components in (27) it can be measured in field experiments.

EXAMPLE 8

A bicycle wheel rotates about its hub at a constant clockwise angular speed of 10 rad/s. A bead on a spoke approaches the hub with a constant speed of 5 m/s. Determine the acceleration of the bead as a function of the distance from the hub to the head.

Solution

The given information involves an angular speed and a rate of approach to a fixed axis. In addition, the solution desired is a function of the radial distance from that axis, and all the motion takes place in a plane. It follows quite naturally that we formulate the problem in terms of the special case of cylindrical coordinates in a plane, namely polar coordinates.

The accompanying sketch shows R and ϕ and the corresponding unit vectors for a general position. In general we know that

$$\bar{a}_P = (\ddot{R} - R\dot{\phi}^2)\bar{e}_R + (R\ddot{\phi} + 2\dot{R}\dot{\phi})\bar{e}_\phi$$

One important feature of the sketch is that $\bar{e}_R$ always points outward and $\bar{e}_\phi$ is perpendicular to $\bar{e}_R$ in the direction of increasing ϕ. The angular speed is the absolute value of $\dot{\phi}$, and the fact that the rotation is clockwise means that for the situation shown in the diagram ϕ is decreasing. Thus $\dot{\phi} = -10$ rad/s. It is given that the angular speed is constant, therefore $\ddot{\phi} = 0$. We are also told that the bead is approaching the hub at 5 m/s. The radial component of velocity $v_R = \dot{R}$ is the rate that a particle is moving away from the axis, so in this case $\dot{R} = -5$ m/s. Because this is a constant rate, $\ddot{R} = 0$. Thus the equation for $\bar{a}_P$ reduces to

$$\bar{a}_P = -R(10^2)\bar{e}_R + 2(-5)(-10)e_\phi$$

$$= -100R\bar{e}_R + 100\bar{e}_\phi \text{ m/s}^2$$

This answer could be converted to X and Y components by simply expressing the unit vectors $\bar{e}_R$ and $\bar{e}_\phi$ in terms of their components in the $\bar{I}$ and $\bar{J}$ directions, just as we did for the path variable unit vectors $\bar{e}_t$ and $\bar{e}_n$. However, a solution that shows the unit vectors for the descriptive method you choose and gives the magnitude of the components will usually be sufficient. Before leaving this problem, can you see what portions of the solution would have changed if we had defined ϕ as the clockwise angle?

EXAMPLE 9

A particle slides along a path that resembles a circular helix of radius R and vertical displacement $h = bt^2/2$. The velocity vector has a constant known magnitude v_P. (a) Find the acceleration. (b) Find the maximum time T for which such a motion is possible.

Solution
Part a
In general we know that

$$\bar{a}_P = (\ddot{R} - R\dot{\phi}^2)\bar{e}_R + (R\ddot{\phi} + 2\dot{R}\dot{\phi})\bar{e}_\phi + \ddot{h}\bar{e}_h$$

From the given information R is constant, so $\dot{R} = \ddot{R} = 0$. Therefore $\bar{a}_P$ reduces to

$$\bar{a}_P = -R\dot{\phi}^2\bar{e}_R + R\ddot{\phi}\bar{e}_\phi + \ddot{h}\bar{e}_h$$

We were given $h(t)$ and the fact that the speed is constant. We do not as yet know $\dot{\phi}$ and $\ddot{\phi}$. In terms of cylindrical coordinates we have

$$\bar{v}_P = \dot{R}\bar{e}_R + R\dot{\phi}\bar{e}_\phi + \dot{h}\bar{e}_h$$
$$v_P = \left|\bar{v}_P\right| = \sqrt{\bar{v}_P \cdot \bar{v}_P} = \sqrt{(\dot{R})^2 + (R\dot{\phi})^2 + (\dot{h})^2}$$

For the given motion we know that $\dot{h} = bt$, so that

$$v_P = \sqrt{(R\dot{\phi})^2 + b^2t^2}$$

Now, solving for $\dot{\phi}$ gives

$$\dot{\phi} = \frac{1}{R}\sqrt{v_P{}^2 - b^2t^2}$$

To find $\ddot{\phi}$ we differentiate this expression with respect to time.

$$\ddot{\phi} = -\frac{b^2t}{R\sqrt{v_P{}^2 - b^2t^2}}$$

We now can substitute the expressions for $\dot{\phi}$, $\ddot{\phi}$, and $\ddot{h}$ ($\ddot{h} = b$) into the equation for $\bar{a}_P$, with the result that

$$\bar{a}_P = -R\left(\frac{v_P{}^2 - b^2t^2}{R^2}\right)\bar{e}_R - \frac{b^2t}{\sqrt{v_P{}^2 - b^2t^2}}\bar{e}_\phi + b\bar{e}_n \qquad \triangleleft$$

Part b
In the solution for $\bar{a}_P$ in Part a, we saw that the transverse component of acceleration a_ϕ contains the square root of a number that decreases with time, specifically $\sqrt{v_P{}^2 - b^2t^2}$. The acceleration will cease to be a real

quantity when $v_P{}^2 - b^2t^2 < 0$. Thus the largest allowable value $t = T$ is the one that gives

$$v_P{}^2 - b^2T^2 = 0$$

so that

$$T = \frac{v_P}{b}$$

The reason this situation occurs is that we required the particle to have a constant speed v_P, but the axial component $\dot{h}$ increases with time according to $\dot{h} = bt$. At $t = T (= v_P/b)$ we find that $\dot{h} = v_P$, so that at this instant the transverse component of velocity is zero. For $t > T$, the value of $\dot{h}$ would exceed this value of v_P, which is obviously an impossible situation because a component of a vector can never exceed the magnitude of the vector.

Prob. II.25

Prob. II.27

Prob. II.28

HOMEWORK PROBLEMS

II.25 The sliding block C is known to have the instantaneous velocity and acceleration illustrated in the diagram. Determine the corresponding values of $\dot{R}$, $\ddot{R}$, $\dot{\phi}$, and $\ddot{\phi}$ in this position.

II.26 The cylindrical coordinates (R, ϕ, h) of a particle are given by $R = 10t^2$, $\phi = 5t^3$, $h = 2t^4$, where R and h are expressed in meters, ϕ is in radians, and t is in seconds. (a) Determine the expressions for the velocity and acceleration of the particle. (b) Draw a sketch of the path.

II.27 The pendulum shown is moving to the right and slowing down. When $\theta = 15°$ it has a speed of 16 in./s and an acceleration of 20 in./s². Determine the angular speed $\dot{\theta}$ and angular acceleration $\ddot{\theta}$ at this instant.

II.28 The rotating arm AB guides pin P through the fixed spiral, which is defined by $R = 4\phi$, where R has units of inches and ϕ has units of radians. The arm has a constant counterclockwise angular speed of 16 rad/s. Determine the velocity and acceleration of the pin when $\phi = 90°$.

II.29 For the system of Problem II.28, determine the horizontal and vertical components of the velocity and acceleration when $\phi = 45°$.

II.30 When the ball is at the midpoint C of the ramp AB, it is falling at a speed of 10 ft/s and is gaining speed at the rate of 6 ft/s². For the coordinate system shown in the figure at the top of page 55, determine the cylindrical components of the position, velocity, and acceleration vectors.

Prob. II.30

Prob. II.31

II.31 A package is descending a helical chute that is defined by $R = 0.8$ m, $h = 0.4\theta/2\pi$ m, where θ is measured in radians. The motion of the package is defined by $\theta = 0.5t + t^2$, where t is in seconds. Determine the acceleration of the package when $t = 1$ s.

II.32 The velocity and acceleration of a particle in planar motion are fully known in terms of cylindrical coordinates. (a) Show that the tangent and normal unit vectors are given by $\bar{e}_t = (v_R\bar{e}_R + v_\phi\bar{e}_\phi)/(v_R{}^2 + v_\phi{}^2)^{1/2}$ and $\bar{e}_n = (v_R\bar{e}_\phi - v_\phi\bar{e}_R)/(v_R{}^2 + v_\phi{}^2)^{1/2}$. (b) From the expressions in part (a), determine the expressions for the normal and tangential components of acceleration.

Prob. II.33

II.33 A governor consists of two heavy balls A on the end of 25-mm rods that are hinged to the vertical shaft at B. The position of the collar C on the vertical shaft may be adjusted to change the angle θ. The entire device rotates at a constant rate $\omega = 100$ rpm about the vertical. The angular speed $\dot{\theta}$ is a constant 0.2 rad/s. Determine the velocity and acceleration of each ball for $\theta = 30°$.

C. STRAIGHT AND CIRCULAR PATHS

Here we wish to consider the simple particle paths of a straight line and a circle. These paths are an important study unto themselves because of the frequency with which they occur in problems. Further, this study will help increase our understanding of the three kinematical descriptions we have previously presented.

1 Straight Path

Classically, the straight path case has the fancy title *rectilinear motion*. The simplest way of describing the motion, if the given information suits it, is to use path variables. We choose some reference point for $s = 0$ and

Figure 13

take $\bar{e}_t$ to be tangent to the straight line in the direction of increasing s, Figure 13. Because the straight line has an infinite radius of curvature, there is no center of curvature and we can choose $\bar{e}_n$ to be in any convenient direction perpendicular to $\bar{e}_t$, as is done in Figure 13. Then either from the figure and the fact that $\bar{e}_t$ is oriented in a fixed direction, or from the formulas derived for path variables, we find

$$\bar{r}_{P/O} = s\bar{e}_t$$

$$\bar{v}_P = \dot{\bar{r}}_{P/O} = \dot{s}\bar{e}_t$$

$$\bar{a}_P = \dot{\bar{v}}_P = \ddot{s}\bar{e}_t \tag{29}$$

from which we obtain the scalar identities

$$v_P = \dot{s} \qquad a_P = \dot{v}_P = \ddot{s} = v_P \frac{dv_P}{ds} \tag{30}$$

From equation (29) we may conclude that for rectilinear motion, the acceleration, as well as the velocity, is tangent to the path. It is for this reason that $a_P = \dot{v}_P$ in rectilinear motion, whereas for a general path we must use the full path variable equation, (13).

EXAMPLE 10

A car accelerates along a straight, level road from rest to a speed of 100 km/hr in a distance of 400 m. If the speed of the car is proportional to the distance the car has traveled, find the position along the 400-m distance where the acceleration is a maximum. Determine this maximum value.

Solution

We know that the path is a straight line. Information is also given about speed and distance traveled, specifically, $v_P = ks$, where k is a constant to be determined. This type of information indicates the use of path variables. For a straight path we know that the acceleration we seek may be written as $\bar{a}_P = \dot{v}_P\bar{e}_t$. We also know that

$$v_P = ks$$

so

$$\dot{v}_P = v_P \frac{dv_P}{ds} = v_P k = k^2 s$$

From the equation for v_P we can find k, because we were given that at $s = 400$ m,

$$v_P = 100 \ \frac{km}{hr} \left(\frac{1000 \ m}{1 \ km} \right) \left(\frac{1 \ hr}{3600 \ s} \right)$$

$$= 27.78 \ m/s$$

Hence,

$$27.78 = k(400)$$

or

$$k = 0.0695 \ 1/\text{seconds}$$

Then, substituting this value for k into the equation for $\dot{v}_P$, we find that

$$\dot{v}_P = (0.0695)^2 s \ \text{m/s}^2$$

From this we see that $\dot{v}_P$ is a maximum when $s = 400$ m and

$$\dot{v}_{P(\max)} = (0.0695)^2 400 = 1.932 \ \text{m/s}^2 \qquad \triangleleft$$

In the previous discussion we assumed that the information given for rectilinear motion best suited path variables. This, however, may not always be the case. For instance, we may have an inclined straight path in the vertical plane with information given about a horizontal or vertical component of velocity as a function of time, as shown in Figure 14. This information is most naturally suited for Cartesian coordinates.

To treat a problem like this we express $\bar{v}_P(t)$ by its Cartesian components,

$$\bar{v}_P(t) = v_P(t)(\cos \alpha \bar{I} + \sin \alpha \bar{K})$$

and then relate the velocity equation to the information given in the problem. Note too that if the constant angle $\alpha = 0$ or $90°$, the Cartesian unit vectors will be parallel to $\bar{e}_t$ and $\bar{e}_n$.

Similarly, we may have a situation of rectilinear motion with information that fits polar coordinates. Such a case is shown in Figure 15, where the information given for the same inclined path problem would be the angle $\theta(t)$ locating the position vector from the X axis. We then can follow the method derived for cylindrical coordinates in planar motion (changing the symbol ϕ in the formulas to θ). The appropriate unit vectors are shown in Figure 15, and equations (21), (25), and (27) may be used to express the position, velocity, and acceleration.

Figure 14

Figure 15

2 Circular Path

The first question that arises when encountering a problem containing a circular path is what method of description we should use. Those problems that are best suited for Cartesian coordinates will be quite obvious, because the given information will involve the Cartesian components of position, velocity, or acceleration. Therefore let us now consider how to determine when path variables or cylindrical coordinates are most suitable.

We have already treated the use of path variables for circular paths in Section II.B.4. As is true for all problems, the fact that the problem is

Figure 16

Figure 17

posed in terms of distance traveled or speed will indicate that path variables are preferable. The key for the use of cylindrical coordinates in circular motion will be that the significant parameters are radial distance from a fixed point to the particle and an angle or angular speed for this radial line. A subtle aspect of the foregoing is that the fixed point need not necessarily be the center of the circle, as is illustrated in Figure 16.

If we were to draw the path variable unit vectors $\bar{e}_t$ and $\bar{e}_n$, for point P in Figure 16, $\bar{e}_t$ would be tangent to the circle and $\bar{e}_n$ would be perpendicular to $\bar{e}_t$ pointing toward the geometric center of the circle C. Thus, in this case neither of the path variable unit vectors would be parallel to the cylindrical coordinate unit vectors. However, if in some problem the fixed point O is the center C, we have the situation shown in Figure 17, where the sense of the direction of $\bar{e}_t$ is dictated by the way the arc length s is measured. Hence, in this case, $\bar{e}_t \equiv \bar{e}_\theta$ and $\bar{e}_n \equiv -\bar{e}_R$.

EXAMPLE 11

A particle moves on the circular path shown with constant angular speed $\dot{\theta}$, where θ is defined in the sketch. Given that $v_P = v_O$ when the particle passes through point A, determine the velocity and acceleration in terms of the angle θ.

Solution

In this case the fact that we are given information regarding the angle θ, specifically that $\dot{\theta}$ is constant, indicates that we can use cylindrical coordinates. We have shown R, $\bar{e}_R$, and $\bar{e}_\theta$ in the sketch accompanying the problem. From this diagram we may write

$$\bar{r}_{P/O} = R\bar{e}_R$$

and because angle OPA is a right angle,

$$\cos \theta = \frac{R}{D} \quad \text{or} \quad R = D \cos \theta$$

As always in problem solving, we have the choice of substituting into the derived formula for velocity or differentiating the position vector using the formulas for the rates of change of the unit vectors. Usually the former approach is faster, but for variety let us follow the latter, recalling that $\dot{\bar{e}}_R = \dot{\theta}\bar{e}_\theta$, and $\dot{\bar{e}}_\theta = -\dot{\theta}\bar{e}_R$. In either approach we shall need to use the chain rule because time is given implicitly. That is, R is a function of θ and θ is a function of time.

$$\bar{v}_P = \dot{\bar{r}}_{P/O} = \frac{d}{dt}(D \cos \theta \bar{e}_R) = -(D \sin \theta)\dot{\theta}\bar{e}_R + D \cos \theta \dot{\bar{e}}_R$$

$$= \dot{\theta}D(-\sin \theta \bar{e}_R + \cos \theta \bar{e}_\theta)$$

Noting that $\dot{\theta}$ is a constant, we then find that $\ddot{\theta} = 0$. Thus when we differentiate $\bar{v}_P$ to find $\bar{a}_P$, we obtain

$$\bar{a}_P = \dot{\bar{v}}_P = \dot{\theta}D[-(\cos\theta)\dot{\theta}\bar{e}_R - \sin\theta\dot{\bar{e}}_R$$
$$- (\sin\theta)\dot{\theta}\bar{e}_\theta + \cos\theta\dot{\bar{e}}_\theta]$$
$$= -2\dot{\theta}^2 D(\cos\theta\bar{e}_R + \sin\theta\bar{e}_\theta)$$

We still must determine $\dot{\theta}$. A piece of information we have not used is that $v_P = v_0$ at point A. From our result for $\bar{v}_P$ we can form

$$v_P = |\bar{v}_P| = \dot{\theta}D\sqrt{\sin^2\theta + \cos^2\theta} = \dot{\theta}D$$

From this we find that v_P is constant because $\dot{\theta}$ and D are both constants. It was given that $v_P = v_0$ at point A, so we can solve for $\dot{\theta}$.

$$\dot{\theta} = \frac{v_0}{D}$$

Placing this into our result for $\bar{a}_P$ gives

$$\bar{a}_P = -\frac{2v_0^2}{D}(\cos\theta\bar{e}_R + \sin\theta\bar{e}_\theta) \qquad \triangleleft$$

Let us check this result by using path variables. For constant speed,

$$\bar{a}_P = \frac{v_P^2}{\rho}\bar{e}_n = \frac{v_0^2}{\rho}\bar{e}_n$$

For a circle, the radius of curvature is the radius of the circle, so that $\rho = D/2$. From the sketch we have

$$\bar{e}_n = -\cos\theta\bar{e}_R - \sin\theta\bar{e}_\theta$$

Substituting for ρ and $\bar{e}_n$ in the path variable expression gives

$$\bar{a}_P = \frac{v_0^2}{(D/2)}(-\cos\theta\bar{e}_R - \sin\theta\bar{e}_\theta)$$
$$= -\frac{2v_0^2}{D}(\cos\theta\bar{e}_R + \sin\theta\bar{e}_\theta)$$

This is the same result as the one we obtained using cylindrical coordinates. If you were skeptical earlier when we said that the resulting *vectors* for velocity and acceleration are the same regardless of what kinematical description is used, you should believe it now!

D. PROBLEM SOLVING

By now you should have a reasonable feel for the meaning of the variables and unit vectors that appear in the three kinematical descriptions presented. Problem solving using these descriptions (i.e., picking the right one) is not an art, but rather a professional decision made by recognizing the proper keys. In what follows we present a logical way of thinking to identify the keys. It is not meant for memorization, but rather as a guide while working problems. Work enough problems *correctly* and the key will become part of your thinking process.

1 Read the problem carefully and express all quantitative information in equation form, at the same time converting the information to a consistent set of units. For instance, if a problem states "After the particle has traveled 700 ft, it has a speed of 30 miles/hr," you should write down

$$\text{when } s = 700 \text{ ft,}$$

$$v_P = 30 \frac{\text{miles}}{\text{hr}} \left(\frac{88 \text{ ft/s}}{60 \text{ miles/hr}} \right) = 44 \text{ ft/s}$$

2 If the path is defined but no diagram is provided, make a sketch of the path.
3 Choose path variables, Cartesian coordinates, or cylindrical coordinates. Make this choice based on which formulation most closely fits the information given in the problem and, equally important, based on the solution variables whose values are sought. Select the coordinate parameters and unit vectors and show them in the sketch. Now, in terms of these coordinates, write general geometric relationships for the lengths and angles.
4 Write down the formulas in the kinematical description you have chosen that relate the given information to the variables of the solution. After studying the formulas and the given information, from the sketch write any additional information needed to relate the system's geometry.
5 Solve the appropriate equations. In the illustrative problems, we shall refer to these steps by their appropriate numbers.

ILLUSTRATIVE PROBLEM 1

A slider moves in the vertical plane along a wire that is bent into a hyperbolic shape according to the equation $XZ = 36$ m^2. Determine the horizontal and vertical components of the acceleration of the slider when it is at $X = 6$ m if (a) the slider has a constant speed of 10 m/s directed as shown, and (b) the slider has a constant vertical component of velocity of 10 m/s downward.

Solution

In both parts of this problem we know the path and desire the Cartesian components of the acceleration. However, in (a) the speed is given, which suggests the use of path variables. In (b), a Cartesian component of velocity is given, which suggests the use of Cartesian coordinates.

Part a

Step 1 $XZ = 36$ m², find a_X and a_Z when $X = 6$ m, given $v_P = 10$ m/s, and $\dot{v}_P = 0$ because v_p is constant.

Steps 2 and 3 Use path variables.

$$Z = \frac{36}{X}$$

$$\alpha = \tan^{-1}\left|\frac{dZ}{dX}\right|$$

Step 4

$$\bar{a} = \dot{v}_P \bar{e}_t + \frac{v_P{}^2}{\rho}\bar{e}_n$$

We know v_P and $\dot{v}_P$, but we must find the curvature $1/\rho$. This is motion in a plane, so

$$\frac{1}{\rho} = \frac{\left|d^2Z/dX^2\right|}{[1 + (dZ/dX)^2]^{3/2}}$$

Step 5

$$Z = \frac{36}{X} \qquad \frac{dZ}{dX} = \frac{-36}{X^2} \qquad \frac{d^2Z}{dX^2} = \frac{72}{X^3}$$

When $X = 6$,

$$\frac{1}{\rho} = \frac{(72/6^3)}{[1 + (36/6^2)^2]^{3/2}} = 0.118 \ 1/m$$

$$\bar{a}_P = 0\bar{e}_t + 10^2(0.1180)\bar{e}_n = 11.80\bar{e}_n \ m/s^2$$

In most cases we may consider the solution to be completed if we have shown the unit vectors in a diagram and determined the corresponding components. However, in this problem we are requested to express $\bar{a}_P$ by its $\bar{I}$ and $\bar{K}$ components. To do this, we express $\bar{e}_n$ (and $\bar{e}_t$ if it were necessary) by its Cartesian components. From the diagram (Steps 2 and 3 above) we have

$$\bar{e}_n = \cos(90° - \alpha)\bar{I} + \sin(90° - \alpha)\bar{K}$$

but

$$\alpha = \tan^{-1}\left|\frac{-36}{6^2}\right| = \tan^{-1}(1.0) = 45°$$

so that

$$\bar{e}_n = 0.707\bar{I} + 0.707\bar{K}$$

$$\bar{a}_P = 11.80 \times 0.707(\bar{I} + \bar{K}) = 8.34(\bar{I} + \bar{K})$$

Thus

$$a_X = 8.34 \text{ m/s}^2 \qquad a_Z = 8.34 \text{ m/s}^2$$

As an aside, you can verify the transformation to Cartesian components by checking to see that the magnitude of the transformed vector is the same as that of the original vector. In this problem the path variable expression yielded $|\bar{a}_P| = 11.80$, whereas the Cartesian expression gives

$$|\bar{a}_P| = \sqrt{(8.34)^2(2)} = 11.80$$

Thus, the transformation checks.

Part b

Step 1 $XZ = 36$ m², find a_X and a_Z when $X = 6$ m, given $v_Z = \dot{Z} = -10$ m/s. $\dot{Z}$ is negative because the particle is moving down and Z is decreasing, and $\ddot{Z} = 0$ because $\dot{Z}$ is constant.

Steps 2 and 3 Use Cartesian coordinates.

Step 4 $\bar{a}_P = \ddot{X}\bar{I} + \ddot{Z}\bar{K}$, $\dot{Z} = -100$ m/s², $\ddot{Z} = 0$. From the geometry of the problem we may write $X = 36/Z$. Differentiating twice yields

Step 5

$$\dot{X} = -\frac{36}{Z^2}\dot{Z} \qquad \ddot{X} = \frac{72}{Z^3}(\dot{Z})^2 - \frac{36}{Z^2}\ddot{Z}$$

Thus, when $X = 6$ m, $Z = 36/6 = 6$ m, $\ddot{X} = 72(10^2/6^3) = 33.3$ m/s² and

$$\bar{a}_P = 33.3 \text{ m/s}^2\ \bar{I}$$

Note that the difference between the accelerations in Parts a and b is substantial, even though the given information seems so similar.

ILLUSTRATIVE PROBLEM 2

A high-speed railroad train travels along a straight track at a speed of 50 m/s. A photographer stands 10 m from the track and wishes to pan (rotate) the camera so that the camera always points directly at the locomotive. Determine what values for the rate of rotation ω of the camera and rate of increase of ω are necessary to accomplish this.

Solution

Here two methods come to mind. One obvious method to solve the problem is to express ϕ in terms of the distance traveled by the locomotive, and then differentiate the expression to find expressions for $\omega(=\dot{\phi})$ and $\dot{\omega}$ as a function of ϕ. This approach is fine, but the derivatives will be complicated. Another approach, the one we shall follow here, does not require differentiation.

Step 1 $v_P = 50$ m/s, $\dot{v}_P = 0$ along a straight path. The distance from fixed point O (photographer's location) to the path is 10 m. Find ω and $\dot{\omega}$ for the radial line from point O to the locomotive.

Steps 2 and 3   Use cylindrical coordinates with $h = 0$ (plane polar coordinates), because an angular speed and angular acceleration of a radial line are desired.

Step 4

$$\bar{v}_P = \dot{R}\bar{e}_R + R\dot{\phi}\bar{e}_\phi$$

$$\bar{a}_P = (\ddot{R} - R\dot{\phi}^2)\bar{e}_R + (R\ddot{\phi} + 2\dot{R}\dot{\phi})\bar{e}_\phi$$

In terms of the problem variables, $\dot{\phi} = \omega$, $\ddot{\phi} = \dot{\omega}$.

Step 5   We know that $\bar{v}_P$ is tangent to the path and that $\bar{a}_P = \bar{0}$ (because $\bar{v}_P$ is constant in magnitude and direction). From the sketch we can express $\bar{v}_P$ by its polar components.

$$\bar{v}_P = 50(\sin \phi \bar{e}_R + \cos \phi \bar{e}_\phi)$$

But the polar coordinates formula gives

$$\bar{v}_P = \dot{R}\bar{e}_R + R\omega\bar{e}_\phi$$

Thus matching components yields

$$\bar{e}_R \text{ direction:} \qquad \dot{R} = 50 \sin \phi$$

$$\bar{e}_\phi \text{ direction:} \qquad R\omega = 50 \cos \phi$$

From step 3 we have $R = 10/\cos \phi$, so that

$$\dot{R} = 50 \sin \phi \text{ m/s}$$

and

$$\omega = \frac{50 \cos \phi}{R} = 5 \cos^2 \phi \text{ rad/s}$$

Next, we know that $\bar{a}_P = \bar{0}$, so that the transverse component of $\bar{a}_P$ is zero. Hence from the polar coordinate formula we have

$$a_R = R\dot{\omega} + 2\dot{R}\omega = 0$$

$$\dot{\omega} = -\frac{2\dot{R}\omega}{R} = \frac{-2 \times 50 \sin \phi \times 5 \cos^2 \phi}{(10/\cos \phi)}$$

$$\dot{\omega} = -50 \sin \phi \cos^3 \phi \text{ rad/s}^2$$

ILLUSTRATIVE PROBLEM 3

A race car traveled at 200 miles/hr along a straight path and then entered a 1000-ft radius curve. Immediately upon entering the curve, the driver applied his brakes and the speed decreased at a 2g rate ($g = 32.17$ ft/s²). Determine the maximum total acceleration of the car and where this maximum occurred. Draw a sketch of this maximum acceleration.

Solution

Step 1 Initially,

$$v = 200 \frac{\text{miles}}{\text{hr}} \left(\frac{88 \text{ ft/s}}{60 \text{ miles/hr}} \right) = 293.3 \text{ ft/s}$$

Also, $\dot{v} = -2g = -64.34$ ft/s² ($\dot{v}$ is negative, because v is decreasing). The path has a constant radius of 1000 ft, so it is a circle. Find $|\bar{a}_P|_{\max}$ and the corresponding location, and draw a sketch of $\bar{a}_P$.

Steps 2 and 3 We use path variables because we know the path and have information about the speed and rate of increase of the speed.

Step 4

$$\bar{a}_P = \dot{v}_P \bar{e}_t + \frac{v_P{}^2}{\rho} \bar{e}_n$$

Step 5 $\dot{v}_P = -64.34$ ft/s² and $\rho = 1000$ ft. Then the magnitude of the acceleration is

$$|\bar{a}_P| = \sqrt{\dot{\bar{v}}_P \cdot \dot{\bar{v}}_P} = \sqrt{(\dot{v}_P)^2 + \left(\frac{v_P{}^2}{\rho} \right)^2}$$

$$= \sqrt{(64.34)^2 + \left(\frac{v_P{}^2}{1000} \right)^2}$$

Clearly $|\bar{a}_P|$ is a maximum when v_P is a maximum. This occurs at the beginning of the curve, where $v_P = 293.3$ ft/s. Thus

$$a_n = \frac{v_P{}^2}{\rho} = \frac{(293.3)^2}{1000} = 86.02 \text{ ft/s}^2$$

$$a_t = \dot{v}_P = -64.34 \text{ ft/s}^2$$

$$\left| \bar{a}_P \right| = \sqrt{(64.34)^2 + (86.02)^2} = 107.4 \text{ ft/s}^2$$

ILLUSTRATIVE PROBLEM 4

The voltage applied to a cathode-ray tube is adjusted so that the position coordinates of an electron vary with time according to $Z = At$, $X = A \cos pt$, $Y = A \sin bt$. Show (a) that for this motion the acceleration of the electron is always parallel to the perpendicular line from the electron to the Z axis, and (b) that the magnitude of the acceleration is proportional to the length of this line.

Solution

Step 1 We are given $X(t)$, $Y(t)$, and $Z(t)$ and wish to determine some properties of $\bar{a}_P$.

Steps 2 and 3 We have the Cartesian coordinates of the electron, so clearly we shall adopt them for the kinematical description.

Step 4 $\bar{r}_{P/O} = X\bar{I} + Y\bar{J} + Z\bar{K}$ with $X(t)$, $Y(t)$, and $Z(t)$ given. We want to show that $\bar{a}_P$ is parallel to $\bar{r}_{P/O'}$ and that $\left| \bar{a}_P \right|$ is proportional to $\left| \bar{r}_{P/O'} \right|$. Hence the kinematical expression we want can be obtained by straightforward differentiation of the position vector, bearing in mind that $\bar{I}$, $\bar{J}$, and $\bar{K}$ are constant.

Step 5 Differentiating twice yields

$$\bar{r}_{P/O} = X\bar{I} + Y\bar{J} + Z\bar{K}$$

$$= A \cos pt\bar{I} + A \sin pt\bar{J} + At\bar{K}$$

$$\bar{v}_P = -Ap \sin pt\bar{I} + Ap \cos pt\bar{J} + A\bar{K}$$

$$\bar{a}_P = -Ap^2 \cos pt\bar{I} - Ap^2 \sin pt\bar{J}$$

Part a
From the sketch we see that

$$\bar{r}_{P/O'} = X\bar{I} + Y\bar{J}$$

$$= A \cos pt\bar{I} + A \sin pt\bar{J}$$

Therefore

$$\bar{a}_P = -p^2\bar{r}_{P/O'}$$

This shows that $\bar{a}_P$ is in the opposite direction from $\bar{r}_{P/O'}$.

Part b
Taking the absolute value of both sides of the result for $\bar{a}_P$, we have

$$\left|\bar{a}_P\right| = p^2\left|\bar{r}_{P/O'}\right|$$

and we see that the magnitude of the acceleration is proportional to $\left|\bar{r}_{P/O'}\right|$.

   This last portion of the problem may seem trivial to you, but it is important because it is one of the things that differentiates the technician from the engineer: We did not leave the answer in the form of a number, but went to the further effort of interpreting the result.

___

ILLUSTRATIVE PROBLEM 5
A helicopter descends at a speed of 10 m/s with an acceleration of 0.5g. The main rotor blades are 5 m long, oriented horizontally, and are rotating at a constant rate of 450 rpm. What is the speed and magnitude of the acceleration of a point on the tip of a blade?

$v = 10$ m/s
$a = 0.5g$

$\dot{\varphi} = 15\pi$ rad/s

Solution

Step 1   Given $v_h = 10$ m/s; $a_h = 0.5g[(9.806 \text{ m/s}^2)/1g] = 4.903$ m/s², both down, $\omega_{\text{rotor}} = 450$ rpm $(2\pi$ rad/rev) $(1$ min/60 s) $= 15\pi$ rad/s, $\dot{\omega} = 0$. (Note that the angular units for rotation must be radians). Find $\left|\bar{v}_P\right|$ and $\left|\bar{a}_P\right|$.

Steps 2 and 3   The motion of the axis of the rotor shaft is the same as that of the helicopter. Therefore, we know the rate of rotation of a radial line from the rotor axis to a blade tip and the motion of the tip parallel to the rotor shaft axis. This is ideal for cylindrical coordinates. If we measure h positive downward, then $\bar{e}_h$ is parallel to the velocity and acceleration of the helicopter. To produce a right-handed system we now measure positive ϕ in the direction of rotation.

Step 4

$$\bar{v}_P = \dot{R}\bar{e}_R + R\dot{\phi}\bar{e}_\phi + \dot{h}\bar{e}_h$$

$$\left|\bar{v}_P\right| = \sqrt{\bar{v}_P \cdot \bar{v}_P}$$

$$\bar{a}_P = (\ddot{R} - R\dot{\phi}^2)\bar{e}_R + (R\ddot{\phi} + 2\dot{R}\dot{\phi})\bar{e}_\phi + \ddot{h}\bar{e}_h$$

$$\left|\bar{a}_P\right| = \sqrt{\bar{a}_P \cdot \bar{a}_P}$$

Step 5

$$R = 5 \text{ m} \qquad \dot{R} = \ddot{R} = 0$$

$$\dot{\phi} = 15\pi \text{ rad/s}, \; \ddot{\phi} = 0$$

$$\dot{h} = 10 \text{ m/s} \qquad \ddot{h} = 4.90 \text{ m/s}^2$$

$$\bar{v}_P = 5(15\pi)\bar{e}_\phi + 10\bar{e}_h, \; |\bar{v}_P| = \sqrt{(75\pi)^2 + 10^2}$$

$$\qquad = 235.8 \text{ m/s}$$

$$\bar{a}_P = -5(15\pi)^2\bar{e}_R + 4.90\bar{e}_h$$

$$|\bar{a}_P| = \sqrt{(1125\pi^2)^2 + (4.90)^2} = 1.110(10^4) \text{ m/s}^2$$

ILLUSTRATIVE PROBLEM 6

In order to ease the shock on packages moving around a curve, a conveyor belt is designed with a transition curve in which the radius of curvature gets smaller as the package moves along the curve. The radius of curvature is given in graphical form in the diagram as a function of the distance s the package has traveled along the conveyor belt. Draw a graph of the acceleration in g's that the package undergoes as a function of s when the conveyor belt is moving at 100 in./s.

Solution

Step 1 $v_P = 100$ in./s, $\dot{v}_P = 0$, $\rho(s)$ given. Find $|\bar{a}_P|$ in units of g.

Steps 2 and 3 We are given the speed and radius of curvature as a function of distance traveled (arc length). This information fits path variables.

Step 4

$$\bar{a}_P = \dot{v}_P\bar{e}_t + \frac{v_P^2}{\rho}\bar{e}_n$$

Step 5

$$\bar{a}_P = 0\bar{e}_t + \frac{(100)^2}{\rho}\bar{e}_n \text{ in./s}^2$$

To find $|\bar{a}_P|$ in units of g, divide by

$$g = 32.17 \ \frac{ft}{s^2} \left(\frac{12 \ in.}{1 \ ft} \right) = 386 \ in./s^2$$

Then

$$|\bar{a}_P| = \frac{10,000}{386\rho} g \qquad \text{(units for } \rho \text{ are inches)}$$

This results in the graph shown.

ILLUSTRATIVE PROBLEM 7

A portion of a machine consists of a block that is free to slide along a rotating rod. The block is tied by means of an inextensible string that is wrapped around a fixed drum, as shown. The length L of the string shown varies according to $L = 5 + 0.3\theta$ (L in meters and θ in radians). If θ varies according to the formula $\theta = (\pi/6)(3 - 2t^2)$, what is the acceleration of point A on the block when $t = 1$ s?

Solution

Step 1 $L = 5 + 0.3\theta$, $\theta = (\pi/6)(3 - 2t^2)$, plane motion. Find $\bar{a}_P$.

Steps 2 and 3 We are given the radial distance from a fixed point and the angle of rotation of the radial line. Therefore we can use polar coordinates, changing the symbols for these coordinates from R, ϕ to L, θ.

Step 4

$$\bar{a}_P = (\ddot{L} - L\dot{\theta}^2)\bar{e}_L + (L\ddot{\theta} + 2\dot{L}\dot{\theta})\bar{e}_\theta$$

Step 5 We have $L(\theta)$ and $\theta(t)$, which suggests the use of the chain rule to find $\dot{L}$, $\ddot{L}$. For instance,

$$\dot{L} = \dot{\theta}\frac{dL}{d\theta}$$

A more direct alternative is to substitute $\theta(t)$ into $L(\theta)$ to find

$$L(t) = 5 + 0.3 \times \frac{\pi}{6}(3 - 2t^2)$$

Continuing with the latter,

$$\dot{L} = -0.2\pi t \qquad \ddot{L} = -0.2\pi$$

$$\dot{\theta} = -\frac{2\pi}{3}t \qquad \ddot{\theta} = -\frac{2\pi}{3}$$

When $t = 1$ s, $L = 5 + 0.05\pi = 5.157$ m, $\dot{L} = -0.2\pi$ m/s, $\ddot{L} = -0.2\pi$ m/s², $\theta = \pi/6$ rad $= 30°$, $\dot{\theta} = -2\pi/3$ rad/s, $\ddot{\theta} = -2\pi/3$ rad/s².

$$\bar{a}_P = a_L \bar{e}_L + a_\theta \bar{e}_\theta$$

$$\bar{a}_P = \left[-0.2\pi - 5.157\left(\frac{2\pi}{3}\right)^2 \right] \bar{e}_L + \left[5.157\left(\frac{-2\pi}{3}\right) \right.$$

$$\left. + 2(-0.2\pi)\left(-\frac{2\pi}{3}\right) \right] \bar{e}_\theta$$

$$= -23.25\bar{e}_L - 8.17\bar{e}_\theta \text{ m/s}^2$$

To convert to Cartesian components, if so desired, we can express (from the sketch) $\bar{e}_L$ and $\bar{e}_\theta$ in terms of their Cartesian components.

$$\bar{e}_L = \cos 30°\bar{I} + \sin 30°\bar{J}$$

$$\bar{e}_\theta = -\sin 30°\bar{I} + \cos 30°\bar{J}$$

$$\bar{a}_P = (a_L \cos 30° - a_\theta \sin 30°)\bar{I}$$

$$+ (a_L \sin 30° + a_\theta \cos 30°)\bar{J}$$

As we have pointed out a number of times, unless expressly asked for, either form will suffice.

ILLUSTRATIVE PROBLEM 8

A ball is thrown at an initial angle of elevation α_0 with an initial speed v_0. The horizontal and vertical distances traveled by the ball in terms of the coordinates shown are given by

$$X = (v_0 \cos \alpha_0)t \qquad Y = (v_0 \sin \alpha_0)t - \tfrac{1}{2}gt^2$$

(We shall show how to obtain this result in the next module.) Determine the radius of curvature of the path and the rate of increase of the speed at the instant of release ($t = 0$).

Solution

Step 1 $X = (v_0 \cos \alpha_0)t$, $Y = (v_0 \sin \alpha_0)t - \tfrac{1}{2}gt^2$. Find ρ and $\dot{v}_P$ when $t = 0$.

Steps 2 and 3   If we wished, we could approach this problem by a scalar method as follows. To find ρ we could determine the equation of the path, that is, Y as a function of X, and then use the formula for ρ. To find $\dot{v}_P$ we could write a formula for v_P as a function of time (using $v_P = \sqrt{(\dot{X})^2 + (\dot{Y})^2}$), and then differentiate.

Such a solution would result in a cumbersome set of calculations. We can obtain an easier solution if we note that $\dot{v}_P$ and ρ both occur in the acceleration formula for path variables. In this approach we relate the acceleration of the particle in Cartesian coordinates to the description in path variables. Let us pursue this method.

Step 4

$$\bar{v}_P = \dot{X}\bar{I} + \dot{Y}\bar{J} = v_P\bar{e}_t$$

$$\bar{a}_P = \ddot{X}\bar{I} + \ddot{Y}\bar{J} = \dot{v}_P\bar{e}_t + \frac{v_P^2}{\rho}\bar{e}_n$$

Step 5

$$\dot{X} = v_0 \cos \alpha_0 \qquad \dot{Y} = v_0 \sin \alpha_0 - gt$$

$$\ddot{X} = 0 \qquad \ddot{Y} = -g$$

Thus

$$\bar{v}_P = v_0 \cos \alpha_0 \bar{I} + (v_0 \sin \alpha_0 - gt)\bar{J}$$

$$\bar{a}_P = -g\bar{J}$$

$$v_P = |\bar{v}_P| = \sqrt{(v_0 \cos \alpha_0)^2 + (v_0 \sin \alpha_0 - gt)^2}$$

$$= \sqrt{v_0^2 - 2v_0 gt \sin \alpha_0 + g^2 t^2}$$

We can now find the tangent and normal components of $\bar{a}_P$ with the aid of the sketch shown above.

$$\bar{J} = \sin \alpha \bar{e}_t - \cos a \bar{e}_n$$

$$\bar{a}_P = -g(\sin \alpha \bar{e}_t - \cos \alpha \bar{e}_n) = -g \sin \alpha \bar{e}_t + g \cos \alpha \bar{e}_n$$

$$= \dot{v}_P\bar{e}_t + \frac{v_P^2}{\rho}\bar{e}_n$$

Thus, by equating components,

$$\bar{e}_t \text{ direction:} \qquad \dot{v}_P = -g \sin \alpha$$

$$\bar{e}_n \text{ direction:} \qquad \frac{v_P^2}{\rho} = g \cos \alpha$$

At the initial position $\alpha = \alpha_0$ and $v_P = v_0$; the results are

$$\dot{v}_P = -g \sin \alpha_0 \qquad \rho = \frac{v_0^2}{g \cos \alpha_0} \qquad \triangleleft$$

If the results at a general instant were desired, the angle α could be found by expressing $\bar{e}_t$ by its Cartesian components, that is, $\bar{e}_t = \cos \alpha \bar{I} + \sin \alpha \bar{J}$, so that

$$\bar{v}_P = v_P \bar{e}_t = v_P \cos \alpha \bar{I} + v_P \sin \alpha \bar{J} = \dot{X}\bar{I} + \dot{Y}\bar{J}$$

$$\tan \alpha = \frac{v_P \sin \alpha}{v_P \cos \alpha} = \frac{\dot{Y}}{\dot{X}} = \frac{v_0 \sin \alpha_0 - gt}{v_0 \cos \alpha_0}$$

This expression for α can be checked by setting $t = 0$, from which we obtain an identity for tan α_0. Noting that the value of v_P can be found from the derived result, in general we have

$$\dot{v}_P = -g \sin \alpha$$

$$\rho = \frac{v_P{}^2}{g \cos \alpha}$$

Prob. II.34

Prob. II.35

Prob. II.36

Prob. II.39

HOMEWORK PROBLEMS

II.34 A pin slides in two yokes as shown. Yoke A has a velocity of 50 mm/s to the right and an acceleration of 20 mm/s² to the left whereas yoke B has a constant downward velocity of 10 mm/s. For the position shown determine the radius of curvature of the pin's path and sketch where the center of curvature is located.

II.35 A *lazy tongs* consisting of bars 1 m long is pivoted at a fixed point O. The pin A is constrained to move along OA. Find the velocity and acceleration of pin A when $\theta = 47°$ and $\dot{\theta}$ is a constant 3 rad/s clockwise. *Hint:* Express the distance from point O to point A in terms of the angle θ.

II.36 A hydraulic piston inside arm AB causes the length of the arm to increase at the constant rate of 3 in./s. Determine (a) the velocity and acceleration of the pin where $\theta = 90°$, (b) the corresponding angular velocity $\dot{\phi}$ and the angular acceleration $\ddot{\phi}$ of arm AB.

II.37 Solve Problem II.36 for the position where $\theta = 120°$.

II.38 The acceleration of a satellite in a circular orbit of radius R around the earth is $g(R_e/R)^2$, where $R_e = 6370$ km is the radius of the earth and $g = 9.806$ m/s². Determine the value of R for which the satellite is in a synchronous orbit, which is an orbit that requires 1 day for a complete revolution. Also determine the corresponding speed of the satellite. (Consider a day to be 24 hr.)

II.39 The angle of rotation of arm OA about the eccentric pivot point O is defined by $\theta = 200t$, where θ is in radians and t is in seconds. A spring within the telescoping arm keeps end A in contact with the circular hole. Determine the velocity and acceleration of point A when $\theta = 90°$.

II.40 Solve Problem II.39 for the instant when $\theta = 30°$.

II.41   At the instant shown the telescoping passenger ramp is 45 m long and is being extended at the constant rate of 0.5 m/s. What is the constant rotation rate $\dot\phi$ that will produce an acceleration of 1 m/s² at end A at this instant? What is the corresponding speed of end A?

Prob. II.41

II.42   It is known that the speed of a rocket sled on a straight track is related to the distance it travels by $v = 2000s - 500s^2$, where v is in kilometers per hour and s is in km. Determine (a) the maximum speed and the location where it occurs, (b) the maximum acceleration and the location where it occurs, (c) the location where the sled comes to rest and the acceleration in that position.

Prob. II.42

II.43   A block slides along the bent road $ABCD$ such that the distance traveled is given by $s = 30t + 5t^2$, where s has units of inches and t is seconds. Determine the acceleration of the block as a function of t. Draw a graph showing the results for each acceleration component.

Prob. II.43

Prob. II.44

II.44   Two bars are welded to a nut that advances along a bolt whose pitch is 4 mm. The 80-mm bar-nut system rotates at the constant rate of 30 rpm. What is the velocity and acceleration of a point at the free end of one of the bars?

II.45 A pipe rotates about the vertical axis at 30 rpm as water flows through it at 20 m/s. Determine the velocity, acceleration, and radius of curvature of the path of the particle of fluid that is at $R = 0.5$ m from the vertical axis.

Prob. II.45

Prob. II.46

II.46 The propeller on the toy shown is pushed along a threaded shaft so that it flies into the air while rotating. The pitch of the thread causes the propeller to revolve once every tenth of an inch. If the hub of the propeller is moved along the shaft with a speed t^2 in./s, determine (a) how long it takes the hub to move the length of the shaft, and (b) the magnitude of the acceleration of the tip of the propeller blade when the propeller leaves the shaft.

II.47 A spring of unstretched length l is attached to the block. Functionally, the speed of the block is

$$v = \pm \sqrt{v_0^2 - K_1 x^2 - \tfrac{1}{2} K_2 x^4}$$

where the sign for v is chosen to fit the direction of the motion. Determine the acceleration of the block as a function of x.

Prob. II.47

II.48 The acceleration of a satellite in an elliptical orbit has a magnitude of $g(R_e/R)^2$, where $g = 9.806$ m/s² and $R_e = 6370$ km is the radius of the earth. Also, the acceleration is always directed toward the center of the earth. At point A the satellite has a speed of 25,000 km/hr. What is the rate of change of the speed and the radius of curvature of the path at this location?

Prob. II.48

Prob. II.49

Prob. II.51

II.49 The block is pulled outward along the slot in the rotating arm AB by the cable whose length is 3 ft. The arm AB has a constant angular speed of 4 rad/s. Determine the speed and magnitude of the acceleration of the block when $\theta = 36.87°$. *Hint:* The rate of increase of the distances R and l are identical because the cable has a constant length.

II.50 Determine the horizontal and vertical components of the velocity and acceleration of the block in Problem II.49 when $\theta = 30°$.

II.51 The amusement park ride rotates about its vertical axis at the constant rate of 10 rpm as the arms swing in and out. At the instant when $\theta = 30°$ the arms are swinging outward at $\dot{\theta} = 0.3$ rad/s and this rate is decreasing at 0.1 rad/s². Determine the acceleration of the cabin at this instant.

II.52 The pin is guided through the fixed straight groove AB by arm CD, which moves horizontally to the right according to $X = 200(1 - e^{-t/2})$ mm. Determine the speed and acceleration of the pin along groove AB in the position where $X = 100$ mm.

Prob. II.52

Prob. II.53

II.53 The diagram shows an elliptic curve forming a portion of an automobile race track. For maximum performance the driver operates the vehicle such that the normal component of acceleration is constant at $1.2g$. Given that $x^2 + 4y^2 = 4(10^4)$ m², determine the speed of the vehicle at points A, B, and C.

Prob. II.54

II.54 An observer at point O on the ground follows an airplane in level flight at a speed of 900 km/hr. For the position where $\phi = 0$, determine the rates of change of the angle ϕ and the distance d.

II.55 Determine $\ddot{d}$ and $\ddot{\phi}$ for the system of Problem II.54.

II.56 A block is pushed up the circular hill; its horizontal component of

Prob. II.56

velocity is constant at 200 mm/s. Determine the velocity and acceleration of the block when $\theta = 0$ and $\theta = 60°$.

II.57   A cycloidal path is defined in parametric form by $X = 0.5(\alpha - \sin \alpha)$ ft, $Y = -0.5 \cos \alpha$ ft, where α is the parameter. In a certain case it is known that the motion of a particle following this path is given by $\alpha = 4t^2$. Determine expressions for the velocity and acceleration of this point.

Prob. II.57

II.58   For the motion in Problem II.57, determine the radius of curvature of the path and the rate of change of the speed when the particle is at point A.

II.59   A compression wave in the air contained in the hollow tube causes the particles of air to follow a rectilinear motion in the X direction defined by $v = A \sin[\pi(X - ct)/L]$, where A and c are constant. Determine an expression for the acceleration of a particle of air.

Prob. II.59

II.60   The rectangular Cartesian coordinates of a particle are given by $X = 200 \sin \pi t/10$, $Y = 200 \cos \pi t/10$, $Z = 400 \cos \pi t/20$, where X, Y, and Z are expressed in millimeters and t is in seconds. Determine the velocity and acceleration vectors and show that the path lies on the surface of a circular cylinder.

II.61   For the motion in Problem II.60, determine expressions for the speed, the rate of increase of the speed, and the radius of curvature, all as functions of time.

II.62   By transforming the unit vectors, express the solutions to Problem II.61 in terms of cylindrical coordinates for which the Z axis is the axial direction. From this result determine the corresponding values of the angular velocity and angular acceleration.

II.63   The particle at the end of the flexible bar follows an elliptic path whose equation is

$$R = \frac{ab}{[b^2 + (a^2 - b^2) \cos^2 \phi]^{1/2}}$$

Prob. II.64

Prob. II.65

It is known that the angular velocity for the motion is given by $\dot\phi = K/R^2$. Find an expression for the acceleration of the particle in an arbitrary position.

II.64   Along a portion of a railroad line the transition curve from a straight track to a curve is defined by $Y = X^2/2000$, where X and Y are in meters. Measuring the distance s from the position where the locomotive is at $X = 0$, the motion of the train is defined by $s = t^2/10$, where s is the arc length in meters and t is in seconds. What is the speed, the rate of change of the speed, and the magnitude of the acceleration of the locomotive when $X = 1000$ m?

II.65   For sufficiently large values of the radial distance R from the centerline of a *vortex* in a fluid, the cylindrical components of the velocity of a particle of fluid are $v_R = 0$, $v_\phi = A/R$, $v_Z = 0$, where A is a constant. Determine an expression for the acceleration of a particle. What is the path of a particle of fluid?

II.66   An important type of flow in the study of fluids is the *doublet,* for which the velocity components in cylindrical coordinates are $v_R = (A \cos \phi)/R^2$, $v_\phi = (A \sin \phi)/R^2$, $v_Z = 0$, where A is a constant. Determine an expression for the acceleration of a particle.

E.   RELATIVE MOTION
1   Translating Frame of Reference
The kinematical equations developed so far describe how we see the motion of a particle when our own position is a fixed point in space, that is, when the origin of the reference frame is a fixed point. Clearly, if we were to view the same particle motion from the window of a moving vehicle, we would not be viewing it from a fixed frame of reference. However, in order to convey our observations to someone not in the vehicle with us, we must be able to relate these observations to those viewed from a fixed point in space. This can be accomplished by the concept of relative motion. Such a situation is depicted in Figure 18.

The XYZ frame is a fixed coordinate system. The location of particle P with respect to the fixed point O is $\bar r_{P/O}$. Let point Q denote the position of an observer in a moving vehicle. Hence the observer's position with respect to the fixed point O is $\bar r_{Q/O}$.

We now choose a second reference frame xyz whose origin is the moving point Q. To aid in describing the kinematical properties that point P has relative to the observer, we ascribe fixed directions to the x, y, and z coordinate axes. That is, these axes are chosen to remain parallel to their initial orientation. Hence, xyz is a reference frame executing *translational*

Figure 18

motion. Our convention, throughout this text, will be to use lowercase letters to indicate that a reference frame is moving.

From Figure 18 we see that $\bar{r}_{P/Q}$ (read *r* of *P* relative to *Q*) is the position of point *P* as seen by the observer at point *Q*, and further, we can write

$$\bar{r}_{P/O} = \bar{r}_{Q/O} + \bar{r}_{P/Q} \qquad (31)$$

We shall refer to any variable that is observed from a fixed point as an absolute quantity. Thus, equation (31) states that the absolute position of point *P* is the sum of the absolute position of point *Q* and the relative position of point *P* with respect to point *Q*.

The next kinematical quantity of interest is velocity. The time derivative of the relative position vector $\bar{r}_{P/Q}$ is the relative velocity $\bar{v}_{P/Q}$ of point *P* with respect to point *Q*. That is,

$$\frac{d}{dt}(\bar{r}_{P/Q}) = \dot{\bar{r}}_{P/Q} = \bar{v}_{P/Q} \qquad (32)$$

Then, differentiating equation (31), we have the following relationship between the absolute and relative velocities:

$$\dot{\bar{r}}_{P/O} = \dot{\bar{r}}_{Q/O} + \dot{\bar{r}}_{P/Q}$$

$$\bar{v}_P = \bar{v}_Q + \bar{v}_{P/Q} \qquad (33)$$

In a similar manner we can reason that the acceleration of point *P* as seen by the observer at *Q* must be

$$\ddot{\bar{r}}_{P/Q} = \dot{\bar{v}}_{P/Q} = \bar{a}_{P/Q} \qquad (34)$$

Then, by differentiating equation (33), we find that

$$\bar{a}_P = \bar{a}_Q + \bar{a}_{P/Q} \qquad (35)$$

Notice the similarity in the form of equations (31), (33), and (35). In their present form they are symbolic. As will be seen in the examples that follow, when combined with our knowledge of kinematics with respect to a fixed point, these equations become a valuable computational tool. One immediate result arises in the special case where *Q* is also a fixed point, for then $\bar{v}_Q = \bar{a}_Q = 0$. This means that

$$\bar{v}_P = \bar{v}_{P/O} \quad \text{and} \quad \bar{a}_P = \bar{a}_{P/Q}$$

thus proving that the absolute velocity and acceleration are independent of the location of the fixed point of observation.

Note that in using a translating reference frame we are still describing motions, be they absolute or relative, in terms of fixed directions. In Module VI we shall relax this restriction and allow the moving frame to rotate. This will enable us to create a method for handling a broader class of problems.

Recall that vector components can be added only when all the vectors are written with respect to the same set of unit vectors. When using a moving reference frame xyz, dynamicists tend to perform computations by describing all vectors in terms of their $(\bar{\imath}, \bar{\jmath}, \bar{k})$ components with respect to xyz. We shall adhere to this procedure.

EXAMPLE 12

A rowboat has a speed of 8 ft/s with respect to a river that is flowing with a constant speed of 4 ft/s. What is the direction of the relative velocity vector, and what is the absolute speed of the boat if it is following a straight path from point A to point B?

Solution

To understand the motion of the rowboat, suppose that the river was at rest. Then the velocity of the boat relative to the river would be the true velocity of the boat. Here, the river is flowing. Hence the boat must head in a somewhat cross-stream direction to maintain its absolute direction along AB.

Using equation (33), in the terms of our problem,

$$\bar{v}_{\text{boat}} = \bar{v}_{\text{water}} + \bar{v}_{\text{boat/water}}$$

The given information is that $\bar{v}_{\text{water}}$ is 4 ft/s downstream and that $\bar{v}_{\text{boat/water}}$ has a magnitude of 8 ft/s. Also, $\bar{v}_{\text{boat}}$ is tangent to line AB, because this is the true path of the boat. To assist in describing these vectors, we draw a sketch and define the axes of the xyz coordinate system. The angle α in this sketch, which defines the orientation of $\bar{v}_{\text{boat/water}}$, also describes the orientation of the longitudinal axis of the boat.

From this sketch we have

$$\bar{v}_{\text{boat}} = v_{\text{boat}} (\sin 15° \, \bar{\imath} + \cos 15° \bar{\jmath})$$

Because the axes of the moving and fixed reference frames are parallel, we have

$$\bar{v}_{\text{water}} = 4\bar{I} \equiv 4\bar{\imath}$$

$$\bar{v}_{\text{boat/water}} = 8(- \sin \alpha \bar{\imath} + \cos \alpha \bar{\jmath})$$

Thus, the relative velocity equation gives

$$v_{boat}(\sin 15°\bar{\imath} + \cos 15°\bar{\jmath}) = 4\bar{\imath} + 8(-\sin \alpha\bar{\imath} + \cos \alpha\bar{\jmath})$$

Equating like components in this vector equation yields

$\bar{\imath}$ component: $v_{boat} \sin 15° = 4 - 8 \sin \alpha$

$\bar{\jmath}$ component: $v_{boat} \cos 15° = 8 \cos \alpha$

Hence

$$8 \sin \alpha = 4 - 0.2588 v_{boat}$$

$$8 \cos \alpha = 0.9659 v_{boat}$$

Squaring each of these equations and adding yields

$$8^2(\sin^2 \alpha + \cos^2 \alpha) \equiv 64 = (4 - 0.2588 v_{boat})^2 + (0.9659 v_{boat})^2$$

$$64 = 16 - 2.070 v_{boat} + v_{boat}{}^2$$

$$v_{boat} = 8.040, -5.970$$

The positive value of v_{boat} is the only one with meaning in this problem (the negative value corresponds to motion from point B to point A).

Substituting v_{boat} into the scalar equations then gives

$$\sin \alpha = \frac{4 - 0.2588(8.040)}{8} = 0.2399$$

$$\cos \alpha = \frac{0.9659(8.040)}{8} = 0.9707$$

Thus, because $\sin \alpha$ and $\cos \alpha$ are both positive, α is positive and we find

$$\alpha = 13.89°$$

EXAMPLE 13

At the instant depicted in the diagram, car A is on the east-west overpass and has a constant speed of 36 km/hr whereas car B has a constant speed of 72 km/hr along the highway. The radius of curvature of the highway is 100 m. Determine the velocity and acceleration of car A as viewed from car B.

Solution

This problem requires the determination of the motion of car A relative to car B, so we place the origin of the moving reference frame at car B. Then, because we know the speed along its path for each car, we describe the absolute velocities and accelerations by using path variables.

Car A is following a straight path at a constant speed of 36 km/hr = 10 m/s, whereas car B has a constant speed of 72 km/hr = 20 m/s along the curved road. Thus

$$\bar{v}_A = 10(\bar{e}_t)_A \qquad \bar{v}_B = 20(\bar{e}_t)_B \text{ m/s}^2$$

$$\bar{a}_A = \bar{0} \qquad \bar{a}_B = \frac{(20)^2}{100}(\bar{e}_n)_B = 4(\bar{e}_n)_B \text{ m/s}^2$$

Before we can substitute these vectors into the relative motion equations, we must express them in terms of their components with respect to the *xyz* reference frame. Following the standard procedure, we transform the unit vectors. From the sketch we see that

$$(\bar{e}_t)_A = -\bar{\imath} \quad \text{and} \quad (\bar{e}_t)_B = -\sin 30°\bar{\imath} + \cos 30°\bar{\jmath}$$

$$(\bar{e}_n)_B = \cos 30°\bar{\imath} + \sin 30°\bar{\jmath}$$

Thus

$$\bar{v}_A = -10\bar{\imath} \qquad \bar{v}_B = 20(-0.5\bar{\imath} + 0.8660\bar{\jmath}) = -10.0\bar{\imath} + 17.32\bar{\jmath}$$

$$\bar{a}_A = \bar{0} \qquad \bar{a}_B = 4(0.8660\bar{\imath} + 0.50\bar{\jmath}) = 3.464\bar{\imath} + 2\bar{\jmath}$$

Finally, from equations (33) and (35) we find that

$$\bar{v}_{A/B} = \bar{v}_A - \bar{v}_B = -17.32\bar{\jmath} \text{ m/s}^2$$

and

$$\bar{a}_{A/B} = \bar{a}_A - \bar{a}_B = -3.464\bar{\imath} - 2.0\bar{\jmath} \text{ m/s}^2$$

Thus we see that, to an observer in car B, at this instant car A seems to be moving south at 17.32 m/s, with acceleration components of 3.464 m/s² east and 2.0 m/s² south.

HOMEWORK PROBLEMS

II.67 The nose of a small airplane is oriented 15° south of west and the airplane is traveling west. Knowing that the wind is from the southwest at 100 km/hr, determine the airspeed and groundspeed of the airplane.

Prob. II.67

Prob. II.68

Prob. II.69

Prob. II.70

II.68 Passengers looking out the side window of an automobile traveling on a level road at a speed v_0 observe rain falling at an angle α forward from the vertical, whereas an observer at rest sees it falling at a forward angle β. Knowing that the absolute speed of a raindrop is 25 m/s, obtain an expression for α in terms of v_0 and β.

II.69 Two automobiles are following straight and level roads as they approach an intersection. Vehicle A has a speed of 100 km/hr and vehicle B has a speed of 60 km/hr, both constant. Determine the values of $\dot{R}$ and $\ddot{R}$ for the position shown. *Hint:* Evaluate the relative motion and express it in terms of polar coordinates having origin at vehicle B.

II.70 Two automobiles are traveling in opposite directions along a divided highway. Both vehicles have a speed of 55 miles/hr. Determine the velocity and acceleration of vehicle A with respect to vehicle B when the angle of separation $\phi = 0°$.

II.71 Solve Problem II.70 for $\phi = 30°$.

II.72 The pin P is constrained to follow the grooves in the two sliding arms AB and CD. In the position depicted in the diagram, arm AB has a constant speed of 80 mm/s, whereas arm CD has a speed of 40 mm/s and is gaining speed at the rate of 10 mm/s². For this position determine (a) the velocity of the pin, (b) the acceleration of the pin.

II.73 In order to avoid a collision, ships A and B maneuver in circular paths. At the instant shown ship A has a speed of 4 m/s and is decelerating at the rate of 0.4 m/s², whereas ship B has a constant speed of 8 m/s. What

Prob. II.72

is the relative position, velocity, and acceleration of ship B with respect to ship A at this instant?

Prob. II.73

II.74 A pin at the end of the 3-ft-long arm AB follows the circular groove in arm CD, which is moving to the right with a constant speed of 60 ft/s. For the instant when $\phi = 45°$, determine (a) the velocity of the pin, and (b) the acceleration of the pin.

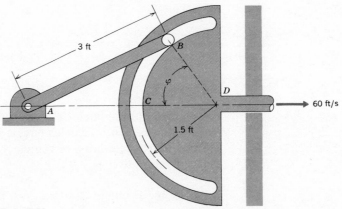

Prob. II.74

II.75 An airplane that can fly at v km/hr in still air is to travel to a point $\bar{r}$ $= R(\cos\theta\bar{I} + \sin\theta\bar{J})$ km from its starting point at the origin in a wind of velocity $v_w\bar{I}$ km/hr. Determine
(a) the direction in which the plane should be headed,
(b) the ground speed v_0 of the airplane,
(c) the time it will take to make the flight,
(d) the wind speed v_{max} for which it becomes impossible for the flight to be made in a straight-line path.

II.76 In the position shown, airplane A, which is diving in a vertical circle, has a speed of 200 miles/hr and is gaining speed at the rate of 5 ft/s². It is being observed by airplane B, which is flying in a straight path at a constant speed of 300 miles/hr. Determine the velocity of airplane A with respect to airplane B.

Prob. II.76

II.77 In Problem II.76, determine the acceleration of airplane B with respect to airplane A.

2 Cable-Pulley Systems

When cables and pulleys are utilized to connect bodies, we are concerned with the motion of more than one point, just as we were in the preceding study on relative motion. In rigid body mechanics cables are generally modeled as inextensible, massless bodies. Hence the bodies connected to the cable are constrained to move with respect to each other in a way that can easily be determined. As will be seen, this is accomplished by writing an equation for the constrained length of the cable.

Consider the two masses A and B in Figure 19 which are connected by an inextensible cable. The distances s_A and s_B locate the masses with respect to the fixed reference point O. The length of the cable is

$$l = l_A + \pi R + l_B$$

where πR is the length of the cable in contact with the pulley. From Figure 19 we may write

$$l_A = s_A - C_A \quad \text{and} \quad l_B = s_B - C_B$$

where C_A and C_B are constants whose magnitudes depend on the fixed distance d and the dimensions of the masses. Replacing l_A and l_B in the equation for l gives

$$l = s_A - C_A + \pi R + s_B - C_B$$

Figure 19

or

$$s_A + s_B = C$$

where C is a combined single constant whose value will prove to be of no importance to us. Differentiating this relationship with respect to time yields

$$\frac{d}{dt}(s_A + s_B) \equiv \dot{s}_A + \dot{s}_B = 0$$

and

$$\ddot{s}_A + \ddot{s}_B = 0$$

For rectilinear motion,

$$\bar{v}_A = \dot{s}_A(\bar{e}_t)_A \qquad \bar{v}_B = \dot{s}_B(\bar{e}_t)_B$$

and

$$\bar{a}_A = \ddot{s}_A(\bar{e}_t)_A \qquad \bar{a}_B = \ddot{s}_B(\bar{e}_t)_B$$

where *the unit vectors,* shown in Figure 19, *are in the direction of increasing s* by definition. Thus, expressing the results in terms of the motion of block A, we have the following constraint equations

$$\bar{v}_A = \dot{s}_A(\bar{e}_t)_A \quad \text{and} \quad \bar{v}_B = -\dot{s}_A(\bar{e}_t)_B$$

$$\bar{a}_A = \ddot{s}_A(\bar{e}_t)_A \quad \text{and} \quad \bar{a}_B = -\ddot{s}_A(\bar{e}_t)_B$$

For the system of Figure 19, $(\bar{e}_t)_B \equiv (\bar{e}_t)_A$. Thus, these constraint equations state that blocks A and B have the same speed and magnitude of acceleration, but the velocity and acceleration of mass B are opposite to those of mass A.

For an arbitrary cable-pulley system we may generalize this approach for obtaining the constraint equations as follows. First, locate each of the connected bodies by its distance from a fixed point of reference. Then sum the length of each free portion of the cable, ignoring any constant values, and equate this sum to a constant. Finally, differentiate this sum to obtain the speed and magnitude of acceleration relationships.

EXAMPLE 14

Block A is moving to the left at 4 ft/s and decelerating at the rate of 12 ft/s². Determine the velocity and acceleration of block B.

Solution

We begin with a sketch in which we define the distance s_A and s_B for the location of the blocks. Remember that these distances must be measured from a fixed point. We have chosen point O.

In terms of the unit vectors shown in the sketch,

$$\bar{v}_A = \dot{s}_A(\bar{e}_t)_A = 4(\bar{e}_t)_A \text{ ft/s} \qquad \bar{v}_B = \dot{s}_B(\bar{e}_t)_B$$

$$\bar{a}_A = \ddot{s}_A(\bar{e}_t)_A = -12(\bar{e}_t)_A \text{ ft/s}^2 \qquad \bar{a}_B = \ddot{s}_B(\bar{e}_t)_B$$

We now construct an equation for the length of the cable. There are four vertical lengths of cable and one horizontal length, so

$$4s_B + s_A = C$$

Differentiating gives

$$4\dot{s}_B + \dot{s}_A = 0 \qquad 4\ddot{s}_B + \ddot{s}_A = 0$$

Then, because the preceding equations for velocity and acceleration give $\dot{s}_A = 4$ ft/s, $\ddot{s}_A = -12$ ft/s^2, we find

$$\dot{s}_B = -1 \text{ ft/s} \qquad \ddot{s}_B = 3 \text{ ft/s}^2$$

Thus

$$\bar{v}_B = -1(\bar{e}_t)_B \text{ ft/s} \qquad \bar{a}_B = 3(\bar{e}_t)_B \text{ ft/s}^2$$

HOMEWORK PROBLEMS

II.78 Block A is given an acceleration of $0.5g$ to the right. Determine (a) the acceleration of block B and (b) the acceleration of block B with respect to block A.

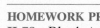

Prob. II.78

II.79 The tow truck A pulls the automobile B out of a ravine by driving forward. Starting from the position shown, the distance the truck moves

Prob. II.79

is given by $s = t^2/2 - t^3/100$, where s is measured in meters and t in seconds. Determine how far the automobile is pulled after 10 seconds and its velocity at the end of this time interval.

II.80 Block B is moving downward with a speed v_B. What is the velocity of block A?

Prob. II.80

II.81 The electrically powered winch pulls the cable upward at the constant speed of 1 m/s. What is the velocity of the platform?

Prob. II.81

II.82 Two alternative designs for an elevator are shown in the diagram. In design A the pulley attached to the motor pulls the cable in with an absolute speed v, whereas in design B the speed v with which the cable is pulled in is with respect to the elevator. Which elevator is faster, and by what percentage is it faster?

Prob. II.82

Prob. II.83

II.83 Block A has a constant velocity of 48 in./s to the left. In the position shown, determine (a) the velocity of block B, and (b) the acceleration of block B. (Note: The radius of the pulley may be neglected.)

MODULE III
KINETICS OF
A PARTICLE

In this module we shall apply our knowledge of the kinematics of a particle to determine how the forces exerted upon the particle affect its motion. The foundation for this study is Newton's second law of motion.

Letting $\Sigma \bar{F}$ denote the resultant of the set of forces acting on a particle P, we may write

$$\Sigma \bar{F} = m\bar{a}_P \tag{1}$$

When employing this relationship we shall express the forces and accelerations in terms of their components with respect to either Cartesian coordinates, cylindrical coordinates, or path variables. As one of our objectives, we shall learn how to recognize which of these coordinate descriptions is best suited for the formulation and solution of particular classes of dynamics problems.

Before continuing, it will prove beneficial to place Newton's two other laws of motion in proper perspective. The first law is embodied in the second law, reaffirming the fact that $\bar{a}_P = \bar{0}$ when $\Sigma \bar{F} = \bar{0}$. In effect, by stating this law separately, we are considering the study of statics to be independent of the study of dynamics. The third law proves most useful when we model physical systems as interacting, but independent, bodies. For such instances the action and reaction will be needed to represent the interaction forces correctly.

The foregoing brings to mind the concept of a free body diagram, in which we isolate a body from its surroundings, showing all forces exerted on the body of interest. This diagram serves two purposes. Drawing the diagram helps us to understand how the forces affect the body, and the resulting diagram aids us to account consistently for all the forces. Because the free body diagram is a primary tool in solving statics problems, we shall not dwell on its construction here.

An important feature that arises in analyzing the motion of a system concerns the restrictions, called *constraints,* that are often placed on the motion. In the case of particle motion, the most common constraint is that the particle must follow a specific path. Such constraint conditions can be thought of as being imposed by the application of constraint forces. For example, consider the particle in Figure 1a, which is moving under the influence of a given force $\bar{F}_1$ along a smooth curved path in a *horizontal plane.* The free body diagram of the particle when it is at a general point P must show the constraint force $\bar{N}$ that is normal to the path. This force has the effect of restraining the velocity of the particle to be tangent to the path. Without this constraint force the particle would follow a different path from the prescribed one. Thus, Figure 1b is the free body diagram for this problem.

For ease in reading Figure 1 we have accentuated the arrows depicting forces to differentiate them from the arrows that describe unit vectors. Also note that the forces normal to the plane of motion, in this case the

Figure 1a   **Figure 1b**

vertical forces, are not shown in Figure 1b. Included in these forces is the constraint force necessary to prevent motion (due to gravity) perpendicular to the plane. If we were interested in this force, we would draw a free body diagram showing the system as it is seen when viewed horizontally.

In the kinematics study in Module II, we concentrated on systems where we knew the position and/or velocity of a particle; the acceleration was obtained by differentiation of the position vector. No problem involved the determination of position and velocity from given information regarding the acceleration. Our reason for presenting these two seemingly similar types of problems separately is motivated by the fact that there are essentially two broad classes of practical problems in dynamics.

The first class involves situations where the motion of the physical system is given and we seek the forces producing this motion. For a particle, this reduces to a kinematical study using the methods in Module II to determine $\bar{a}_P$. Then, a free body diagram and $\Sigma \bar{F} = m\bar{a}_P$ will yield the necessary information about the forces.

In the second class of problems, the forces exerted on a physical system are given and we seek to determine the system's motion. For the case of a particle, we may determine its acceleration from $\bar{a}_P = \Sigma \bar{F}/m$. It is then necessary to follow an integration process (the opposite of differentiation) to determine the velocity and position vectors. (Clearly, a third class of problems is a combination of the foregoing.)

In this module we shall study how to determine the motion of a particle under the action of a given force system. In order to understand this process clearly, we shall first develop procedures for problem solving by considering the case of rectilinear (straight-line) motion. We shall then study a spectrum of problems in curvilinear motion, treating situations where forces are given, where the motion is specified, and where combinations of the two exist.

A.   EQUATIONS OF RECTILINEAR MOTION

We begin the study of the motion of a particle along a straight path by drawing a free body diagram that shows *all* forces acting on the particle. To complete the sketch we also need a set of unit vectors, which of course involves a decision as to which kinematical formulation we should select to describe the motion. As always, this decision depends on how the information given in the problem is presented. Let us assume that the given forces are functions of the path variable quantities, which are the distance s (traveled along the path), the speed v, and the time t. With this type of given information we normally use path variables with unit vectors $\bar{e}_t$, $\bar{e}_n$, and $\bar{e}_b$ when setting up the free body diagram. Recall that $\bar{e}_b$ is perpendicular to the plane of $\bar{e}_t$ and $\bar{e}_n$ and is defined by $\bar{e}_t \times \bar{e}_n = \bar{e}_b$. Thus,

Figure 2

for the motion of a particle along a straight path, letting $\bar{F}_1$, $\bar{F}_2$, etc., denote the set of forces acting on the particle, we have the general free body diagram of Figure 2. Note that $\bar{e}_t$ is shown tangent to the path and that the plane formed by $\bar{e}_n$ and $\bar{e}_b$ is perpendicular to the path.

We now wish to apply Newton's second law to this problem. To do this we resolve all vectors into their components with respect to the directions of the unit vectors. The resultant force $\Sigma \bar{F}$ is the sum of the vectors $\bar{F}_i$,

$$\Sigma \bar{F} = \Sigma \bar{F}_i \equiv (\Sigma F_t)\bar{e}_t + (\Sigma F_n)\bar{e}_n + (\Sigma F_b)\bar{e}_b \tag{2}$$

where ΣF_t, ΣF_n, and ΣF_b are the sum of the force components in the tangent and two normal directions, respectively. From kinematics we know that for rectilinear motion

$$\bar{a}_P = \dot{v}_P \bar{e}_t \tag{3}$$

Thus, because two vectors are equal only when their components are equal, the vector relation $\Sigma \bar{F} = m\bar{a}_P$ reduces to the scalar equations

$$\boxed{\begin{array}{l} \Sigma F_t = m\dot{v}_P \\[1mm] \Sigma F_n = 0 \\[1mm] \Sigma F_b = 0 \end{array}} \tag{4}$$

Equations (4) are the basic equations for rectilinear motion. The components perpendicular to the direction of motion express the fact that there must be a balance of forces in these directions, whereas the tangential component relates the forces to the rate of increase of the speed.

Equations (4) may be used to determine the forces for a known value of $\dot{v}_P$ or to determine $\dot{v}_P$ for a known set of forces. In the first case we have three algebraic equations for the force components, which can be solved by conventional methods. Here we shall concentrate on the latter case, in that it gives rise to an interesting situation. Let us assume that we know the functional dependence of the forces. Then, when we substitute the functions describing these forces into equations (4), we obtain a *differential equation,* that is, an equation relating the derivative of a function to other functions. The solution of this equation will tell us how the speed is related to the elapsed time and to the distance traveled and, equally important, how the distance traveled depends on the time. These data are defined by the functions $v_P(t)$, $v_P(s)$, and $s(t)$.

It will soon be obvious to you that we cannot determine these characteristics of the motion unless we know what state the particle was in prior to the application of the forces. (Remember, we require a force to alter a state of motion.) To define the initial state, we need *initial conditions,* which are the time t_0 when the motion was initiated, the distance s_0 describing the initial location of the particle, and the initial speed v_0.

B. SEPARATION OF VARIABLES

In general, there are many methods, both analytical and computerized, for solving the differential equation of motion obtained from equation (4). An effective approach is the method of separating variables. This method, which is applicable in situations where the differential equation contains only two variables, aims at forming integrable terms. In what follows we shall consider situations where this method can be employed.

1   Force as a Function of Time

In this case we have $\dot{v}_P = f(t)$, where $f(t) \equiv \Sigma\, F_t/m$ represents a known function of time. Then the differential equation given by the first of equations (4) can be placed into a form suitable for separating variables by using $\dot{v}_P \equiv dv_P/dt$. Hence equation (4) becomes

$$\dot{v}_P = \frac{dv_P}{dt} = f(t) \tag{5}$$

The term "separating variables" refers to the process of manipulating the differential equation so that only one variable appears on each side of the equation. To obtain this result for equation (5), we multiply through by dt, which gives

$$dv_P = f(t)\, dt$$

Knowing $f(t)$, this equation can be integrated. For dynamics problems with given initial conditions, a definite integral can be formed. We therefore have

$$\int_{v_0}^{v_P} dv_P \equiv v_P - v_0 = \int_{t_0}^{t} f(t)\, dt \tag{6}$$

Notice that the lower limits in the integrals are the initial values of the variables, and the upper limits are the values associated with point P. As an aside, remember that a valuable aid in the integration process is standard tables of integrals.

The foregoing equation yields the solution for $v_P(t)$, which can then be integrated to determine $s(t)$. Toward this end, we write

$$\dot{s} = \frac{ds}{dt} = v_P(t)$$

Separating variables and integrating yields

$$\int_{s_0}^{s} ds \equiv s - s_0 = \int_{t_0}^{t} v_P(t) \, dt \tag{7}$$

Finally, if we desire to relate v_P to s, we can treat t as a parameter that is to be eliminated between equations (6) and (7).

EXAMPLE 1

An aircraft lands with a speed of 200 km/hr and the pilot then gradually brakes the aircraft, producing the deceleration rate shown in the graph. How far will the aircraft travel prior to coming to rest?

Solution

The given initial conditions for $t_0 = 0$ are

$$v_0 = 200 \, \frac{km}{hr} \left(\frac{1 \, hr}{3600 \, s} \right) \left(\frac{1000 \, m}{km} \right) = 55.56 \, m/s$$

$$s_0 = 0$$

The graph shows that v_P is defined by two functions. For $0 \leq t \leq 10$ seconds, we have

$$\dot{v}_P = -\tfrac{1}{2}t \, m/s^2$$

and for $t \geq 10$ seconds, we have

$$\dot{v}_P = -5 \, m/s^2$$

We first solve for the motion during the initial time period. Thus, for $0 \leq t \leq 10$ seconds, the acceleration is

$$\dot{v}_P \equiv \frac{dv_P}{dt} = -\tfrac{1}{2}t$$

Upon separating variables and integrating, this yields

$$\int_{55.56}^{v_P} dv_P \equiv v_P - 55.56 = -\tfrac{1}{2}\int_{0}^{t} t \, dt = -\tfrac{1}{4}t^2$$

$$v_P = 55.56 - \tfrac{1}{4}t^2 \, m/s$$

To find $s(t)$, we write

$$\dot{s} \equiv \frac{ds}{dt} \equiv v_P = 55.56 - \tfrac{1}{4}t^2$$

$$\int_{0}^{s} ds \equiv s = \int_{0}^{t} (55.56 - \tfrac{1}{4}t^2) \, ds$$

$$s = 55.56t - \tfrac{1}{12}t^3 \, m$$

We can now determine if the aircraft will come to rest within this 10 second time interval by substituting $t = 10$ seconds. This gives

$$v_P(10) = 30.36 \text{ m/s}$$

$$s(10) = 472.3 \text{ m}$$

Hence the aircraft is still moving at the end of the first time period. We next consider the time interval $t \geq 10$ seconds. The initial conditions for this time interval are the values of v_P and s at the end of the initial time interval. Repeating the foregoing steps,

$$\dot{v}_P \equiv \frac{dv_P}{dt} = -5$$

$$\int_{30.56}^{v_P} dv_P \equiv v_P - 30.56 = -5 \int_{10}^{t} dt = -5(t - 10)$$

$$v_P = 80.56 - 5t \text{ m/s}$$

Now, to find $s(t)$ we write

$$\dot{s} \equiv \frac{ds}{dt} \equiv v_P = 80.56 - 5t$$

$$\int_{472.3}^{s} ds \equiv s - 472.3 = \int_{10}^{t} (80.56 - 5t) \, dt$$

$$s - 472.3 = 80.56(t - 10) - 2.5(t^2 - 10^2)$$

$$= -83.3 + 80.56t - 2.5t^2 \text{ m}$$

The value of t at which the aircraft comes to rest is obtained by setting $v_P = 0$. Thus

$$0 = 80.56 - 5t, \text{ so } \quad t = 16.11 \text{ seconds}$$

Finally, we can now obtain the distance required for the airplane to come to rest.

$$s = -83.3 + 80.56(16.11) - 2.5(16.11)^2 = 565.7 \text{ m}$$

If we were interested in how the motion varies with time, we could present the results for $v_P(t)$ and $s(t)$ graphically as shown in the diagram.

HOMEWORK PROBLEMS

III.1 A particle in rectilinear motion is given the acceleration plotted in the diagram. When $t = 0$ the particle is at rest at the origin. Sketch diagrams showing how the speed and distance traveled vary with time.

Prob. III.1

Prob. III.2

Prob. III.3

Prob. III.5

III.2 Experimental measurement of the contact force F exerted between a golf ball and a club indicates that the force-time plot is as shown. However, because of an error the scale factor for the force was lost, so the value of F_{max} is unknown. If the velocity of the ball after the impact was 70 m/s, determine the value of F_{max} in terms of the mass m of the ball.

III.3 The thrust T of a toy rocket that follows a straight vertical path is shown in the diagram. The toy weighs 2 lb and rests on its platform until the thrust is sufficient to launch it. Determine the maximum speed of the toy and the maximum height it attains.

III.4 The acceleration of a particle starting from rest along a straight line is known to follow the relation

$$a = k_n t^n$$

where k_n is a constant that depends on the nonnegative integer n. It is also known that the distance $s = l$ when $t = \tau$ seconds. Determine (a) an expression for k_n in terms of l, τ, and n, and (b) the speed of the particle when $t = \tau$. Does this speed increase or decrease with increasing values of n?

III.5 In order to prevent an automobile from skidding on ice when coming to a stop, it is common practice to "pump" the brakes. A driver followed this procedure, resulting in the acceleration-time plot shown. (a) Letting v_n and s_n denote the speed and distance traveled after n seconds, derive expressions for v_{n+1} and s_{n+1} in terms of v_n and s_n. (b) Use the results of part (a) to determine how long it takes to bring the automobile to rest from an initial speed of 20 miles/hr, and also determine the corresponding distance traveled.

III.6 A particle is rectilinear motion passes the origin traveling at 400 m/s when it is given a deceleration of $50t$ mm/s², where t is in units of seconds. Determine (a) the length of time required to bring the particle to rest, and (b) the distance of the particle from the origin and the actual distance it has traveled after 6 seconds.

2 Force as a Function of Position

When force is a function of position, $\dot{v}_P = \Sigma F_t/m = f(s)$, we cannot separate variables by writing $\dot{v}_P = dv_P/dt$, in that we would then have three variables (v_P, s, and t) in the differential equation. The s variable must be retained because it appears in the force term. Hence, we shall eliminate time from the equation. This is achieved by employing the chain rule, specifically,

$$\dot{v}_P \equiv \frac{dv_P}{dt} \equiv \frac{ds}{dt}\frac{dv_P}{ds} \equiv v_P\frac{dv_P}{ds} = f(s)$$

We now can separate variables and integrate. Thus

$$\int_{v_0}^{v_P} v_P \, dv_P \equiv \tfrac{1}{2}(v_P{}^2 - v_0{}^2) = \int_{s_0}^{s} f(s) \, ds \tag{8}$$

Equation (8) yields the function $v_P(s)$. To determine how the distance s depends on time, we write

$$\dot{s} \equiv \frac{ds}{dt} = v_P(s)$$

We then separate variables and integrate, with the result that

$$\int_{s_0}^{s} \frac{ds}{v_P(s)} = \int_{t_0}^{t} dt \equiv t - t_0 \tag{9}$$

Upon evaluating the integral on the left side of equation (9), we can solve for the value of s at any instant t. If the value of v_P at a specific instant is required, we can substitute the result for s at this instant into equation (8) and solve for the corresponding value of v_P.

EXAMPLE 2
A cable is attached to a 0.5-kg block that slides over the smooth rigid horizontal rod AB. The diagram shown depicts the vertical plane. The cable tension is a constant value T. At point C the speed of the block is 6 m/s to the left and at point D the speed is 1 m/s to the left. Determine the value of T.

Solution
In order to relate the speed of the block at points C and D, we sketch a free body diagram of the block at a general position. (Do not forget the normal force N that constrains the block to follow rod AB.)
We let s be the distance the block has traveled from point C and show $\bar{e}_t$ in the direction of increasing s.

Summing forces in the normal direction, we have

$$\Sigma F_n = N + T \sin\alpha - 4.903 = 0.5a_n = 0$$

Summing forces in the tangential direction yields

$$\Sigma F_t = -T \cos\alpha = 0.5a_t = 0.5\dot{v}_P$$

The first equation gives

$$N = 4.903 - T \sin \alpha$$

In this problem the expression for N is not needed. The second equation shows that the acceleration is a function of the angle α, but from the right triangle illustrated in the free body diagram we find that

$$\cos \alpha = \frac{s}{\sqrt{s^2 + 3^2}}$$

Hence, the tangential force equation will give $\dot{v}_P$ as a function of s. We therefore write

$$\dot{v}_P \equiv v_P \frac{dv_P}{ds} = -\frac{T}{(0.5)} \frac{s}{\sqrt{s^2 + 9}}$$

The initial conditions are that $s = 0$ and $v_P = 6$ m/s when $t = 0$. Therefore, separating variables and integrating yields

$$\int_6^{v_P} v_P \, dv_P = -2T \int_0^s \frac{s \, ds}{\sqrt{s^2 + 9}}$$

The integrand on the right side has the form $\int x^{-1/2} \, dx$. Hence

$$\tfrac{1}{2}(v_P^2 - 6^2) = -2T[\sqrt{s^2 + 9} - \sqrt{9}]$$

$$v_P^2 = 36 - 4T[\sqrt{s^2 + 9} - 3]$$

The foregoing defines $v_P(s)$ for a given value of T. To find T we use the fact that $v_P = 1$ m/s when $s = 4$ m, which yields

$$1^2 = 36 - 4T[\sqrt{4^2 + 3^2} - 3]$$

$$T = \frac{35}{4(2)} = 4.375 \text{ newtons}$$

Before leaving this problem, let us note the problem confronting us if we wish to follow the analogous steps to those in equation (9). The form of the function $v_P(s)$ that was obtained above would not allow for a closed-form evaluation of the integral on the left side of this equation. For such a case the integral could be evaluated numerically with the aid of a computer.

HOMEWORK PROBLEMS

III.7   The 10-kg block is pulled upward from the rest position shown by a

Prob. III.7

constant cable tension of 400 N. Determine the velocity of the block when it reaches the elevation of the pulley (a) if friction is negligible, (b) if the coefficient of friction between the block and its guide is $\mu = 0.1$.

III.8   A steel block weighing 800 lb is to be picked up by an electromagnet whose attractive force in pounds is given by the inverse square formula $F = 3(10^4)/s^2$ lb, where s is the distance in feet. The block was initially at $s = 5$ ft. Determine the velocity with which the block strikes the electromagnet.

Prob. III.8

Prob. III.9

III.9   The inverse square law for the gravitational attraction exerted on a particle of mass m is

$$F = \frac{mg}{(1 + h/R_e)^2}$$

where $g = 32.17$ ft/s², h is the altitude of the particle above the surface of the earth, and $R_e = 2.091(10^7)$ ft is the mean radius of the earth. In a hypothetical situation, a particle is released from rest at a height $h = 10,000$ ft. Neglecting air resistance, determine the velocity of the particle when it hits the earth. Compare this result to the value obtained by considering the gravity force to be simply mg.

Prob. III.10

III.10   The block of mass m is attached to a spring whose stiffness is k and slides on the smooth horizontal surface. Initially the block is held at rest at a distance s_0 from the position where the spring exerts no force. (a) Derive an expression for $v(s)$. (b) Use the result from part (a) to determine $s(t)$.

Prob. III.11

III.11   A 0.25-kg collar slides over a horizontal rod under the action of a linear spring whose constant is 20 N/m. There is no force in the spring when the collar is at point A. In a certain experiment the collar is released from rest at point B. Determine the velocity of the collar when it passes point A. Friction is negligible.

Cylinder, cross-sectional area A

Piston →

x

Prob. III.12

III.12 The 0.5-kg piston slides within a cylinder whose cross-sectional area $A = 5000$ mm². The gas pressure p within the cylinder is inversely proportional to the volume of gas Ax, where x is as shown. Also, when $x = 50$ mm, the pressure p equals the constant atmospheric pressure $p_0 = 1.0133(10^5)$ pascal. If the piston passes the position $x = 50$ mm with a velocity to the right of 500 mm/s, determine the distance the piston travels before coming to rest.

3 Force as a Function of Speed

In the case where $\dot{v}_P = \Sigma\,F_t/m = f(v_P)$, we have an alternative in the approaches available. If we desire to determine $v_P(t)$ and/or $s(t)$, we may write

$$\dot{v}_P \equiv \frac{dv_P}{dt} = f(v_P)$$

Then separating variables and integrating yields

$$\int_{v_0}^{v_P} \frac{dv_P}{f(v_P)} = \int_{t_0}^{t} dt \equiv t - t_0 \tag{10}$$

After evaluating the integral on the left side of equation (10), we can solve for $v_P(t)$. To determine $s(t)$, we write

$$\dot{s} \equiv \frac{ds}{dt} = v_P(t)$$

so that

$$\int_{s_0}^{s} ds \equiv s - s_0 = \int_{t_0}^{t} v_P(t)\,dt \tag{11}$$

The alternative method involves the situation where we desire to know the value of v_P corresponding to a value of s. Here the chain rules yields

$$\dot{v}_P = \frac{dv_P}{dt} = \frac{ds}{dt}\frac{dv_P}{ds} = v_P \frac{dv_P}{ds} = f(v_P)$$

which, upon separating variables and integrating, gives

$$\int_{v_0}^{v_P} \frac{v_P\,dv_P}{f(v_P)} = \int_{s_0}^{s} ds \equiv s - s_0 \tag{12}$$

Note that we now have two ways in which $s(t)$ can be evaluated. The first is to follow equation (11), whereas the second is to treat v_P as a parameter that is to be eliminated between equations (10) and (12).

tan^{-1}(0.1)

EXAMPLE 3

The brakes of an automobile whose mass is 1500 kg are released and the automobile rolls down a 10% grade. The air resistance is $kv_P{}^2$, where $k = 2.5$ newtons-seconds^2/meter^2, and the rolling resistance of the tires is 0.01 N, where N is the normal force exerted by the ground. Determine (a) the speed of the automobile after it has rolled 100 m, and (b) the length of time required for the car to reach this position.

Solution

As shown in the free body diagram, the forces acting on the automobile are the frictional air and rolling resistances (opposing the velocity of the automobile), the weight force, and the normal force exerted by the ground.

$\bar{e}_t$ $\bar{e}_n$

kv_P^2

0.01N

$\alpha = \tan^{-1}(0.10) = 5.711°$

mg N

Summing forces in the $\bar{e}_n$ direction yields

$$\Sigma F_n = N - mg \cos \alpha = 0$$

whereas the force sum in the direction of $\bar{e}_t$ gives

$$\Sigma F_t = mg \sin \alpha - kv_P{}^2 - 0.01N = m\dot{v}_P$$

Using the expression for N from the first equation to solve the second equation for $\dot{v}_P$, we obtain

$$\dot{v}_P = g(\sin \alpha - 0.01 \cos \alpha) - \frac{k}{m} v_P{}^2$$

This can be written in the more convenient form

$$\dot{v}_P = \frac{k}{m}(c^2 - v_P{}^2), \text{where} \qquad c = \sqrt{\frac{mg}{k}(\sin \alpha - 0.01 \cos \alpha)}$$

The distance s is measured from the position where the car is released, so that the initial conditions are $v_P = 0$ and $s = 0$ when $t = 0$. To determine the speed at a specified position, namely $s = 100$ m, we write

$$\dot{v}_P \equiv v_P \frac{dv_P}{ds} = \frac{k}{m}(c^2 - v_P{}^2)$$

Upon separating variables and integrating, we get

$$\int_0^{v_P} \frac{v_P\, dv_P}{c^2 - v_P^2} = \frac{k}{m}\int_0^s ds$$

From this result v_P may be obtained as follows:

$$-\tfrac{1}{2}\ln(c^2 - v_P^2) + \tfrac{1}{2}\ln(c^2) = \frac{k}{m}s$$

$$c^2 - v_P^2 = c^2 \exp\!\left(\frac{-2ks}{m}\right)$$

$$v_P = c\left[1 - \exp\!\left(\frac{-2ks}{m}\right)\right]^{1/2}$$

Now, substituting the parameters and letting $s = 100$ m yields

$$c = 22.95 \text{ m/s}$$

$$v_P = 12.22 \text{ m/s} \qquad\qquad \triangleleft$$

Having determined the value of v_P corresponding to $s = 100$ m, the most direct approach for determining the time required to reach this position is to obtain $v_P(t)$ from the original differential equation. To achieve this we write

$$\dot{v}_P \equiv \frac{dv_P}{dt} = \frac{k}{m}(c^2 - v_P^2)$$

$$\int_0^{v_P} \frac{dv_P}{(c^2 - v_P^2)} = \frac{k}{m}\int_0^t dt \equiv \frac{k}{m}t$$

With the aid of a table of integrals we find that

$$t = \frac{1}{2c}\frac{m}{k}\ln\!\left(\frac{c + v_P}{c - v_P}\right)$$

Hence, for $v_P = 12.22$ m/s, we get

$$t = 15.52 \text{ seconds} \qquad\qquad \triangleleft$$

For completeness, the derived equations relating v_P to t, and v_P to s, are presented here graphically. If we desired to relate s to t, the solution for v_P as a function of s could be substituted into the time equation.

The dashed lines in the graph show that v_P approaches a maximum value as s and $t \to \infty$. This maximum value, which is called the *terminal velocity*, can be determined without actually solving the differential equation of motion by simply setting $\dot{v}_P = 0$. Thus, for this problem we have

$$\dot{v}_P = 0 = \frac{k}{m} \left[c^2 - (v_P)^2_{\text{terminal}} \right]$$

$$(v_P)_{\text{terminal}} = c = 22.95 \text{ m/s}$$

HOMEWORK PROBLEMS

III.13 The viscous friction between a 30° inclined plane coated by a layer of lubricant and a 10-lb block is given by $f = 0.25v$, where f is expressed in pounds and v is in feet per second. The block is released from rest. Determine (a) the time required for the block to attain a speed of 10 ft/s, and (b) the corresponding distance traveled by the block.

Prob. III.13

Ball bearing

Prob. III.14

III.14 The frictional resistance on a ball bearing falling through oil is know to be proportional to the speed of the ball bearing, $f = kv$. It is also known that the terminal speed is 800 mm/s. The ball bearing is released from rest. Determine (a) the time required for the ball bearing to reach one-half its terminal speed, and (b) the speed of the ball bearing at twice the elapsed time found in part (a).

III.15 A block of mass m sliding on a horizontal surface hits a shock absorbing device (called a *linear dashpot*), which slows the block by exerting a retarding force $f = kv$, where v is the speed of the block and k is a constant. Letting s be the distance the dashpot is compressed, derive expressions for $v(t)$, $v(s)$, and $s(t)$ if $v = v_0$ and $s = 0$ initially.

Prob. III.15

III.16 The air resistance on a 200-g ball in free flight is given by $f = 2(10^{-4})v^2$, where f is in newtons and v is in meters per second. If the ball is dropped from rest 500 m above the ground, determine the speed at which it hits the ground. What percentage of the terminal speed is the result?

III.17 A 200-g ball is thrown vertically upward from the ground with a speed of 25 m/s. The air resistance is $f = 2(10^{-4})v^2$, where f is in newtons and v is in meters per second. Determine (a) the maximum height attained by the ball, (b) the speed of the ball when it returns to the ground, (c) the total elapsed time.

III.18 The resistance of the water to the motion of a $5(10^4)$-ton ship is known to be proportional to the square of the speed. It is also known that the propellers must produce a thrust of a quarter of a million pounds to give the ship a speed of 40 ft/s. The ship is traveling at 40 ft/s when the engine is turned off and the ship slows to a speed of 10 ft/s. Determine (a) the distance the ship travels, and (b) the elapsed time.

4 Force as a Constant

If the forces acting on a particle in rectilinear motion have constant magnitude and direction, then equation (9) shows that $\dot{v}_P$ is constant. We could consider this case to be a special case of any of the three preceding ones. However, let us treat it separately because this situation arises quite frequently. We first determine v_P as a function of time by writing $\dot{v}_P \equiv dv_P/dt$, separating variables and integrating as follows.

$$\int_{v_0}^{v_P} dv_P = \dot{v}_P \int_{t_0}^{t} dt$$

$$v_P = v_0 + \dot{v}_P(t - t_0) \tag{13}$$

To find $s(t)$ we then have

$$v_P \equiv \frac{ds}{dt} = v_0 + \dot{v}_P(t - t_0)$$

$$\int_{s_0}^{s} ds = \int_{t_0}^{t} [v_0 + \dot{v}_P(t - t_0)]\, dt$$

$$s = s_0 + v_0(t - t_0) + \tfrac{1}{2}\dot{v}_P(t - t_0)^2 \tag{14}$$

Finally, we find $v_P(s)$ by using the chain rule.

$$v_P \frac{dv_P}{ds} = \dot{v}_P$$

$$\int_{v_0}^{v_P} v_P\, dv_P = \dot{v}_P \int_{s_0}^{s} ds$$

$$(v_P)^2 = (v_0)^2 + 2\dot{v}_P(s - s_0) \tag{15}$$

Equations (13)–(15) are commonly employed in elementary physics courses, but their derivation is sufficiently easy to make it unnecessary to memorize them. However, if you should choose to memorize them, take care not to apply these formulas in situations where $\dot{v}_P$ is not constant.

EXAMPLE 4

The two blocks shown in the sketch are released from rest. It is observed that block A has a downward velocity of 6 ft/s after falling 2 ft. Knowing that block A weighs 50 lb, determine the weight of block B. Each pulley has negligible mass.

Solution

We draw a free body diagram of each block, and also of the pulley D, because it supports block B. In drawing these diagrams we use the fact that the tension in a cable is unchanged as it passes over ideal (massless and frictionless) pulleys.

← Inextensible cable

We have used the same tangent unit vector for each free body diagram, so we may relate the motion of the blocks by expressing their distances from a fixed datum. Thus, following the steps discussed in Module II, we draw a sketch of the pulley system that shows the various distances.

The length of the cable is constant, so that

$$2s_B + s_A = \text{constant}$$

Differentiation of this expression gives

$$2\dot{s}_B + \dot{s}_A = 0 \quad \text{and} \quad 2\ddot{s}_B + \ddot{s}_A = 0$$

so

$$v_B = -\tfrac{1}{2}v_A \quad \text{and} \quad \dot{v}_B = -\tfrac{1}{2}\dot{v}_A$$

These relationships may now be used in the equations of motion. Summing forces in the $\bar{e}_t$ direction for each block yields

$$\text{Block } A: \qquad \Sigma F_t = W_A - T = \frac{W_A}{g}\dot{v}_A$$

$$\text{Block } B: \qquad \Sigma F_t = W_B - F_B = \frac{W_B}{g}\dot{v}_B = -\tfrac{1}{2}\frac{W_B}{g}\dot{v}_A$$

The value F_B is determined by summing the forces on massless pulley D. Thus,

$$\text{Pulley } D: \qquad \Sigma F_t = F_B - 2T = \frac{W_{\text{pulley}}}{g}\dot{v}_{\text{pulley}} = 0$$

As a result, the equations of motion become

$$W_A - T = \frac{W_A}{g}\dot{v}_A$$

$$W_B - 2T = -\tfrac{1}{2}\frac{W_B}{g}\dot{v}_A$$

There are three unknowns in the foregoing two equations (W_A was given). The third relation needed is obtained from the given information about the motion of block A, which states that the speed of this block increases from zero to 6 ft/s as the block travels 2 ft. Because the equations of motion show that $\dot{v}_A$ is a constant, we may use equation (15) to get

$$v_A^2 = v_0^2 + 2\dot{v}_A(s - s_0)$$

$$6^2 = 0 + 2\dot{v}_A(2)$$

$$\dot{v}_A = 9.0 \text{ ft/s}^2$$

Substituting this value into the equations of motion gives

$$W_A - T = \frac{W_A}{g}(9.0)$$

$$W_B - 2T = -\frac{W_B}{g}(4.5)$$

from which we find

$$T = W_A\left(1 - \frac{9.0}{g}\right)$$

$$W_B - 2W_A\left(1 - \frac{9.0}{g}\right) = -\frac{W_B}{g}(4.5)$$

$$W_B = 2W_A\left(\frac{g - 9}{g + 4.5}\right) = 63.2 \text{ lb}$$

In conclusion, note that the tension force is not equal to the weight of the block A, for only in the case of static equilibrium is $T = W_A$. The difference is attributable to the inertia of the block.

HOMEWORK PROBLEMS

III.19   A rock is thrown downward from a bridge with a speed of 5 m/s and hits the water 4 s later. Determine (a) the height of the bridge and (b) the speed with which the rock hits the water.

III.20   A 3000-lb automobile accelerates at a constant rate from rest to 60 miles/hr in 10 s. The rolling resistance attributable to the undriven wheels is 20 lb. What is the tractive force exerted by the drive wheels on the ground during the acceleration? The air resistance is negligible.

III.21   A rider on an open elevator that is descending at 3 m/s throws a ball vertically upward with a speed of 9 m/s relative to the elevator. How long will it be before that ball returns to the platform, and how far will the platform have descended in this time interval?

III.22   The 100-lb box was initially descending at 8 ft/s. What constant force $\bar{F}$ must be applied to the 20-lb counterweight to bring the box to rest in 4 s? How far does the counterweight move in this time interval?

III.23   The driver of an automobile traveling 40 miles/hr sees a traffic light 1500 ft ahead turn red. The light is timed to be red for 15 s and green

20 lb
counterweight

F

100-lb box

Prob. III.22

Prob. III.23

for 30 s. (a) What constant acceleration will enable the automobile to pass the light at the instant it turns green, and what is the corresponding speed as it passes the light? (b) What is the maximum constant deceleration the automobile can have and still pass the light in the green cycle? Also, find the speed of the automobile as it passes the light for this deceleration rate.

III.24   A motorist in an automobile traveling 72 km/hr at a distance of 60 m behind a truck going at the same speed wishes to pass the truck. In order to accomplish this maneuver safely, the automobile must accelerate and reach a position 60 m in front of the truck in no more than 30 s. (a) Determine the minimum constant acceleration of the automobile that is safe. (b) Determine the corresponding speed of the car at the end of the passing maneuver.

Prob. III.24

5   Force as a General Function

Force is a general function when the tangential component of the resultant force depends on more than one of the kinematic variables t, s, and v_P. Then, in the most general case, we find $\dot{v}_P = \Sigma\, F_t/m = f(t, s, v_P)$. This equation is not easy to solve, for we *cannot* use the method of separating variables to reduce the problem to a series of differential equations in first derivatives. Instead, we must form a second-order differential equation by using $v_P \equiv \dot{s}$ to obtain

$$\ddot{s} = f(t, s, \dot{s}) \tag{16}$$

Equations of this type can be solved, but in many cases the solution is possible only with the aid of an electronic computer. In certain situations equation (16) may form a linear differential equation, methods for the solution of which are the subject of the section on vibrations in Module VIII. However, obtaining solutions for the case of general functions $f(t, s, \dot{s})$ is beyond the scope of this text.

C.   EQUATIONS OF CURVILINEAR MOTION

With some basic techniques for solving differential equations established, we can now proceed to study the motion of a particle as it follows a curved path. Scalar equations of motion can be obtained from the components of Newton's second law, equation (1). To do this we must first choose the unit vectors associated with a particular coordinate system.

This choice is an important facet of the process of problem solving, in that a poor choice can unnecessarily complicate the scalar equations whose solutions we require. Consequently, we shall now discuss the form of the equations of motion that arise when each of the kinematical descriptions (path variables, rectangular Cartesian coordinates, and cylindrical coordinates) is considered.

1 Path Variables

The path variable approach can be employed when the particle path is known. Letting s be the arc length traversed by the particle, we know that

$$\bar{v}_P = \dot{s}\bar{e}_t \quad \text{and} \quad \bar{a}_P = \dot{v}_P\bar{e}_t + \frac{v_P{}^2}{\rho}\,\bar{e}_n$$

Equating components of $\Sigma\,\bar{F}$ to their corresponding components of $m\bar{a}_P$, we obtain

$$\boxed{\begin{array}{l} \Sigma\,F_t = ma_t \equiv m\dot{v}_P \\[2mm] \Sigma\,F_n = \;\; ma_n \equiv m\dfrac{v_P{}^2}{\rho} \end{array}} \tag{17}$$

where $\Sigma\,F_t$ and $\Sigma\,F_n$ are the force components in the direction of the tangent unit vector $\bar{e}_t$ and normal unit vector $\bar{e}_n$, respectively. Equations (17) are sufficient for treating planar problems. However, as we saw in the case of rectilinear motion, an added equation is available for solving three-dimensional problems. In such a case we use the fact that there is no acceleration in the $\bar{e}_b$ (binormal) direction. Recalling that the binormal is perpendicular to the osculating plane formed by $\bar{e}_t$ and $\bar{e}_n$, there is no acceleration in that direction. Hence, there is a force balance in that direction, as there was in equations (4). The importance of this fact will be demonstrated in Example 6.

    The similarity of the first of equations (4) to the first of equations (17) suggests that path variables should be employed in situations where the particle is following a given path *and* the significant kinematical quantities for describing the forces and motion are t, s, and/or v_P. We shall see that situations that fail to meet these two criteria can be treated more readily by using the other kinematical descriptions.

EXAMPLE 5

A rocket is programmed to gain speed as it moves along the path shown. Is the depicted situation possible?

Solution

An obvious error in the diagram is that $\bar{F}$ and $\bar{a}_P$ are not parallel, which violates Newton's second law. Also, we know that

$$\bar{v}_P = v_P \bar{e}_t$$

so that the velocity vector must be tangent to the path. Finally, we can determine the general orientation of the acceleration vector using

$$\bar{a}_P = \dot{v}_P \bar{e}_t + \frac{v_P^2}{\rho}\,\bar{e}_n$$

From the given information we know that v_P is increasing, so that $\dot{v}_P$ must be positive. The centripetal acceleration is directed toward the center of curvature. Hence the correct situation is as shown below.

EXAMPLE 6

A 3000-lb automobile is traveling at 54 miles/hr along a level highway whose radius of curvature is 500 ft. The highway is banked at 10° from the horizontal. What are the normal and frictional forces that must be exerted between the tires and the road if the automobile is to remain in its lane?

Solution

Let us begin by assuming that the automobile would skid outward if there were no friction, in which case the friction force $\bar{f}$ is inward. Recall that for two surfaces that are not sliding over each other, the only restriction on the friction force is that it must be less than the coefficient of static friction times the normal force. Thus, the free body diagram is as shown at left.

We choose path variables for the formulation, noting that the automobile is following a circular path in the horizontal plane. Thus $\bar{e}_n$ is horizontal, oriented toward the center of curvature, and $\bar{e}_t$ is perpendicular to the plane of the free body diagram. We must therefore define a unit vector $\bar{e}_b$ to describe the vertical direction.

From the given information we have

$$v_P = 54\,\frac{\text{miles}}{\text{hr}} \times \left(\frac{88\ \text{ft/s}}{60\ \text{miles/hr}}\right) = 79.20\ \text{ft/s}$$

$$\dot{v}_P = 0 \qquad \rho = 500\ \text{ft}$$

Thus, the acceleration of the automobile is

$$\bar{a}_P = \dot{v}_P \bar{e}_t + \frac{v_P^2}{\rho}\,\bar{e}_n = 12.545 \bar{e}_n\ \text{ft/s}^2$$

The force sums in the directions illustrated in the free body diagram then yield, with $g = 32.17\ \text{ft/s}^2$,

$$\Sigma F_n = N \sin 10° + f \cos 10° = ma_n = \frac{3000}{32.17} (12.545)$$

$$\Sigma F_b = N \cos 10° - f \sin 10° - 3000 = 0$$

The solution of these equations is

$$N = 3158 \text{ lb}$$

$$f = 631 \text{ lb}$$

Because f is positive, our initial assumption that the vehicle would slide outward in the absence of friction was correct. What would we do if this assumption were wrong?

Prob. III.26

Prob. III.27

Prob. III.28

HOMEWORK PROBLEMS

III.25 The longitudinal axis of a rocket ship is aligned tangent to the path, at an angle of 30° from the vertical. The thrust is 1,000,000 lb and the weight of the rocket is 200,000 lb. If the speed at this instant is 4000 ft/s, determine (a) the rate at which the rocket is gaining speed, and (b) the radius of curvature of the path.

III.26 A 50-g sphere is initially held by two cables in the position shown. The left cable is cut. Determine the tension in the right cable just before and just after the cable is cut. Also determine the initial acceleration of the sphere.

III.27 A small sphere of mass m is attached to a cord of length l. The ball rotates in a horizontal circle for which the angle between the cord and the vertical axis is θ. Derive expressions for the tension in the cord and the constant speed of the ball in terms of l and θ.

III.28 The driver of a miniature go-cart wants to perform a stunt by driving along a circular track that is banked at 90°, so that the surface of the track is vertical. The top speed of the car is 45 miles/hr and the coefficient of friction between the tires and the ground is 0.8. What is the maximum radius r that the track can have?

III.29 An automobile is slowed at a constant rate from 108 km/hr to 72 km/hr in a distance of 600 m as the vehicle follows a level curve of 500-m radius. What are the magnitude and angle relative to the road of the horizontal force exerted between the automobile and the ground at a position 400 m from the location where the braking was begun?

III.30 The 0.5-kg collar is pulled down the curved rod by a spring whose stiffness is 20 N/m and whose unstretched length is 500 mm. Friction is negligible. If the collar has a speed of 10 m/s in the position shown, determine the corresponding acceleration of the collar and the force the collar exerts on the rod.

60°

500 mm

←500 mm→

Prob. III.30

2 Rectangular Cartesian Coordinates

A general feature of extrinsic (path-independent) coordinates is that they may be applied to problems where the particle path is to be determined, as well as to situations where the particle path is given. The specific decision to use rectangular Cartesian coordinates is usually based on noticing that the forces *and* acceleration are most easily described by components in fixed directions, for example, horizontal or vertical.

Let (X, Y, Z) denote the rectangular coordinates of a particle. The acceleration of the particle is

$$\bar{a}_P = \ddot{X}\bar{I} + \ddot{Y}\bar{J} + \ddot{Z}\bar{K}$$

Hence, matching the components of $\Sigma \bar{F}$ and $m\bar{a}_P$, as required by equation (1), yields

$$\boxed{\begin{aligned} \Sigma F_X &= m\ddot{X} \\ \Sigma F_Y &= m\ddot{Y} \\ \Sigma F_Z &= m\ddot{Z} \end{aligned}} \tag{18}$$

The following example demonstrates a common application of equation (18).

EXAMPLE 7

Firefighters are trying to put out a fire on the roof of a warehouse that is 21 m tall. To do this they set up their hose 15 m from the base of the building, holding the nozzle 1 m above the ground. They aim the hose at an angle of elevation α, so as just to clear the edge of the roof. The speed of the water as it leaves the nozzle of the hose is 25 m/s. Neglecting the effects of air resistance, (a) determine the value of the angle α, and (b) locate where the water will land on the level roof.

Solution

Let us first determine the general relationships for the motion of a particle in freefall near the earth's surface. We begin by drawing a free body diagram of the particle in a general position, using rectangular Cartesian coordinates.

The decision to use Cartesian coordinates is based primarily on the fact that the weight force is in the (constant) vertical direction, and also on the fact that the particle path is not given. The latter makes path variables unsuitable. For convenience, the origin of the XYZ coordinate system was chosen to coincide with a known point, in this case the launching point of the particle. Also for convenience the XY plane was chosen such that the initial velocity vector $\bar{v}_0$ is contained in it.

In general, for the chosen coordinate system, we have

$$\bar{a}_P = \ddot{X}\bar{I} + \ddot{Y}\bar{J} + \ddot{Z}\bar{K}$$

The initial velocity and position vectors are

$$\bar{v}_0 = v_0(\cos \alpha \bar{I} + \sin \alpha \bar{I}) \qquad \bar{r}_0 = \bar{0}$$

Summing force components yields

$$\Sigma F_x = 0 = ma_x = m\ddot{X}$$

$$\Sigma F_y = -mg = ma_y = m\ddot{Y}$$

$$\Sigma F_z = 0 = m\ddot{Z}$$

Thus, the acceleration components reduce to

$$\ddot{X} = 0 \qquad \ddot{Y} = -g \qquad \ddot{Z} = 0$$

In words, the acceleration downward is constant and there is no component of acceleration in either horizontal direction.

We now integrate the equations of motion, using equations (13)–(15) for constant acceleration, which give

$$X = X_0 + \dot{X}_0 t$$

$$Y = Y_0 + \dot{Y}_0 t - \tfrac{1}{2}gt^2$$

$$Z = Z_0 + \dot{Z}_0 t$$

In the foregoing X_0, Y_0, Z_0, and $\dot{X}_0$, $\dot{Y}_0$, $\dot{Z}_0$ represent the components of the initial position and velocity vectors, $\bar{r}_0$ and $\bar{v}_0$, respectively. Therefore, using the initial conditions, we have

$$X = (v_0 \cos \alpha)t \qquad Y = (v_0 \sin \alpha)t - \tfrac{1}{2}gt^2 \qquad Z = 0$$

Because $Z = 0$, we can conclude that the particle path is planar. In Homework Problem III.33 we shall eliminate t between the X and Y equations, to find that Y is a quadratic function of X. Hence, the path is a parabola. However, this knowledge is not essential for solving the problem at hand.

To determine α we use the fact that the edge of the roof is located at $X = 15$ m, $Y = 20$ m. Thus, upon substituting $v_0 = 25$ m/s and $g = 9.806$ m/s², we find

$$15 = (25 \cos \alpha)t$$

$$20 = (25 \sin \alpha)t - \tfrac{1}{2}(9.806)t^2$$

Solving the first equation for t and placing it into the second, we get

$$t = \frac{0.6}{\cos \alpha}$$

$$20 = (25 \sin \alpha)\left(\frac{0.6}{\cos \alpha}\right) - 4.903\left(\frac{0.6}{\cos \alpha}\right)^2$$

Using the trigonometric identity, $\sin \alpha = (1 - \cos^2 \alpha)^{1/2}$, the latter equation becomes

$$20 = 15(1 - \cos^2 \alpha)^{1/2}\left(\frac{1}{\cos \alpha}\right) - 1.7651\left(\frac{1}{\cos^2 \alpha}\right)$$

Simplifying this expression yields

$$625.0 \cos^4 \alpha - 154.40 \cos^2 \alpha + 3.116 = 0$$

We now have a quadratic equation for $\cos^2 \alpha$. The resulting values of the angle are $\alpha = 64.69°$ and $\alpha = 81.44°$; the latter value corresponds to the case where the stream hits the corner of the roof on the way down. Hence, the solution to part (a) is

$$\alpha = 64.69°$$

Now part (b) can be solved by substituting this value for α into the equations for $X(t)$ and $Y(t)$. From the sketch provided with the problem, we see that we need the larger value of X corresponding to $Y = 20$ m. Thus

$$X = (25 \cos 64.69°)t$$

$$20 = (25 \sin 64.69°)t - \tfrac{1}{2}(9.806)t^2$$

From this we find that a particle will be at the elevation of the roof at both 1.194 s and 3.415 s after leaving the nozzle. The first value of time corresponds to the position at the near corner of the roof. Using the second value of time, $t = 3.415$ s, we get

$$X = (25 \cos 64.69°)(3.415) = 37.28 \text{ m}$$

from which we see that the stream of water will fall on the roof 22.28 m from the near corner.

HOMEWORK PROBLEMS

III.31 A basketball is thrown from the balcony of a building that is 4 m above the ground. It is thrown at an angle of elevation of 45° with an initial speed of 12 m/s. Determine the maximum height attained by the ball and the location where the ball strikes the ground.

III.32 A pitching machine can throw baseballs at any angle of elevation with a speed of 120 ft/s. A ball player standing 180 ft away catches the ball at the same height above the level ground as that from which it was thrown. Neglecting air resistance, determine (a) the two possible angles of elevation for the throw, and (b) the final speed of the ball and the time duration of the flight for both possible angles.

Prob. III.31

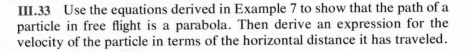

Prob. III.32

III.33 Use the equations derived in Example 7 to show that the path of a particle in free flight is a parabola. Then derive an expression for the velocity of the particle in terms of the horizontal distance it has traveled.

III.34 A ball is projected from a hill with a speed v_0 at an angle of elevation α. Determine the distance d from the launch point to the location where the ball hits the hill (a) using the xyz coordinate system shown, and (b) using the $x'y'z'$ coordinate system shown.

Prob. III.34

Prob. III.35

III.35 A 200-g pin is pulled through a circular groove by the grooved vertical arm, which is moving to the right at the steady speed of 20 m/s. Determine the normal force exerted by each groove on the pin in the position where $\theta = 36.87°$.

III.36 A 1-oz steel ball bearing in free flight passes through a magnetic field that exerts a constant horizontal pull of 2 oz in the X direction. Initially, the ball bearing is at a point 1 ft above the ground and its velocity is $20\bar{J}$ ft/s. Determine (a) the location of point B where the ball bearing hits the XY plane, and (b) the velocity of the ball bearing when it reaches point B.

Prob. III.36

3 Cylindrical Coordinates

The acceleration of a particle in cylindrical coordinates is

$$\bar{a}_P = (\ddot{R} - R\dot{\phi}^2)\bar{e}_R + (R\ddot{\phi} + 2\dot{R}\dot{\phi})\bar{e}_\phi + \ddot{h}\bar{e}_h$$

We follow the approach of the previous cases by matching the radial, transverse, and axial components of $\Sigma \bar{F}$ to those of $m\bar{a}_P$. Hence, the scalar equations of motion are

$$
\begin{array}{ll}
\Sigma F_R = m(\ddot{R} - R\dot{\phi}^2) & \quad\quad (19) \\
\Sigma F_\phi = m(R\ddot{\phi} + 2\dot{R}\dot{\phi}) & \\
\Sigma F_h = m\ddot{h} &
\end{array}
$$

As is true for Cartesian coordinates, cylindrical coordinates can be applied in cases where the particle path is not given. Key factors in choosing this kinematical description are that the forces are described with respect to a fixed axis or point, or equally important, the angle of rotation and radial distance to a fixed axis (or point) are the significant kinematical quantities.

EXAMPLE 8

A sharpshooter is tracking a moving target. In so doing, the rifle barrel is elevated 30° above the horizontal and rotated about the vertical axis at a constant rate of 0.6 rad/s. A bullet weighing 4 oz leaves the barrel with a muzzle velocity of 2000 ft/s. Determine the horizontal component of the force the bullet exerts on the barrel immediately before leaving the barrel.

Solution

The forces acting on the bullet are its weight (0.25 lb), the components N_ϕ and N_v of the contact force perpendicular to the barrel in the horizontal and vertical planes, and the explosive force F tangent to the barrel. The value of N_ϕ represents the solution of the problem. To illustrate this three-dimensional situation let us draw free body diagrams in side and top views.

Because the rate of rotation about a fixed axis is given, we choose cylindrical coordinates defined with respect to this axis. It follows that $\bar{e}_R$ is perpendicular to the axis. To describe the motion of the bullet, we note that the muzzle velocity, which is tangent to the rifle barrel, represents the radial and axial components of the total velocity of the bullet. Thus

$$\bar{v}_B = \dot{R}\bar{e}_R + R\dot{\phi}\bar{e}_\phi + \dot{h}\bar{e}_h$$
$$= 2000(\cos 30°\bar{e}_R + \sin 30°\bar{e}_h) + 0.6R\bar{e}_\phi$$

Therefore, by equating coefficients of the unit vectors, we find

$$\dot{R} = 2000 \cos 30° = 1732 \text{ ft/s}$$
$$\dot{h} = 2000 \sin 30° = 1000 \text{ ft/s}$$

Also, notice that the length of the rifle is not given, so we cannot determine the value of R. (We continue, hoping that this will not affect the computations for the force N_h.)

We next determine the acceleration of the bullet. The values of $\dot{R}$, $\dot{\phi}$, and $\dot{h}$ are known; it is also known that $\ddot{\phi} = 0$ because $\dot{\phi}$ is constant. Hence

$$\bar{a}_B = (\ddot{R} - R\dot{\phi}^2)\bar{e}_R + (R\ddot{\phi} + 2\dot{R}\dot{\phi})\bar{e}_\phi + \ddot{h}\bar{e}_h$$
$$= [\ddot{R} - R(0.6)^2]\bar{e}_R + [0 + 2(1732)(0.6)]\bar{e}_\phi + \ddot{h}\bar{e}_h$$

The value of N_ϕ is determined directly from the force sum in the transverse direction, which gives

$$\Sigma F_\phi = N_\phi = ma_\phi = \left(\frac{0.25}{32.17}\right)(2)(1732)(0.6)$$

$$N_\phi = 16.15 \text{ lb}$$

Does this force assist or oppose the rotation of the rifle barrel?

Prob. III.37

HOMEWORK PROBLEMS

III.37 A block is held in position within a groove in a rotating disk by a cable. Derive expressions for the tension force in the cable and the normal force exerted by the groove on the block in terms of the constant angular speed ω of the disk and the dimensions R and α shown. The motion occurs in the horizontal plane.

III.38 The 2-kg block slides over the rigid arm AB, which is rotating in the vertical plane at 2 rad/s counterclockwise. A radial force F gives the block a constant inward speed of 1 m/s. When $\theta = 30°$, the radial distance R is 400 mm. Determine the corresponding value of F if the coefficient of friction is 0.10.

III.39 Passengers on an amusement park ride sit in seats that are suspended by 25-ft-long cables from the horizontal arm AB, as shown. Arm AB rotates about the vertical axis at a constant angular speed ω. If $\theta = 36.87°$, determine the corresponding value of ω.

Prob. III.38

Prob. III.40

Prob. III.39

III.40 An 8-oz sphere is attached to a spring whose stiffness is 10 lb/ft as it swings in the vertical plane. The unstretched length of the spring is 18 in. When $\theta = 30°$ it is observed that $l = 20$ in. and that the velocity of the sphere is 80 in./s horizontally to the right. Determine the corresponding values of $\ddot{\theta}$ and $\ddot{l}$.

III.41 The speed governer shown in cross section consists of a 200-g plunger that is restrained by a spring whose stiffness is 30 N/mm. When the angular speed is zero the compressive force in the spring is 200 N and the plunger is in the position shown. Determine the constant angular speed ω that will cause the plunger to move the 10 mm required to trip switch A.

Prob. III.41

III.42   Collars A and B, each having mass m, are tied together by a cable as they slide over their guides. The entire assembly is rotating about the vertical axis at 30 rpm. In the position shown block A is descending at 4 m/s. Determine the acceleration of collar B at this instant. Friction is negligible.

D.   PROBLEM SOLVING

The steps for formulating and solving the equations of motion for a body when it is modeled as a particle are probably apparent to you as a result of the eight examples we have considered. Nevertheless, let us present them formally.

1   Draw a free body diagram. If several bodies are interconnected, draw a free body diagram of each body, being sure to satisfy Newton's third law of action and reaction. As an aid in constructing these diagrams, remember that each constraint (restriction) upon the motion of the body can only be imposed by a force.

2   Write down or depict in the free body diagram all given information about the forces and kinematical variables. List the quantities required for the solution.

3   Choose a coordinate system for describing vector components, and show the associated unit vectors in the free body diagram. This choice should be based on suiting the kinematical quantities and the forces associated with the system.

4   Following the kinematical methods in Module II, express the acceleration of each body in terms of the coordinate system chosen in the preceding step.

5   Sum the force components in the direction of each unit vector and equate each term to the corresponding component of $m\bar{a}_P$. Check that the number of algebraic and differential equations match the number of unknowns. Common reasons for these two numbers being unequal are a force that was not properly accounted for, a given quantity that was not used, or a kinematical constraint that was not considered.

6   Solve the equations of motion.

ILLUSTRATIVE PROBLEM 1

A conveyor belt moves with a constant speed of 0.9 m/s up a 20° incline and passes over an idler wheel whose radius is 300 mm. The coefficient of static friction between a package and the conveyor belt is 0.4. Determine the maximum deceleration rate that can be imparted to the belt without causing packages to slip as they pass over the idler wheel. How does this result compare to the maximum deceleraton rate for which packages will not slip when they are on the incline?

$\bar{e}_n$ mg

20° $\bar{e}_t$

f

N

$r = 300$ mm

Solution

Step 1 The critical position in regard to slipping of the packages is the location where the packages first begin to follow the idler, for it is at this position along the idler that the component of gravity opposite the direction of motion is largest. The friction force $\bar{f}$ is oriented as shown in the free body diagram, because this force opposes the direction in which the package would slip if the deceleration rate were too large.

Note that if we set $r = \infty$, the diagram then describes the situation when the package is approaching the idler on the incline.

Step 2 Given $v_P = 0.9$ m/s and $\mu_s = 0.4$, find the maximum value of $\dot{v}_P$ for the condition of no slipping at the idler and on the incline.

Step 3 As shown in the free body diagram, we chose to use path variables based on the fact that the speed is the principal kinematical quantity in this problem.

Step 4 The acceleration of the package is easily shown to be

$$\bar{a}_P = \dot{v}_P \bar{e}_t + \frac{v_P{}^2}{\rho}\, \bar{e}_n = \dot{v}_P \bar{e}_t + 2.7 \bar{e}_n \text{ m/s}^2$$

Step 5 The maximum deceleration is obtained when $f = \mu_s N$. Thus, the equations of motion are

$$\Sigma\, F_t = -0.4N - mg \sin 20° = ma_t = m\dot{v}_P$$

$$\Sigma\, F_n = mg \cos 20° - N = ma_n = m(2.7)$$

Considering m as a parameter, we can solve these equations for N and $\dot{v}_P$.

Step 6 Using $g = 9.806$ m/s², we find

$$N = m(9.806 \cos 20° - 2.7) = 6.515m \text{ newtons}$$

$$\dot{v}_P = -0.4\,\frac{N}{m} - 9.806 \sin 20° = -5.960 \text{ m/s}^2$$

Next we consider the situation when the package is on the inclined ramp. Here, the maximum deceleration rate is easily determined, because the free body diagram for this situation is the same as the one shown earlier. Then, because this is rectilinear motion, we set the centripetal acceleration to zero, from which it follows that

$$N = m(9.806 \cos 20°)$$

$$\dot{v}_P = -0.4\,\frac{N}{m} - 9.806 \sin 20° = -7.040 \text{ m/s}^2$$

The maximum deceleration rate is smaller for the position on the roller because the centripetal acceleration at this location decreases the value of the normal force, which, in turn, has the effect of lowering the friction force.

ILLUSTRATIVE PROBLEM 2

A small ball bearing moves downward over the surface of a smooth semicircular cylinder after being released from rest at the top. Determine the location at which the ball bearing will cease to be in contact with the cylinder.

Solution

Step 1   The ball bearing loses contact with the cylinder when the normal force $\bar{N}$ becomes zero. Therefore, we draw a free body diagram with the aim of determining this force.

Step 2   We must find the value of the angle ϕ defined in the free body diagram at which $N = 0$, knowing that $\bar{v} = \bar{0}$ when $\phi = 0$.

Step 3   Because the ball bearing follows a circular path, we could use path variables to formulate the equations of motion. However, as we wish to relate the force $\bar{N}$ to the angle ϕ, a more direct approach is to utilize polar coordinates, as illustrated in the free body diagram.

Step 4   For polar coordinates we have

$$\bar{a}_P = (\ddot{R} - R\dot{\phi}^2)\bar{e}_R + (R\ddot{\phi} + 2\dot{R}\dot{\phi})\bar{e}_\phi$$

For the coordinate system chosen, $R = b$. However, the only information given about ϕ is that $\dot{\phi} = 0$ when $\phi = 0$. Therefore, we have

$$\bar{a}_P = -b\dot{\phi}^2\bar{e}_R + b\ddot{\phi}\bar{e}_\phi$$

Step 5   Newton's second law gives

$$\Sigma F_R = N - mg \cos \phi = ma_R = -mb\dot{\phi}^2$$

$$\Sigma F_\phi = mg \sin \phi = ma_\phi = mb\ddot{\phi}$$

The second of these two equations is a differential equation for $\phi(t)$, whereas the first serves as an algebraic equation that gives N. We therefore proceed to the solution of these equations.

Step 6   We want to find the value of ϕ for which $N = 0$. Hence, set $N = 0$ in the first equation of motion to obtain

$$0 - mg \cos \phi = mb\dot{\phi}^2$$

In other words, $N = 0$ when $\dot{\phi}^2 = (g/b) \cos \phi$. We must now use the second equation of motion to determine how $\dot{\phi}$ depends on ϕ. To achieve this we write

$$\ddot{\phi} = \frac{g}{b} \sin \phi$$

Notice that this differential equation is identical in form to that obtained in rectilinear motion when acceleration is a function of position. To separate variables we write

$$\ddot{\phi} \equiv \frac{d}{dt}\dot{\phi} \equiv \frac{d\phi}{dt}\frac{d}{d\phi}\dot{\phi} \equiv \dot{\phi}\frac{d\dot{\phi}}{d\phi} = \frac{g}{b} \sin \phi$$

Using the initial condition that $\dot{\phi} = 0$ when $\phi = 0$, we have

$$\int_0^{\phi} \dot{\phi} \, d\dot{\phi} = \frac{g}{b}\int_0^{\phi} \sin \phi \, d\phi$$

$$\tfrac{1}{2}\dot{\phi}^2 = \frac{g}{b}(1 - \cos \phi)$$

By substituting the expression for the value of $\dot{\phi}^2$ at which $N = 0$ into this equation, we find

$$\tfrac{1}{2}\frac{g}{b} \cos \phi = \frac{g}{b}(1 - \cos \phi)$$

$$\cos \phi = \tfrac{2}{3}, \text{ so } \qquad \phi = 48.18°$$

When the ball bearing reaches this position it flies off tangent to the cylinder and follows a free-fall trajectory until it hits the ground. We mention in closing that the expression for the value of $\dot{\phi}^2$ at any angle ϕ can more easily be determined by energy principles. This approach, which will be presented later in this module, avoids the need to solve a differential equation.

ILLUSTRATIVE PROBLEM 3

A horizontal force $\bar{F}$ pushes a 5-kg block up a smooth parabolic hill whose shape is defined by $Y = 4X^2$, where X and Y are measured in meters. Determine how $\bar{F}$ must vary with the location of the block in order for the horizontal component of the velocity of the block to be a constant value c.

Solution

Step 1 There is no friction force because the hill is smooth. Thus, we have the following free body diagram.

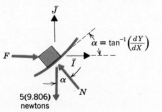

Note that the normal force $\bar{N}$ is perpendicular to the local tangent to the parabola.

Step 2 We are told that the equation of the path is $Y = 4X^2$ and that $\dot{X} = c, \ddot{X} = 0$. We wish to determine the value of F.

Step 3 As indicated in the free body diagram, we shall use the XY coordinate system to formulate the equations of motion. This approach is chosen because the given kinematical information involves the velocity component in the X direction.

Step 4 In terms of rectangular Cartesian components, the acceleration of the block is

$$\bar{a}_P = \ddot{X}\bar{I} + \ddot{Y}\bar{J} + \ddot{Z}\bar{K}$$

but for planar motion $\ddot{Z} = 0$. Furthermore, because we know that $\dot{X} = c$ and $\ddot{X} = 0$, we can determine $\ddot{Y}$ by differentiating the equation for Y twice with respect to time. Thus

$$\dot{Y} \equiv \frac{dY}{dt} = \frac{d}{dt}(4X^2) = 8X\dot{X} = 8cX$$

$$\ddot{Y} = 8c\dot{X} = 8c^2$$

Therefore

$$\bar{a}_P = 8c^2\bar{J}$$

Step 5 Referring to the free body diagram, we have

$$\Sigma F_X = F - N \sin \alpha = ma_X = 0$$

$$\Sigma F_Y = N \cos \alpha - 5(9.806) = ma_Y = 5(8c^2)$$

Because the value of the angle α can be determined from the slope of the parabola, the two equations of motion have only two unknowns.

Step 6 We solve the first equation for N, and substitute this result into the second equation, with the following results.

$$N = \frac{F}{\sin \alpha} \qquad \left(\frac{F}{\sin \alpha}\right) \cos \alpha = 49.03 + 40c^2$$

$$F = (49.03 + 40c^2) \tan \alpha \text{ newtons}$$

Noting that

$$\tan \alpha = \frac{dY}{dX} = 8X$$

we find that

$$F = (392.2 + 320c^2) X \text{ newtons}$$

You probably did not expect it to be necessary to increase the force $\bar{F}$ in proportion to the horizontal distance traveled by the block. Can you give a physical explanation of why a constant value of F does not produce the desired motion?

ILLUSTRATIVE PROBLEM 4

The slotted arm rotates in a horizontal plane at a constant clockwise rate $\omega = 4$ rad/s as it guides the pin P over the fixed cam. The cam is defined by $R = 16 + 8$ cos ϕ, where R is the distance in inches from the fixed point O to a point on the perimeter of the cam. The compression force in the spring is 3.0 lb when $\phi = 0°$ and 1.0 lb when $\phi = 180°$. Calculate the value of the contact force exerted between the 12-oz pin and the cam for the position where $\phi = 60°$.

Solution

Step 1 There are three forces acting on the pin in the horizontal plane: the spring force $\bar{S}$ which acts radially inward, a force $\bar{F}$ exerted by the slotted arm which is pependicular to the walls of the groove, and a normal force $\bar{N}$ exerted by the cam, perpendicular to the local tangent to the cam. Because the value of N is desired when $\phi = 60°$, we draw a free body diagram of the pin in that position.

Step 2 It is given that $\omega = 4$ rad/s and $\dot{\omega} = 0$. Also, it is known that $S = 3$ lb when $\phi = 0$ and $S = 1$ lb when $\phi = 180°$, and that the weight of the pin is $\frac{12}{16}$ lb. We wish to find the value of N when $\phi = 60°$.

Step 3 It is logical to use cylindrical coordinates whose origin is point O, because the given kinematical information is the angular speed of the radial line from point O to the pin. Recalling that the unit vector $\bar{e}_\phi$ is oriented in the direction of increasing ϕ, the unit vectors are as shown in the free body diagram.

Step 4 For cylindrical coordinates we have

$$\bar{a}_P = (\ddot{R} - R\dot{\phi}^2)\bar{e}_R + (R\ddot{\phi} + 2\dot{R}\dot{\phi})\bar{e}_\phi + \ddot{h}\bar{e}_h$$

For planar motion $\ddot{h} = 0$, and for clockwise rotation ϕ is decreasing, so that

$$\dot{\phi} = -\omega = -4 \text{ rad/s} \qquad \ddot{\phi} = -\dot{\omega} = 0$$

The distance R can be determined by using the given equation defining the shape of the cam. Thus

$$R = 16 + 8 \cos \phi$$

$$\dot{R} = -8\dot{\phi} \sin \phi$$

$$\ddot{R} = -8\dot{\phi}^2 \cos \phi - 8\ddot{\phi} \sin \phi$$

Solving for these values at the position where $\phi = 60°$ yields

$$R = 20.0 \text{ in.} \qquad \dot{R} = 27.71 \text{ in./s} \qquad \ddot{R} = -64.0 \text{ in./s}^2$$

Therefore

$$\bar{a}_P = [-64.0 - 20(4)^2]\bar{e}_R + [0 + 2(27.71)(-4)]\bar{e}_\phi$$

$$= -384.0\bar{e}_R - 221.7\bar{e}_\phi \text{ in./s}^2$$

Another quantity that is related to the kinematics of the motion is the angle α between the normal force $\bar{N}$ and the radial line, as shown in the free body diagram. The accompanying sketch shows that α is also the angle between the local tangent to the cam and the transverse line.

Hence, to determine α we can use the fact that the velocity $\bar{v}_P$ is tangent to the path. From the sketch we have

$$\alpha = \tan^{-1}\left(-\frac{v_R}{v_\phi}\right)$$

Note that the negative sign in this equation arises because the vector $\bar{v}_P$ illustrated in the sketch has a negative transverse component. Thus, evaluating $\bar{v}_P$ at $\phi = 60°$, we find

$$\bar{v}_P = \dot{R}\bar{e}_R + R\dot{\phi}\bar{e}_\phi = 27.71\bar{e}_R + 20.0(-4)\bar{e}_\phi$$

$$\alpha = \tan^{-1}\left(\frac{27.71}{80.0}\right) = 19.10°$$

Step 5   Referring to the free body diagram in step 1, the equations of motion are

$$\Sigma F_R = N \cos \alpha - S = \frac{0.75}{g}a_R = \frac{0.75}{g}(-384.0)$$

$$\Sigma F_\phi = N \sin \alpha + F = \frac{0.75}{g}a_\phi = \frac{0.75}{g}(-221.7)$$

where $g = 386.0$ in./s², because the units of length are inches.

These two equations have three unknowns (N, S, and F), so we need another equation. This is obtained by using the given information about the spring force. For a linearly elastic spring $S = k\Delta$, where Δ is the amount by which the length of the spring has been decreased from the unstretched value. Let Δ_0 be this value at $\phi = 0$. Then, because $R = (16+8)$ in. at $\phi = 0$, $R = 20.0$ in. at $\phi = 60°$, and $R = (16 - 8)$ in. at $\phi = 180°$, it follows that

$$S_0 = 3 \text{ lb} = k\Delta_0$$

$$S_{60} = k[\Delta_0 - (24.0 - 20.0)]$$

$$S_{180} = 1 \text{ lb} = k[\Delta_0 - (24.0 - 8.0)]$$

Solving these equations, we find

$$1 = k\Delta_0 - 16k = 3 - 16k, \text{ so } \quad k = 0.125 \text{ lb/in.}$$

$$\Delta_0 = 3/k = 24.0 \text{ in.}$$

$$S_{60} \doteq 0.125[24.0 - (24.0 - 20.0)] = 2.50 \text{ lb}$$

Step 6 We may now find N at $\phi = 60°$ by solving the equation for $\Sigma \, F_R$ in the preceding step, using the value of α found in step 4.

$$N = \frac{1}{\cos 19.10°}[2.50 - \frac{0.75}{386.0}(384.0)] = 1.856 \text{ lb}$$

ILLUSTRATIVE PROBLEM 5

The magnitude of the air resistance on a projectile is known to be a specific function of the speed of the projectile, $f(v_P)$. For example, in many situations $f(v_P) = kv_P{}^2$. Determine the equations describing the motion of a projectile near the surface of the earth whose solution would yield the location of the projectile at any instant. The initial velocity of the projectile is v_0 at an angle of elevation θ.

Solution

Step 1 The air resistance is a frictional force opposite the velocity of the projectile. Because the path is planar in the absence of air resistance, it follows that the motion here is also planar. Hence, the free body diagram is as shown below.

Step 2 We wish to obtain the equations for the position of a projectile for a known $f(v_P)$.

Step 3 As indicated in the free body diagram, we choose rectangular Cartesian coordinates for the formulation. To a certain extent this choice is by default, in that we cannot employ path variables because the path is unknown, whereas polar coordinates are clearly not applicable. Also, the chosen approach is consistent with that employed in Example 7 to treat projectile motion in the absence of air resistance.

Step 4 For rectangular Cartesian coordinates,

$$\bar{a}_P = \ddot{X}\bar{I} + \ddot{Y}\bar{J} + 0\bar{K}$$

We also need the angle α between the velocity vector and the horizontal line in order to compute the sum of the forces. Toward this end, we draw a sketch of the velocity vector $\bar{v}_P = \dot{X}\bar{I} + \dot{Y}\bar{J}$.

From this sketch we see that

$$\tan \alpha = \frac{\dot{Y}}{\dot{X}} \qquad \sin \alpha = \frac{\dot{Y}}{\sqrt{\dot{X}^2 + \dot{Y}^2}} \qquad \cos \alpha = \frac{\dot{X}}{\sqrt{\dot{X}^2 + \dot{Y}^2}}$$

Step 5 Using the free body diagram to sum forces, we have

$$m\ddot{X} = -[f(v_P)] \cos \alpha$$

$$m\ddot{Y} = -mg - [f(v_P)] \sin \alpha$$

We now express all terms in these equations by their Cartesian coordinate equivalents. Thus

$$\ddot{X} + \frac{1}{m}[f(\sqrt{\dot{X}^2 + \dot{Y}^2})]\frac{\dot{X}}{\sqrt{\dot{X}^2 + \dot{Y}^2}} = 0$$

$$\ddot{Y} + \frac{1}{m}[f(\sqrt{\dot{X}^2 + \dot{Y}^2})]\frac{\dot{Y}}{\sqrt{\dot{X}^2 + \dot{Y}^2}} = -g$$

Accompanying these equations are the initial conditions that

$$\bar{r}_{P/O}(t = 0) = X_0\bar{I} + Y_0\bar{J} \text{ and } \bar{v}_P(t = 0) = v_0(\cos \theta \bar{I} + \sin \theta \bar{J})$$

where X_0 and Y_0 are the initial coordinates of the projectile.

The foregoing equations are, in general, complicated differential equations whose solutions are intertwined. As a specific illustration, consider the case when $f(v_P) = kv_P^2$. We then have

$$f(\sqrt{\dot{X}^2 + \dot{Y}^2}) = k[\sqrt{\dot{X}^2 + \dot{Y}^2}]^2 = k(\dot{X}^2 + \dot{Y}^2)$$

so that the equations of motion become

$$\ddot{X} + \frac{k}{m}\dot{X}\sqrt{\dot{X}^2 + \dot{Y}^2} = 0$$

$$\ddot{Y} + \frac{k}{m}\dot{Y}\sqrt{\dot{X}^2 + \dot{Y}^2} = -g$$

Clearly, the solution of these equations is beyond the scope of an elementary course. In fact, accurate solutions are possible only with the aid of an electronic computer. However, if the air resistance were proportional to the speed, we could solve the resulting equations of motion. This is the subject of Homework Problem III.82.

HOMEWORK PROBLEMS

III.43 A 3-kg block moving down the inclined rod is brought to a stop by a cord that is held horizontally. The tension in the cord is 100 N and the coefficient of friction between the block and the rod is 0.4. If the block was initially sliding at 2 m/s, determine the distance the block slides before coming to rest.

III.44 A pendulum composed of a 2-lb sphere and a 3-ft-long cable is mounted from a post on a cart whose weight is 4 lb. If the cart is pulled by

Prob. III.43

Prob. III.44

Prob. III.45

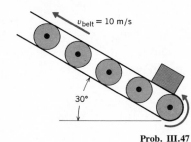

Prob. III.47

a horizontal force $\bar{F}$, determine an expression for the constant angle θ between the cable of the pendulum and the vertical direction.

III.45 A crate weighing 1 ton rests on the flat bed of a 5-ton truck traveling at 55 miles/hr. The coefficients of static and kinetic friction between the crate and the truck are $\mu_s = 0.50$ and $\mu_k = 0.45$, respectively. Determine the minimum length of time in which the truck may be brought to rest without causing the crate to shift relative to the truck. What is the total braking force that the tires must produce for this deceleration?

III.46 The truck in Problem III.45 decelerates at 20 ft/s². Determine the deceleration rate of the crate and the required braking force for this deceleration.

III.47 A package is at rest when it is placed on a conveyor belt that is moving up a 30° incline at a constant speed of 10 m/s. The coefficient of kinetic friction between the package and the belt is 0.75. Determine how far a package will move up the incline before it ceases to slip relative to the conveyor belt. Also determine the corresponding elapsed time.

III.48 Each car of a commuter train is self-propelled and has a mass of 10,000 kg. The maximum forward traction force from the electric motors in each car is 10,000 N and the maximum braking force is 20,000 N. What is the minimum length of time in which the train can depart from one station and come to a stop at the next station 6 km away without exceeding a speed 90 km/hr? What is the average speed of the train between stops?

Prob. III.48

III.49 A locomotive pulls a two-car train with a constant acceleration from rest to 20 m/s in 1 min along a level track. The mass of each car is 10,000 kg and the mass of the locomotive is 20,000 kg. The frictional resistance on each car and the locomotive attributable to the undriven wheels is 1000 N each. Determine the traction force of the locomotive and the force exerted within the couplers between the cars during the acceleration.

Prob. III.49

III.50   A 1000-kg motorboat is pointed upstream in a river flowing at 3 m/s and the thrust of the outboard motor is 400 N, with the result that the boat holds its position relative to the riverbank. The frictional force of the water is proportional to the speed of the boat relative to the water. Determine (a) the true speed of the boat 30 s after the engine is turned off, and (b) the distance the boat will drift 30 s after the motor is turned off.

Prob. III.50

III.51   The 70-kg sign painter applies a tensile force of 220 N to the free end of the cable. Determine (a) the acceleration of the painter, and (b) the force exerted between the painter and the seat.

III.52   Each block has a mass of 20 kg. What force $\bar{F}$ must be applied to block A to accelerate block B at 5 m/s²? What is the corresponding tension force in the cable? Friction is negligible.

III.53   The weights of the blocks are $W_A = 10$ lb, $W_B = 5$ lb, and $W_C = 20$ lb. The coefficient of friction for blocks A and B is $\mu = 0.20$. Determine how far block C will descend in 4 s, starting from rest.

Prob. III.51

Prob. III.52

Prob. III.53

Prob. III.55

III.54   When $t = 0$, block A is moving down the incline at 2 m/s. Block A has a mass of 4 kg, block B has a mass of 8 kg, and the coefficient of friction between the block and the incline is $\mu = 0.2$. How far will block A move down the incline before coming to rest?

III.55   The crate is raised by a tensile force $\bar{F}$, which pulls the end of the cord downward at the constant speed v_0. Derive an expression for the value of F in terms of the speed v_0, the mass of the crate m, and the dimensions h and L.

Prob. III.56

III.56 Block A weighs 3 lb and block B weighs 6 lb. The two blocks are tied together by a cord and slide with negligible resistance from friction. Block A is released at the elevation of block B with the cord taut. Derive the differential equation of motion governing the distance block A descends.

III.57 A 150-m-radius turn in a highway is designed such that 120 km/hr is the maximum safe speed of an automobile whose tires have a coefficient of friction of 0.40 for sliding (sideways) on the road. What is the angle of banking for the road?

Prob. III.57

III.58 Hollow plastic spheres weighing 2 oz are tested within the centrifuge device whose cross section is shown. Determine the normal forces exerted between a sphere and each of the smooth walls for a constant rate of rotation of 3600 rpm.

III.59 A small ball bearing follows a horizontal circle as it rolls at a constant speed v on the interior of a hemispherical bowl of radius r. Determine the relationship between v and the depth h of the ball below the rim of the bowl.

Prob. III.58

Prob. III.59

III.60 Two cords are attached to a 1-kg sphere that revolves in a horizontal circle of 500-mm radius at a constant speed of 2.5 m/s. Determine the tension force in each cord.

III.61 For what range of values of the speed of the sphere in Problem III.60 will both cords be taut?

Prob. III.60

III.62 Two blocks rest on a turntable that is rotating about the vertical axis at a constant angular speed ω. The blocks are connected by an

600 mm

500 mm

Prob. III.62

inextensible cable. The masses of blocks A and B are 1.0 and 1.5 kg, respectively, and the coefficient of friction is 0.30. Determine the maximum angular speed for which the blocks will not slide. Which way will the blocks slide if ω exceeds this value?

III.63   A train follows a level unbanked curve of 5-km radius. Starting from rest, the locomotive has a constant acceleration of 0.25 m/s². The mass of the locomotive is $2(10^4)$ kg and the mass of each of the four cars it is pulling is $2(10^5)$ kg. Also, the frictional resistance on each car is (10^4) N. Determine the components of the total reaction force between the locomotive and the track after it has traveled 1000 m. Assume that the coupler force between the locomotive and the cars being pulled is tangent to the longitudinal axis of the locomotive.

$\rho = 5$ km

Prob. III.63

III.64   Solve Problem III.63 if the track is banked at 10°.

III.65   An automobile is traveling at a constant speed over a hill. At the top the radius of curvature is 500 ft. (a) Determine the maximum speed the automobile can have in this positon and still remain in contact with the road. (b) If the automobile is traveling at 60 miles/hr and the brakes are suddenly applied at the top of the hill, determine the maximum possible deceleration rate. The coefficient of friction between the tires and the ground is 0.70.

III.66   The vertical elevation of a straight road is defined by $Y = 2 \sin(\pi X/20)$, where X and Y are measured in meters. Determine the maximum deceleration rate of an automobile traveling at 45 km/hr at (a) point A, (b) point B, (c) point C. The coefficient of friction between the tires and the ground is 0.60.

$\rho = 500$ ft

Prob. III.65

45 km/hr

A B 2 m

20 m C

Prob. III.66

III.67 The curve of a level unbanked entrance ramp on a highway is defined by $Y = X^2/1600$, where X and Y are expressed in feet. A 3000-lb automobile, starting from rest at point A, accelerates at a constant rate and passes point B with a speed of 50 miles/hr. What are the components of the frictional force between the tires and the ground as the automobile passes point B? *Hint:* First determine the distance the automobile travels between points A and B.

Prob. III.67

Prob. III.68

Prob. III.69 and Prob. III.70

III.68 A 20-kg block of ice slides within a chute. Section AB forms a level 90° turn of 5-m radius. The viscous friction force between the ice and the metal chute is given by $f = 0.2v$ N, where v is the speed in meters per second. At position A the ice block is entering the turn at a speed of 10 m/s. Determine (a) the maximum and (b) the minimum normal forces exerted by the block of ice on the side wall of the chute in section AB.

III.69 A package descends a spiral chute that is defined by $R = 2$ and $h = c\theta$, where R and c are expressed in meters and θ is in radians. There is no friction. Determine the speed of the package after it has fallen 4 m with (a) $c = 1$ m, and (b) $c = 2$ m. Also determine the time required for the descent.

III.70 A package descends a spiral chute defined by $R = 2$, $h = c\theta$, where R and c are measured in meters and θ is in radians. Because of friction the package has a constant speed of 4 m/s. The coefficient of friction between the package and the side wall is the same value μ as that between the package and the floor of the chute. What is the value of μ?

III.71 A pendulum is formed from a 400-g sphere at the end of a 2-m cable. If the sphere is released from rest at $\theta = 60°$, determine the tension in the cable (a) when $\theta = 0°$, (b) when $\theta = 30°$.

Prob. III.71

III.72 A satellite following an ellipitical orbit about the earth has the

$v = 8$ km/s

Earth

45°

$y = 8000$ km

R_e

Prob. III.72

instantaneous position and velocity shown. The inverse square law for gravitation is $F = mgR_e^2/(R_e + y)^2$, where $R_e = 6730$ km is the mean radius of the earth. Determine (a) the value of $\ddot{y}$, (b) the radius of curvature of the orbit at this location.

III.73 A particle follows a spiral in the horizontal plane defined by $R = 200\pi/\theta$, where R is in millimeters and θ is in radians. The tangential force $\bar{F}_t$ is varied with the angle θ so as to maintain $\dot{\theta}$ constant at 4 rad/s. Determine the tangential and normal forces acting on the particle when $\theta = 120°$.

Prob. III.73

Prob. III.74

III.74 Arm AB rotates counterclockwise at the constant rate of 120 rpm as it guides an 8-oz pin over the fixed cam shaft. The equation of the camshaft is $R = 10 + 2\cos 4\theta$, where R is in inches and θ is in radians. The maximum compressive force in the spring is 8 lb and the minimum force is 4 lb. What is the force that the pin exerts on the cam (a) at point A, (b) at point B?

Y

X

A

B

4 m/s

Prob. III.75

III.75 Guide AB moves to the right at a constant speed of 4 m/s as it guides the 250-g pin through the fixed parabolic groove lying in the horizontal plane. The equation of the groove is $X = Y^2$, where X and Y have units of meters. Determine the force the pin exerts on guide AB in terms of the distance X.

III.76 The eccentric arm AB rotates clockwise at the steady rate of 180 rpm, causing the 100-g pin to follow the fixed circular groove whose radius is 200 mm. Friction is negligible. Determine the normal force exerted between arm AB and the pin when $\theta = 45°$. Which side of the groove in arm AB is in contact with the pin? *Hint:* Use the law of cosines to express the radial distance R in terms of θ.

R

B

θ

A

200 mm

100 mm

Prob. III.76

III.77 The hollow tube rotates about a vertical axis with a constant

1 m

v_R

10 rad/s

Prob. III.77

angular speed of 10 rad/s, as a rubber ball rolls inside the tube. (a) Derive the equation defining the distance of the ball from the axis of rotation. (b) If a ball is introduced at the origin with a radial speed of 500 mm/s, determine the radial speed and the absolute velocity of the ball as it emerges from the tube.

III.78 The collar slides over bar *AB,* which is rotating about the vertical axis at the constant rate of 10 rad/s. Friction is negligible. (a) Derive the equation governing the distance *l* of the collar from point *A*. (b) If the collar was held at *l* = 1 ft and then released, determine the absolute velocity of the collar at *l* = 3 ft.

III.79 A rider on an open platform that is descending at the constant speed of 3 m/s throws a ball. Relative to the platform, the ball's initial velocity is horizontal at 12 m/s. The ground is 10 m below the location where the ball is thrown. (a) Where does the ball hit the ground? (b) How long after the ball hits the ground does the platform reach ground level?

A

53.13°

10 rad/s

l

B

Prob. III.78

3 m/s

10 m

Prob. III.79

III.80 Mail parcels are projected from a horizontal conveyor belt moving at 8 ft/s and are to be caught by bins running on a parallel conveyor belt 4 ft below. The speed of this conveyor belt is 2 ft/s. In order to catch package *B* in bin *A,* what are the range of values for the distance *d* that the bin *A* should be ahead of package *B* when the package is projected from the upper conveyor belt?

8 ft/s

B

4 ft

d

3 ft

2 ft/s

A

Prob. III.80

Prob. III.81

III.81 A person standing on the bed of a pickup truck traveling at 30 miles/hr on a straight road throws a rock horizontally, aiming the throw perpendicular to the velocity of the truck, with the result that the rock hits a tree at point A, as shown in the top view. Knowing that the rock was thrown at position B, 15 ft before the location of the tree, determine (a) the true initial velocity of the rock, (b) the location of the rider when the rock hits the tree, and (c) the distance the rock had fallen when it hits the tree.

III.82 The air resistance on a slow-moving body is essentially proportional to its velocity, that is, $\bar{f} = -k\bar{v}_P$. Use the equations derived in Illustrative Problem III.5 to derive expressions for $X(t)$ and $Y(t)$ for a body in free flight. The object had an initial speed v_0 at an angle of elevation α.

Prob. III.82

III.83 In Problem III.82 the value of k for a 200-g ball is 0.05 N-s/m. The ball has an initial speed of 20 m/s at an angle of elevation of 45°. Determine where the ball returns to its initial elevation and compare this result to the result for $k = 0$.

Prob. III.84

III.84 The longitudinal axis of a 1000-lb guided missile remains horizontal as the engine produces a constant thrust of 2000 lb. If the missile was launched from an aircraft traveling at 900 miles/hr in level flight at an altitude of 1000 ft, determine (a) the horizontal distance the missile will travel before it hits the ground, (b) the velocity of the missile when it hits the ground. Neglect air resistance.

III.85 A steel ball bearing rolls on a 30° incline. Its initial speed at point A was 5 m/s horizontally, as shown. Determine (a) the length of time required for the ball bearing to reach point B at the base of the incline, and (b) the distance d locating point B.

Prob. III.85

III.86 An electron is introduced at rest midway between the two parallel charged plates shown, which are separated by a distance D. The force on the electron, whose mass is m, is eV/D, where e is the electron charge and V is the voltage difference. When $t = 0$ the voltage is varied sinusoidally according to $V = V_0 \cos \omega t$. (a) What is the maximum value of V_0 for

Prob. III.86

Prob. III.87

which the electron does not hit either charged plate? (b) What is the resulting position of the electron as a function of time for this maximum value of V_0?

III.87   The gravitational force on an electron of mass m and charge e in a cathode-ray tube is negligible, so the path of an electron is essentially a straight line except when it passes between the charged plates. The electron is attracted to the positively charged plate by a force eV/b perpendicular to the plates, where V is the voltage difference and b is the distance between the plates. When an electron is emitted from the cathode and passes through the anode, its speed is u. Derive an expression for the deflection d of the electron for a constant voltage $V = V_0$.

III.88   Solve Problem III.87 if the voltage is varied according to $V = V_0(ut/l)$.

E.   NEWTON'S EQUATION OF MOTION — SYSTEM OF PARTICLES
In all of the examples and problems discussed thus far in this text we have treated some large bodies, such as automobiles and airplanes, as if they were particles. To understand why this may be done, and also to derive an important result for our later studies, let us consider a general system of particles.

A primary property of such a system is the position of its center of mass. Therefore, let us start by reviewing how this point is located. In Figure 3a we see a system of N particles, in which gravity acts in the negative Z direction.

Figure 3a   **Figure 3b**

The center of mass is found by noting that the resultant of the gravitational forces in Figure 3a is equivalent to the total weight force acting at the center of mass G, as illustrated in Figure 3b. Recall that in statics it was shown that equivalence of the moments of the force systems in the figures leads to an equation relating the *first moments of mass*.

To detail this result we denote the coordinates of the ith particle by $(X_i,\ Y_i,\ Z_i)$ and the coordinates of the center of mass by $(X_G,\ Y_G,\ Z_G)$. Then, for the orientation of the gravity force in Figures 3, moment equivalence about the X and Y axes gives, respectively,

$$\left(\sum_{i=1}^{N} m_i\right)X_G = \sum_{i=1}^{N} m_i X_i$$

$$\left(\sum_{i=1}^{N} m_i\right)Y_G = \sum_{i=1}^{N} m_i Y_i$$

These two equations may be used to determine the values of X_G and Y_G. To determine Z_G we can imagine realigning the gravitational field in Figures 3; for instance, let gravity act in the negative Y direction. In such a case, moment equivalence about the X axis requires

$$\left(\sum_{i=1}^{N} m_i\right)Z_G = \sum_{i=1}^{N} m_i Z_i$$

These relations may be summarized in vector form by

$$\boxed{\;m\bar{r}_{G/O} = \sum_{i=1}^{N} m_i \bar{r}_{i/O}\;} \tag{20}$$

where m is the total mass of the body and the vectors are as defined in Figures 3a and 3b.

Before considering Newton's laws for the system of particles, let us categorize the individual forces acting on each particle in the system according to their source. We denote by $\bar{F}_i$ the resultant force on particle i from sources external to the system. In addition to the external forces, there will be interaction forces $\bar{f}_{ij}$ ($\bar{i} \neq \bar{j}$) on each particle i due to every other particle j in the system.

Examples of external forces are gravity and contact forces exerted between the earth and the particles being studied. Interaction forces can arise from particles within the system that are in contact, and from the gravitational attractions between particles. A simple case is illustrated in Figure 4 for a system of three particles. Notice that each pair of interaction forces satisfies Newton's third law. Extending Figure 4 to the case of a system of N particles, we see that Newton's second law for particle i is

$$\bar{F}_i + \sum_{\substack{j=1 \\ \bar{i} \neq \bar{j}}}^{N} \bar{f}_{ij} = m_i \bar{a}_i = m_i \ddot{\bar{r}}_{i/A} \tag{21}$$

Figure 4

where A is an arbitrary point in the system.

Let us now compute the total resultant of all forces. As can be seen from the system of Figure 4, the forces $\bar{f}_{ij}$ and $\bar{f}_{ji}$ associated with any two particles i and j cancel each other. This occurs because the forces are equal in magnitude but opposite in direction. Therefore, when we sum equation (21) for each particle, we find

$$\sum_{i=1}^{N} \bar{F}_i = \sum_{i=1}^{N} m_i \ddot{\bar{r}}_{i/A} = \frac{d^2}{dt^2} \sum_{i=1}^{N} m_i \bar{r}_{i/A}$$

This expression gains greater significance when we note that the summation in the right side defines the center of mass of the system according to equation (20). Hence

$$\sum_{i=1}^{N} \bar{F}_i = \frac{d^2}{dt^2} (m \bar{r}_{G/A}) = m \ddot{\bar{r}}_{G/A}$$

$$\boxed{\sum_{i=1}^{N} \bar{F}_i = m \bar{a}_G} \qquad (22)$$

In words, equation (22) states that the acceleration of the center of mass is affected only by the total external force, and, equally important, it is unaffected by the interaction between the particles of the system.

The primary reason for presenting equation (22) at this point in our studies is the direct analogy between this equation and Newton's second law. This shows that *when we model the collection of atoms composing a physical body as a single particle, we are studying the motion of the center of mass of the body.*

Equation (22) is essential to the study of kinetics of rigid bodies, which we shall begin in Module V. We shall also use it later in this module to investigate interacting particles further.

F. WORK-ENERGY RELATIONS

We have encountered a number of problems where the application of Newton's laws resulted in differential equations relating the acceleration of a particle in a general position to the forces. The work-energy principle, which follows, represents a standard integral of such differential equations. Basically, it relates the motion of a particle at two different positions along its path.

Because the work-energy principle circumvents the need to solve a differential equation, it is a powerful tool for solving problems. However, as we shall see, the work-energy relations cannot be applied in all situations. Therefore, they complement, but do not replace, the formulation and solution of equations of motion.

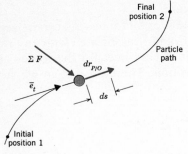

Figure 5

1 Work Performed by Forces

Let us follow the particle in Figure 5 as it travels an infinitesimal arc length ds along its path under the action of the resultant force $\Sigma \bar{F}$.

We define the *work dU* performed by this resultant force in moving the particle this infinitesimal amount as the product of the component of the force in the direction of motion and the infinitesimal distance traveled. Mathematically, this definition is

$$dU = \Sigma F_t \, ds$$

Thus, the total work done by the resultant force in moving the particle from point 1 to point 2, as indicated in Figure 5, is

$$U_{1\rightarrow 2} \equiv \int_{s_1}^{s_2} \Sigma F_t \, ds \tag{23a}$$

The evaluation of this expression requires that we express the resultant force in terms of its tangent and normal components. However, it may be preferable in some problems to write $\bar{F}$ in terms of a different set of unit vectors. Then, a more general definition of the work done can be obtained by referring to the infinitesimal change in the position vector of the particle $d\bar{r}_{P/O}$, as illustrated in Figure 5. Specifically, we have

$$U_{1\rightarrow 2} = \int_{r_1}^{r_2} \Sigma \bar{F} \cdot d\bar{r}_{P/O} \tag{23b}$$

where r_1 and r_2 denote the coordinates of the initial and final positions of point P.

Before considering some examples of the evaluation of the work done by forces in moving a particle, we can make a few general observations about this determination.

1 Notice that the work done by a force is a scalar quantity, so the total work performed by a system of forces is the scalar sum of the work done by each force in the system.

2 Any force that is normal to the direction of motion, such as a reaction force, does no work because it has no component tangent to the direction of motion.

3 In order to compute the work done by a force, its component in the direction of motion *must be a function only of the position* of the particle.

The last statement is particularly significant, because it states that we should *not* try to apply the work-energy principle to problems where the forces doing work depend on either time or the particle velocity. An example of such a force is the air resistance in Illustrative Problem 5.

EXAMPLE 9

The 4-kg block shown in the sketch is pulled up the smooth incline by applying a 40-N force to the end of the cable. Compute the total work done in moving the block from position A to position B on the incline.

Solution

To see which forces do work in moving the block, we draw a free body diagram of the block in a general position at a distance s from point A. From this diagram we see that work is done by gravity and the cable force, but not by the normal force $\bar{N}$.

To compute the work done by gravity, we note that the component of gravity in the direction of motion is the constant value $-4g \sin 30°$. Thus

$$(U_{A \to B})_{\text{gravity}} = \int_{s_A=0}^{s_B=5} (-4g \sin 30°)\, ds$$

$$= -4(9.806)(0.50)(S_B - S_A)$$

$$= -98.06 \text{ N-m}$$

$$\equiv -98.06 \text{ joules}$$

Note that the SI units of work are joules (J), a joule being defined as a newton-meter. Also, the work done by gravity in this problem is negative because it is opposite to the direction of motion.

To determine the work done by the cable force, we can express the component of $\bar{F}$ tangent to the incline in terms of the angle α, which gives

$$(U_{A \to B})_{\text{cable}} = \int_0^5 (40 \cos \alpha)\, ds$$

Then, from the dashed triangle in the free body diagram we find

$$\cos \alpha = \frac{5.866 - s}{\sqrt{(1.5)^2 + (5.866 - s)^2}}$$

so that

$$(U_{A \to B})_{cable} = 40 \int_0^5 \frac{(5.866 - s)}{\sqrt{2.25 + (5.866 - s)^2}} \, ds$$

$$= -40[\sqrt{2.25 + (5.866 - s)^2}] \Big|_{s=0}^{s=5}$$

$$= -40[1.732 - 6.055] = 172.91 \; J$$

An alternative way to obtain the work done by the cable is to consider the work done in pulling the end of the cable. Note that the length of the cable from the block to the pulley, when the block is at point A, is $((5.866)^2 + (1.5)^2)^{1/2} = 6.055$ m, and that it is $1.5 \sin 60° = 1.732$ m at point B. Hence, in moving the block from point P to point B, the end of the cable is pulled 4.323 m, so that the work done is $40(4.323) = 172.91$ J.

The solution is now completed by adding the individual work terms to obtain

$$W_{A \to B} = 172.91 - 98.06 = 74.85 \text{ J}$$

EXAMPLE 10

One end of a linear elastic spring whose unstretched length is L_0 is attached to a fixed point O, and the other end is attached to collar C which slides on a smooth curved bar, as shown in the diagram. The spring and collar are coplanar. Determine the work done by the spring force when the collar C is moved from point A to point B.

Solution

We formulate the solution in terms of cylindrical coordinates having origin at point O because the force in the spring depends on the deformed length of the spring. Also, the spring force has only a radial component. Therefore, if the spring is elongated by an amount Δ from its undeformed length, we have the following free body diagram of the collar.

We use equation (23b) to compute the work done by the spring. The infinitesimal change in the position vector is the vector sum of the radial movement dR and the transverse movement $Rd\phi$ of the collar C. Thus, assuming that the spring force is inward,

$$U_{1 \to 2} = \int_{r_1}^{r_2} \bar{F}_{sp} \cdot d\bar{r}_{P/O}$$

$$= \int_{r_1}^{r_2} (-k\Delta \bar{e}_R) \cdot (dR\bar{e}_R + Rd\phi\bar{e}_\phi)$$

$$= -\int_{R_1}^{R_2} k\Delta \, dR$$

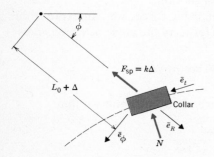

Notice that the spring does work only if its length is changed. Then, because the value of dR is the amount of additional elongation in the spring, $dR \equiv d\Delta$, so

$$U_{1 \to 2} = -k \int_{\Delta_1}^{\Delta_2} \Delta d\Delta \equiv \tfrac{1}{2}k(\Delta_1{}^2 - \Delta_2{}^2)$$

We can gain added insight to the meaning of this result by examining a graph of the force exerted by the spring as a function of its elongation, as shown below.

When the motion of the particle results in increasing the elongation of the spring from Δ_1 to Δ_2, the force applied by the spring is opposite the motion of the particle, so that

$$U_{1 \to 2} = -\int_{\Delta_1}^{\Delta_2} F_{sp}\, d\Delta = \tfrac{1}{2}k(\Delta_1{}^2 - \Delta_2{}^2)$$

This integral is the lightly shaded trapezoidal area in the foregoing sketch. Hence

$$U_{1 \to 2} = -\tfrac{1}{2}[(F_{sp})_1 + (F_{sp})_2](\Delta_2 - \Delta_1)$$
$$= -\tfrac{1}{2}[k\Delta_1 + k\Delta_2](\Delta_2 - \Delta_1) = \tfrac{1}{2}k(\Delta_1{}^2 - \Delta_2{}^2)$$

This, of course, is the same result as was obtained earlier by integration.

It should be noted that the result obtained in this example shows that the value of the work depends only on the magnitude of the initial and final deformation, and not on whether these deformations are positive (elongation) or negative (compression).

HOMEWORK PROBLEMS

III.89 As the 50-kg package descends the incline, it is caught by a bumper (spring) at point A. The stiffness of the spring, which is initially held at 50 N compression by the two tensioned cables, is 100 N/m. The coefficient of friction is 0.20. Determine the work done by all the forces acting on the package if the package compresses the spring 10 mm.

Prob. III.89

III.90 The throttle of a 600-lb motorcycle carrying a 175-lb rider is held steady, resulting in a constant traction force of 150 lb. Neglecting friction, determine the total work done by all the forces in the system when the motorcyclist moves from point A to point B. The center of mass for the motorcycle and rider is 3 ft above the ground.

Prob. III.90

III.91 A ball is thrown from the ground at 80 ft/s at an angle of elevation

Prob. III.91

Prob. III.92

Prob. III.93

of 45°. Determine the work done by gravity as the ball moves from its initial location to the highest point in its trajectory.

III.92   A 2-kg block is tied to a cable as it slides on a smooth table in the horizontal plane. The cable is passed through a hole in the table at point O, and a 40-N force is applied to its free end. This force causes the block to spiral inward, from point A where $R = 500$ mm, to point B where $R = 400$ mm. Determine the work done by the cable (a) by considering the force applied to the block, (b) by considering the force applied to the free end of the cable.

III.93   A 0.5-kg collar is pushed along the rigid bar by the constant 10-N force shown. Determine the work done by all the forces acting on the block moving the block from position A to position B. The Z axis is vertical, and friction is negligible.

III.94   A constant tension force of 100 lb at the end of the cable raises the suspended 40-lb crate 1.75 ft from the position shown. Determine the total work done in raising the crate.

Prob. III.94

2 Potential Energy and Conservative Forces

A subtlety contained in Example 10 is that we did not need to know the curve formed by the bar in order to compute the work done by the spring. Any force that has the characteristic of doing the same amount of work in moving a particle between two points, independent of the path the particle follows, is called a *conservative force*. A consequence of this property is that the total work done by a conservative force when it moves a particle along a closed path back to its starting point is zero. For example, suppose that the spring in Example 10 were freed from the constraint of the curved bar and went from position 1 to position 2 along *any* arc C_1 and returned along any other arc C_2, as shown in Figure 6.

Then, from the result of Example 10 we may write

$$U_{1\rightarrow2} = \tfrac{1}{2}k\Delta_1{}^2 - \tfrac{1}{2}k\Delta_2{}^2$$

so that interchanging the subscripts gives

$$U_{2\rightarrow1} = \tfrac{1}{2}k\Delta_2{}^2 - \tfrac{1}{2}k\Delta_1{}^2 = -U_{1\rightarrow2}$$

Figure 6

Hence the total work performed by the spring in following the closed path is

$$U_{1\rightarrow2\rightarrow1} = U_{1\rightarrow2} + U_{2\rightarrow1} \equiv 0 \tag{24}$$

Thus, if we suppose that the spring force does negative work in moving a particle from point 1 to point 2 because it opposes the motion, this work is not lost from the system; it is stored within the spring and can be used to return the spring to position 1.

Because the work done by a conservative force depends only on the initial and final positions of the particle, it follows that the work can be expressed as the difference in the value of a function that depends only on the position of the particle. That is, for such forces we can create *potential energy* functions. A potential energy function, V, is defined to be a position-dependent function that satisfies

$$\boxed{U_{1\rightarrow2} = V_1 - V_2} \tag{25}$$

To understand this further, consider equations (24) and (25) in the special case where the potential energy at position 1 is zero. Then

$$U_{2\rightarrow1} = -U_{1\rightarrow2} = V_2$$

Thus, potential energy may be regarded as a measure of the work a conservative force must perform to return the particle to the reference position of zero potential energy.

From the foregoing it is readily verified that in the case of a linear

spring,

$$V_{sp} = \tfrac{1}{2}k\Delta^2 \qquad (26)$$

Note that for all cases (compression or extension) V_{sp} is positive if $\Delta \neq 0$.

In addition to the linearly elastic spring whose potential energy is given by equation (26), another conservative force that is frequently encountered is gravity. The determination of the potential energy of gravity is the topic of the next example.

EXAMPLE 11
Determine the potential energy of the gravity force acting on (a) a ball that we have thrown from the surface of the earth, and (b) an artificial satellite that is in orbit about the earth.

Solution
Part a
In this case we may neglect the curvature of the surface of the earth and thus consider the vertical direction to be fixed. This results in the following free body diagram.

To determine the work done by the gravity force, we use equation (23b) and express all quantities in terms of rectangular Cartesian coordinates. Thus,

$$d\bar{r}_{P/O} = dX\bar{I} + dY\bar{J} + dZ\bar{K}$$

$$\bar{F} = -mg\bar{K}$$

$$U_{1 \to 2} = \int_{r_1}^{r_2} \bar{F} \cdot d\bar{r}_{P/O} = -mg \int_{Z_1}^{Z_2} dZ$$

$$= mgZ_1 - mgZ_2$$

Comparing this result with equation (25), we have $V_1 = mgZ_1$, $V_2 = mgZ_2$. Hence

$$V_{gr} = mgZ \qquad (27)$$

where Z is the elevation of the body above an *arbitrary* reference position, which we call the *datum*.
Part b
In this case we must account for the fact that the gravitational force is

always oriented toward the center of the earth. Also, we must use the inverse square law to describe the magnitude of the force. This law is

$$F_{gr} = G\frac{Mm}{R^2}$$

where G is the universal gravitational constant, M is the mass of the earth, and R is the distance from the body to the center of the earth. The value of GM may be estimated by noting that $F_{gr} = mg$ at the earth's surface. Substituting for F_{gr} in the inverse square law yields

$$GM = gR_e^2$$

where $R_e = 3960$ miles $= 6370$ km is the radius of the earth when the earth is considered to be a sphere.

We select cylindrical coordinates with origin at the center of the earth, because it facilitates the description of $\bar{F}_{gr}$. This results in the free body diagram shown. The sketch implies that the satellite follows a planar path; this fact will be established when we study orbital motion in Module VIII.

We now use equation (23b) to compute the work done by $\bar{F}_{gr}$. Thus

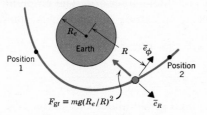

$$d\bar{r}_{P/O} = dR\bar{e}_R + Rd\phi\bar{e}_\phi$$

$$\bar{F}_{gr} = -mg\frac{R_e^2}{R^2}\bar{e}_R$$

$$U_{1\to2} = \int_{r_1}^{r_2} \bar{F}_{gr} \cdot d\bar{r}_{P/O}$$

$$= -mgR_e^2\int_{R_1}^{R_2}\frac{dR}{R^2} = mgR_e^2\left(\frac{1}{R_2} - \frac{1}{R_1}\right)$$

Comparing this result with equation (25), we have $V_1 = -mgR_e^2/R_1$, $V_2 = -mgR_e^2/R_2$, so that

$$\boxed{V_{gr} = -mg\frac{R_e^2}{R}} \qquad (28)$$

This result does not seem to resemble that obtained in part a, even though both results describe the same physical phenomenon. To treat this matter let us specialize equation (28) to the case of motion very close to the earth's surface. Letting $R = R_e + Z$, where Z is the elevation above the surface, an infinite series expansion yields

$$V = -mg\frac{R_e^2}{R_e + Z} = -mg(R_e - Z + \frac{Z^2}{R_e} - \ldots)$$

$$= -mgR_e + mgZ - \ldots$$

Thus, as long as $Z/R_e \ll 1$, the series converges rapidly and the two expressions for the potential energy differ only by a constant. In general, the addition of an arbitrary constant to the result for the potential energy of any conservative force has no effect, for the work done by a conservative force only depends on the *change* in the value of the potential energy. An additive constant does not affect the difference of the potential energy values.

A more mathematical treatment of potential energy can be established by proving that a conservative force is defined by the *gradient* of its potential energy. For rectangular Cartesian coordinates this operation is given by

$$\bar{F} = -\bar{\nabla}V \equiv -\frac{\partial V}{\partial X}\bar{I} - \frac{\partial V}{\partial Y}\bar{J} - \frac{\partial V}{\partial Z}\bar{K}$$

where the symbol $\bar{\nabla}$ is called the "grad" or "del" operator. The gradient formulation has many useful applications in the study of mechanics, particularly fluid mechanics.

HOMEWORK PROBLEMS

III.95 A 1-kg collar slides over the inclined guide under the action of a 100-N tangential force and an elastic spring whose stiffness is 200 N/m. The spring's unstretched length is 400 mm. The diagram depicts the vertical plane. Determine the total work done by all the forces acting on the collar in moving the collar from point A to point B.

III.96 A variable-stiffness bumper for packages on a level ramp is composed of two springs. As shown, the unstretched length of spring 1, of stiffness 10 lb/in., is 3 in. greater than the length of spring 2, of stiffness 20 lb/in. In terms of the distance x that spring 1 is compressed by a package, derive an expression for the total potential energy stored in the springs (a) if $x < 3$ in., (b) if $x > 3$ in.

III.97 The elevation of a 5-kg object above the surface of the earth is increased by 1 km. Determine the increase in the potential energy of the gravitational force acting on this body (a) if the object was initially at the surface of the earth, (b) if the object was initially 6370 km above the earth's surface.

III.98 A type of slingshot is formed by stretching a rubber band horizontally between pegs A and B with an intial tension force $\bar{F}_0$. An object is pulled by the horizontal force $\bar{P}$ and held at point D. The stiffness of the rubber band is k, (a) Derive an expression for the potential energy V_{rb} in

500 mm
100 N
45°
Prob. III.95

3 in.
v_0
Spring 1
Spring 2
Prob. III.96

Prob. III.98

Rigid,
massless bar

Sphere

Prob. III.99

terms of k, F_0, and the dimensions x and l shown. (b) How much work is done by the force $\bar{P}$ in moving the object from point C to point D? (c) When the rubber band is released and the object is projected past point C, how much work does the rubber band do?

III.99 Derive an expression for the potential energy V_{gr} of the gravitational force acting on the sphere in terms of the mass m, the length l, and the angle θ. Plot the results for $-180° \leq \theta \leq 180°$. What is the significance of the locations where V_{gr} is a maximum or a minimum?

III.100 The tensile force in a nonlinear spring is given by $F = k_1\Delta + k_2\Delta^3$, where Δ is the elongation. As shown, the spring is said to be of the hardening or softening type according to whether k_2 is positive or negative. (a) Derive an expression for the potential energy V of such a spring. (b) For the case where it is of the softening type, determine the values of Δ where $F = 0$ and where $V = 0$. Why don't the two coincide?

F

$k_2 > 0$: hardening

$k_2 = 0$: linear

$k_2 < 0$: softening

Δ

Prob. III.100

3 General Principle of Work and Energy

The work done by the resultant force in moving a particle is related by Newton's second law to the kinematic variables describing the motion of the particle. We begin by using the path variables description to relate the work done by the resultant force moving along arc length ds to the motion variables. This gives

$$\Sigma F_t \, ds = m\dot{v}_P \, ds$$

From the chain rule we have $\dot{v}_P \equiv v_P \, dv_P/ds$, so the foregoing becomes

$$\Sigma F_t \, ds = mv_P \, dv_P$$

Integrating this expression from its initial position to its final position along the path yields

$$\int_{s_1}^{s_2} \Sigma F_t \, ds = m \int_{v_1}^{v_2} v_P \, dv_P \equiv \tfrac{1}{2}m(v_2{}^2 - v_1{}^2) \tag{29}$$

From equation (23a) we see that the left side of equation (29) is the total work done by the forces in moving the particle from position 1 to position 2. We now define *kinetic energy*, denoted by the symbol T, to be

$$\boxed{T = \tfrac{1}{2}mv_P{}^2 \equiv \tfrac{1}{2}m\bar{v}_P \cdot \bar{v}_P} \qquad (30)$$

so equation (29) may be rewritten as

$$T_2 - T_1 = U_{1\to2} \qquad (31)$$

In other words, the change in the kinetic energy of the particle equals the total work done by the forces acting on the particle. Note that the work-energy equation (31) is a scalar equation.

Applying equation (31) to solve problems requires that, for each problem, we compute the work done by all forces. The computations can be considerably reduced if we use the concept of potential energy to evaluate the work done by conservative forces. Therefore, let us decompose the total work $U_{1\to2}$ into the contributions of the conservative forces, denoted as $U_{1\to2}^{(c)}$, and the nonconservative forces, denoted as $U_{1\to2}^{(nc)}$. Equation (31) then becomes

$$T_2 - T_1 = U_{1\to2}^{(c)} + U_{1\to2}^{(nc)}$$

Now let V *be the total potential energy,* that is, the sum of the potential energy associated with each conservative force. From equation (25) we then have

$$T_2 - T_1 = V_1 - V_2 + U_{1\to2}^{(nc)}$$

Rearranging the terms in this expression yields

$$\boxed{T_2 + V_2 = T_1 + V_1 + U_{1\to2}^{(nc)}} \qquad (32)$$

This is the form of the work-energy relation that we shall employ to solve problems. It can be utilized to relate the speed of the particle at any two positions, provided that the work done by the nonconservative forces can be evaluated. The significance of the latter requirement will be obvious in the second of the following examples. However, before we study some examples, let us consider the implication of this work-energy equation.

First, if all forces acting on the particle are nonconservative, then there is no potential energy and equation (32) reduced to equation (31). On the other hand, if all of the forces are conservative, we then have

$$\boxed{T_2 + V_2 = T_1 + V_1} \qquad (33)$$

This means that the sum of potential and kinetic energy must remain constant in the motion. Equation (33) is usually referred to as the principle of *conservation of mechanical energy.*

There are other forms of energy in addition to mechanical energy, such as heat and acoustic energy. When we account for these, we find that equation (32) is equivalent to the first law of thermodynamics, which states that the sum of all forms of energy in a system is constant. For example, the friction force on a sliding body always opposes the motion, and therefore does negative work. This results in a dissipation in the value of $T + V$, where the lost mechanical energy is manifested mainly in the production of heat.

EXAMPLE 12

The 10-lb block shown in the sketch slides over the smooth vertical rod until it is caught at point A by a ring that is attached to two linearly elastic springs. When the springs are in position A they are each elongated by 6 in. and the tensile force within them is 40 lb. Knowing that the block has a velocity of 15 ft/s downward when the block is 3 ft above point A, determine the maximum distance the block will fall below point A.

Solution

We begin with a free body diagram of the block. The spring forces are depicted by dashed arrows in order to indicate that these forces act in only a portion of the motion.

From this diagram we see that the only forces doing work are the gravity and spring forces. ($\bar{N}$ is always perpendicular to the motion; in fact, $\bar{N} \equiv 0$ in this problem.) We therefore apply the work-energy equation (32), using the concept of potential energy to describe the work done by the springs and gravity. Thus

$$T_2 + V_2 = T_1 + V_1 + U_{1 \to 2}^{(nc)}$$

$$V = V_{gr} + V_{sp}, \quad U_{1 \to 2}^{(nc)} = 0$$

We are assuming that there is no loss of mechanical energy when the block collides with the ring. This assumption allows us to relate directly the motion of the block at the initial and final positions, without being concerned about the collision process.

We find the properties of the springs from the information given in the problem, which states that $F_{sp} = 40$ lb and $\Delta = 6$ in. $= \frac{1}{2}$ ft at position A. For a linear spring, $F_{sp} = k\Delta$, so in this case

$$k = \frac{40}{\frac{1}{2}} = 80 \text{ lb/ft}$$

To determine the elongation of each spring at any position other than point A, we need to know the unstretched lengths. Noting that the springs are each 2 ft long at point A where they are elongated by $\frac{1}{2}$ ft, we have

$$L_0 = 2.0 - 0.5 = 1.5 \text{ ft}$$

We can now evaluate the system's potential energy. To do this let s be the distance the block descends below point A and choose point A to be the datum for gravity, as shown in the following sketch.
Considering that there are two springs, equations (23) and (24) for potential energy give

$$V = 2 \left(\tfrac{1}{2}k\Delta^2\right) + mgh = 80\Delta^2 + 10h$$

At the initial position the block is 3 ft above the datum ($h = +3$ ft) and $\Delta = 0.5$ ft, hence

$$V_1 = 80(0.5)^2 + 10(3) = 50 \text{ lb-ft}$$

From the sketch we see that the length of the spring in the final position is $(s^2 + 2^2)^{1/2}$, so that

$$\Delta_2 = \sqrt{s^2 + 2^2} - L_0 = \sqrt{s^2 + 4} - 1.5$$

Then, because the block is below the datum in the final position, $h = -s$, so that

$$V_2 = 80(\sqrt{s^2 + 4} - 1.5)^2 - 10s$$

The kinetic energy in the initial position is

$$T_1 = \tfrac{1}{2}\left(\frac{10}{32.17}\right)(15)^2 = 34.97 \text{ lb-ft}$$

In the final position $T_2 = 0$ because the value of s is a maximum when the block comes to rest ($\dot{s} = 0$). Thus the work-energy equation gives

$$0 + 80(\sqrt{s^2 + 4} - 1.5)^2 - 10s = 34.97 + 50$$

Removing the radical sign from this expression results in a quartic polynomial for s. The real roots of this polynomial are

$$s = 1.707 \text{ ft} \qquad s = -1.400 \text{ ft}$$

The positive answer is the solution we seek. The negative value of s represents the maximum elevation above point A to which the block would rebound if it were actually attached to the springs.

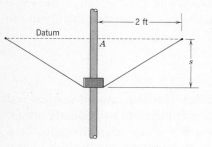

EXAMPLE 13

A skier descends a 40° slope and enters a circular transition to a level grade at a speed of 50 km/hr. The transition has a radius of 140 m. Determine the speed of the skier at the bottom of the transition curve (a) when friction is neglected, and (b) when friction between the snow and skies is accounted for. The coefficient of sliding friction is $\mu = 0.10$.

Solution

In case a the friction force on the skier is negligible, whereas in case b the frictional force obviously opposes the velocity. Thus a free body diagram that includes both cases is the following.

Case a For this case we set $f \equiv 0$, and determine the individual terms in the work-energy equation (32). At point A the speed is 50 km/hr, so that the kinetic energies at the initial and final positions are

$$T_1 = \tfrac{1}{2}m\left[(50)\,\frac{1000}{3600}\right]^2 = 96.45(m) \text{ J} \qquad T_2 = \tfrac{1}{2}mv_B{}^2$$

Turning to the free body diagram, we see that the normal force $\bar{N}$

does no work and that the only other force is gravity, which is conservative. Hence, $U_{1\to2} \equiv 0$, so that equation (32) reduces to the conservation of energy equation (33). Let us choose the end of the slope, point B as the datum elevation. To compute the elevation of the skier at point A, we sketch the geometry of the slope, considering the skier as a particle.

From this sketch we see that

$$h = 140 - 140 \cos 40°$$

so that

$$V_1 = mg(140)(1 - \cos 40°) = 32.75mg$$

$$V_2 = 0$$

Equation (33) now gives

$$\tfrac{1}{2}mv_B{}^2 + 0 = 96.45m + 32.75mg$$

or

$$v_B = 28.90 \text{ m/s} = 104.0 \text{ km/hr}$$

Case b In this case $f = \mu N$. Thus, in order to compute the work done by the friction force we must determine the value of N. Using the path variable unit vectors shown in the free body diagram, the acceleration of the skier is

$$\bar{a}_P = \dot{v}\bar{e}_t + \frac{v^2}{140}\bar{e}_n$$

so that

$$\Sigma F_n = N - mg \cos \theta = m\frac{v^2}{140}$$

$$N = m\left(g \cos \theta + \frac{v^2}{140}\right)$$

For the work done by the friction force we therefore have

$$U_{1\to2}^{(nc)} = -\int_{s_A}^{s_B} \mu N \, ds = -\mu m \int_{s_A}^{s_B} \left(g \cos \theta + \frac{v^2}{140}\right) ds$$

The evaluation of this integral requires that we know how the speed depends on the value of s. Because this information is unknown, and in fact represents the solution we desire, we must use a different method to solve the problem. Remember that at the outset we indicated that the work-energy relations cannot be used in situations where forces doing work depend on the speed of the particle. The real purpose of case b, then, is to present a situation where the method of work-energy is not an effective tool.

To complete the problem we return to the free body diagram and write the equations of motion.

$$\Sigma F_t = mg \sin \theta - \mu N = m\dot{v}$$

$$\Sigma F_n = N - mg \cos \theta = m \frac{v^2}{R}$$

We leave it as a homework problem to show that the solution to these equations is

$$v^2 = Ae^{2\,\mu\theta} + \frac{2gR}{1 + 4\mu^2} [3\mu \sin \theta + (1 - 2\mu^2) \cos \theta]$$

where the constant A may be found from the initial conditions: $v = 50$ km/hr when $\theta = 40°$. Carrying out the computations yields $v = 74.64$ km/hr at the bottom point B. Comparing this to the result in part a, we see the sizable effect of friction in this problem.

HOMEWORK PROBLEMS

III.101 Solve Problem III.43 using the work-energy principle.

III.102 A 3000-lb automobile descends a 5% grade at 45 miles/hr. The coefficient of friction between the tires and the ground is 0.60. What is the minimum distance required to reduce the speed (a) to 15 miles/hr, (b) to zero. Neglect the resistance of the air.

Prob. III.102

III.103 The sliding collar in Problem III.93 is at rest at point A. Determine its speed when it reaches point B.

III.104 An object is at rest with respect to the center of the earth at an altitude of 2000 km above the earth's surface. Determine the speed at which it will hit the earth if it falls, neglecting air resistance.

Prob. III.104

III.105   In Problem III.90, the motorcycle starts from rest at point A. Determine the speed of the motorcycle when it reaches point B.

III.106   The maximum speed of a bicyclist on level ground is 10 m/s. What is the largest radius R of a vertical loop that the rider may safely negotiate without pedaling the bicycle? The center of mass of the bicyclist and bicycle is 1 m above the ground.

Prob. III.106

III.107   At $\theta = 10°$ the pendulum is returning to its static equilibrium position at 2.5 m/s. Determine (a) the speed of the sphere when it passes the vertical position, (b) the maximum value of the angle θ during the motion.

Prob. III.107

Prob. III.108

III.108   A simple pendulum of mass m and length l is given an initial velocity v_0 to the right at its lowest point. Determine the minimum value of v_0 necessary to cause the mass to follow a full circle if the mass is connected to the pivot A (a) by a massless rigid bar, (b) by a cord.

III.109   A simple pendulum of mass m and length l is released from rest in the horizontal position. The cord hits peg B, which is directly below the pivot A, and begins to wrap around it. Determine the smallest value of the distance d for which the mass m will follow a complete circle about peg B.

Prob. III.109

III.110 A daredevil toboggan rider proposes to descend the ice-covered slope from point A, fly into the air at point B, and then land on the ledge C when the toboggan is at the top of its trajectory. The take-off angle at point B is 30°. Neglecting friction and air resistance, determine (a) the speed v_B of the toboggan at take-off, (b) the dimensions h and d where the ledge C should be located, (c) the landing speed v_C.

Prob. III.110

III.111 Verify by substitution that the expression for $v(\theta)$ in case (b) of Example 13 satisfies the equations of motion given there. Then use this result to drive an expression for the work done by the friction force as a function of θ.

III.112 What constant tension force $\bar{F}$ must be applied to the end of the cable to give the 20-lb block a downward velocity of 2 ft/s after the block descends 4 ft from rest?

Prob. III.112

III.113 Solve Problem III.7a using the work-energy principle.

III.114 The 40-lb crate is raised from point A where it is at rest by a constant tensile force of 35 lb. Determine (a) the speed of the package after it has been raised 1.75 ft, and (b) the maximum height the crate will attain.

Prob. III.114

III.115 A 10-lb block is dropped from a height of 2 ft onto a spring whose stiffness is 2 lb/in. Determine the maximum deformation of the spring and the maximum compressive force it sustains.

III.116 A 10-kg package slides down from point A and is brought to rest at point B by a spring whose stiffness is 5 kN/m. Neglecting frction, determine the maximum compression of the spring.

Prob. III.115

Prob. III.116

III.117 The 50-kg package descends the incline and is caught by the spring bumper. The stiffness of the spring is 10 kN/m and the spring is precompressed by the two tensioned cables, which apply a force of 50 N. The coefficient of friction is 0.20. If the package was initially 2 m from the bumper when it was released from rest, determine the maximum compression force in the spring.

Prob. III.117

III.118 Determine the maximum distance to which the package in Problem III.117 rebounds.

III.119 In Problem III.96, determine the maximum compression of spring 1 if a 4-lb package hits it (a) at 50 in./s, (b) at 200 in./s.

III.120 A 2-kg collar slides over the rigid guide AB. The spring has a stiffness of 5 N/m and an unstretched length of 100 mm. If the collar is released from rest at point A, what is its speed when it reaches point B?

Prob. III.120

Prob. III.121

Prob. III.122

Prob. III.123

III.121 The diagram shows a schematic view of an arresting gear to slow automobiles in a crash test, in which a hook on the vehicle is caught at point C. When the cable is stretched between the vertical posts A and B, the tension within the springs is 10,000 N. The stiffness of each spring is $k = 2(10^4)$ N/m. After hitting the cable, the hook on the automobile moves a distance of 6 m, coming to rest at point D. Determine the speed of the automobile at point C. The mass of the vehicle is 1200 kg, and the vehicle is rolling freely.

III.122 The 0.5-kg collar is pulled down the smooth curved rod by gravity and a spring whose stiffness is 20 N/m. At point A the collar is at rest and the tension force in the spring is 16 N. Determine the force exerted by the collar on the rod when the collar is on the curved segment (a) at point B, (b) at point C.

III.123 An elastic cord whose stiffness is 24 lb/ft and whose unstretched length is 16 in. is attached to a 2-lb collar. The collar rides on the circular guide $ABCD$ in a horizontal plane. When the collar is at point A it has a velocity of 10 ft/s to the left. Determine (a) the maximum speed of the collar, (b) whether the collar reaches point D, and if so, its speed at this location. *Hint:* Recall that a cord cannot sustain a compressive force.

III.124 Determine the minimum speed that the collar in Problem III.122 should have at point A in order to reach point D.

III.125 The 5-kg sphere is attached to arm AB, which may be considered to be massless and rigid. The stiffness of the spring is 800 N/m and the force in the spring is zero when arm AB is horizontal. Knowing that the system was released from rest in the horizontal position, determine the speed of the block when the arm falls to the vertical position.

Prob. III.125

4   Work and Energy for a System of Particles

The work-energy equation for a system of N particles is obtained by summing the corresponding equations for each particle. Let T be the total kinetic energy of the system. Then

$$T = \tfrac{1}{2} \sum_{i=1}^{N} m_i v_i^2 \tag{34}$$

Now redefine V to be the total potential energy of all conservative forces acting on the particles of the system. The foregoing definitions lead to the result that *equation (32) is also the work-energy relation for a system of particles*. That this result follows so readily stems from the fact that the work done by the individual forces is additive as scalars.

EXAMPLE 14

The system shown in the sketch is released from rest. Determine the velocity of both blocks when block B has descended 1.5 m. The masses of the blocks are $m_A = 12$ kg, $m_B = 6$ kg, and the coefficient of kinetic friction between block A and the horizontal surface is 0.20.

Solution

To relate the energy at the initial and final positions, we begin by drawing free body diagrams of both blocks.

An analysis of the cable-pulley system shows that block A moves 0.75 m to the right when block B descends 1.50 m, and also that $v_A = v_B/2$. The displacement information is used to express the work done by the forces acting on both bodies. Because each of the tangential components of force is constant, the total work is

$$U_{1 \to 2} = [6(9.806) - F](1.50) + [2F - 0.2N_A](0.75) = 88.25 - 0.15N_A \text{ J}$$

Notice that we chose not to use the fact that gravity is a conservative force, because for this comparatively uncomplicated problem the work done by this force is simple to determine. In essence, we are ignoring equation (32) and using the more fundamental relationship, equation (31).

The evaluation of the work term is completed by determining N_A. The equilibrium equation for block A normal to its direction of motion yields

$$N_A = 12(9.806) = 117.67 \text{ newtons}$$

so that

$$U_{1 \to 2} = 88.25 - 0.15(117.67) = 70.60 \text{ J}$$

Finally, the total kinetic energy is

$$T = \tfrac{1}{2}m_A v_A{}^2 + \tfrac{1}{2}m_B v_B{}^2 = 6v_A{}^2 + 3v_B{}^2$$

Because the system is initially at rest and $v_A = v_B/2$, we have

$$T_1 = 0 \qquad T_2 = 6\left[\frac{(v_B)_2}{2}\right]^2 + 3(v_B)_2{}^2 = 4.5(v_B)_2{}^2$$

Thus from equation (30) we obtain

$$T_2 = T_1 + U_{1 \to 2} \qquad 4.5(v_B)_2{}^2 = 0 + 70.60$$

$$(v_B)_2 = 3.96 \text{ m/s}$$

A significant feature of this system, common to *all* systems in which cables and pulleys are considered to have ideal properties (massless, inextensible, and frictionless), is that the cable tension does not contribute to the total work done in moving the connected bodies, so that the work done by the cable forces may be ignored. This subtlety is displayed in the formulation of $U_{1 \to 2}$ in the foregoing example.

HOMEWORK PROBLEMS

III.126 Determine the force $\bar{F}$ that must be applied to the 20-kg block A

to make the 20-kg block B attain a speed of 4 m/s after moving 2 m to the right from rest. Neglect friction.

Prob. III.126

III.127 Solve Problem III.126 if the coefficient of friction between the block and the ground is 0.20.

III.128 When released from rest, block A attains a speed of 1 m/s after rising 2 m. Knowing that the coefficient of friction between block B and the incline is 0.40 and that the mass of block B is 5 kg, determine the mass of block A.

Prob. III.128

III.129 Two spheres A and B are attached to opposite ends of the rigid bar, which pivots about point C. The masses of the spheres are $m_A = 2$ kg, $m_B = 1$ kg, and the mass of the bar is negligible. The bar is released from rest in the horizontal position. Determine the angular velocity of the bar when it passes the vertical position.

Prob. III.129

III.130 A 4-oz pin is guided through the circular groove by the grooved vertical arm, which is pulled to the right by a 2-lb force. The arm weighs 1 lb. The pin starts from rest at point A. Determine the speed of the pin when it has traveled 8 in. to the right.

Prob. III.130   **Prob. III.131**

III.131 Sliders A and B are connected by a rigid bar of negligible mass in a vertical plane. If block A is released from rest in the position shown, determine its velocity after it falls (a) 2 ft, (b) 4 ft. Both blocks weigh 2 lb, and friction is negligible.

←—40 in.—→

B

A

Prob. III.132

III.132 Block *A* is released from rest at the same elevation as block *B* with the cable taut. Neglecting friction, determine the velocity of block *B* after block *A* has fallen 30 in. Both blocks weigh 1 lb.

III.133 Solve Problem III.132 for the case where there is frictional resistance to the motion of block *B*, the coefficient of friction being 0.20.

G. POWER

The concept of power is intimately tied to the energy concepts we have just studied. Basically, the *mechanical power* supplied to a system by a force is the time rate at which this force does work; that is,

$$\text{Power} = \frac{dU}{dt} \tag{35}$$

A more specific relation for power is obtained by considering the definition of the work performed by a force in an infinitesimal displacement, which is

$$dU = \bar{F} \cdot d\bar{r}_{P/O}$$

Dividing this expression by dt yields

$$\frac{dU}{dt} = \bar{F} \cdot \frac{d\bar{r}_{P/O}}{dt}$$

$$\text{Power} = \bar{F} \cdot \bar{v}_P \tag{36}$$

The unit of power in the SI system is a watt (W), which is equal to 1 J/s, which is 1 N-m/s. In the British system the basic unit is a horsepower (hp) one horsepower being defined as 550 ft-lb/s. (One horsepower is 776 W.)

For a different viewpoint of the power developed by a force, let us consider equation (32), rewritten as follows:

$$U_{1\to2}^{(nc)} = (T_2 - T_1) + (V_2 - V_1)$$

In the situation where position 2 is different from position 1 by a differential amount, this equation becomes

$$dU^{(nc)} = dT + dV$$

Dividing this result by dt gives

$$\text{Power}^{(nc)} \equiv \frac{d}{dt}(T + V) \tag{37}$$

Thus, we see that the total power supplied to a system by a nonconservative force equals the rate of increase of the potential and kinetic energy of the system.

A common type of nonconservative force is friction, which generally does negative work. Thus, the power associated with a friction force is negative, representing a dissipation of energy essentially in the form of heat. In this book we are concerned only with mechanical energy; the science of thermodynamics is concerned with all forms of energy and the efficiency with which this energy is used.

EXAMPLE 15

The power output of the motor of an electric car is constant over its entire speed range for a fixed setting of the motor control. The maximum power that can be delivered to the drive wheels is 60 hp. The automobile weighs 2500 lb and the air resistance in pounds is $0.012v^2$, where v is in units of feet per second. Consider the rolling resistance to be negligible. Determine the maximum speed of the vehicle and the minimum time required to accelerate the car from rest to a speed of 60 miles/hr, both on level ground.

Solution

In drawing a free body diagram we must remember that the friction force $\bar{f}$ on the drive wheels pushes the automobile forward.

The power of the motor is transmitted to the drive wheels. The friction force $\bar{f}$ is in the direction of motion. Therefore, using power $= fv$, for maximum acceleration

$$f = \frac{60(550)}{v} = \frac{33,000}{v} \text{ lb}$$

We now sum forces in the tangential direction.

$$\Sigma F_t = f - 0.012v^2 = m\dot{v} = \frac{2500}{32.17}\dot{v}$$

$$\frac{33,000}{v} - 0.012v^2 = 77.71\dot{v}$$

This is a differential equation defining acceleration as a function of position. To determine the maximum speed we set $\dot{v} = 0$ and find

$$\frac{33,000}{v_{max}} - 0.012v_{max}^2 = 0$$

$$v_{max} = 140.10 \text{ ft/s} = 95.5 \text{ miles/hr} \qquad \triangleleft$$

The solution to the second part of the problem is more complicated, because it requires that we solve the differential equation of motion. To determine the time required to attain a speed of 60 miles/hr (= 88 ft/s), we write $\dot{v} = dv/dt$ and separate variables.

$$\frac{77.71 \, dv}{(33,000/v) - 0.012v^2} = dt$$

Using the initial conditions we obtain the following integral

$$\frac{77.71}{0.012}\int_0^{88} \frac{v \, dv}{2.75(10^6) - v^3} = \int_0^t dt \equiv t$$

From a table of integrals we find

$$t = \frac{77.71}{0.012}\frac{1}{3k}\left[\tfrac{1}{2}\ln\left(\frac{k^2 - kv + v^2}{k^2 + 2kv + v^2}\right) + \sqrt{3}\tan^{-1}\left(\frac{2v - k}{k\sqrt{3}}\right)\right]_0^{88}$$

where

$$k = [2.75(10^6)]^{1/3} = 140.10$$

Evaluating this expression for t yields

$$t = 8.33 \text{ seconds} \qquad \triangleleft$$

It may seem unlikely to you that an automobile whose top speed is only 95 miles/hr should be able to accelerate this rapidly. The result is obtained because the electric motor is capable of high power output at low speed. A gasoline engine does not have this characteristic. In fact, it is for this reason that automobiles with gasoline engines require transmissions.

HOMEWORK PROBLEMS
III.134 A 30,000-kg truck has a constant speed of 40 km/hr up a 10% grade. The air resistance is $F = 40.0v^2$, where F is in newtons and v is the speed of the vehicle in meters per second. It is known that 75% of the

power produced by the engine is actually used to propel the truck. Determine the power produced by the engine.

Prob. III.134

III.135 In Problem III.134, what acceleration will the truck have when it reaches level ground if its engine continues to produce the same power?

III.136 Once a motorboat exceeds 3 m/s, its hull begins to *plane* over the water. It is known that the water resistance is then given by $F_w = k_1/v^2 + k_2 v^2$, where v is the speed of the boat. It is also known that 5 kW are required to propel the boat at 15 m/s, and that a speed of 5 m/s requires the minimum power output from the engine. (a) Determine the values of the coefficients k_1 and k_2. (b) At what speed is the water resistance a minimum? Why isn't this speed the same as that for which the power is a minimum?

III.137 Four identical diesel-electric locomotives weighing 30 tons each accelerate a freight train from rest. There are 150 cars in the train, the average weight of each car being 60 tons. The rolling resistance on each car is 0.2% of its weight, and air resistance is negligible. Each locomotive is producing 2000 hp of usable mechanical power. Determine (a) the top speed of the train, (b) the time required to attain a speed of 50 ft/s.

Prob. III.137

III.138 An electrically powered winch is utilized to move the 50-kg crate with a constant acceleration from rest to 5 m/s in a distance of 10 m. The inclined ramp consists of a series of roller bearings, so friction is negligible. Determine (a) the maximum horsepower produced by the motor, and (b) the horsepower produced in the position where the crate has moved 5 m from its initial position.

36.87°

Prob. III.138

**Prob. III.139
and Prob. III.140**

III.139   The electric motor raises block A at the constant speed $v_A = 6$ m/s. Both blocks have a mass of 2 kg. Determine the power required of the motor.

III.140   The motor is switched on when block A is descending at 6 m/s, immediately producing 200 watts to slow the system. Determine the deceleration rate of block A when the motor is turned on.

III.141   The electric motor pulls the cable in at the constant rate v. Derive an expression for the power required from the motor in terms of v, the mass m of the package being raised, and the dimensions l and d shown.

Prob. III.141

III.142   The escalator carries people upward at 4 ft/s. What is the average power that the motors must furnish in order to transport 45 people simultaneously, if the average weight of a person is 170 lb? Also, determine how many people are being transported per minute.

Prob. III.142

H.   LINEAR IMPULSE-MOMENTUM RELATIONS

The momentum principles that we shall study are additional standard integrals of the differential equations of motion. In contrast to the energy principle, which treats situations where forces vary with the position of the particle, we *consider the momentum of the particle when the forces depend only on time.*

The derivation is straightforward. We begin with Newton's second law and note that when the forces depend only on t, we can separate variables in the vector equation by writing $\bar{a}_P \equiv \dot{\bar{v}}_P = d\bar{v}_P/dt$. Thus

$$m \, d\bar{v}_P = \Sigma \bar{F} \, dt$$

Integrating over an arbitrary time interval from t_1 to t_2 yields

$$m \int_{\bar{v}_1}^{\bar{v}_2} d\bar{v}_P = \int_{t_1}^{t_2} \Sigma \bar{F} \, dt$$

$$\boxed{m\bar{v}_2 - m\bar{v}_1 = \int_{t_1}^{t_2} \Sigma \bar{F} \, dt} \tag{38}$$

where $\bar{v}_1$ and $\bar{v}_2$ are the initial and final velocities of the particle.

The quantity $m\bar{v}_P$ is the *linear momentum* of the particle. The integral in the right side of equation (38) is called the *linear impulse* of the force resultant. In words, equation (38) states that the change in the linear momentum equals the linear impulse. Notice that this is a vector relation representing three scalar equations.

Let us now consider how we can apply equation (38) in conjunction with the several kinematical descriptions. In the case of rectangular Cartesian coordinates, we have

$$\Sigma \bar{F} = \Sigma F_X \bar{I} + \Sigma F_Y \bar{J} + \Sigma F_Z \bar{K}$$

Then, because the unit vectors are constant, the linear impulse is

$$\int_{t_1}^{t_2} \Sigma \bar{F} \, dt = \int_{t_1}^{t_2} \Sigma F_X \, dt \, \bar{I} + \int_{t_1}^{t_2} \Sigma F_Y \, dt \, \bar{J} + \int_{t_1}^{t_2} \Sigma F_Z \, dt \, \bar{K}$$

This tells us that the linear impulse may be evaluated whenever we know how the XYZ components of the forces vary with time. If we so desire, we can then obtain scalar equations equating corresponding components in equation (38). For instance, the $\bar{I}$ component gives

$$m(v_X)_2 - m(v_X)_1 = \int_{t_1}^{t_2} \Sigma F_X \, dt$$

We should note that this result can also be obtained with equal ease from Newton's second law whenever we know how ΣF_X depends on time by writing

$$\Sigma F_X = ma_X \equiv m\ddot{X} = m\dot{v}_X$$

Separating variables and integrating then gives the same result as that obtained in the momentum development.

Let us now consider what is involved in evaluating the linear impulse in situations where we are using either path variables or cylindrical coordinates. The unit vectors in these coordinate descriptions vary, so they cannot be treated as constants in the integral for the impulse. For example, in the case of path variables we have

$$\int_{t_1}^{t_2} \Sigma \, \bar{F} \, dt = \int_{t_1}^{t_2} (\Sigma \, F_t \, \bar{e}_t) \, dt + \int_{t_1}^{t_2} (\Sigma \, F_n \, \bar{e}_n) \, dt$$

For a general motion these integrals can be evaluated only by resolving the unit vectors into their $(\bar{I}, \bar{J}, \bar{K})$ components. Therefore, we make the general statement that the principle of linear momentum and impulse should always be formulated in terms of rectangular Cartesian coordinates.

One situation where a momentum formulation is very useful occurs when we are interested only in the average value of the forces in a time interval, because in that case the linear impulse is

$$\int_{t_1}^{t_2} \Sigma \, \bar{F} \, dt = (\Sigma \, \bar{F}_{\text{average}})(t_2 - t_1) \tag{39}$$

Other applications of the impulse-momentum principles are for systems of interacting particles and for impulsive forces, such as those exerted between colliding bodies. These are the next topics we shall study.

1 Interacting Particles

Let us begin with the equation for the acceleration of the center of mass G of a system of N particles, equation (22),

$$\sum_{i=1}^{N} \bar{F}_i = m\bar{a}_G$$

where $\Sigma \, \bar{F}_i$ is the resultant of the external forces acting on the system. From equation (2) we know that the position of the center of mass is defined by

$$m\bar{r}_{G/O} = \sum_{i=1}^{N} m_i \, \bar{r}_{i/O}$$

Therefore

$$m\bar{a}_G \equiv m\ddot{\bar{r}}_{G/O} = \sum_{i=1}^{N} m_i \, \ddot{\bar{r}}_{i/O} = \sum_{i=1}^{N} \frac{d}{dt}(m_i \, \dot{\bar{r}}_{i/O})$$

$$= \sum_{i=1}^{N} \frac{d}{dt}(m_i \bar{v}_i)$$

Assuming that we know how the external forces vary with time, separation of variables yields

$$\sum_{i=1}^{N} d(m_i \bar{v}_i) = \sum_{i=1}^{N} \bar{F}_i \, dt$$

We now integrate over the time interval $t_1 \le t \le t_2$ to obtain

$$\sum_{i=1}^{N} \int_{(\bar{v}_i)_1}^{(\bar{v}_i)_2} d(m_i \bar{v}_i) = \sum_{i=1}^{N} \int_{t_2}^{t_1} \bar{F}_i \, dt$$

$$\boxed{\sum_{i=1}^{N} m_i(\bar{v}_i)_2 - \sum_{i=1}^{N} m_i(\bar{v}_i)_1 = \sum_{i=1}^{N} \int_{t_1}^{t_2} \bar{F}_i \, dt} \qquad (40)$$

In words, equation (40) shows that the total linear momentum is changed only by the impulse of the external forces.

Occasionally, we shall encounter a situation where the external forces exert a negligible impulse in comparison to the impulse of the interaction forces. In such a case equation (40) yields

$$\sum_{i=1}^{N} m_i(\bar{v}_i)_2 = \sum_{i=1}^{N} m_i(\bar{v}_i)_1$$

In conventional terminology this is the principle of *conservation of linear momentum*.

The primary feature of equation (40) in general is that it provides a relationship for the motion of a system that does not require knowledge of the interactions occurring within the system. In addition to its application to conventional mechanical systems, we shall employ this formulation in Module VIII to treat some problems in fluid mechanics.

EXAMPLE 16

A truck towing a trailer starts from rest and accelerates along level ground to a speed of 40 km/hr in 20 s. The truck has a mass of 5000 kg and the trailer has a mass of 15,000 kg. The frictional resistances on the truck and the trailer attributable to the undriven wheels are 200 N each. Determine the average traction force produced by the drive wheels of the truck during the acceleration and determine the average pull exerted by the truck on the trailer.

Solution

We draw free body diagrams of the truck and of the trailer, letting $\bar{H}$ be the horizontal pull force exerted by the truck on the trailer and $\bar{F}$ be the tractive force.

As you can see, the interaction between the truck and the trailer also involves the vertical force $\bar{V}$ needed to establish equilibrium for the trailer. The determination of the value of this force is beyond the scope of this module, because such a determination requires that we first study the kinetics of rigid bodies.

We begin by considering the truck and trailer as a system in order to eliminate the unknown interaction force. We refer to the free body diagrams to obtain the external forces acting on the system and use equation (39) to formulate the impulse in terms of average forces. Hence

$$(m_{\text{truck}} + m_{\text{trailer}})(\bar{v}_2 - \bar{v}_1) = \Sigma \, \bar{F}_{\text{average}}(\Delta t)$$

$$20,000[40\left(\frac{1,000}{3,600}\right)\bar{I} - \bar{0}] = \{(F - 200 - 200)\bar{I}$$

$$+ [-(15,000 + 5,000)(9.806)$$

$$+ N_A + N_B]\bar{J}\}(20)$$

Equating the $\bar{I}$ components in this equation gives

$$20,000(40)\left(\frac{10}{36}\right) = (F - 400)(20)$$

$$F = 1.151(10^4) \text{ N}$$

To determine the force $\bar{H}$ we consider either the truck or the trailer individually, so that the effect of this force does not cancel in the interaction. Treating the trailer, we write

$$15,000[40\left(\frac{1,000}{3,600}\right)\bar{I} - \bar{0}] = \{(H - 200)\bar{I} + [N_B + V - 15,000(9.806)]\bar{J}\}(20)$$

Equating the $\bar{I}$ components yields

$$15{,}000(40)\left(\frac{10}{36}\right) = (H - 200)(20)$$

$$H = 8530 \text{ N}$$

Can you explain why H and F are not equal?

HOMEWORK PROBLEMS

III.143 With the aid of a catapult, a 20-ton aircraft is launched in an easterly direction from the deck of an aircraft carrier. The wind is calm and the aircraft has an airspeed of 150 miles/hr when launched, attained after being accelerated for 4 s. If the engines provide 40,000 lb of thrust, determine the force provided by the catapult, (a) if the aircraft carrier is at rest, (b) if the aircraft carrier is heading east at a constant speed of 30 miles/hr.

III.144 A three-car commuter train slows from an initial speed of 72 km/hr on a level track. Each car has a mass of (10^4) kg, and the coefficient of friction between the wheels and the track is 0.20. Determine the minimum amount of time required to stop the train and the corresponding force within the couplers (a) if the brakes in all cars are functioning, (b) if the brakes in the middle car fail. Air resistance is negligible.

Prob. III.144

III.145 The 1500-kg automobile accelerates from rest and attains a speed of 5 m/s relative to the 6000-kg barge in 5 s. Determine the average force in the mooring cable that is required to hold the barge stationary and the average traction force exerted by the tires on the barge.

Prob. III.145 and Prob. III.146

III.146 The 1500-kg automobile starts from rest and attains a speed of 5 m/s relative to the 6000-kg barge in 5 s. Because someone forgot to attach the mooring cable, the barge moves away from the pier. Determine (a) the velocity of the barge after 5 s, (b) the average traction force exerted by the tires on the barge, and (c) the distance the barge moves in this 5-s interval.

III.147 The 20-lb block slides down the smooth ramp of the 100-lb cart. The system was at rest initially. Determine (a) the velocity of the block and of the cart when the box reaches the bottom of the ramp, and (b) the distance the cart shifts during this motion.

Prob. III.147

III.148 A 40-lb piece of luggage has a speed of 8 ft/s when it is projected from the conveyor belt onto the ramp. It then slides down the smooth ramp and onto the 80-lb cart. The coefficient of friction between the piece of luggage and the cart is 0.50. Determine (a) the velocity of the cart when the luggage comes to rest on it, (b) the time required for the luggage to come to rest on the cart, and (c) the total distance the luggage slides. Neglect rolling resistance.

Prob. III.148

III.149 As the 1000-kg automobile descends the hill, the total friction force $\bar{f}$ exerted between the tires and the ground varies as shown, where a negative value represents braking. The initial speed of the car was 72 km/hr. Determine the speed of the car (a) after 15 s, (b) after 30 s. Neglect air resistance.

Prob. III.149

Prob. III.150

III.150 A 0.5-kg ball is thrown at 10 m/s as shown. The XY plane is horizontal. The ball is subjected to a constant 1-N force from a wind that is blowing in the negative Y direction. How long will it be before the ball is traveling parallel to the XZ plane? Determine the velocity of the ball at this time.

III.151   The initial and final velocities of a 200-g particle over a 1-min time interval are $\bar{v}_1 = 6\bar{I} - 6\bar{J} + 3\bar{K}$ m/s and $\bar{v}_2 = -8\bar{I} + 9\bar{J} + 12\bar{K}$ m/s, respectively. (a) Determine the average force acting on the particle. (b) If the force on the particle is known to vary with time according to $\bar{F} = C_1\bar{I} + C_2\bar{J} + C_3t^2\bar{K}$, where t is in seconds and $\bar{F}$ is in newtons, determine the maximum magnitude of $\bar{F}$ in the 1-min interval.

2   Impulsive Forces

To investigate impulsive forces, let us consider the example shown in Figure 7a where a block initially held at rest on an incline by friction is struck by a hammer at instant t_1. The force $\bar{F}$ resulting from the hammer blow is displayed in Figure 7b. We say that the force $\bar{F}$ is an *impulsive force* if it significantly increases the speed of the block in the small time interval from t_1 to t_2. More specifically, an impulsive force acts over a small time duration and imparts an acceleration of many g's to a body.

Figure 7a   **Figure 7b**

From Figure 7b we can determine the magnitude of the impulse of $\bar{F}$, because $\int F\,dt$ is the area A under the $F-t$ curve. We shall assume that during the time interval $t_1 \leq t \leq t_2$ the hammer force exceeds the friction force $\bar{f}$ by a significant amount, so that we can approximate the velocity of the block at t_2 by the linear impulse-momentum relation

$$m(v_2\bar{I} - \bar{0}) = \int_{t_1}^{t_2} F(t)\,dt\,\bar{I} = A\bar{I}$$

$$v_2 = \frac{A}{m}$$

After the hammer blow is completed, the block moves up the incline, decelerating uniformly due to the effects of gravity and friction. Thus the graph of $v(t)$ is as shown in Figure 8.

From the shape of the graph we observed that a reasonable approximation to the velocity history of the block can be obtained by considering the time instants t_1 and t_2 to be identical, which is equivalent to considering the velocity of the particle to change instantaneously. From Figure 8

Figure 8

Figure 9

we can construct a graph of $X(t)$, because for rectilinear motion along the X axis,

$$X(t) = X(0) + \int_0^t v(t)\, dt$$

The qualitative character of $X(t)$, depicted in Figure 9, is obtained by considering the integral to be the area under the $v(t)$ graph. Notice that the value of X is shown close to zero at instant t_2, which is consistent with the approximation that the position of the block does not change during the period of the hammer blow.

The foregoing assertions about the action of a particle acted upon by an impulsive force can be generalized as follows:

1   The change in the velocity of a particle

$$\Delta \bar{v} = \frac{1}{m} \int_{t_1}^{t_2} \bar{F}(t)\, dt$$

occurs, in essence, instantaneously.

2   The change in position of the particle during the time interval the impulsive force acts is negligible.

3   All nonimpulsive forces are negligible compared to the impulsive force during the time interval in which the impulsive force acts.

EXAMPLE 17

The last stage of a guided missile shuts down when the system attains an orbital speed of 27,000 km/hr. To separate the 3000-kg payload from the expended 2000-kg booster, six explosive bolts are fired. The explosives in each bolt generate 400 J of energy, 40% of which converts to mechanical energy. The charges are placed such that the payload continues to move parallel to the initial velocity. Determine the speed of the payload and of the booster immediately after the firing of the explosive bolts and the separation distance between the two bodies 1 min after the firing.

Solution

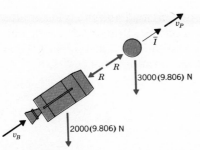

We begin with free body diagrams of the payload and of the booster. Note that the impulsive interaction force $\bar{R}$ generated by the explosion is depicted parallel to the direction of the original velocity in order to maintain this direction. Also, we are using the standard value of g because no information is given about the altitude.

Neglecting the gravity force in comparison to the explosive force, the principle of conservation of linear momentum for the system yields

$$3000(7500 + \Delta v_P) + 2000(7500 + \Delta v_B) = 5000(7500) \text{ kg-m/s}$$

or

$$3000\, \Delta v_P + 2000\, \Delta v_B = 0$$

where 7500 m/s is the given initial speed and Δv_P and Δv_B represent the increase in the speeds of the payload and booster, respectively.

It is given that 40% of the 6(400) J produced by the explosion is converted to mechanical energy. Because we neglect any change in position during the short duration of the explosion, the work-energy equation (32) yields

$$T_2 = T_1 + 0.40(2400)$$

$$\tfrac{1}{2}(3{,}000)(7{,}500 + \Delta v_P)^2 + \tfrac{1}{2}(2{,}000)(7{,}500 + \Delta v_B)^2$$

$$= \tfrac{1}{2}(5{,}000)(7{,}500)^2 + 960$$

This expression simplifies to

$$1{,}500[15{,}000\, \Delta v_P + (\Delta v_P)^2] + 1{,}000[15{,}000\, \Delta v_B + (\Delta v_B)^2] = 960$$

We now have two equations for the two unknown increments in speed, which we solve as follows.

$$\Delta v_B = -\frac{3{,}000}{2{,}000}\, \Delta v_P = -1.5\, \Delta v_P$$

$$1{,}500[15{,}000\, \Delta v_P + (\Delta v_P)^2] + 1{,}000[15{,}000(-1.5\, \Delta v_P)$$

$$+ (1.5\, \Delta v_P)^2] = 960$$

$$3{,}750(\Delta v_P)^2 = 960$$

$$\Delta v_P = 0.5060 \text{ m/s}$$

$$\Delta v_B = -1.5(0.5060) = -0.7590 \text{ m/s}$$

Hence the speeds are v (initial) $+ \Delta v$, or

$$(v_P)_2 = 7500 + 0.5060 \text{ m/s}$$

$$(v_B)_2 = 7500 - 0.7590 \text{ m/s}$$

To determine the distance between the payload and the booster after separation, we consider their relative motions. Both bodies are only under the influence of gravity, which is essentially constant in a 1-min interval. They therefore have the same acceleration, that is

$$\bar{a}_{P/B} = \bar{a}_P - \bar{a}_B = \bar{0}$$

or

$$\bar{v}_{P/B} = \bar{v}_P - \bar{v}_B = (0.5060 + 0.7590)\bar{I}$$

$$= 1.2650\bar{I} \text{ m/s}$$

Considering the two bodies to be adjacent to each other at the end of the explosion, we find for $t = 60$ s,

$$\bar{r}_{P/B} = 1.2650(t)\bar{I} = 1.265(60)\bar{I}$$

$$= 75.4\bar{I} \text{ meters} \qquad \triangleleft$$

or, in words the payload is 75.4 m ahead of the booster at the end of 1 min.

150 miles/hr

30°

90 miles/hr

Prob. III.152

Inextensible cable

45°

4 ft

1300 ft/s

Prob. III.153

HOMEWORK PROBLEMS

III.152 The initial and final velocities of a 5-oz baseball being hit by a bat are shown. Knowing that the ball and bat are in contact for 0.05 s, determine the average force exerted between the ball and the bat.

III.153 A 2-oz bullet traveling at 1300 ft/s in the direction shown hits and is embedded in the 8-lb sphere. The impact is completed in 0.05 s. Determine (a) the average increase in the tension of the cable during the impact, (b) the speed of the sphere immediately after the impact, and (c) the height to which the sphere rises after the impact.

III.154 The velocities of a 20-lb block of wood and a 1.5-oz bullet immediately before they collide are shown. The bullet is embedded in the block 0.01 s after the first contact. Determine the final velocity of the block of wood and the average force exerted on the block by the bullet during the collison.

$v_{block} = 10$ ft/s

30°

$v_{bullet} = 2000$ ft/s

30°

Z

Y

X

500 m/s

Prob. III.155

Prob. III.154

III.155 A 50-g bullet traveling at 500 m/s hits the 5-kg block and embeds in it. The block is then slowed by the spring whose stiffness is 100 N/mm. Determine the maximum distance the spring is compressed.

III.156 A schematic view of a cannon and its recoil mechanism is shown. A 20-kg shell has a muzzle velocity of 500 m/s. The shell emerges 1 ms after being fired. The mass of the barrel and its movable support is 600 kg. Also, the spring has a stiffness of 1000 kN/m. Determine (a) the average increase in the reaction force on the rails during the firing, (b) the recoil velocity of the barrel, and (c) the maximum distance that the spring is compressed.

Prob. III.156

III.157 Different weight bullets are fired from a rifle with the same kinetic energy of 2500 J. Each bullet emerges from the barrel 20 ms after being fired. Determine the average force required to hold the rifle stationary for (a) a 25-g bullet, (b) a 50-g bullet.

III.158 Three identical $3(10^4)$-kg freight cars are to be coupled. The coupling process is completed in 0.50 s. Car A is at rest and the coupled cars B and C are traveling at 1 m/s. Determine the speed of the cars after the coupling and the average force in each of the two couplers during the coupling.

Prob. III.158

III.159 Solve Problem III.158 if cars B and C are uncoupled. Car C is sufficiently far behind car B to allow cars A and B to couple before hitting car C.

3   Central Impact

A common way in which impulsive forces are generated is through the collision of bodies. Figure 10a depicts a general condition of a collision (*impact*) of two spheres, where $(\bar{v}_A)_1$ and $(\bar{v}_B)_1$ are the velocities of the spheres immediately before impact, and $(\bar{v}_A)_2$ and $(\bar{v}_B)_2$ are the velocities immediately after the impact. To study this system we choose a coordinate system xyz such that the x axis is normal to the plane of contact (the yz plane) of the spheres, as shown in the figure.

The situation depicted in Figure 10a is called *central impact*. This is the case when the centers of mass of both bodies and the point of contact

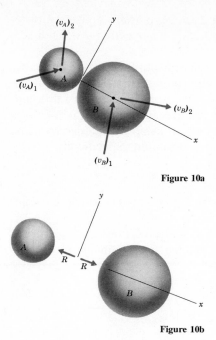

Figure 10a

Figure 10b

are all situated along a common line normal to the plane of contact. The case where these three points are not all situated along the normal is called *eccentric impact*. The latter is a topic in rigid body kinetics, treated in Module V.

We assume that the spheres are perfectly smooth, so that the only interaction force between the bodies is the normal force $\bar{R}$ illustrated in Figure 10b. At an instant when time equals t_1, the bodies first contact and begin to deform under the influence of this force. After attaining their maximum deformation, at time t_{max}, the bodies begin to regain their shapes, until ultimately they lose contact at time t_2. To a certain extent the collision process resembles the sudden compression and release of a spring, but it is much more complex.

In order to avoid the necessity of studying the deformation characteristics of the bodies, let us consider the results of a number of experiments in which we measure the final velocities of the two spheres for many different initial velocities. We find in these experiments that the ratio of the initial and final relative velocity components in the x direction (normal to the plane of contact) is essentially constant for a range of velocities of engineering significance. In order for the particles to collide and then separate, we must have

$$(v_{Ax})_1 > (v_{Bx})_1$$

$$(v_{Ax})_2 \leq (v_{Bx})_2$$

Therefore, the experiments tell us that

$$\epsilon = -\frac{(v_{Bx})_2 - (v_{Ax})_2}{(v_{Bx})_1 - (v_{Ax})_1} \tag{41}$$

where ϵ is a positive constant called the *coefficient of restitution*.

Equation (41) also has meaning in regard to the interaction force $\bar{R}$. Let us define the period of deformation by $t_1 \leq t \leq t_{max}$ and the period of restitution by $t_{max} \leq t \leq t_2$. We can then show that

$$\epsilon = \frac{\int_{t_{max}}^{t_2} R\,dt}{\int_{t_1}^{t_{max}} R\,dt} \tag{42}$$

In other words, the ratio of the impulses of the contact force in the two time intervals is the coefficient of restitution.

A collision generally involves a loss of mechanical energy, the energy normally being expended in the form of heat and sound. Only when $\epsilon = 1$, which is the case of a *perfectly elastic impact,* is there conservation of mechanical energy. The case $\epsilon = 0$, which is a *perfectly plastic impact,* represents a maximum loss of energy.

Equation (41) represents only one relationship for the velocity components $(v_{Ax})_2$ and $(v_{Bx})_2$. Clearly we require another relationship to determine the velocities of both bodies after the collision. Toward this end, we use the free body diagram of Figure 10b. Because the interaction force $\bar{R}$ is the only significant force, the other relationship is obtained by using conservation of linear momentum in the x direction. Specifically,

$$m_A(v_{Ax})_2 + m_B(v_{Bx})_2 = m_A(v_{Ax})_1 + m_B(v_{Bx})_1 \qquad (43)$$

Solving this equation simultaneously with equation (41) yields the x components of the velocities.

The components of velocity in the plane of contact remain unchanged, because the impact does not produce any significant forces acting tangent to the contact plane. Hence

$$(v_{Ay})_2 = (v_{Ay})_1 \qquad (v_{By})_2 = (v_{By})_1 \qquad (44)$$

$$(v_{Az})_2 = (v_{Az})_1 \qquad (v_{Bz})_2 = (v_{Bz})_1$$

We should note the limitations of the foregoing development. Equation (41), for the coefficient of restitution, is applicable to all cases of central impact. On the other hand, equations (43) and (44) were obtained for the special case where the impulsive interaction force $\bar{R}$ was the only force, Figure 10b. In some problems the bodies will be restrained by impulsive external forces that may invalidate either equation (43) or equation (44). Situations where this occurs are featured in some of the homework problems that follow.

Finally, consider the case when a small body collides with the earth. In such a situation the enormous mass of the earth results in a negligible change in the earth's velocity. Therefore, to study such problems we consider the earth to be fixed, $\bar{v}_e \equiv \bar{0}$, and determine the motion of the small body from equations (41) and (44).

EXAMPLE 18
Disk A on an air hockey game board collides with disk B, which is initially at rest. The disks are of identical mass and the coefficient of restitution is ϵ. Determine (a) the velocities with which the disks rebound from the collision, and (b) the percentage of the initial kinetic energy that is dissipated in the collision.

Solution
The study of the collision process requires that we first establish the contact plane. Therefore, we make a sketch that shows the two disks in contact.

$$\alpha = \sin^{-1}\left(\tfrac{3}{5}\right) = 36.87°$$

Choosing the x axis normal to the contact plane as shown, the initial velocities are

$$(\bar{v}_B)_1 = \bar{0}$$

$$(\bar{v}_A)_1 = 20(\cos \alpha \bar{\imath} + \sin \alpha \bar{\jmath}) = 16\bar{\imath} + 12\bar{\jmath} \text{ ft/s}$$

Because the contact force is in the x direction, the velocity components in the y direction are unchanged. Hence

$$(v_{By})_2 = 0 \qquad (v_{Ay})_2 = 12$$

The motion in the x direction can be determined from the equation for the coefficient of restitution and from conservation of linear momentum. Thus, for $m_A = m_B = m$,

$$\epsilon = -\frac{(v_{Bx})_2 - (v_{Ax})_2}{0 - 16}$$

$$m(v_{Ax})_2 + m(v_{Bx})_2 = m(16) + 0$$

Solving these equations simultaneously gives

$$(v_{Ax})_2 = 8(1 - \epsilon) \quad (v_{Bx})_2 = 8(1 + \epsilon)$$

Therefore the final velocities are

$$(\bar{v}_A)_2 = 8(1 - \epsilon)\bar{\imath} + 12\bar{\jmath} \text{ ft/s}$$

$$(\bar{v}_B)_2 = 8(1 + \epsilon)\bar{\imath} \text{ ft/s}$$

These vectors are illustrated in the following sketch.

The decrease in kinetic energy is now computed as

$$T_1 = \tfrac{1}{2}m_A(v_A)_1{}^2 + \tfrac{1}{2}m_B(v_B)_1{}^2 = \tfrac{1}{2}m(20)^2 = \tfrac{1}{2}m(400)$$

$$T_2 = \tfrac{1}{2}m_A(v_A)_2{}^2 + \tfrac{1}{2}m_B(v_B)_2{}^2$$

$$= \tfrac{1}{2}m[8^2(1 - \epsilon)^2 + 12^2 + 8^2(1 + \epsilon)^2]$$

$$= \tfrac{1}{2}m[272 + 128\epsilon^2]$$

Thus

$$\Delta T(\%) = \frac{T_2 - T_1}{T_1}(100) = \frac{272 + 128\epsilon^2 - 400}{400}(100)$$

$$= -32(1 - \epsilon^2) \%$$

From this result we can deduce that 32% of the kinetic energy is dissipated in a perfectly plastic collision and no energy is lost in a perfectly elastic collision.

In closing, we call attention to a common misconception about a perfectly plastic collision; that is, that the two bodies are joined after the collision. This need not be so, as can be seen in this problem by setting $\epsilon = 0$ in $(\bar{v}_A)_2$ and $(\bar{v}_B)_2$.

EXAMPLE 19

A steel ball bearing is dropped from a height of 1.60 m onto a steel surface that is inclined at an angle of 15°. The coefficient of restitution for the collision is 0.60. Determine the maximum height to which the ball bearing will rebound.

Solution

There are three time intervals involved in this problem. The first is the free-fall time of the ball bearing until it hits the steel surface. Then there is a short time interval in which the collision occurs. Finally, the ball bearing rebounds, once again under the influence of gravity.

For the initial free fall we have rectilinear motion from rest with constant acceleration g. Hence the ball bearing impacts the surface at the speed of

$$v_1 = \sqrt{v_0^2 + 2gh} = \sqrt{0 + 2(9.806)(1.60)} = 5.602 \text{ m/s}$$

To study the collision we sketch the impact, showing the x axis normal to the contact plane.

From this diagram we may write

$$\bar{v}_1 = 5.602(-\cos 15°\bar{i} + \sin 15°\bar{j})$$

$$= -5.411\bar{i} + 1.450\bar{j} \text{ m/s}$$

Treating the steel surface as a massive fixed body, from equation (41) we have

$$\epsilon = 0.60 = -\frac{(v_x)_2}{(v_x)_1} = -\frac{(v_x)_2}{(-5.411)}$$

$$(v_x)_2 = 3.247 \text{ m/s}$$

The velocity components tangent to the contact plane are unchanged, so

$$(v_y)_2 = (v_y)_1 = 1.450$$

Therefore

$$\bar{v}_2 = 3.247\bar{i} + 1.450\,\bar{j} \text{ m/s}$$

We next study the motion after the collision by computing the horizontal and vertical components of $\bar{v}_2$. This computation is conveniently performed by choosing another coordinate system $x'y'z'$ in which x' is horizontal and y' is vertical, as shown in the sketch, and then transforming the unit vectors. (In essence, we are setting up a new problem and hence choosing a more convenient set of coordinate axes.) From this diagram we have

$$\bar{i} = \sin 15°\,\bar{i}' + \cos 15°\,\bar{j}' = 0.2588\bar{i}' + 0.9659\bar{j}'$$

$$\bar{j} = \cos 15°\,\bar{i}' - \sin 15°\,\bar{j}' = 0.9659\bar{i}' - 0.2588\bar{j}'$$

Thus

$$\bar{v}_2 = 3.247(0.2588\bar{i}' + 0.9659\bar{j}') + 1.450(0.9659\,\bar{i}' - 0.2588\bar{j}')$$

$$= 2.241\bar{i}' + 2.761\bar{j}' \text{ m/s}$$

In this problem we require the value of $h_{\max}$, which is the maximum upward distance traveled after the collision. The motion in the vertical direction involves a constant downward deceleration g with the initial upward speed of 2.761 m/s. Thus, with h as the vertical distance, we have

$$(v_{y'})^2 = (2.761)^2 - 2(9.806)h$$

The value of h is a maximum when the velocity in the vertical direction is zero, which results in

$$h_{\max} = \frac{(2.761)^2}{2(9.806)} = 0.389 \text{ m}$$

HOMEWORK PROBLEMS

III.160 Two automobiles collide head-on. Vehicle A has a mass of 1200 kg and an initial speed of 3 m/s, whereas vehicle B has a mass of 1500 kg and an initial speed of 2 m/s. The coefficient of restitution is 0.20. Determine (a) the velocities of the two vehicles following the collision, and (b) the amount of mechanical energy dissipated by the collision.

Prob. III.160

Prob. III.161

III.161 Three identical collars slide over a smooth horizontal rod. Initially, block A is moving to the left at speed v and the other blocks are at rest. Determine how many collisions will occur and the velocity of each block after all collisions are over. Let the coefficient of restitution between the collars be ϵ, where $0 < \epsilon < 1$.

III.162 Spheres A and B have identical mass m and radius r. The coefficient of restitution is ϵ. If sphere A is released from a height h_1, determine the height to which the spheres rebound after colliding.

Prob. III.162

III.163 The velocities of a 2-lb block and a 1-lb disk sliding on a horizontal table are shown. Determine their velocities after they collide, if the coefficient of restitution is 0.40.

Prob. III.163

Prob. III.164

III.164 Two disks sliding over a horizontal surface collide in the position shown. The masses of the disks are $m_A = 800$-g and $m_B = 200$-g, and the coefficient of restitution is 0.80. As shown in the diagram, the initial velocity of disk B and the final velocity of disk A are unknown. Determine these unknown quantities.

Prob. III.165

III.165 Show that, in the absence of friction, a ball rolling on a horizontal plane will emerge from a corner such that its final velocity $\bar{v}_2$ is $-\epsilon\bar{v}_1$, where $\bar{v}_1$ is the initial velocity. Also explain what would happen in the case where $\epsilon = 0$.

Prob. III.166

III.166 A ball is dropped vertically onto the inclined plane, hitting it at a speed v_1. It is desired to have the ball rebound horizontally. Derive expressions for the required angle θ for the incline and the corresponding rebound speed v_2 in terms of the coefficient of restitution ϵ.

Prob. III.167

III.167 A ball bearing is thrown horizontally at a speed v_1 at a distance h_1 above the ground. Derive expressions for the distances d and h_2 locating the highest point attained by the ball bearing after it hits the ground. Let the coefficient of restitution be ϵ.

III.168 The pendulum is released from rest in the horizontal position and strikes the inclined surface when it is in the vertical position. Determine the velocity of the sphere as it rebounds from the collision if the sphere is connected to the pivot by (a) a cable, (b) a rigid, massless bar.

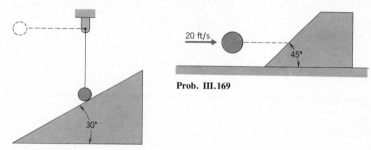

Prob. III.169

Prob. III.168

III.169 A 4-oz ball bearing hits the 2-lb steel angle block with a horizontal velocity of 20 ft/s. Assuming that all surfaces are smooth, determine the velocities of both bodies after the collision. The coefficient of restitution is 0.75.

III.170 The unit vector normal to the shaded plane is $\bar{e} = (2\bar{I} + 2\bar{J} + \bar{K})/3$. The velocity of a ball before it hits the plane is $\bar{v} = -40\bar{I} - 30K$ ft/s, and the coefficient of restitution is 0.80. Determine the velocity of the ball after it hits the plane.

Prob. III.171

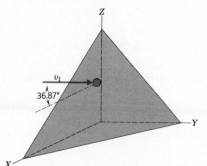

Prob. III.170

III.171 The 900-kg hammer of a pile driver falls from a height of 2 m above a 150-kg pile. The collision between the hammer and the pile is perfectly plastic, driving the pile 150 mm into the ground. Determine the average resistance of the ground to the motion of the pile.

III.72 The 4-kg head of a sledge hammer strikes the 20-kg anvil at 10 m/s. The mass of the handle of the sledge hammer is negligible. The coefficient of restitution is 0.50 and the collision is completed in 20 ms. (a) Determine the average impulsive force exerted between the hammer and the anvil if the stiffness of the two springs supporting the anvil is $k = 20$ kN/m each. (b) Solve part (a) if the springs are rigid ($k = \infty$). (c) Determine the maximum additional distance the springs are compressed after the collision in case (a).

I. ANGULAR IMPULSE-MOMENTUM RELATIONS

The relationships to be developed in this section are additional standard integrals of the differential equations of motion involving the linear momentum of a particle. However, in contrast to the linear impulse-momentum principles, the result obtained here will be relationships useful for systems that are rotating about an axis or a point. In addition to their applications to particle systems, the principles we derive will be fundamental to further studies, such as the kinetics of rigid bodies. As was done previously, we shall first study the motion of a single particle, and then consider a system of particles.

1 Single Particle

Let us compute the moment of the resultant force $\bar{F}$ exerted on the particle P in Figure 11 about the fixed point O. In general, the moment of a force about an arbitrary point may be written as

$$\bar{M}_O \equiv \bar{r}_{P/O} \times \bar{F}$$

Using Newton's second law to replace $\bar{F}$ gives

$$\bar{M}_O = \bar{r}_{P/O} \times m\bar{a}_P$$

Figure 11

We can now manipulate this equation by using the fact that $\bar{a}_P = \dot{\bar{v}}_P = \ddot{\bar{r}}_{P/O}$, and applying the rule for the time derivative of a cross product, as follows.

$$\bar{M}_O = \bar{r}_{P/O} \times m\dot{\bar{v}}_P \equiv \frac{d}{dt}(\bar{r}_{P/O} \times m\bar{v}_P) - \dot{\bar{r}}_{P/O} \times m\bar{v}_P$$

$$= \frac{d}{dt}(\bar{r}_{P/O} \times m\bar{v}_P) - \bar{v}_P \times m\bar{v}_P$$

$$= \frac{d}{dt}(\bar{r}_{P/O} \times m\bar{v}_P)$$

The quantity $\bar{r}_{P/O} \times m\bar{v}_P$ is called the *angular momentum* with respect to point O and is denoted by the symbol $\bar{H}_O$. As can be seen from this definition, $\bar{H}_O$ is the moment of the linear momentum vector about point O. In general then, we have

$$\boxed{\begin{array}{l} \bar{H}_O = \bar{r}_{P/O} \times m\bar{v}_P \\[1ex] \bar{M}_O = \dot{\bar{H}}_O \end{array}} \tag{45}$$

The derivation of the angular momentum relationship for a single particle can now be completed by assuming that we know how the moment $\bar{M}_O$ varies with time. Knowing this, we can separate variables in the second of equations (45) to obtain

$$\int_{(\bar{H}_O)_1}^{(\bar{H}_O)_2} d\bar{H}_O = \int_{t_1}^{t_2} \bar{M}_O \, dt$$

$$\boxed{(\bar{H}_O)_2 = (\bar{H}_O)_1 + \int_{t_1}^{t_2} \bar{M}_O \, dt} \tag{46}$$

Note the similarity to equation (38) for linear momentum. Hence it makes sense to call $\int \bar{M}_O \, dt$ the *angular impulse*. In situations where the resultant force exerts no moment about the origin, equation (46) tells us that $(\bar{H}_O)_1 = (\bar{H}_O)_2$, which is the statement of the principle of *conservation of angular momentum*.

To gain further insight about the angular momentum principles let us examine one scalar component of the vector relationships given in equations (45) and (46). Let the direction of the chosen component be the axial direction $\bar{e}_h = \bar{K}$ of the set of cylindrical coordinates shown in Figure 12.

We begin by writing the vectors in terms of their cylindrical coordinate components. Thus

$$\bar{r}_{P/O} = R\bar{e}_R + h\bar{e}_h$$

$$\bar{v}_P = \dot{R}\bar{e}_R + R\dot{\phi}\,\bar{e}_P + \dot{h}\bar{e}_h$$

$$\bar{F} = F_R\,\bar{e}_R + F_\phi\bar{e}_\phi + F_h\,\bar{e}_h$$

Figure 12

Now we compute

$$\bar{H}_O = \bar{r}_{P/O} \times m\bar{v}_P = m[-hR\dot{\phi}\,\bar{e}_R + (h\,\dot{R} - R\dot{h})\bar{e}_\phi + R^2\dot{\phi}\,\bar{e}_h]$$

$$\bar{M}_O = \bar{r}_{P/O} \times \bar{F} = -hF_\phi\,\bar{e}_R + (hF_R - RF_h)\bar{e}_\phi + RF_\phi\,\bar{e}_h$$

As stated above, our interest is with the axial components of $\bar{H}_O$ and $\bar{M}_O$. Noting that $\bar{e}_h \equiv \bar{K}$, these components are

$$(H_O)_z = mR^2\dot{\phi}, \qquad (M_O)_z = RF_\phi$$

Consider first the axial component equation (45), that is
$$(M_O)_z = (\dot{H}_O)_z$$

Upon substitution of the expressions for $(H_O)_z$ and $(M_O)_z$, this becomes

$$RF_\phi = \frac{d}{dt}\,(mR^2\dot{\phi})$$

Performing the differentiation yields

$$RF_\phi = m(2R\dot{R}\dot{\phi} + R^2\ddot{\phi})$$

The interesting aspect of this result is that, when a factor R is cancelled, we have $F_\phi = ma_\phi$. In other words, the axial component of equation (45) is identical to the transverse component of Newton's second law. We therefore can conclude that equation (45) does not represent a significantly new viewpoint for treating the motion of a particle.

The utility of the concept of angular momentum lies in the application of the impulse-momentum equation (46), not the basic equation (45). Using the foregoing results to express the axial components of the vectors in equation (46) gives

$$(H_O)_{Z_2} = (H_O)_{Z_1} + \int_{t_1}^{t_2} (M_O)_z\,dt$$

or

$$mR_2^2\dot{\phi}_2 = mR_1^2\dot{\phi}_1 + \int_{t_1}^{t_2} RF_\phi\,dt \tag{47}$$

The application of equation (47) requires that we know how the product RF_ϕ varies with time, in order to evaluate the angular impulse. Such situations occur infrequently. However, there is an important class of physical problems that have the special property, $F_\phi \equiv 0$. Equation (47) then tells us that $R^2\dot{\phi} = (H_O)_z/m$ is constant in the motion, which is the statement of conservation of angular momentum about the Z axis.

A common but not unique case in which $F_\phi = 0$ arises when the resultant force acting on a particle is always directed toward or away from a fixed point. We refer to the force in such a situation as a *central force,* as typified by the problem of orbital motion of a space vehicle under the influence of the gravitational attraction of the earth. This problem is treated separately in Module VIII because of its practical importance.

Summarizing the foregoing discussion, we can make the following generalization. The concept of angular momentum proves important to study the motion of a *single particle* only when the particle moves about an axis under the influence of a *central force,* in which case we have *conservation of angular momentum* about that axis. Denoting the axis as Z, equation (47) then yields

$$(H_O)_Z = R_2{}^2\dot\phi_2 = R_1{}^2\dot\phi_1 \tag{48}$$

or equivalently

$$(H_O)_Z = R_2(v_\phi)_2 = R_1(v_\phi)_1$$

EXAMPLE 20

A particle slides down the smooth interior surface of a hemisphere of radius r. It was released from the rim with a velocity $\bar v_1$ directed at an angle of 20° with the horizontal, as shown. Determine the magnitude of v_1 that will result in the lowest position attained by the particle being an elevation of $0.8r$ below the rim.

Solution

The motion will consist of a spiralling motion about the vertical axis of the hemisphere. Hence we choose cylindrical coordinates, defined with respect to this axis. Because the surface is smooth, none of the forces acting on the particle have transverse components, as indicated in the free body diagram.

Neither of the forces acting on the particle exert a moment about the Z axis. Therefore, there is conservation of angular momentum about this axis. Also, noting that only gravity does work, we can use the work-energy principle to relate the initial and final positions.

The velocity of the particle is always tangent to the hemispherical surface. Thus, at the rim the velocity is in the vertical plane, so that

$$\bar v_1 = v_1(\sin 20°\, \bar e_h + \cos 20°\, \bar e_\phi)$$

The specific position we wish to consider is $h = 0.8r$, where $\dot h \equiv v_h = 0$. This means that $v_R = 0$ also, because

$$R = \sqrt{r^2 - h^2} \qquad v_R = \dot R = -\frac{h}{\sqrt{r^2 - h^2}}$$

Thus $\bar{v}_2$ is in the transverse direction.

$$\bar{v}_2 = v_2 \bar{e}_\phi$$

Hence, using equation (48) to describe the conservation of angular momentum, we write

$$R_2(v_\phi)_2 = R_1(v_\phi)_1$$

For the spherical surface $h^2 + R^2 = r^2$, so $R_1 = r$ and $R_2 = 0.6r$. Hence

$$0.6r(v_2) = r(v_1 \cos 20°)$$

$$v_2 = \frac{\cos 20°}{0.6} v_1 = 1.5662v_1$$

This is one relation between the initial and final speeds. As was mentioned previously, we can also use the work-energy principle to relate v_1 and v_2. Let us arbitrarily choose the elevation of the rim as the datum for the potential energy of gravity, in which case

$$V_1 = 0 \qquad V_2 = mg(-h_2) = -0.8mgr$$

Then, because only gravity does work in the motion of the particle, we have

$$T_2 + V_2 = T_1 + V_1$$

$$\tfrac{1}{2}mv_2{}^2 - 0.8mgr = \tfrac{1}{2}mv_1{}^2 + 0$$

$$v_1{}^2 = v_2{}^2 - 1.6gr$$

We now have two equations in the two unknowns v_1 and v_2. Solving for v_1 yields

$$v_1{}^2 = (1.5662v_1)^2 - 1.6gr$$

$$v_1 = 1.049 \sqrt{gr}$$

HOMEWORK PROBLEMS

III.173   Use the relationship between the moment about point O and the corresponding angular momentum to derive the differential equation of motion for the angle θ.

Prob. III.173

III.174   An artificial satellite makes its closest approach to the surface of the earth (perigee) in an elliptical orbit. It has an altitude of 400 km and a

Prob. III.174

speed of 33,400 km/hr. Determine the altitude and speed of the satellite when it is farthest from the earth (apogee).

III.175 A cable is attached to an 8-oz block that may slide on a smooth horizontal surface. The cable is then passed through a hole at point O and a tensile force $\bar{F}$ is applied to it. Initially, the block is following a circular path at $R = 1$ ft corresponding to $F = 12$ lb. (a) Determine the initial speed of the block. (b) If the value of F is suddenly decreased to a constant value of 4 lb, determine the velocity of the block after it has spiralled outward to $R = 1.5$ ft.

Prob. III.175

III.176 In Problem III.175, the tensile force is decreased from 12 lb to some unknown value of F, with the result that the maximum distance attained by the block in its spiralling motion about point O is $R = 2$ ft. Determine (a) the velocity of the block at $R = 2$ ft, and (b) the corresponding value of F. (c) Do the answers to parts (a) and (b) satisfy circular motion for the block at a 2-ft radius?

III.177 A cable that is attached to a mooring post at point A is thrown to the person in the rowboat B. When the cable is caught the rowboat has the position and velocity shown. The person immediately pulls on the cable with a force of 100 N, in order to approach the dock. Determine the velocity of the rowboat after the cable is pulled in by 1.5 m. The combined mass of the passenger and the boat is 120 kg, and the water resistance is negligible.

Prob. III.177

Prob. III.178

III.178 A 200-g disk on a smooth horizontal plane is tied to a rubberband whose opposite end is tied to a peg at point A. The disk is set into motion at point B, as shown. The stiffness of the rubberband is 8 N/m and the rubberband is 600 mm long when it is slack. Determine the maximum distance between the disk and end A of the rubberband.

Prob. III.179

III.179 A ball bearing is projected tangentially along the rim of the cone shown at a speed v_1. At its lowest point the ball bearing is 12 in. above the apex of the cone. Determine (a) the initial speed v_1 of the ball bearing, and (b) the speed of the ball bearing at its lowest point. (c) Is the solution to part (b) the same as the speed for circular motion on the interior surface of the cone at that elevation?

III.180 An inextensible cable is passed through the hollow vertical post and then tied to a 0.5-kg sphere. The sphere is made to revolve in a horizontal circle for which $l_1 = 2$ m and $\theta_1 = 30°$. The cable is then drawn in very slowly until the sphere follows a horizontal circle for which $\theta_2 = 60°$. Determine (a) the initial speed v_1 of the sphere, (b) the final speed v_2, and (c) the amount of work done by the cable in pulling the sphere inward.

Prob. III.180

2 System of Particles

Figure 13a shows a typical set of forces acting on a three-particle system, where the $\bar{F}_i$ represent external forces and the $\bar{f}_{ij}$ represent the interaction forces exerted between pairs of particles.

Summing moments about an arbitrary point A, we find that the moment of each force $\bar{f}_{ij}$ cancels the moment of the corresponding force $\bar{f}_{ji}$ because the two forces are collinear, equal in magnitude, and opposite in direction. Hence, using relative position vectors such as those shown in Figure 13b, we can extend this analysis to conclude that the total moment about point A for a system of N particles is

$$\Sigma \bar{M}_A \equiv \sum_{i=1}^{N} \bar{r}_{i/A} \times \left(\bar{F}_i + \sum_{j=1}^{N} \bar{f}_{ij}\right) = \sum_{i=1}^{N} \bar{r}_{i/A} \times \bar{F}_i \qquad (49)$$

In other words, *the total moment acting on the system is the sum of the moments of the external forces.* This statement is basic to the study of statics.

Figure 13a

Figure 13b

Newton's second law for the ith particle is

$$\bar{F}_i + \sum_{i=1}^{N} \bar{f}_{ij} = m_i\bar{a}_i$$

Substituting this expression into equation (49) yields

$$\Sigma \bar{M}_A = \sum_{i=1}^{N} \bar{r}_{i/A} \times (m_i\bar{a}_i) \qquad (50)$$

This expression can be placed into a more convenient form by using the concept of relative motion with respect to point A. From Figure 13b we can write

$$\bar{r}_{i/O} = \bar{r}_{A/O} + \bar{r}_{i/A}$$

so that

$$\bar{v}_i = \bar{v}_A + \bar{v}_{i/A}$$

$$\bar{a}_i = \bar{a}_A + \bar{a}_{i/A}$$

Therefore, equation (50) becomes

$$\Sigma \bar{M}_A = \sum_{i=1}^{N} \bar{r}_{i/A} \times [m_i(\bar{a}_A + \bar{a}_{i/A})]$$

$$= \left(\sum_{i=1}^{N} m_i\bar{r}_{i/A}\right) \times \bar{a}_A + \sum_{i=1}^{N} \bar{r}_{i/A} \times m_i\bar{a}_{i/A} \qquad (51)$$

The first term on the right-hand side of equation (51) involves the definition of the location of the center of mass G of the system with respect to point A [refer to equation (20)]. Hence

$$\left(\sum_{i=1}^{N} m_i\bar{r}_{i/A}\right) \times \bar{a}_A \equiv \bar{r}_{G/A} \times m\bar{a}_A$$

We now rewrite the last term in equation (51) as the moment of a momentum vector by using the rule for the derivative of a vector product. This yields

$$\bar{r}_{i/A} \times m_i\bar{a}_{i/A} = \bar{r}_{i/A} \times [\frac{d}{dt}(m_i\bar{v}_{i/A})]$$

$$= \frac{d}{dt}(\bar{r}_{i/A} \times m_i\bar{v}_{i/A}) - \frac{d}{dt}(\bar{r}_{i/A}) \times m_i\bar{v}_{i/A}$$

$$= \frac{d}{dt}(\bar{r}_{i/A} \times m_i\bar{v}_{i/A}) - \bar{v}_{i/A} \times m_i\bar{v}_{i/A}$$

$$= \frac{d}{dt}(\bar{r}_{i/A} \times m_i\bar{v}_{i/A})$$

Using the results obtained above, equation (51) reduces to

$$\Sigma \, \bar{M}_A = \bar{r}_{G/A} \times m\bar{a}_A + \frac{d}{dt}\left(\sum_{i=1}^{N} \bar{r}_{i/A} \times m_i\bar{v}_{i/A}\right) \tag{52}$$

Let us consider the last term in equation (52) more carefully. If we think of $m_i\bar{v}_{i/A}$ as being the linear *momentum of particle i relative to point A*, then $\bar{r}_{i/A} \times m_i\bar{v}_{i/A}$ is the moment about point A of the momentum of particle i relative to point A. Therefore, let the vector $\bar{H}_A$ denote the *angular momentum of the system of particles relative to point A;* this vector is defined as

$$\bar{H}_A = \sum_{i=1}^{N} \bar{r}_{i/A} \times m_i\bar{v}_{i/A} \tag{53}$$

Thus, if point A is an arbitrary point, the moment equation is

$$\Sigma \, \bar{M}_A = \bar{r}_{G/A} \times m\bar{a}_A + \dot{\bar{H}}_A$$

A significant difference between this equation and the simple relation $\bar{M}_O = \dot{\bar{H}}_O$ for a single particle is that, if point A is arbitrarily chosen, it is necessary to determine the acceleration of this point in addition to determining the rate of change of the angular momentum $\bar{H}_A$.

Let us investigate the conditions required for this equation to reduce to the simpler form

$$\boxed{\Sigma \, \bar{M}_A = \dot{\bar{H}}_A} \tag{54}$$

For this to occur, point A must fit one of three special cases that reduces the term $\bar{r}_{G/A} \times m\bar{a}_A$ to zero.

1   If point A is fixed, or has constant velocity, then A is an inertial reference point and $\bar{a}_A \equiv 0$.
2   If point A is the center of mass, then $\bar{r}_{G/A} \equiv 0$.
3   If point A is accelerating directly toward or away from the center of mass, then $\bar{a}_A$ is parallel to $\bar{r}_{G/A}$ and their cross product vanishes.

Restricting ourselves to formulating the angular momentum with respect to one of these special points in no way limits the scope of the problems we can investigate.

We now complete the derivation of angular momentum principles by considering the situation where the moment of the forces is a known function of time. For such a case equation (54) can be integrated to obtain

$$\boxed{(\bar{H}_A)_2 = (\bar{H}_A)_1 + \int_{t_1}^{t_2} \Sigma \, \bar{M}_A \, dt} \tag{55}$$

This is the angular impulse-momentum equation for a system of particles.

As was true for a single particle, the concept of angular momentum for a system of independent particles is useful for problems involving conservation of angular momentum with respect to a specific axis. Equation (55) then shows that it is only necessary that the external forces exert no moment about the axis to have a constant value for the corresponding component of $\bar{H}_A$. Another important use of the concept of angular momentum is for treating the motion of rigid bodies. It will be shown in Modules V and VII that equation (54) represents one of the fundamental equations of motion for rigid bodies.

EXAMPLE 21
Blocks A and B, which are tied together by a cord, slide without friction along the horizontal arms of the cross bar shown while the entire system rotates freely about the vertical axis at the angular speed ω. Blocks A and B have masses of 2.0 kg and 0.5 kg, respectively, whereas the mass of the crossbar and the cord connecting the blocks is negligible. When block A is 600 mm from the vertical axis, it is moving outward with a speed of 40 mm/s and the angular speed of the system is $\omega = 0.5$ rad/s. Determine the value of $\dot{\omega}$ in this position, and also determine whether the radial speed of block A is increasing or decreasing.

Solution

The information desired in this problem involves acceleration quantities, so we begin with free body diagrams of each block which will enable us to apply Newton's second law. Because the motion of each block is in the horizontal plane, the vertical component of the reaction force on each block merely counterbalances gravity. Hence, we only need show top views of the system illustrating the forces that act in the horizontal plane.

The transverse component of the reaction forces $\bar{N}_A$ and $\bar{N}_B$ are a result of the interaction of the blocks and the crossbar. If, instead of considering the two blocks and the crossbar individually, as we have done in drawing the free body diagrams, we consider them to form a system that also contains the cord connecting the two blocks, then all forces in the free body diagrams are internal to the system. It then follows that the angular momentum of this system about the vertical axis is constant, because the crossbar is rotating freely.

When we compute the total angular momentum of this system, we find that it is simply the sum of the values for the two blocks, because the other bodies have negligible mass. Hence, letting the Z axis coincide with the vertical axis of rotation, we have

$$(H_O)_Z = m_A R_A{}^2 \omega + m_B R_B{}^2 \omega$$

In order to obtain a relation for $\dot{\omega}$, we must use the derivative form of the angular momentum equation, that is, equation (55).

$$(M_O)_Z \equiv 0 = \frac{d}{dt}(H_O)_Z$$

$$= 2(m_A R_A \dot{R}_A + m_B R_B \dot{R}_B)\omega + (m_A R_A{}^2 + m_B R_B{}^2)\dot{\omega}$$

At the instant described in the problem, we know that $R_A = 0.6\,\text{m}$, $\dot{R}_A = +0.4$ m/s, and $\omega = 0.5$ rad/s. From the constraint that the length of the cord is constant we have

$$R_B = 1.0 - R_A = 0.4 \text{ m}$$

We differentiate this expression and use the given value of $\dot{R}_A$ to obtain

$$\dot{R}_B = -\dot{R}_A = -0.4 \text{ m/s}$$

Differentiating again yields $\ddot{R}_B = -\ddot{R}_A$. We now solve the above equations for $(M_O)_Z$ to find

$$\dot{\omega} = -\frac{2(m_A R_A \dot{R}_A + m_B R_B \dot{R}_B)}{m_A R_A{}^2 + m_B R_B{}^2} \qquad \omega = -0.50 \text{ rad/s}^2 \qquad \triangleleft$$

Here the negative sign means that the angular speed is decreasing.

Having considered the overall motion of the system, we are still free to consider the individual motion of each block. In view of the free body diagrams, the equations of motion for block A are

$$\Sigma F_R = -F = m_A(\ddot{R}_A - R_A \omega^2) = 2[\ddot{R}_A - 0.6(0.5)^2]$$

$$\Sigma F_\phi = N_{A\phi} = m_A(R_A \dot{\omega} + 2\dot{R}_A \omega) = 2[0.6(-0.5) + 2(0.4)(0.5)]$$

Because $\ddot{R}_B = -\ddot{R}_A$, the corresponding equations for block B are

$$\Sigma F_R = -F = m_B(\ddot{R}_B - R_B\omega^2) = 0.5[-\ddot{R}_A - 0.4(0.5)^2]$$

$$\Sigma F_\phi = N_{B\phi} = m_B(R_B\dot{\omega} + 2\dot{R}_B\omega)$$

$$= 0.5[0.4(-0.5) + 2(-0.4)(0.5)]$$

The solution of these equations is

$$N_{A\phi} = 0.20 \text{ newtons} \qquad N_{B\phi} = -0.30 \text{ newtons}$$

$$\ddot{R}_A = 0.10 \text{ m/s}^2$$

$$F = 0.10 \text{ newtons}$$

We conclude that the radial speed of block A is increasing becauce $\ddot{R}_A$ is positive. Finally, can you explain why the reaction forces above satisfy the equation $R_A N_{A\phi} + R_B N_{B\phi} = 0$?

HOMEWORK PROBLEMS

III.181 Two identical 1-kg disks of negligible diameter are fastened to the rigid bar AC, which is free to pivot in the vertical plane about point C. The system is at rest when disk A is hit by a 100-g bullet that is traveling at 600 m/s as shown. The bullet becomes embedded in the disk. Determine the angular velocity of the system immediately after the collision is completed.

Prob. III.181

Prob. III.182

III.182 Two identical 1-kg disks of negligible diameter are fastened to the rigid bar AC, which is free to pivot in the vertical plane about point C. The system is at rest when disk A is hit by another 1-kg disk that is

Prob. III.183

traveling at 6 m/s, as shown. The coefficient of restitution for the collision is 0.50. Determine the angular velocity of bar AC after the collision is completed.

III.183 The collar of mass m can slide over the smooth rigid crossbar as the bar rotates about the vertical axis. The mass of the bar is negligible. A constant counterclockwise torque M is applied to the bar, causing it to rotate, and also causing the collar to slide along it. Initially, the system is at rest and $R = R_0$. Derive a single differential equation governing the radial distance R in terms of time.

III.184 Each of the three collars weighs 4 lb. The bar assembly on which they ride has negligible mass and rotates freely about the vertical axis. Initially, collars A and B are at rest relative to the bars at $R = 9$ in., and the angular speed of the assembly is $\omega = 2$ rad/s. Determine the velocity of each of the collars when $R = 18$ in.

III.185 Each of the three collars weighs 4 lb. The bar assembly on which they ride has negligible mass and rotates freely about the vertical axis. When $R = 15$ in., the collars are sliding outward relative to the bars at a speed of 9 in./s, with $\omega = 1$ rad/s. Determine the angular acceleration $\dot{\omega}$ and the acceleration of collar C at this instant.

III.186 Arm AB of the angular speed governor can be moved up and down by the force $\bar{F}$, causing the angle θ to change. In one case the system is rotating at $\omega = 80$ rad/s and $\theta = 60°$, when this force is changed causing the angle θ to change, very slowly, to $\theta = 30°$. Determine (a) the angular speed ω of the governor in its final position, and (b) the average value of F that caused this change in the value of θ. Each of the spheres has a mass of 400-g, and the mass of the rigid bars is negligible.

Prob. III.184 and III.185

Prob. III.186

III.187 The assembly rotates in the vertical plane about the pivot at point A. The mass of each of the four disks is m, and the mass of the rigid bars is negligible. Determine the system's differential equation of motion in terms of the angle θ.

Prob. III.188

Prob. III.187

III.188 A constant clockwise couple M is applied to the bar AB, causing the system to rotate in the vertical plane about the pin at point A. This causes the block of mass m to slide along the bar. Considering bar AB to be smooth and massless, derive the differential equations of motion for the radial distance R and the angle θ.

J.   SUMMARY OF PARTICLE DYNAMICS

The fundamental tool for a complete investigation of the motion of particle systems is Newton's second law expressed in terms of the appropriate kinematical description. Here we concentrated on path variables, rectangular Cartesian coordinates, and cylindrical coordinates for the kinematical description. These methods are sufficient for problems where only a force or an acceleration value are to be determined.

In some problems we must relate the initial and final velocity parameters of the system. One method for achieving this is to derive and solve the differential equation of motion by formulating Newton's second law for a general position. There are many techniques for solving differential equations of motion, the most elementary being the method of separating variables.

This approach for relating the initial and final velocities is complemented by the principles of work and energy, linear impulse and momentum, and angular impulse and momentum. All of these principles are standard solutions of the differential equations of motion. The key features to look for in applying these principles are summarized below.

1   Energy considerations are generally useful when the forces depend only on the position of the particle.

2   Linear momentum is suitable to situations where the rectangular components of the forces are functions of time, particularly, for impulsive forces.

3   Angular momentum is suitable to situations where bodies are rotating about a specific axis.

Finally, it should be noted that these principles may be utilized jointly to solve a problem, or individually to treat separate intervals in the motion of a body.

MODULE IV
KINEMATICS OF A
RIGID BODY IN
PLANAR MOTION

Now that we have completed a study of the dynamics of particles, we turn our attention to rigid bodies. Our aim in this module is to describe the motion of a rigid body when each point of the body is confined to move parallel to a prescribed plane. Hence, this module has the classic title, "kinematics of rigid bodies in plane motion," where kinematics is defined as the study of the motion of a physical system without regard to the forces causing the motion.

Simply stated, a rigid body is a collection of particles that are always at a fixed distance from each other. Consider the two-particle systems shown in Figure 1.

Figure 1   Two-particle systems.

Figure 1a shows the two particles connected by an elastic spring. If the spring is extended or shortened, the particles will no longer be at their original distance from each other; hence the system is not a rigid body. In Figure 1b, if we neglect the secondary (small) effects of the deformation of the steel bar, the particles are always at a fixed distance apart. Thus, we regard the bar to be a rigid body. Clearly, then, the concept of a rigid body is a mathematical modeling tool in that no real material can remain undeformed under the influence of forces. What we are saying is that for the class of problems we shall study, the microscopic deformation effects will not influence the results. Obviously the foregoing arguments can be extended to a system of particles (atoms) that compose a general body. Incidentally, if we did not neglect the deformation of the steel bar, we would be concerned with the dynamics of flexible bodies, which is an important topic (an integral part of the design process) that you will encounter in your later years of study.

Now let us direct attention to the phrase "plane motion." For the most part we shall use the plane of the paper to represent the plane of motion. A simple example of such motion is a disk rolling in a straight line. More generally, a rigid body executes plane motion if the path of each point in the body lies wholly in a plane, with the planes for all points being parallel to each other. An example of this is a body that is constrained to rotate about a fixed shaft.

Before we develop the basic kinematical model for the general motion of a rigid body, let us look more closely at the categories of motion we shall consider.

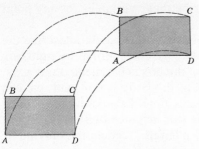

Figure 2

A. TYPES OF MOTION

1 Translation

When a body moves in a manner that results in every line in the body always being parallel to its original position, the body is said to be translating. Figure 2 shows a typical translation motion for a body. Note that in translation the points in the body can move in a curvilinear path. It is a common misconception to think that translation is equivalent to rectilinear motion for each point in the body.

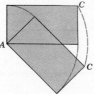

Figure 3

2 Pure Rotation

If a rigid body moves in a manner such that one point in the body is fixed in space, then the body is said to be in pure rotation. Thus, in Figure 3 the box is executing a pure rotation about point *A* about an axis passing through *A* perpendicular to the plane of the diagram. An example of a pure rotation is the rotor of an electric motor.

3 General Motion

If the motion of a body is not a translation or a pure rotation, the body is in general motion. Later, we shall show that general motion consists of a combination of translation and rotation. An illustration of general motion for the box of Figure 2 is shown in Figure 4. An automobile traveling over a roadway with varying radius of curvature is in general motion.

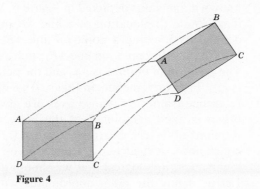

Figure 4

B. KINEMATICAL EQUATIONS OF PLANE MOTION

The important kinematic variables for rigid body motion are the same as those for particle motion, namely position, velocity, and acceleration.

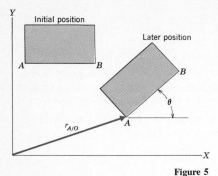

Figure 5

However, now we must be able to describe these variables for any point in the body.

To help in our thinking about the information needed to describe the position of all points in a body, let us start by considering the rigid box shown in Figure 5. Initially the edge AB of the box is parallel to the X axis. Because the distance between all points of the body is constant, we can determine the position of the box at a later time if we know the dimensions of the box and the location of a line (for instance, AB) of the box at that time. Line AB is a rigid line of known length. Hence, the only information required to describe the position of line AB is a position vector of any point on the line with respect to point O, for instance, $\bar{r}_{A/O}$, and the angle θ through which line AB was rotated.

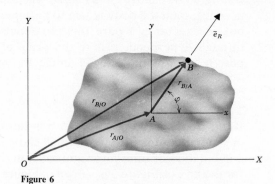

Figure 6

Let us generalize these concepts by considering the arbitrary body of known dimensions depicted in Figure 6. We know the position of the body if we know the location of a line AB in the body. This position can be defined by locating a point on line AB, for instance point A, and then locating another point on line AB, such as point B, with respect to point A. The position of point A is $\bar{r}_{A/O}$ and the position of point B with respect to A is $\bar{r}_{B/A}$. For convenience, in Figure 6 we introduced an auxiliary translating reference frame xyz whose axes are always parallel to XYZ, and whose origin is fixed at point A of the body. Then at any instant, all that is required to describe $\bar{r}_{B/A}$ is the angle φ, because the magnitude of $\bar{r}_{B/A}$ is constant for a rigid body. Reflecting on the tools we learned in particle kinematics, it is logical that we should describe $\bar{r}_{B/A}$ is polar coordinates. Let us define the polar coordinates with respect to the translating xyz reference frame. Thus

$$\bar{r}_{B/A} = \left| \bar{r}_{B/A} \right| \bar{e}_R$$

From Figure 6 we have the simple vector relationship

$$\bar{r}_{B/O} = \bar{r}_{A/O} + \bar{r}_{B/A} \qquad (1)$$

We now have a formula to define the location of *any* point in the rigid body (in this case point B), which can be used when we know the absolute position of some other arbitrary point in the body (in this case point A), the distance from point A to point B, and the angle associated with the radial unit vector $\bar{e}_R$.

Let us now consider the matter of velocities. Velocity is the time derivative of the position vector. Therefore

$$\frac{d\bar{r}_{B/O}}{dt} \equiv \dot{\bar{r}}_{B/O} = \dot{\bar{r}}_{A/O} + \dot{\bar{r}}_{B/A} \tag{2}$$

But $\dot{\bar{r}}_{B/O} \equiv \bar{v}_B$ and $\dot{\bar{r}}_{A/O} \equiv \bar{v}_A$. Equation (2) may now be written as

$$\bar{v}_B = \bar{v}_A + \bar{v}_{B/A} \tag{3}$$

where $\bar{v}_{B/A}$ is the velocity of point B with respect to point A. An alternative way of considering $\bar{v}_{B/A}$ is to think of yourself as being stationed on the xyz coordinate system at point A (oblivious to the fact that this coordinate system is translating) watching point B move, in which case you would observe the velocity of B with respect to A.

We now require a way to describe the term $\bar{v}_{B/A}$ mathematically. Consider the two configurations shown in Figure 7. They show a body at some initial instant t and a slightly later time $t + \Delta t$.

Because $|\bar{r}_{B/A}|$ is a constant, we have

$$\bar{v}_{B/A} = \frac{d}{dt}(\bar{r}_{B/A}) = \frac{d}{dt}(|\bar{r}_{B/A}|\bar{e}_R) = |\bar{r}_{B/A}|\frac{d\bar{e}_R}{dt}$$

$$= |\bar{r}_{B/A}|\dot{\bar{e}}_R$$

In the time interval Δt, the unit vector $\bar{e}_R$ has rotated through an angle $\Delta\varphi$. Thus using the chain rule and the results from cylindrical coordinates for the derivative of the unit vector $\bar{e}_R$, we have

$$\frac{d\bar{e}_R}{dt} = \frac{d\bar{e}_R}{d\varphi}\frac{d\varphi}{dt} = \bar{e}_\varphi\dot{\varphi}$$

so that the equation for $v_{B/A}$ may now be written as

$$\bar{v}_{B/A} = |\bar{r}_{B/A}|\dot{\varphi}\bar{e}_\varphi \tag{4}$$

Equation (4) will prove to be useful. Hence, let us write this pure rotational effect in a more general form. If the plane of the page is the plane of motion, then for the rotational effect associated with $\varphi(t)$, it is easy to visualize the rotation as taking place about an axis perpendicular to the page. With this in mind let us define the angular velocity of line AB by a vector whose magnitude is $\dot{\varphi}$ and whose direction is normal to the

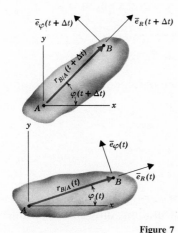

Figure 7

plane of motion. The sense of this normal is according to the right-hand rule, with the fingers of the right hand curling in the direction of increasing φ. Thus, the vector is in the direction of both the Z and z axes; that is,

$$\bar{\omega} \equiv \dot{\varphi}\bar{K} = \dot{\varphi}\bar{k} \tag{5}$$

The unit vectors of cylindrical coordinates are $\bar{e}_R, \bar{e}_\varphi, \bar{e}_h (\equiv \bar{k})$, from which we have

$$\bar{k} \times \bar{e}_R = \bar{e}_\varphi$$

This allows us to write equation (4) as

$$\bar{v}_{B/A} = \dot{\varphi}|\bar{r}_{B/A}|(\bar{k} \times \bar{e}_R) = \dot{\varphi}\bar{k} \times |\bar{r}_{B/A}|\bar{e}_R$$

or more simply,

$$\bar{v}_{B/A} = \bar{\omega} \times \bar{r}_{B/A} \tag{6}$$

Then from equation (3) we have the resulting vector equation for the velocity of any point on a rigid body in general motion:

$$\bar{v}_B = \bar{v}_A + \bar{\omega} \times \bar{r}_{B/A} \tag{7}$$

The acceleration equation may now be obtained by differentiating the velocity equation.

$$\dot{\bar{v}}_P = \dot{\bar{v}}_A + \dot{\bar{v}}_{B/A}$$
$$= \dot{\bar{v}}_A + \dot{\bar{\omega}} \times \bar{r}_{B/A} + \bar{\omega} \times \dot{\bar{r}}_{B/A}$$
$$= \dot{\bar{v}}_A + \dot{\bar{\omega}} \times \bar{r}_{B/A} + \bar{\omega} \times \bar{v}_{B/A} \tag{8}$$

The quantity $\dot{\bar{\omega}}$ is called the *angular acceleration* $\bar{\alpha}$ of the line AB. Because $\bar{\omega}$ is always normal to the plane of motion, only its magnitude can vary in planar motion. Thus

$$\bar{\alpha} \equiv \dot{\bar{\omega}} = \ddot{\varphi}\bar{k} \tag{9}$$

The time rate of change of velocity is acceleration. Hence substitution of equation (7) into equation (9) yields

$$\bar{a}_B = \bar{a}_A + \bar{a}_{B/A}$$
$$= \bar{a}_A + \bar{\alpha} \times \bar{r}_{B/A} + \bar{\omega} \times (\bar{\omega} \times \bar{r}_{B/A}) \tag{10}$$

Incidentally, in three-dimensional motion, the angular velocity can also change its direction. This becomes a major complication in such problems. However, the basic equations for rigid body motion in three dimensions will have the same form as those given by equations (7) and (10).

We used $\bar{\omega}$ and $\bar{\alpha}$ as properties of a line AB in the development of equations (7) and (10) for rigid body motion. Let us now show that at any instant of time $\bar{\omega}$ and $\bar{\alpha}$ are properties of the entire rigid body, so that the equations hold for any line in the rigid body.

Consider the rigid body shown in Figure 8. Arbitrarily choose two intersecting lines AB and CD that are fixed in the rigid body. Now measure their angular orientation with respect to a fixed line, say the $+X$ axis. Let γ be the fixed angle between lines AB and CD. We now have

Figure 8

$$\phi = \psi + \gamma \tag{11}$$

The time derivative of equation (11) gives

$$\dot{\phi} = \dot{\psi} + \dot{\gamma}$$

However, because γ is constant, $\dot{\gamma} \equiv 0$. Therefore

$$\dot{\phi} = \dot{\psi} \equiv \omega$$

and we conclude that at any given instant ω is the rate of rotation of all lines in a rigid body. Differentiating again, we have

$$\ddot{\phi} = \ddot{\psi} \equiv \alpha$$

from which we conclude that the angular acceleration is also a property of all lines in a rigid body.

The application of the basic kinematical equations (1), (7), and (10) requires that we express the vectors in terms of their components. These components may be referred to the axes of either the fixed XYZ or moving xyz reference frames. Consistent with the procedure we adopted for treating relative motion in Module Two, we shall use the latter. Hence, the unit vectors will be $(\bar{\imath}, \bar{\jmath}, \bar{k})$. Recalling that vector components do not depend on the location of the origin of the reference frame, we shall specify only the orientation of the axes when solving kinematics problems. In the next module, in the process of formulating the kinetics principles for rigid bodies in planar motion, we shall develop rules governing the location of the origin.

C. PHYSICAL INTERPRETATION
OF KINEMATICAL EQUATIONS

In order to understand better the physical significance of equations (7) and (10), let us use cylindrical coordinates to represent the vectors, rather than the rectangular Cartesian representation that we shall employ when solving problems. Thus, because

$$\bar{r}_{B/A} = |\bar{r}_{B/A}|\bar{e}_R, \quad \bar{\omega} = \dot{\varphi}\bar{k}, \quad \text{and} \quad \bar{\alpha} = \ddot{\varphi}\bar{k}$$

we have

$$\bar{v}_B = \bar{v}_A + \dot{\varphi}|\bar{r}_{B/A}|\bar{e}_\varphi \tag{12}$$

$$\bar{a}_B = \bar{a}_A + \ddot{\varphi}|\bar{r}_{B/A}|\bar{e}_\varphi - \dot{\varphi}^2|\bar{r}_{B/A}|\bar{e}_R \tag{13}$$

First, suppose that we had a problem where a body is in pure translation. Then $\dot{\varphi} = \ddot{\varphi} \equiv 0$, and our intuition tells us that all points in the body should have the same velocity and acceleration. This expectation is confirmed by equations (12) and (13).

Next, suppose that we had a problem where a body is in pure rotation about an axis perpendicular to the plane of the body passing through a point A. Then $\bar{v}_A = \bar{a}_A \equiv \bar{0}$ and all points move in circular paths about the axis of rotation. Note that the angular velocity and angular acceleration are parallel to the axis of rotation. For this case equation (12) tells us that the velocity of a point is tangent to its circular path. The acceleration is seen to have two components for this case. One is tangent to the path and the other is directed toward the center of curvature of the circular path. These properties are, of course, identical to those obtained when considering particle motion on a curved path.

In the case of general plane motion, we can regard $\bar{v}_A$ and $\bar{a}_A$ as translational effects with respect to the motion of point A and the remaining terms may be regarded as being caused by a rotation about an axis through point A normal to the plane of motion. Thus general motion is a superposition of translational and pure rotational motion. This is known as *Chasles' theorem*.

Note that if we had replaced point A with a different point, say point C, the translational portion of the motion would be that of point C, and the rotation would now be about a normal axis through C. However, as shown above, for either point A or point C, the angular velocity $\bar{\omega} \equiv \dot{\varphi}\bar{k}$ is the same, because $\bar{\omega}$ is a property of the body, and not of the points whose motion is being described.

EXAMPLE 1

Bar OA rotates in a plane about the fixed point O. At the instant when $\theta = 30°$, the angular velocity is 4 rad/s clockwise and the angular acceleration is 6 rad/s counterclockwise. Find (a) the velocity and (b) the acceleration of point A at this instant.

Solution
Part a
In general we know that

$$\bar{v}_A = \bar{v}_O + \bar{\omega} \times \bar{r}_{A/O}$$

Using the given coordinate system and the given information, we know that at the desired instant $\theta = 30°$, $\bar{\omega} = -4\bar{k}$ (note that the negative sign arises because of the right-hand rule), and $\bar{\alpha} = 6\bar{k}$. From the geometry of the figure, for $\theta = 30°$ we have

$$\bar{r}_{A/O} = l(\sin\theta\bar{i} + \cos\theta\bar{j}) = l(0.5\bar{i} + 0.866\bar{j})$$

Also, $\bar{v}_O = \bar{0}$ because point O is fixed.

The rigid body velocity formula then gives

$$\bar{v}_A = \bar{0} + (-4\bar{k}) \times l(0.5\bar{i} + 0.866\bar{j})$$

$$= l(3.46\bar{i} - 2\bar{j})$$

Part b

In general we know that

$$\bar{a}_A = \bar{a}_O + \bar{\alpha} \times \bar{r}_{A/O} + \bar{\omega} \times (\bar{\omega} \times \bar{r}_{A/O})$$

From the given information we have $\bar{\alpha} = 6\bar{k}$. We also know $\bar{\omega}$ and $\bar{r}_{A/O}$, and in part a we determined $\bar{\omega} \times \bar{r}_{A/O}$. Thus the acceleration formula yields

$$\bar{a}_A = \bar{0} + (+6\bar{k}) \times l(0.5\bar{i} + 0.866\bar{j})$$

$$+ (-4\bar{k}) \times l(3.46\bar{i} - 2\bar{j})$$

$$= l(-5.20\bar{i} + 3\bar{j}) + l(-8\bar{i} - 13.84\bar{j})$$

$$= l(-13.20\bar{i} - 10.84\bar{j})$$

Alternative Method

In working any kinematical problem one can always revert to the fundamental definitions. Namely, the velocity is the time derivative of the position vector. As is true for any function, we must differentiate the general position vector, not the one evaluated at a specific instant. Hence

$$\bar{r}_{A/O} = l(\sin\theta\bar{i} + \cos\theta\bar{j})$$

As the coordinate axes have fixed directions, for the velocity we find

$$\bar{v}_A = \frac{d}{dt}(\bar{r}_{A/O}) \equiv \dot{\bar{r}}_{A/O} = \frac{d}{dt}[l(\sin\theta\,\bar{i} + \cos\theta\,\bar{j})]$$

$$= l\dot{\theta}(\cos\theta\bar{i} - \sin\theta\bar{j})$$

From the given information, at the position of interest $\theta = 30°$ and $\dot{\theta} = 4$ (note that in this example, clockwise rotation means θ is increasing so that $\dot{\theta}$ is positive). Thus, we have

$$\bar{v}_A = l(3.46\bar{i} - 2\bar{j}) \qquad \text{when } \theta = 30°$$

Next, we find the acceleration from the time derivative of the velocity. To do this we must be certain to differentiate the general expression for velocity. Hence

$$\bar{a}_A = \frac{d}{dt}(\bar{v}_A) = \frac{d}{dt}[l\dot{\theta}(\cos\theta\,\bar{i} - \sin\theta\,\bar{j})]$$

$$= l\ddot{\theta}(\cos\theta\,\bar{i} - \sin\theta\,\bar{j}) + l\dot{\theta}^2(-\sin\theta\,\bar{i} - \cos\theta\,\bar{j})$$

Using the given information that $\theta = 30°$, $\dot{\theta} = 4$, and $\ddot{\theta} \equiv -6$ (counterclockwise angular acceleration means that $\dot{\theta}$ is decreasing), we have

$$\bar{a}_A = l(-5.20\bar{i} + 3\bar{j}) + l(-8\bar{i} - 13.84\bar{j})$$

$$= l(-13.20\bar{i} - 10.84\bar{j}) \qquad \text{when } \theta = 30°$$

The key point in this alternative method is that one must work symbolically. For instance, if we inadvertently attempted to differentiate the velocity vector evaluated at $\theta = 30°$ to find the acceleration, we would have erroneously obtained zero acceleration, because we then would not have differentiated the vector velocity function.

Which method is easier? It doesn't really matter; however, our experience is that most people feel more comfortable using the derived formulas to solve kinematical problems, rather than performing differentiations according to the alternative method.

EXAMPLE 2

The steel fire door shown in the diagram is supported by two parallel arms AB and CD, each 1.5 m long. If the equal arms AB and CD are each rotating at the constant rate of $\omega = 10$ rad/s, what is the velocity and acceleration of the center of mass G of the door?

Solution

Points A, C, and G are part of the same body (the door), so we can use the rigid body equations. Because points A and C have the same motion, we can use either point. Let us use point A.

$$\bar{v}_G = \bar{v}_A + \bar{\omega}_{\text{door}} \times \bar{r}_{G/A}$$

$$\bar{a}_G = \bar{a}_A + \bar{\alpha}_{\text{door}} \times \bar{r}_{G/A} + \bar{\omega}_{\text{door}} \times (\bar{\omega}_{\text{door}} \times \bar{r}_{G/A})$$

However, the polygon formed by $ABDC$ is a parallelogram, so that line AC is always horizontal and the door only translates. Thus $\bar{\omega}_{\text{door}} \equiv \bar{0}$ and

$$\bar{v}_G = \bar{v}_A \qquad \bar{a}_G = \bar{a}_A$$

Now, the problem of finding $\bar{v}_A$ and $\bar{a}_A$ is similar to that in the previous problem, because arm AB is in pure rotation about point B.

$$\bar{v}_A = \bar{v}_B + \bar{\omega}_{AB} \times \bar{r}_{A/B}$$

$$\bar{a}_A = \bar{a}_B + \bar{\alpha}_{AB} \times \bar{r}_{A/B} + \bar{\omega}_{AB} \times (\bar{\omega}_{AB} \times \bar{r}_{A/B})$$

As shown in the sketch, the coordinate system we choose for describing the vector components has the x axis horizontal and the y axis vertical. We are given $\bar{v}_B = \bar{a}_B = \bar{0}$, $\bar{\omega}_{A/B} = 10\bar{k}$, $\bar{\alpha}_{AB} = \bar{0}$. From the sketch we have

$$\bar{r}_{A/B} = 1.5(-\cos\theta\,\bar{i} - \sin\theta\,\bar{j})$$

Thus,

$$\bar{v}_G = \bar{v}_A = \bar{0} + 10\bar{k} \times 1.5(-\cos\theta\,\bar{i} - \sin\theta\,\bar{j})$$

$$= 15(-\cos\theta\,\bar{j} + \sin\theta\,\bar{i})\ \text{m/s}$$

$$\bar{a}_G = \bar{a}_A = \bar{0} + \bar{0} + 10\bar{k} \times [10\bar{k} \times 1.5(-\cos\theta\,\bar{i} - \sin\theta\,\bar{j})]$$

$$= 150(\cos\theta\,\bar{i} + \sin\theta\,\bar{j})\ \text{m/s}^2$$

An important thing to note in this example is that the system was composed of three rigid bodies. Hence we had the possibility of having three different angular velocities and three different angular accelerations. It was the geometrical configuration of the system that reduced those quantities to simply one nonzero angular velocity.

EXAMPLE 3
A portion of a machine consists of a bar AB, 750 mm long, with smooth sliders on both ends. The bar moves in the vertical plane such that slider A follows a vertical guide and slider B follows a horizontal guide, as shown in the diagram. At the instant when $\theta = 30°$, slider A is driven downward with a speed of 250 mm/s and this downward velocity is decreasing at the rate of 50 mm/s². Find the velocity and acceleration of end B at this instant.

Solution
The approach we shall follow is to use the rigid body motion formulas, expressing all quantities by their vector components at the instant of interest. First we select coordinate axes xy, as shown in the diagram. With respect to these axes $\bar{v}_A = -250\bar{j}$ mm/s, $\bar{a}_A = 50\bar{j}$ mm/s² (because the downward velocity is decreasing, the acceleration is upward). The problem is to find $\bar{v}_B$ and $\bar{a}_B$. We cannot simply use the velocity and acceleration formulas for bar AB because we do not know $\bar{\omega}_{AB}$ and $\bar{\alpha}_{AB}$. There is, however, an additional requirement we have not as yet utilized, which is that slider B can only move horizontally.

The requirements that slider A moves vertically and slider B moves horizontally are called *geometrical constraints,* because they are a result of the geometrical design of the system. The importance of geometrical constraints is that they are conditions the designer can impose on a system to obtain particular types of motion from the system, that is, translation, rotation, or specialized general motion. In general, if we overlook a problem's geometrical constraint condition, we probably shall not be able to solve the problem (even in general terms) because we shall have too many unknowns.

With the knowledge of the geometrical constraint the solution may now be completed. We first determine $\bar{v}_B$. The velocity formula gives

$$\bar{v}_B = \bar{v}_A + \bar{\omega}_{AB} \times \bar{r}_{B/A}$$

The geometrical constraint requires that $\bar{v}_B = v_B\bar{\imath}$, where the speed v_B is unknown. Because this is plane motion, $\bar{\omega}_{AB} = \omega_{AB}\bar{k}$. ($\omega_{AB}$ may be either a positive or negative number. By the right-hand rule, if it is positive the rotation is counterclockwise.) Then from the diagram with $\theta = 30°$, we have

$$\bar{r}_{B/A} = 750(\cos 30° \, \bar{\imath} - \sin 30° \, \bar{\jmath})$$

Thus, we find

$$v_B\bar{\imath} = -250\bar{\jmath} + \omega_{AB}\bar{k} \times 750(\cos 30° \, \bar{\imath} - \sin 30° \, \bar{\jmath})$$

$$= 750\omega_{AB} \sin 30° \, \bar{\imath} + (-250 + 750\omega_{AB} \cos 30°)\bar{\jmath}$$

In order for the vectors on the left and right sides of the equation to be equal, their corresponding components must be equal. Equating components gives

$$\bar{\imath} \text{ direction:} \qquad v_B = 750\omega_{AB} \sin 30°$$

$$\bar{\jmath} \text{ direction:} \qquad 0 = -250 + 750\omega_{AB} \cos 30°$$

Note that we have obtained two scalar equations. This was to be expected because in planar motion the velocity has two components. The results are

$$\omega_{AB} = \frac{250}{750 \cos 30°}$$

$$v_B = 750\left(\frac{250}{750 \cos 30°}\right) \sin 30° = 250 \tan 30°$$

$$\bar{\omega}_{AB} = 0.385\bar{k} \text{ rad/s} \qquad \bar{v}_B = 0.144\bar{\imath} \text{ m/s}$$

Note that in addition to using the solution for $\bar{\omega}_{AB}$ to find $\bar{v}_B$, we shall need $\bar{\omega}_{AB}$ in the solution for $\bar{a}_B$.

$$\bar{a}_B = \bar{a}_A + \bar{\alpha}_{AB} \times \bar{r}_{B/A} + \bar{\omega}_{AB} \times (\bar{\omega}_{AB} \times \bar{r}_{B/A})$$

We were given $\bar{a}_A$, and we have determined $\bar{\omega}_{AB}$ and $\bar{r}_{B/A}$. The geometrical constraint on slider B requires that $\bar{a}_B = a_B \bar{i}$, and because this is plane motion $\bar{\alpha}_{AB} = \alpha_{AB} \bar{k}$. Thus the acceleration formula yields

$$a_B \bar{i} = 50\bar{j} + \alpha_{AB}\bar{k} \times 750(\cos 30^\circ\, \bar{i} - \sin 30^\circ\, \bar{j})$$

$$+ 0.385\bar{k} \times [0.385\bar{k} \times 750(\cos 30^\circ\, \bar{i} - \sin 30^\circ\, \bar{j})]$$

$$= (750\alpha_{AB} \sin 30^\circ - 111.11 \cos 30^\circ)\bar{i}$$

$$+ (50 + 750\alpha_{AB} \cos 30^\circ + 111.11 \sin 30^\circ)\bar{j}$$

Equating components then gives

$\bar{i}$ direction: $a_B = 375.0\alpha_{AB} - 96.22$

$\bar{j}$ direction: $0 = 50 + 649.52\alpha_{AB} + 55.55$

Thus the solutions are

$$\bar{\alpha}_{AB} = -0.1625 \text{ rad/s}^2$$

$$\bar{a}_B = -0.1572 \text{ m/s}^2$$

Before leaving this problem, we should consider a geometrical interpretation of the solution steps presented. In solving for both velocity and acceleration we expressed the general motion of bar AB as a translation with the motion of slider A plus a rotation about the pivot of slider A (with the angular motion of bar AB). Then we noted that the geometrical constraint on slider B requires that the combination of the motions of slider B associated with the translation and rotation must result in slider B having horizontal motion.

HOMEWORK PROBLEMS

IV.1   The angle of rotation of the reel of a tape recorder is defined by $\theta = 1.2t + 0.01 \sin 30\pi t$, where θ is measured in radians and t in seconds. Determine the horizontal and vertical components of the acceleration of point A on the rim of the reel when (a) $t = 0$, (b) $t = \frac{1}{60}$ s, and (c) $t = \frac{1}{45}$ s.

Prob. IV.1

IV.2   The truck shown in the sketch is at rest. The tailgate of the truck is lowered such that the vertical component of the velocity of its free end is

Prob. IV.2

constant at 5 m/s. Determine the angular velocity and angular acceleration of the tailgate for the instant when $\theta = \tan^{-1}(\frac{3}{4})$.

IV.3   The acceleration of slider A is 2 m/s² down to the right, and its velocity is 1 m/s up to the left as the bar passes the horizontal. For this position determine (a) the velocity of slider B, (b) the angular acceleration of the bar, and (c) the acceleration of slider B.

Prob. IV.3

Prob. IV.4

IV.4   A box resting against a wall begins to slide. In the position where θ = 30° the angular velocity is 3.0 rad/s and the angular acceleration is 1.0 rad/s², both counterclockwise. Compute the velocities and accelerations of corners A and C at this instant.

Prob. IV.5

IV.5   In the position shown collar A at the end of a 1.30-m rod is moving to the left at 300 mm/s and is gaining speed at the rate of 100 mm/s². Determine the velocity and acceleration of collar B at this instant.

IV.6   The rigid link ABC is a portion of the control linkage in an automotive carburetor. At the instant shown, the control rod BE has a velocity of 5 in./s and an acceleration of 2 in./s², both to the left. Determine the velocity and acceleration of point C at this instant.

Prob. IV.7

Prob. IV.6

IV.7   Collar A is moving to the right at the constant speed of 10 m/s. Determine the angular velocity and angular acceleration of rod AB in the position shown.

IV.8   Observation of the motion of the circular disk reveals the following information about the instantaneous velocity of points A, B, and C: $v_{Ax} = 20$ m/s, $v_{By} = -15$ m/s, $v_{Cx} = 25$ m/s. Determine the angular velocity of the disk and the velocity of the midpoint O.

Prob. IV.8

IV.9   Bars AB and BC rotate in the vertical plane such that the angles γ and β are identical. Bar AB is rotating clockwise at the constant rate of 20 rev/min. Determine the acceleration of point C when $\gamma = 30°$.

Prob. IV.9

IV.10   Slider A moves along rod OB and is also constrained to move in the vertical slot. Given that the rod is rotating with angular velocity ω and angular acceleration α as shown, derive expressions for the velocity and acceleration of the slider. *Hint:* Write the position vector for point A and differentiate.

Prob. IV.10

Prob. IV.11

IV.11 The collar O slides along the vertical rod with constant velocity v. Determine expressions for the velocity and acceleration of point A on the link OA. *Hint:* Write the position vector for point O and differentiate.

D. APPLICATIONS OF KINEMATICAL EQUATIONS AND SPECIAL CASES

Before looking at some illustrative problems that demonstrate how we can systematically apply the kinematical equations to real systems, let us briefly study three special cases. The results of these discussions will prove to be of great value in formulating many problems.

1 Instantaneous Center of Zero Velocity

In this section we shall consider a rigid body undergoing general planar motion. We shall show that if we can find a point of the body that has zero velocity, it will enable us to simplify greatly the determination of the velocities of other points of the body.

Consider the rigid body of Figure 9. Let us assume that we can measure the velocities at A and B as a function of time. Hence, $\bar{v}_A$ and $\bar{v}_B$ are known. We then can relate the velocity of either point A or B to any other point C. From equation (7) we have

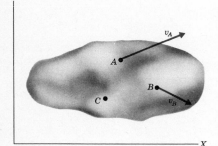

Figure 9

$$\bar{v}_A = \bar{v}_C + \bar{\omega} \times \bar{r}_{A/C} \tag{14}$$

$$\bar{v}_B = \bar{v}_C + \bar{\omega} \times \bar{r}_{B/C} \tag{15}$$

The problem now reduces to two questions. What are the conditions for $\bar{v}_C = \bar{0}$, and how can point C be found? From equations (14) and (15) we see that for $\bar{v}_c = \bar{0}$,

$$\bar{v}_A = \bar{\omega} \times \bar{r}_{A/C} \qquad \bar{v}_B = \bar{\omega} \times \bar{r}_{B/C} \tag{16}$$

From the definition of the cross product we know that $\bar{\omega} \times \bar{r}$ is perpendicular to both $\bar{\omega}$ and $\bar{r}$. Because $\bar{\omega}$ is perpendicular to the plane of motion, it must therefore be true that $\bar{\omega} \times \bar{r}$ is perpendicular to $\bar{r}$, and lies in the XY plane.

In other words, from equation (16) we see that the position vector $\bar{r}_{A/C}$ from the point of zero velocity to point A must be perpendicular to $\bar{v}_A$. Of course, a similar statement is true for point B. When we construct the perpendiculars to $\bar{v}_A$ and $\bar{v}_B$, they will intersect at point C, which is the only point common to both perpendiculars.

In Figure 10 we have shown this construction for the rigid body of Figure 9. The important conclusion is that because the velocity formulas

Figure 10

in equation (16) resemble those for a body in pure rotation, for the purposes of computing *velocity only,* we can think of the body as rotating about point C. Thus if we let D be any point in the body, we can say that $\bar{v}_D$ is perpendicular to $\bar{r}_{D/C}$ in the sense of the rotation of the body. This is shown in Figure 10.

Point C has zero velocity only at the instant for which we have drawn the diagram, because at a later instant the location of points A and B, as well as the velocities of these points, will have changed. The velocity formulas, equations (16), resemble those for a body in pure rotation about point C. Therefore, we call point C an *instantaneous center of zero velocity*. This is frequently shortened to *instant center*.

These statements are true for a general planar motion of a rigid body. If the body is in pure rotational motion, the instant center is the fixed center of rotation. On the other hand, if the body is in pure translation, $\bar{v}_A = \bar{v}_B$ so that the perpendiculars to $\bar{v}_A$ and $\bar{v}_B$ will never intersect; thus there is no instant center for a body in pure translation. Also note that frequently the instant center is not situated on the body. In such cases we can, if we wish, think of the instant center as lying on an imaginary extension of the body. Finally, we emphasize that although the instantaneous center can be located graphically, it can be located more exactly by trigonometry and geometry.

An obvious question to ask now is, can we use the instant center for determining accelerations? With the exception of pure rotation of a rigid body, the instant center is not fixed. Therefore the various points of the rigid body do not follow circular paths, and hence in general we *cannot* use the acceleration equations for circular motion about the instant center.

Our approach for acceleration problems will be to use the vector equation (10). However, because we need to know the angular velocity of the body before we can use equation (10), we shall find that the instant center is a convenient aid.

EXAMPLE 4

Consider the slider mechanism of Example 3. In that example the speed of slider A was 250 mm/s directed downward when $\theta = 30°$. Using the instant center concept, solve for the velocity of slider B when $\theta = 30°$, and show that the solution is the same as that obtained previously.

Solution

Sequentially, the steps in the solution are as follows: We begin an instant center solution by making a sketch that shows the known information about the velocities of the system. In this problem we know $\bar{v}_A$, and we also know that point B is constrained to move horizontally. With regard to $\bar{v}_B$, we need only know that it is horizontal in order to draw the appropriate perpendicular. Eventually the sense of $\bar{\omega}$ will yield the sense of $\bar{v}_B$, which at this stage is an unknown. The point of intersection of the perpendiculars to $\bar{v}_A$ and $\bar{v}_B$ is the instant center. The value of the angular speed ω is found by dividing the known speed v_A by the distance from point A to the instant center. The direction of $\bar{\omega}$ is determined by pretending that point A is actually in circular motion about the instant center. We then can find the magnitude and direction of $\bar{v}_B$ by pretending that point B is also in circular motion about the instant center. All of these statements are illustrated in the following sketch.

In this diagram ω is shown counterclockwise because point A is pictured as moving in a counterclockwise circle about the instant center, point C. Then, because ω is counterclockwise, circular motion for point B dictates that $\bar{v}_B$ is to the right. All that remains is to determine the lengths from C to A and B, respectively. We see that angle ACB is a right angle, so

$$\left|\bar{r}_{A/C}\right| = 750 \cos 30° \qquad \left|\bar{r}_{B/C}\right| = 750 \sin 30°$$

Thus

$$\omega = \frac{250}{750 \cos 30°} = 0.384 \text{ rad/s} \qquad \text{(c.c.w.)}$$

$$v_B = \left(\frac{250}{750 \cos 30°}\right) 750 \sin 30° = 0.144 \text{ m/s} \qquad \text{(to the right)}$$

These results are, of course, the same as those obtained by the two methods used in Example 3.

Among the three methods for solving for velocities in rigid body motion, the instant center method is usually the fastest and easiest. Some people dislike it because it is more geometrical and involves more visualization. The alternative method of differentiating the general position vector becomes progressively more difficult as the system becomes more complicated. The primary advantage of using the vector equation (7) for the determination of the velocity is that the method of solving for acceler-

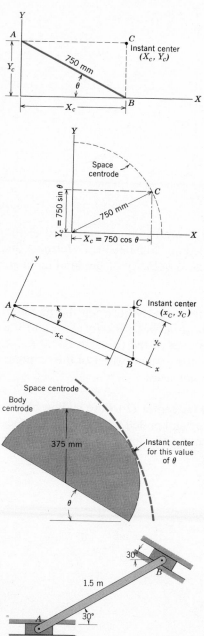

ations is similar to that used in solving for velocities. The decision of which approach to use is yours. However, *do not* forget that the instant center can be used *only* in the velocity portion of the solution.

Additional Comments As we have noted several times, the instant center changes location with time. A curve that is sometimes of interest is the locus of the points in space that the instant center occupies, called the *space centrode*. For the system of Example 4 we can determine the space centrode by finding the coordinates of the instant center X_C and Y_C for an arbitrary θ. We see from the sketch that the coordinates of the instant center with respect to XYZ are

$$X_C = 750 \cos \theta \qquad Y_C = 750 \sin \theta \qquad Z_C = 0$$

Then, because $X_C{}^2 + Y_C{}^2 = (750)^2$ and the range of values of θ is from $0°$ to $90°$, we find that the space centrode is a quarter circle, centered at the origin, of radius 750 mm.

A less obvious curve that we could seek is the locus of points on the body (or its imaginary extension) that are the instant centers, called the *body centrode*. To find the body centrode in Example 4, we resketch the instant center diagram and determine the coordinates x and y that define the location of the instant center with respect to the body. We know that $\left| \vec{r}_{A/C} \right| = 750 \cos \theta$. Then by trigonometry we have

$$x_C = \left| \vec{r}_{A/C} \right| \cos \theta = 750 \cos^2 \theta = 375.0(1 + \cos 2\theta)$$

$$y_C = \left| \vec{r}_{A/C} \right| \sin \theta = 750 \sin \theta \cos \theta = 375.0 \sin 2\theta$$

From this wee see that $(x_C - 375)^2 + y_C{}^2 = (375)^2$. Also, because θ varies between $0°$ and $90°$, x is in the range of values from 0 to 750 mm. Thus, a sketch showing the locus of values of x_C and y_C is a semicircle, centered at the point $(x = 375.0, y = 0)$ of radius 375 mm.

The principal significance of the body and space centrodes is that we can reconstruct the motion of the system by fixing the body centrode to the body and fixing the space centrode in space and allowing the body centrode to roll over the space centrode. These concepts prove of great value in the general area of computer-aided design of mechanisms.

HOMEWORK PROBLEMS

IV.12 At the instant shown, slider A is moving to the left with a speed of 7.5 m/s. Use the method of instant centers to determine the angular velocity of the rod and the velocity of slider B in this position.

IV.13 Determine the velocity of collar B in Problem IV.5 using the method of instant centers.

Prob. IV.12

IV.14   Two people are moving a table. Knowing that $\bar{v}_A$ is 15 in./s parallel to the diagonal line AC and that $\bar{v}_C$ is parallel to edge BC, as shown, determine the velocity that the person at corner B imparts to the table.

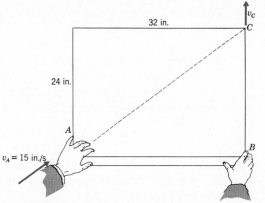

32 in.

24 in.

$v_A = 15$ in./s

Prob. IV.14

IV.15   The forward landing gear of an aircraft is being retracted at the rate of 15 rad/min. The aircraft has a speed of 120 miles/hr. Determine the location of the instantaneous center of zero velocity of the landing gear. Does the result depend on the angle θ?

$v = 120$ miles/hr

2 ft θ ω

Prob. IV.15

IV.16   A hammer is rotated such that the angle γ between the hammer and the forearm is identical to the angle β between the forearm and the horizontal. When $\beta = 45°$ the angular speed of the forearm is 2.0 rad/s. Determine the corresponding location of the instant center of the hammer and the velocity of point D on the hammer head.

50 mm

D

300 mm

β

Fixed

400 mm

Prob. IV.16

IV.17   The stepped pulley is pinned at its center O to the sliding block. Two cords are wrapped around the pulley and pulled with constant velocities, as shown in the diagram. Determine the angular velocity of the pulley and the velocity of the block.

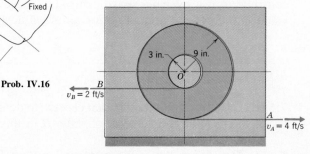

3 in.   9 in.

O

B
$v_B = 2$ ft/s

A
$v_A = 4$ ft/s

Prob. IV.17

Prob. IV.18

IV.18 The hydraulic piston at A is being extended at the rate of 40 mm/s. Determine the angular velocity of the semicircular plate and the velocity of point D at the instant shown.

IV.19 The triangular plate is supported by two taut inextensible cables. In the position shown, cable CD has an angular velocity of 12 rad/s counterclockwise. Determine (a) the corresponding angular velocity of the plate, and (b) the velocity of the centroid G.

IV.20 The hydraulic cylinder actuates the motion of rod AB by moving the follower D to the right with a speed of 2.0 m/s. In the position where $\alpha = \tan^{-1}(\frac{4}{3})$, determine the angular velocity of the rod and the velocity of point A.

Prov. IV.19

Prob. IV.20

2 Rolling Without Slipping

In this section we shall consider the case where the curved surfaces of two bodies are in contact with each other without being physically joined. Thus, as the bodies rotate, their point of contact changes. This type of motion is called rolling. We have illustrated a general rolling situation in Figure 11. Let us draw an x axis parallel to the plane of contact and a y axis perpendicular to that plane. To distinguish between the points of contact on each body, we shall denote the point on body 1 as point C_1 and the point on body 2 as point C_2. Because the bodies are rigid they cannot penetrate each other, so the velocity components of points C_1 and C_2 in the y direction must be identical.

Now let us focus on the expression "without slipping." We have

Figure 11

reasoned that the two surfaces can only move relative to each other along the tangent plane (parallel to the x axis). Such a motion is called *slipping* or *sliding*. If there is no slipping, point C_1 and C_2 will also have the same velocity components parallel to the tangent plane, that is, in the x direction. We therefore conclude that the constraint of no slip means that the x and y components of the velocities of the points of contact are identical, so that

$$\boxed{\bar{v}_{C_1} = \bar{v}_{C_2}} \tag{17}$$

In the special case where one of the bodies, for instance body 2, is stationary, the no slip condition (17) reduces to $\bar{v}_{C_1} = \bar{0}$. For this case it then follows that the contact point is the instant center for body 1.

Two common systems whose motion fits the expression rolling without slipping are a wheel rolling over the ground and a set of gears. In the case of the wheel, friction may prevent the wheel from sliding over the ground. Clearly, the teeth of a gear limit sliding.

Equation (17) represents a constraint on the velocities. Now let us look at the restrictions that no slipping imposes upon the accelerations. If the no slip condition applies, the two points will follow each other in the direction parallel to the plane of contact. Thus the condition of no slip requires that

$$(a_{C_1})_x = (a_{C_2})_x \tag{18}$$

As the bodies roll, the location of the point of contact changes. Thus if we redraw Figure 11 for a slightly later instant, points C_1 and C_2 will have separated. Therefore the y components of the accelerations of points C_1 and C_2 need not be the same.

Finally, it will be seen in the examples that follow that in the special case where the rolling bodies have circular surfaces, we shall be able to solve problems without recourse to equation (18). However, for general surface geometries we would need to use both equations (17) and (18).

EXAMPLE 5
A wheel of diameter 90 mm rolls over the ground to the right so that its center O has a constant speed of 900 mm/s. Determine the velocities and the accelerations of points A, B, and C at the instant shown.

Solution
We know that the center, O, follows a straight path parallel to the ground. Thus, we are given $\bar{v}_O = 900$ mm/s to the right and $\bar{a}_O = \bar{0}$. We wish to find the motion of points A, B, and C. To accomplish this we must first determine the angular velocity and angular acceleration of the wheel. As al-

ways, we begin with angular velocity. Because contact point A is not slipping over the ground and the ground is not moving, the velocity condition for no slip gives

$$\bar{v}_A = \bar{0}$$

Thus point A is the instant center. The instant center method gives

$$\omega = \frac{v_O}{|\bar{r}_{B/A}|} \qquad v_B = \omega|\bar{r}_{B/A}| \qquad v_C = \omega|\bar{r}_{C/A}|$$

From the sketch we have $|\bar{r}_{O/A}| = 45$ mm, $|\bar{r}_{B/A}| = 45/\cos 45° = 45\sqrt{2}$ mm, $|\bar{r}_{C/A}| = 90$ mm. Also from the sketch we see that $\bar{v}_C$ must be parallel to the ground and that $\bar{v}_B$ must be at an upward 45° angle. Thus

$$\omega = \frac{900}{45} = 20 \text{ rad/s}$$

$$\bar{v}_B = (20)45\sqrt{2}(\cos 45° \, \bar{i} + \sin 45° \, \bar{j})$$

$$= 900(\bar{i} + \bar{j}) \text{ mm/s}$$

$$\bar{v}_C = (20) \, 90\bar{i} = 1800\bar{i} \text{ mm/s} = 1.8\bar{i} \text{ m/s}$$

There are two approaches to finding the angular acceleration. One which is valid for a surface of any shape, consists of noting that the x direction shown in the sketch is parallel to the contact plane. Because the ground is at rest, the no slip constraint requires that

$$(a_A)_x = 0$$

However, points A and O are two points on the wheel, so

$$\bar{a}_A = \bar{a}_O + \bar{\alpha} \times \bar{r}_{A/O} + \bar{\omega} \times (\bar{\omega} \times \bar{r}_{A/O})$$

From the velocity solution we know that $\bar{\omega} = -20\bar{k}$ (the negative sign occurs because $\bar{\omega}$ is clockwise). Also, $\bar{r}_{A/O} = -45\bar{j}$ and $\bar{\alpha} = \alpha\bar{k}$ for plane motion. Thus

$$\bar{a}_A = \bar{0} + 45\alpha\bar{i} + 45(20)^2\bar{j}$$

Taking the x component of $\bar{a}_A$ and equating it to zero gives

$$0 = 45\alpha$$

$$\bar{\alpha} = \bar{0}$$

Thus

$$\bar{a}_A = (45)400\bar{j} = 18{,}000\bar{j} \text{ mm/s}^2 = 18\bar{j} \text{ m/s}^2 \qquad \triangleleft$$

The alternative approach for finding α is considerably easier, but it is *only* valid for a *circular* surface. In this approach we note that regardless of the position of the wheel, the sketch for the instant center is unchanged. Thus, it is always true that $\omega = v_O/R$, where R is the radius of the wheel. We can differentiate this formula to find

$$\alpha = \dot{\omega} = \frac{\dot{v}_O}{R}$$

In this problem v_O is constant, so $\dot{v}_O = 0$ and $\alpha = 0$.

Once we have found $\bar{\alpha}$, we can continue on to find $\bar{a}_B$ and $\bar{a}_C$.

$$\bar{a}_B = \bar{a}_O + \bar{\alpha} \times \bar{r}_{B/O} + \bar{\omega} \times (\bar{\omega} \times \bar{r}_{B/O})$$

$$\bar{a}_C = \bar{a}_O + \bar{\alpha} \times \bar{r}_{C/O} + \bar{\omega} \times (\bar{\omega} \times \bar{r}_{C/O})$$

From the diagram we have

$$\bar{r}_{B/O} = -45\bar{i} \qquad \bar{r}_{A/O} = 45\bar{j}$$

so that

$$\bar{a}_B = \bar{0} + \bar{0} + 45(20)^2\bar{i} = 18{,}000\bar{i} \text{ mm/s}^2 = 18\bar{i} \text{ m/s}^2 \qquad \triangleleft$$

$$\bar{a}_C = \bar{0} + \bar{0} - 45(20)^2\bar{j} = -18{,}000\bar{j} \text{ mm/s}^2 = 18\bar{j} \text{ m/s}^2 \qquad \triangleleft$$

Finally, because rolling wheels are an integral part of a multitude of physical systems, let us discuss two more aspects of the general problem of rolling without slipping that relate to the problem we just solved. First, in the derivations we saw that the general motion of a body can be considered to be the superposition of a translation (having the motion of any point on the body) and a pure rotation (about that same point). In the case of the wheel in this problem, we can say that the motion is a superposition of a translation (to the right) of the center point O and a rotation (clockwise) about point O. The effect of the constraint condition of no slipping is to force the rate of rotation to be such that the contact point A has no velocity and only has an acceleration perpendicular to the ground.

With regard to the other aspect we wish to mention, suppose that we drew the path of point A in space as the wheel rolls. You may remember that this path is a cycloid. The cusps (e.g., point O) in the cycloidal path are the points in space where point A is in contact with the ground. As point A approaches a cusp in its path, its velocity is downward and its speed is decreasing. After passing the cusp the velocity has reversed direction and the speed increases, thus at the cusp, point A has zero velocity and its acceleration is upward. The foregoing is an alternative way to describe the "no slip" constraint condition.

EXAMPLE 6
A portion of the steering gear of an automobile consists of the rack and pinion set shown in the sketch. The pinion rotates about a fixed axis passing through point A, and the rack is constrained to translate only. If the pinion is rotating at 0.5 rev/s clockwise and this rate is increasing at 4 rev/s², determine the velocity and acceleration of the rack.

Solution
First we note that for computational purposes the angular units must be changed to radians.

$$\omega = (0.5)2\pi = \pi \text{ rad/s}$$

$$\alpha = (4)2\pi = 8\pi \text{ rad/s}^2$$

The motion of any point on the pinion is known because point A is on the axis of rotation and is fixed. We can therefore find the motion of the contact point on the pinion, and then use of the no slip condition will give the motion of the rack. Point A is at rest, so it is always the instant center.

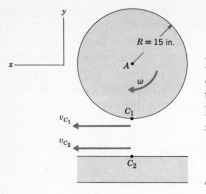

$$v_{C_1} = \omega \left| \bar{r}_{C_1/A} \right| = \omega R = 1.5\pi \text{ in./s}$$

$$\bar{v}_{C_1} = \omega R \bar{i} = 1.5\pi \bar{i}$$

For no slipping $\bar{v}_{C_1} = \bar{v}_{C_2}$ but $\bar{v}_{C_2} = \bar{v}_{\text{rack}}$. Thus $\bar{v}_{\text{rack}} = 1.5\pi\bar{i}$ in./s. To find $\bar{a}_{\text{rack}}$ we could use the no slip condition $(a_{\text{rack}})_x = (a_{C_1})_x$ and subsitute into the acceleration formula, but in this problem — because we have a circular surface — it is easier to note that $v_{\text{rack}} = \omega R$ for all positions of the system. Hence

$$\dot{v}_{\text{rack}} = \alpha R = (8\pi)1.5 = 12\pi \text{ in./s}^2$$

or vectorially,

$$\bar{a}_{\text{rack}} = \dot{\bar{v}}_{\text{rack}} = \dot{v}_{\text{rack}}\bar{i} = 12\pi\bar{i} \text{ in./s}^2 \qquad \triangleleft$$

HOMEWORK PROBLEMS

IV.21   A cylindrical tank of radius 400 mm is rolling backward without slipping of the flatbed of a truck with a speed of 10 m/s relative to the truck. The truck has a forward speed of 20 m/s. Determine the angular velocity of the tank.

Prob. IV.21

IV.22   The stepped disk rolls without slipping on a set of rails. The disk is rotating at the constant rate of 30 rev/min. Determine the acceleration of points A and B.

IV.23   The stepped drum B rolls without slipping. Drum A has an angular velocity of 20 rad/s clockwise and an angular acceleration of 20 rad/s^2 clockwise. Determine (a) the angular velocity of body B, (b) the angular acceleration of body B, (c) the velocity of point C, (d) the acceleration of point O, and (e) the acceleration of point C.

Prob. IV.22

Prob. IV.23

IV.24 A hoist consists of a motor driven gear A which rotates the larger stepped gear B. Determine (a) the gear ratio ω_B/ω_A, and (b) the velocity of the block when $\omega_A = 300$ rev/min.

IV.25 A wheel rolling in a circular path has an angular velocity $\bar{\omega}$ and an angular acceleration $\bar{\alpha}$, both clockwise. There is no slipping at the contact point. Determine the velocity and acceleration of the center point O.

Prob. IV.26

Prob. IV.24

Prob. IV.25

IV.26 A box is placed on rollers and given an acceleration of 4 ft/s² to the right. There is no slippage between the rollers and the crate and between the rollers and the ground. At a specific instant it is known that the total acceleration of the point on the perimeter of a roller that is in contact with the crate is 5 ft/s². Determine the velocity and acceleration of the centers of the rollers at this instant. The radius of the roller is 250 mm.

IV.27 At the instant shown, the elevator is descending at the rate of 4 m/s and is slowing down at 0.5 m/s². The cable does not slip over the pulley. Determine (a) the angular velocity and angular acceleration of the large upper pulley, and (b) the velocity and acceleration of the counterweight.

IV.28 Determine the speed ratio v_A/v_B using the method of instant centers. Compare this solution to that obtained by imposing the kinematical constraint that the lengths of all cables are constant. The cables do not slip as they pass over the pulleys.

Prob. IV.27

Prob. IV.28

IV.29   The semicircular cylinder rolls without slipping. When $\theta = 45°$ the angular velocity is $\bar{\omega}$ and the angular acceleration is $\bar{\alpha}$, both counterclockwise. Determine the velocity and acceleration of the centroid G at this instant.

Prob. IV.29

3   Mechanical Linkages

A mechanical linkage is a mechanism by which motion and power may be transferred. For example, the crankshaft and connecting rod in Figure 12 transform the rectilinear motion of the piston into a rotary motion of the crankshaft. In a linkage, bodies are interconnected so that they cannot move independently of each other. In this module we shall consider only the case where they are connected by pins. Other types of connections will be studied in Module VI.

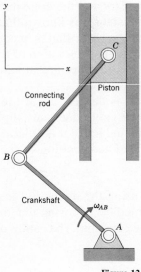

Figure 12

Let us assume that we know the angular velocity $\bar{\omega}$ of the drive shaft AB in Figure 12, and for simplicity assume that $\bar{\alpha}_{AB} = \bar{0}$. The problem is to determine the velocity and acceleration of the piston.

The method of solution differs little from the work already presented. However, as a good summary of what we have already done, and for completeness, let us discuss the analysis.

The motion of rod AB is defined by the given information. That must be our starting point. The piston can only translate, so all points on the piston undergo the same motion as the pin C. Therefore we shall seek information about point C on rod BC, because point C is a point common to the piston and the pin. Furthermore, point B is common to both rods. Thus, in the solution we shall start at A and determine the velocity of point B, considering rod AB. Then we shall consider rod BC using our knowledge of the kinematics of point B.

For link AB we may write

$$\bar{v}_B = \bar{v}_A + \bar{\omega}_{AB} \times \bar{r}_{B/A}$$

Point A is stationary, because it is on the axis of rotation of the crankshaft. Therefore $\bar{v}_A \equiv \bar{0}$. We now have

$$\bar{v}_B = \bar{\omega}_{AB} \times \bar{r}_{B/A}$$

Now, considering $\bar{v}_B$ to be known, we may express the velocity of point C using the velocity equation for the connecting rod.

$$\bar{v}_C = \bar{v}_B + \bar{\omega}_{BC} \times \bar{r}_{C/B}$$

Then substituting for $\bar{v}_B$ gives

$$\bar{v}_C = \bar{\omega}_{AB} \times \bar{r}_{B/A} + \bar{\omega}_{BC} \times \bar{r}_{C/B} \qquad (19)$$

We are dealing with the motion of a point in a plane. Hence equation (19) represents two scalar equations that are obtained by matching the x and y components of the resulting vectors. How many unknowns are there in equation (19)? At this juncture, we do not know $\bar{v}_C$ and $\bar{\omega}_{BC}$. However, we know that we have plane motion, so $\bar{\omega}_{BC} = \omega_{BC}\bar{k}$ and $\bar{v}_C$ can have only two components. Thus we have three unknown scalar quantities, one more than the number of equations presently available. To get the additional equation we use the fact that the piston is constrained to follow the walls of the cylinder, so that $\bar{v}_C = v_C\bar{j}$ (a negative value for v_C will tell us that the piston is moving downward). Now we have two equations and two unknowns, so we can solve for ω_{BC} and v_C.

The acceleration problem may be solved in the same manner, after having first solved the velocity problem. We proceed through the linkage from point A to point B to point C using the relative acceleration equation for each link. For these computations we need $\bar{\omega}_{BC}$, which was determined in the solution to the velocity problem. We then impose the conditions that the system is constrained to move in a plane and that the acceleration of the piston is constrained to be a rectilinear motion in the vertical direction.

We close this discussion by noting that alternatively, the instant center method can be used to find the angular velocity of the crankshaft and the velocity of the piston. However, the vector equation approach must be used for the determination of acceleration. This will be illustrated in the next example.

EXAMPLE 7

The wheel in the linkage shown in the sketch is rotating in a clockwise direction at the constant rate of 40 rev/min. Determine the angular speed and angular acceleration of bar AB for the position shown.

Solution

From the given information we find $\omega_{CD} = (40)(2\pi/60) = (4\pi/3)$ rad/s and $\alpha_{CD} = 0$. Because point D is fixed, we can determine the velocity and acceleration of any point on the wheel. A study of the sketch reveals that we can proceed through the linkage system from point D to point A.

Let us begin by using instant centers to solve the velocity problem.

(The vector approach would work just as well.) Note that because point A is fixed, it is the center of rotation of bar AB. Similarly, point D is the center of rotation of the wheel. It is important that you realize that each rigid body in the system has its own instant center. We now draw a velocity diagram of the linkage.

The instant center for bar BC was located as follows. The velocity of point B must be perpendicular to bar AB because point A is its instant center. Hence the instant center for BC must be along AB or its extension. Similarly, the velocity of point C must be perpendicular to line DC because point D is the instant center of the wheel. Hence, the instant center for BC must be along DC or its extension. It then follows that the instant center of bar BC is at point D, where lines AB and CD intersect. Recall that *each* rigid body has its own instant center. Before we can find ω_{AB} we must first determine the lengths of several lines. Because line OB is collinear with line AB, we know that

$$\beta = \tan^{-1} \tfrac{12}{5}$$

Then, because $\left| \bar{r}_{B/C} \right| = 100$ mm, we find

$$\left| \bar{r}_{C/O} \right| = 100 \tan \beta = 240 \text{ mm}$$

$$\left| \bar{r}_{B/O} \right| = 100 \cos \beta = 260 \text{ mm}$$

From the diagram we have

$$v_C = \left| \bar{r}_{C/D} \right| \omega_{CD} = (60)\frac{4\pi}{3} = 80\pi \text{ mm/s to the right}$$

$$= \left| \bar{r}_{C/O} \right| \omega_{BC} = 240\omega_{BC}$$

Solving for ω_{BC} gives

$$\omega_{BC} = \frac{80\pi}{240} = \frac{\pi}{3} \text{ rad/s counterclockwise}$$

Point B is a common point to bars AB and BC. Thus we find v_B using the instant center O and center of rotation A.

$$v_B = |\bar{r}_{B/O}|\omega_{BC} = (260)\frac{\pi}{3}$$

$$= |\bar{r}_{B/A}|\omega_{AB} = 130\omega_{AB}$$

Solving for ω_{AB} gives

$$\omega_{AB} = \frac{(260)\pi/3}{130} = \frac{2\pi}{3} \text{ rad/s clockwise}$$

We can now solve the acceleration problem using the vector approach. (Instant centers are not valid for accelerations). Starting from fixed point D, $\bar{a}_D = \bar{0}$, we have

$$\bar{a}_C = \bar{0} + \bar{\alpha}_{CD} \times \bar{r}_{C/D} + \bar{\omega}_{CD} \times (\bar{\omega}_{CD} \times \bar{r}_{C/D})$$
$$\bar{a}_B = \bar{a}_C + \bar{\alpha}_{BC} \times \bar{r}_{B/C} + \bar{\omega}_{BC} \times (\bar{\omega}_{BC} \times \bar{r}_{B/C})$$
$$\bar{a}_A = \bar{a}_B + \bar{\alpha}_{AB} \times \bar{r}_{A/B} + \bar{\omega}_{AB} \times (\bar{\omega}_{AB} \times \bar{r}_{A/B})$$

Because point A is fixed, we must also require that $\bar{a}_A = \bar{0}$.

Before we can solve the foregoing equations we must select a coordinate system for describing vector components. Because we have the horizontal and vertical distances for the locations of the points, we use xyz coordinates as shown in the first diagram of the problem. From this diagram we find

$$\bar{r}_{C/D} = 60\bar{j}$$

$$\bar{r}_{B/C} = -100\bar{i}$$

$$\bar{r}_{A/B} = -50\bar{i} - 120\bar{j}$$

From the velocity solution and the fact that for plane motion $\bar{\omega}$ and $\bar{\alpha}$ are perpendicular to the plane, we have (using the right-hand rule)

$$\bar{\omega}_{CD} = \frac{-4\pi}{3}\bar{k} \qquad \bar{\omega}_{BC} = \frac{\pi}{3}\bar{k} \qquad \bar{\omega}_{AB} = \frac{-2\pi}{3}\bar{k} \text{ rad/s}$$

$$\bar{\alpha}_{CD} = \bar{0} \qquad \bar{\alpha}_{BC} = \alpha_{BC}\bar{k} \qquad \bar{\alpha}_{AB} = \alpha_{AB}\bar{k}$$

Substituting these vectors into the acceleration equations gives

$$\bar{a}_C = -\left(\frac{4\pi}{3}\right)^2 60\bar{j} = -\frac{960}{9}\pi^2\bar{j}$$

$$\bar{a}_B = -\frac{960}{9}\pi^2\bar{j} - 100\alpha_{BC}\,\bar{j} + \left(\frac{\pi}{3}\right)^2 100\bar{i}$$

$$= \frac{100}{9}\pi^2\bar{i} - \left(\frac{960}{9}\pi^2 + 100\alpha_{BC}\right)\bar{j}$$

$$\bar{a}_A = \bar{0} = \frac{100}{9}\pi^2\bar{i} - \left(\frac{960}{9}\pi^2 + 100\alpha_{BC}\right)\bar{j}$$

$$+ \alpha_{AB}(-50\bar{j} + 120\bar{i}) + \left(\frac{2\pi}{3}\right)^2(50\bar{i} + 120\bar{j})$$

$$= \left(\frac{100}{9}\pi^2 + 120\alpha_{AB} + \frac{200}{9}\pi^2\right)\bar{i}$$

$$+ \left(-\frac{960}{9}\pi^2 + 100\alpha_{BC} - 50\alpha_{AB} + \frac{480}{9}\pi^2\right)\bar{j}$$

We may now obtain the scalar equations for α_{AB} and α_{BC} by matching components on the left- and right-hand sides of the equation for $\bar{a}_A$. Therefore

$$\bar{i} \text{ direction:} \qquad 0 = \frac{300}{9}\pi^2 + 120\alpha_{AB}$$

$$\bar{j} \text{ direction:} \qquad 0 = -\frac{480}{9}\pi^2 + 100\alpha_{BC} - 50\alpha_{AB}$$

Using these equations the angular accelerations are found to be

$$\bar{\alpha}_{AB} = 2.74\bar{k} \text{ rad/s}^2 \qquad \bar{\alpha}_{BC} = 6.63\bar{k} \text{ rad/s}^2$$

HOMEWORK PROBLEMS

IV.30   Each system shown is a four-bar linkage (the fixed line AD being the fourth bar). In the position shown, bar AB is rotating clockwise at 100 rev/min. Determine the corresponding angular velocity of bar CD.

Prob. IV.30 (a)   Prob. IV.30 (b)

Prob. IV.31

IV.31 Slider A has a downward velocity of 600 mm/s. Determine the angular velocity of rod BC and the velocity of slider C for the position shown in the sketch.

IV.32 A motor rotates the inner gear in the planetary gear transmission at a rate of 3600 rev/min clockwise. (a) Determine the angular velocity of the outer gear when the spider arms are held fixed. (b) Determine the angular velocity of the spider arms when the outer gear is held fixed.

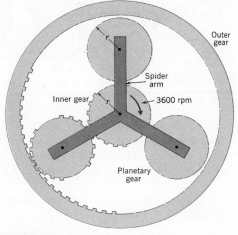

Prob. IV.32

IV.33 The disk rotates at a constant angular velocity $\bar{\omega}$ clockwise. (a) Determine the angular velocity of bar BC as a function of $r, l, \omega,$ and θ. (b) Determine the ratio v_C/v_B as a function of θ for the case when $(l/r) = 2$.

Prob. IV.33

IV.34 End A of rod AB is given a constant velocity of 10 ft/s to the left. The wheel to which its other end is pinned rolls without slipping. Determine the angular velocity of the wheel and the velocity of the center point O (a) when $\theta = 0°$, and (b) when $\theta = 90°$.

IV.35 The driven gear B and the rear wheel of the bicycle shown in the sketch are mounted on the same shaft, with the result that their angular velocities are identical. The diameter of the drive gear A is twice that of

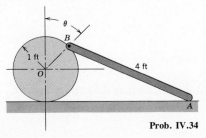

Prob. IV.34

gear B. What rate of rotation ω_A will cause the bicycle to have a speed of 4 m/s? The diameter of each wheel is 700 mm.

Prob. IV.35

IV.36   A hydraulically driven piston moves point C to the right with a constant acceleration of 4.5 m/s². At the instant depicted in the diagram, point C has a velocity of 15 m/s, also to the right. Determine the angular velocity and angular acceleration of both rods at this instant.

Prob. IV.36

IV.37   In the position shown, the angular velocity of bar AB is the constant value $\bar{\omega}$. Determine the angular velocities and angular accelerations of bars BC and CD.

Prob. IV.37

IV.38   Gear A meshes with the horizontal rack as it is driven by bar BC. When $\theta = 60°$, slider C has a velocity of 10 ft/s to the right and its speed is decreasing at the rate of 4 ft/s². Determine the corresponding velocity and acceleration of the center of the gear.

Prob. IV.38

Prob. IV.41

IV.39 Determine the acceleration of point *C* in Problem IV.30 (a) if for the position shown α_{AB} is 2 rad/s² counterclockwise.

IV.40 Determine the acceleration of point *C* in Problem IV.30 (b) if for the position shown α_{AB} is 2 rad/s² clockwise.

IV.41 Bar *AB* is rotating at the constant rate of 3 rad/s counterclockwise. Determine the values of the angular velocity and angular acceleration of bar *BC* for the position shown. The radius of the roller at *C* is negligible.

E. PROBLEM SOLVING

As was done in previous modules, let us develop a series of logical steps that lead to a systematized approach to problem solving. Again, we emphasize that this is strictly to help organize your thoughts. If these steps are just memorized the result will be that you will not truly understand the process of problem solving. As you gain proficiency you will intuitively perform many of the steps delineated below.

In this module we dealt with kinematics of rigid bodies in plane motion. Within the framework of problems in this topic we must always solve a velocity problem. Remember, even if we are only required to find an acceleration, the values of angular and linear velocities were shown to be an integral part of the solution for acceleration. With this in mind we have our starting point, the velocity problem.

In order to formulate any problem properly, we require a pictoral representation from which we can work.

1   The Velocity Problem

1   Draw a simple sketch of the system in the position of interest. Choose an *xyz* coordinate system such that *x* and *y*, which are in the plane of motion, make it as easy as possible to express the position of points in the system. For instance, if horizontal and vertical dimensions are given, choose *x* and *y* such that one is horizontal and the other is vertical.

2   Express the given information in the form of equations and list the quantities you seek.

3   Write down the equations relating the velocities of each of the constrained points of a body to the angular velocity of the body. Do this for each body. Typical constrained points to look for are

a)   fixed points
b)   points pinned to another body
c)   points to follow a specific path
d)   points of contact for bodies that roll

4   Write down the mathematical restrictions imposed on *all* constrained points. For instance, if a point A must follow a straight line in the *x* direction, write down $\bar{v}_A = v_A \bar{\imath}$.

5   Express all position vectors and angular velocity vectors by their *xyz* components. Remember that for plane motion $\bar{\omega} = \pm \omega \bar{k}$.

6   Substitute the constraint conditions from step 4 and the position vectors and angular velocity vectors from step 5 into the velocity equations from step 3. Calculate the cross products.

7   Collect components in the *x* and *y* directions. Count the number of scalar equations you have. Count the number of scalar unknowns in these equations. If the two numbers are not equal, you have probably forgotten to use a given piece of information or forgotten to utilize a constraint condition that the system must obey.

8   Solve the equations.

9   After determining the velocity of any point on the body and the angular velocity of the body, the velocity of unconstrained points may be found from the relative velocity equation.

Alternative to (3–9): The method of instant centers may be used in lieu of the vector equations. This method uses the fact that the velocity $\bar{v}_P$ of any point P on the body must be perpendicular to the line from point P to the instant center C of the body and that $v_P = \omega \left| \bar{r}_{P/C} \right|$.

2   The Acceleration Problem

10–16   If accelerations must be determined, repeat steps 3 through 9, replacing the word velocity by the word acceleration. In these steps consider the solutions for velocities as given information. Remember that instant centers are not valid for determining acceleration.

We shall now investigate a series of illustrative problems, calling attention to the application of these steps as we proceed through the solutions.

ILLUSTRATIVE PROBLEM 1

Two movers lowered a cabinet from the window of an apartment building. They each let their cables out at the same constant rate of 3 ft/s when, as the cabinet neared the ground, they simultaneously tightened their grips to slow it down. In so doing the cable at A was immediately given an upward acceleration of 5 ft/s² and the cable at B was given an upward acceleration of 1 ft/s². Determine the acceleration of the center of the cabinet C and the angular acceleration of the cabinet at the instant the movers tightened their grips.

Solution

Step 1 Draw a sketch labeling all important points and showing xyz.

Step 2 Given $\bar{v}_A = \bar{v}_B = -3\bar{j}$ ft/s, $\bar{a}_A = 5\bar{j}$ ft/s², and $\bar{a}_B = \bar{j}$ ft/s². Find the acceleration of point C. Our thinking at this point is that the sudden change in the motion of the cables will *tend* to cause a rotation of the cabinet as shown. However, we also realize that the time interval during which the movers tightened their grips is so brief that the cabinet will not have time to alter its initial velocity appreciably, so that we should expect the cabinet to have no angular velocity at this instant. If this seems strange to you, think of other situations where the derivative of a quantity undergoes a sudden change. For instance, suppose that you are in a car traveling at 90 km/hr and suddenly apply the brakes. What is your velocity at time zero when you activate the brakes?

Step 3 We know that the formula we shall eventually use for the acceleration requires that we know $\bar{\omega}$, the angular velocity of the cabinet. Let us prove that $\bar{\omega} = \bar{0}$. To do this we relate the velocities of the two points about which we have information. Clearly, in this problem they are points A and B.

$$\bar{v}_B = \bar{v}_A + \bar{\omega} \times \bar{r}_{B/A}$$

Step 4 The constraints we know are that at the instant the cables were seized the velocities of A and B were equal and directed downward. Hence

$$\bar{v}_A = \bar{v}_B = -3\bar{j} \text{ ft/s}$$

Step 5 We know that $\bar{\omega} = -\omega\bar{k}$ for the coordinate system chosen and assumed sense of rotation. The position vector we require in step 3 in

$$\bar{r}_{B/A} = 4\,\bar{i} \text{ ft}$$

Step 6 We may now return to the formula in step 3 and write

$$-3\bar{j} = -3\bar{j} + (-\omega\bar{k}) \times 4\bar{i}$$
$$= -3\bar{j} - 4\omega\bar{j}$$

Step 7 Match components.

$$\bar{j} \text{ direction:} \qquad -3 = -3 - 4\omega$$

Step 8 We have one equation and one unknown, from which

$$\omega \equiv 0$$

Hopefully our foreshadowing made this an expected result. However, it does not follow that the angular acceleration is zero at time zero.

Step 9 We were not asked to find $\bar{v}_C$, but because $\omega = 0$, $\bar{v}_C = \bar{v}_B = \bar{v}_A = -3\bar{j}$.

Step 10 We now wish to determine the acceleration of point C. Begin by repeating step 3. Points A and B are constrained, so we write

$$\bar{a}_B = \bar{a}_A + \bar{\alpha} \times \bar{r}_{B/A} + \bar{\omega} \times (\bar{\omega} \times \bar{r}_{B/A})$$

Because $\bar{\omega} = \bar{0}$ at the initial instant, the equation reduces to

$$\bar{a}_B = \bar{a}_A \times \bar{\alpha} \times \bar{r}_{B/A}$$

Step 11 The constraints given in the problem are, at time zero,

$$\bar{a}_A = 5\bar{j} \quad \text{and} \quad \bar{a}_B = 1\bar{j} \text{ ft/s}^2$$

Step 12

$$\bar{\alpha} = -\alpha\bar{k}$$
$$\bar{r}_{B/A} = 4\bar{j} \text{ ft}$$

Step 13 Place the vectors in steps 11 and 12 into the formula in step 10.

$$\bar{j} = 5\bar{j} - \alpha\bar{k} \times 4\bar{i} = 5\bar{j} - 4\alpha\bar{j}$$

Step 14 Match components.
$$\bar{j} \text{ direction:} \qquad 1 = 5 - 4\alpha$$
Step 15 We have one equation and one unknown, therefore

$$\alpha = 1$$
$$\bar{\alpha} = -1\bar{k} \text{ rad/s}^2$$

Step 16

$$\bar{a}_C = \bar{a}_A + \bar{\alpha} \times \bar{r}_{C/A}$$

$$= 5\bar{j} + (-\bar{k}) \times (2\bar{i} - \bar{j})$$

$$= 5\bar{j} - 2\bar{j} - \bar{i}$$

$$= 3\bar{j} - \bar{i} \text{ ft/s}^2$$

ILLUSTRATIVE PROBLEM 2

A slider mechanism of length L slides with its ends constrained to move in slots that make an angle θ with one another. (a) Prove that when the link is perpendicular to one slot, the velocity of the slider in contact with the other slot vanishes. (b) Show that when the link makes equal angles with the slots, the speeds of the two sliders are the same. (c) Find the speed for part (b) in terms of l, θ, and ω_{AB}.

Solution

Step 1 Because we require two different geometric configurations, let us sketch the system in terms of a general angle α.

Step 2 In terms of the angle α, let us restate the problem. For part (a), show that $v_A \equiv 0$ when $\alpha = 90°$. For part (b), show that $|\bar{v}_A| = |\bar{v}_B|$ when $\alpha = (90° - \theta) + (90° - \alpha)$, that is, when $\alpha = 90° - \theta/2$. Part (c) remains the same, namely show that $|\bar{v}_A| \equiv |\bar{v}_B|$ for the condition of part (b).

Step 3 Clearly there are two points of interest in this problem, and both are constrained. We shall solve the problem using the vector approach. Hence we write

$$\bar{v}_B = \bar{v}_A + \bar{\omega}_{AB} \times \bar{r}_{B/A}$$

Step 4 Points A and B are constrained to move in the slots shown. Hence, assuming $\bar{v}_B$ to be to the left, we may write

$$\bar{v}_A = v_A(\cos \theta \, \bar{i} + \sin \theta \, \bar{j})$$

$$\bar{v}_B = -v_B \bar{i}$$

Step 5 We require the position vector $\bar{r}_{B/A}$.

$$\bar{r}_{B/A} = L(-\cos \alpha \, \bar{i} + \sin \alpha \, \bar{j})$$

The motion takes place in the xy plane. Therefore

$$\bar{\omega}_{AB} = \omega_{AB} \bar{k}$$

Step 6   Place the vector relationships determined in steps 4 and 5 into the velocity formula shown in step 3.

$$-v_B\bar{i} = v_A(\cos\theta\,\bar{i} + \sin\theta\,\bar{j}) + \omega_{AB}\bar{k} \times L(-\cos\alpha\,\bar{i} + \sin\alpha\,\bar{j})$$

$$= v_A(\cos\theta\,\bar{i} + \sin\theta\,\bar{j}) - \omega_{AB}L(\sin\alpha\,\bar{i} + \cos\alpha\,\bar{j})$$

Step 7   These lead to the scalar equations

$$\bar{i}\ \text{direction:}\quad -v_B = v_A\cos\theta - \omega_{AB}L\sin\alpha$$

$$\bar{j}\ \text{direction:}\quad 0 = v_A\sin\theta - \omega_{AB}L\cos\alpha$$

Note that for a problem containing numbers we would count equations and unknowns at this point. Here we have two general equations and must apply the conditions of the specific problem before looking for a solution.

Step 8   Part (a) may be looked upon as, find v_A when $\alpha = 90°$. From the second equation of step 7 we have

$$0 = v_A\sin\theta$$

Because θ is a fixed nonzero angle, we conclude that $v_A = 0$, as was to be shown.

Parts (b) and (c) require that we find v_A and v_B when $\alpha = 90° - \theta/2$. From the second equation in step 7 we have

$$v_A\sin\theta = \omega_{AB}L\cos\left(90° - \frac{\theta}{2}\right)$$

$$\equiv \omega_{AB}L\sin\frac{\theta}{2}$$

Recall the trigonometric identity

$$\sin 2\beta = 2\sin\beta\cos\beta$$

Therefore, letting $2\beta = \theta$, we may write

$$2v_A\sin\frac{\theta}{2}\cos\frac{\theta}{2} = \omega_{AB}L\sin\frac{\theta}{2}$$

or

$$v_A = \frac{\omega_{AB}L}{2\cos(\theta/2)}$$

which is the solution to part (c).

Place this into the first equation in step 7 with $\alpha = 90° - \theta/2$.

$$-v_B = v_A \cos \theta - 2v_A \cos \frac{\theta}{2} \sin \left(90° - \frac{\theta}{2} \right)$$

$$\equiv v_A \cos \theta - 2v_A \left(\cos \frac{\theta}{2} \right)^2$$

We now make use of another trigonometric identity,

$$\cos 2\beta = \cos^2 \beta - \sin^2 \beta = 2 \cos^2\beta - 1$$

Letting $2\beta = \theta$ again, we find

$$-v_B = v_A \left[2\left(\cos \frac{\theta}{2} \right)^2 - 1 \right] - 2v_A \cos \left(\frac{\theta}{2} \right)^2 = -v_A$$

This is the identity we were asked to show in part (b).

3 Alternative Methods

There are two alternate methods that are equally valid to use for a problem such as this. One is, of course, the method of instant centers. The other, a bit more subtle, is to write a position vector, say $\bar{r}_{A/C}$, and differentiate.

ILLUSTRATIVE PROBLEM 3

A winch is being used to move a large spool of inextensible electrical transmission wire. The wire from the spool is wrapped around the drum of the winch and the spool rolls along the ground without slipping. At the instant shown, the winch has an angular speed of 1 rad/s and an angular acceleration of 2 rad/s², both clockwise. Find (a) the velocity of the center of the wire spool, (b) the rate at which the cord is being wound up or let out from the spool, and (c) the acceleration of the center of the wire spool.

Solution

Intuitively, we expect the spool of wire to be pulled to the left. This is useful information to know, but we shall not require it, for the analysis will yield the proper direction of motion even if we begin with an incorrect assumption. Hence, assuming the motion to be to the left, if the contact point on the spool does not slip, the spool will rotate counterclockwise. You may be wondering how the given information could state as a fact that the spool does not slip. The answer to this question is that it is a result that we shall establish in the next module. There, our study of the effects of the system's forces will tell us whether or not slipping occurs. Here, we accept the given premise of rolling without slipping.

We shall solve the problem by the vector approach.

Step 1   Draw a sketch labeling any important points and showing *xyz*.

Step 2   Given $\omega_w = 1$ rad/s, $\alpha_w = 2$ rad/s². Find (a) $\bar{v}_B$, (b) rate of winding or unwinding of cord, (c) $\bar{a}_B$.

Part (b) is unlike anything we have encountered thus far. To motivate our thinking, let us imagine a few hypothetical situations. First, if center point *B* and point *C* have the same velocity, the spool will have no rotation and the wire will neither wind nor unwind. However, if there is no rotation, slipping will occur. Hence, by the constraint of the given information this cannot be. Second, if point *C* is approaching the winch faster than point *B*, then the spool will be rotating clockwise and the cord will unwind. Third, if point *C* is approaching the winch slower than point *B*, then the spool will rotate counterclockwise and the cord will wind around the drum.

From the foregoing we conclude that the rate of unwinding is determined from the difference between $\bar{v}_C$ and $\bar{v}_B$. But

$$\bar{v}_C - \bar{v}_B = \bar{v}_{C/B} = \bar{\omega}_s \times \bar{r}_{C/B}$$

so that we need to find the angular velocity of the spool, $\bar{\omega}_s$.

Step 3   For the winch we know that point *A* is fixed and that point *E* is the contact point between the winch and the wire. We therefore write

$$\bar{v}_E = \bar{v}_A + \bar{\omega}_w \times \bar{r}_{E/A}$$

For the spool we know that point *D* is the contact point with the ground and that point *C* is the contact point between the wire and the spool. Thus,

$$\bar{v}_C = \bar{v}_D + \bar{\omega}_s \times \bar{r}_{C/D}$$

Note that point *B* is an unconstrained point, so we do not consider it in this step.

Step 4   The wire is inextensible, and because it does not slip over either drum, we have the constraint condition

$$\bar{v}_C = \bar{v}_E$$

Point A is fixed, so $\bar{v}_A = \bar{0}$, and the ground is stationary, so the no slip condition gives $\bar{v}_D = \bar{0}$.

Step 5   We know $\bar{\omega}_w = -\omega_w \bar{k} = -1\bar{k}$ rad/s. For plane motion $\bar{\omega}_s = \omega_s \bar{k}$. The position vectors in step 3 are

$$\bar{r}_{E/A} = -1\bar{j} \qquad \bar{r}_{C/D} = 0.5\bar{j} \text{ m}$$

Step 6   Substitute steps 4 and 5 into step 3, leaving $\bar{\omega}_w$ in terms of the algebraic rate ω_w, in lieu of substituting its current numerical value. The reason for this will be obvious shortly.

$$\bar{v}_E = \bar{0} - \omega_w \bar{k} \times -\bar{j} = -\omega_w \bar{i}$$

$$= \bar{v}_C = \bar{0} + \omega_s \bar{k} \times 0.5\bar{j} = -0.5\omega_s \, \bar{i}$$

Step 7   Match components.

$$\bar{i} \text{ direction:} \qquad -\omega_w = -0.5\omega_s$$

We have one equation and one unknown.

Step 8   Solve the equation.

$$\omega_s = \frac{\omega_w}{0.5} = 2\omega_w = 2 \text{ rad/s} \qquad \text{at this instant}$$

Therefore, at this instant,

$$\bar{\omega}_s = 2\bar{k} \text{ rad/s}$$

We now can discuss the reason for leaving ω_w as an algebraic variable. Generally in problems of circular wheels that are rolling without slipping, the geometry of the system does not change as time goes by. We shall therefore be able to differentiate the general formula for the angular speed(s) to obtain a relationship for angular acceleration(s). This may enable us to skip some or all of steps 10–16. For example, in this problem the cord is always horizontal, so the diagram in step 1 applies at all instants of time. Therefore at this instant,

$$\alpha_s = \dot{\omega}_s = 2\dot{\omega}_w = 2\alpha_w = 4 \text{ rad/s}^2$$

so that

$$\bar{\alpha}_s = 4\bar{k} \text{ rad/s}^2$$

Step 9   For part (a) we want $\bar{v}_B$. Clearly, we wish to relate B to a point in the spool of known velocity, which is point D. Therefore

$$\bar{v}_B = \bar{v}_D + \bar{\omega}_s \times \bar{r}_{B/D} = \bar{0} + \omega_s \bar{k} \times 1.5\bar{j}$$

$$= -1.5\omega_s \bar{i}$$

$$= -3\bar{i} \text{ m/s} \qquad \text{at this instant}$$

For part (b) we have

$$\bar{v}_{C/B} = \bar{\omega}_s \times \bar{r}_{C/B} = 2\bar{k} \times -1\bar{j} = 2\bar{i} \text{ m/s}$$

From this result and the answer to part (a), we conclude that point B is moving to the *left* at 2 m/s faster than point C. Therefore the cord is being wound up at the rate of 2 m/s. What conclusion would you reach if the result was $\bar{v}_{C/B} = -2\bar{i}$?

Steps 10–16   From part (a) we have the general equation

$$\bar{v}_B = -1.5\omega_s \bar{i}$$

Because this equation is always true, we can differentiate it. Therefore

$$\bar{a}_B = \dot{\bar{v}}_B = -1.5\dot{\omega}_s \bar{i} = -1.5\,\alpha_s \bar{i}$$

From step 8 we know that $\alpha_s = 4$ rad/s^2, so

$$\bar{a}_B = -6\bar{i} \text{ m/s}^2$$

ILLUSTRATIVE PROBLEM 4

Blocks at both ends of rigid bar AB slide in grooves. At the instant shown, block B is moving upward at a constant speed of 2.5 m/s. Determine the acceleration of the midpoint C of bar AB for this position.

Solution
Step 1

Step 2 Given $\bar{v}_B = 2.5\bar{j}$ m/s, $\bar{a}_B = \bar{0}$. Find $\bar{a}_C$.

Step 3 We shall solve the problem using the vector equation approach. Point B must follow the vertical groove, and point A must follow the circular one. Thus

$$\bar{v}_A = \bar{v}_B + \bar{\omega}_{AB} \times \bar{r}_{A/B}$$

Step 4 We were given $\bar{v}_B = 2.5\bar{j}$. From path variables we know that $\bar{v}_A$ is tangent to the circle, so that

$$\bar{v}_A = v_A \bar{e}_t$$

The unit vectors $\bar{e}_t$ and $\bar{e}_n$ are shown in the sketch of step 1. From the given geometry we may write

$$\bar{e}_t = \cos 30° \, \bar{i} + \sin 30° \, \bar{j} = 0.866\bar{i} + 0.5\bar{j}$$

$$\bar{e}_n = -\sin 30° \, \bar{i} + \cos 30° \, \bar{j} = -0.5\bar{i} + 0.866\bar{j}$$

The unit vectors $\bar{e}_t$ and $\bar{e}_n$ are expressed by their components in the x and y directions because we shall need the components of $\bar{v}_A$ and $\bar{a}_A$ in these directions. Thus

$$\bar{v}_A = v_A(0.866\bar{i} + 0.5\bar{j})$$

Step 5 $\bar{\omega}_{AB} = \omega_{AB}\bar{k}$ (ω_{AB} is positive if counterclockwise).

$$\bar{r}_{A/B} = -0.96\bar{i} - 0.72\bar{j} \text{ m}$$

Step 6 Substitute the vectors from steps 4 and 5 into the velocity equation in step 3.

$$v_A(0.886\bar{\imath} + 0.5\bar{\jmath}) = 2.50\bar{\jmath} + \omega_{AB}\bar{k} \times (-0.96\bar{\imath} - 0.72\bar{\jmath})$$

$$= 0.72\omega_{AB}\,\bar{\imath} + (2.50 - 0.96\omega_{AB})\bar{\jmath}$$

Step 7   Match components.

$$\bar{\imath} \text{ direction:} \qquad 0.866v_A = 0.72\omega_{AB}$$

$$\bar{\jmath} \text{ direction:} \qquad 0.5v_A \;\;= 2.50 - 0.96\omega_{AB}$$

We have two equations and two unknowns.

Step 8   Solve the equations.

$$v_A = \frac{0.72}{0.866}\,\omega_{AB} = 0.831\omega_{AB}$$

$$(0.5)0.831\omega_{AB} = 2.50 - 0.96\omega_{AB}$$

$$1.375\omega_{AB} = 2.50$$

$$\omega_{AB} = \frac{2.50}{1.375} = 1.818 \text{ rad/s}$$

$$v_A = 1.511 \text{ m/s}$$

Step 9   We are not asked for $\bar{v}_C$, so we do not compute it.

Step 10   We repeat the process for accelerations. Points A and B are constrained, so we write

$$\bar{a}_A = \bar{a}_B + \bar{\alpha}_{AB} \times \bar{r}_{A/B} + \bar{\omega}_{AB} \times (\bar{\omega}_{AB} \times \bar{r}_{A/B})$$

Step 11   We were given $\bar{a}_B = \bar{0}$. From path variables we know that

$$\bar{a}_A = \dot{v}_A\bar{e}_t + \frac{v_A{}^2}{R}\,\bar{e}_n$$

From step 3 we know $\bar{e}_t$ and $\bar{e}_n$, and we know v_A from step 8. Thus

$$\bar{a}_A = \dot{v}_A(0.866\bar{\imath} + 0.5\bar{\jmath}) + \frac{(1.511)^2}{0.50}(-0.5\bar{\imath} + 0.866\bar{\jmath})$$

$$= (0.866\dot{v}_A - 2.283)\bar{\imath} + (0.5\dot{v}_A + 3.954)\bar{\jmath}$$

Step 12   The only additional vector needed is

$$\bar{\alpha}_{AB} = \alpha_{AB}\bar{k}$$

Step 13   Substitute the vectors from steps 5, 11, and 12, and the value of ω_{AB} from step 8 into the acceleration equation in step 10.

$$(0.866\dot{v}_A - 2.283)\bar{i} + (0.5\dot{v}_A + 3.954)\bar{j}$$

$$= \bar{0} + \alpha_{AB}\bar{k} \times (-0.96\bar{i} - 0.72\bar{j})$$

$$+ 1.818\bar{k} \times [1.818\bar{k} \times (-0.96\bar{i} - 0.72\bar{j})]$$

$$= \alpha_{AB}(-0.96\bar{j} + 0.72\bar{i})$$

$$+ (1.818)^2(0.96\bar{i} + 0.72\bar{j})$$

$$= (0.72\alpha_{AB} + 3.173)\bar{i} + (-0.96\alpha_{AB} + 2.38)\bar{j}$$

Step 14 Match components.

$\bar{i}$ direction: $\qquad 0.866\dot{v}_A - 2.283 = 0.72\alpha_{AB} + 3.173$

$\bar{j}$ direction: $\qquad 0.5\dot{v}_A + 3.954 = -0.96\alpha_{AB} + 2.38$

We have two equations and two unknowns.

Step 15 Solve the equations. To find α_{AB}, multiply the first equation of step 14 by 0.5 and the second by 0.866.

$$0.5(0.866\dot{v}_A - 2.283) = 0.5(0.72\alpha_{AB} + 3.173)$$

$$0.866(0.5\dot{v}_A + 3.954) = 0.866(-0.96\alpha_{AB} + 2.38)$$

Now we subtract the second from the first.

$$-4.566 = 1.1914\alpha_{AB} - 0.4746$$

$$\alpha_{AB} = -3.434 \text{ rad/s}^2$$

Then solving for $\dot{v}_A$ gives

$$\dot{v}_A = 3.444 \text{ m/s}^2$$

Step 16 We now can find the acceleration of the unconstrained point C by relating it to either point A or point B. In this problem point B is the easier choice because it has no acceleration.

$$\bar{a}_C = \bar{a}_B + \bar{\alpha}_{AB} \times \bar{r}_{C/B} + \bar{\omega}_{AB} \times (\bar{\omega}_{AB} \times \bar{r}_{C/B})$$

From the diagram in step 1,

$$\bar{r}_{C/B} = -0.48\bar{i} - 0.36\bar{j}$$

Then substituting the results for $\bar{\omega}_{AB}$ and $\bar{\alpha}_{AB}$ gives

$$\bar{a}_C = \bar{0} - 3.434\bar{k} \times (-0.48\bar{i} - 0.36\bar{j}) + 1.818\bar{k}$$

$$\times [1.818\bar{k} \times (-0.48\bar{i} - 0.36\bar{j})]$$

$$= 1.648\bar{j} - 1.236\bar{i} + 1.587\bar{i} + 1.190\bar{j}$$

$$= 0.350\bar{i} + 0.284\bar{j} \text{ m/s}^2$$

To reinforce your confidence, we should note that this is one of the more complicated problems you can expect to see in the kinematics of rigid bodies in planar motion. If you followed all of the development, you have established a good level of understanding. If some of the steps gave you trouble, you should go back and review the corresponding material in the text.

HOMEWORK PROBLEMS

IV.42 The circular cam is mounted eccentrically on a fixed axis of rotation passing through point O. Express the velocity and acceleration of the follower in terms of the angle θ. *Hint:* Express the distance y in terms of the angle and the constant angular speed ω θ.

IV.43 The block is moving to the right at a constant speed of 0.5 m/s. Determine the angular velocity and angular acceleration of the 400-mm rod as a function of θ.

IV.44 The lower rack has a constant velocity of 6 in./s to the right. The upper rack has an acceleration of 8 in./s² to the right, and an instantaneous velocity of 2 in./s to the right. Determine (a) the angular velocity and angular acceleration of the gear, (b) the acceleration of point C.

Prob. IV.42

Prob. IV.43

Prob. IV.44

IV.45 The cable is unwrapped from the rim of the stepped disk as the block is lowered. The angular velocity of the disk is 12 rad/s and the angular acceleration is 6 rad/s², both counterclockwise. Assuming that there is no slippage and that the cable remains vertical, determine (a) the velocity of the block, and (b) the acceleration of the block.

IV.46 Block A has fallen 1 ft from rest with a contant acceleration of 16 ft/s². There is no slippage of the cables as they pass over the stepped pulley B. For this instant, determine (a) the angular velocity of pulley B, (b) the relative velocity of block A with respect to the center of pulley B, and (c) the velocity of block C.

Prob. IV.46

1.6 ft

3.2 ft

1.6 ft

θ

Prob. IV.48

IV.47   For the system of Problem IV.46, determine (a) the angular acceleration of pulley B, (b) the relative acceleration of block A with respect to the center of pulley B, and (c) the acceleration of block C.

IV.48   The angle of rotation of the flywheel is given by $\theta = 30t + 180t^2$, where θ is measured in degrees and t in seconds. Determine (a) the velocity of the piston when $t = 1$ s, (b) the acceleration of the piston when $t = 1$ s.

IV.49   Solve Problem IV.48 for $t = \frac{1}{2}$ s.

IV.50   Solve Problem IV.48 for $t = \frac{1}{3}$ s.

IV.51   The angular velocity of bar AB is 5 rad/s. Determine the velocity of slider C at the instant depicted in the diagram.

IV.52   If bar AB in Problem IV.51 has an angular deceleration of 2 rad/s^2 in the position illustrated, and all other parameters are as given, determine the acceleration of slider C.

IV.53   Solve Illustrative Problem 2 by writing the appropriate position vector and differentiating.

IV.54   In the position shown, slider A has a velocity of 20 ft/s to the left. Determine the angular velocities of bars AB and BC.

B

400 mm

5 rad/s

600 mm

75°

45°

A

60°

C

Prob. IV.51

C

B

45°

60°

2 ft

60°

45°

$v_A = 20$ ft

A

Prob. IV.54

IV.55   In Problem IV.54 the acceleration of slider A in the given position is 40 ft/s^2 to the right. Determine the angular accelerations of both bars.

IV.56   A side view of three identical panels from a garage door is shown in the diagram. The bottom is being raised with a speed of 7.50 m/s. For the position depicted, determine the speed of (a) roller B, (b) roller D.

500 mm   500 mm   500 mm

B C D

60°

A

$R = 500$ mm

Prob. IV.56

IV.57 Determine the acceleration of roller D for the system of Problem IV.56.

IV.58 A portion of the mechanism of a typewriter is shown in the diagram. In this position the downward components of the velocity and acceleration of the key are 200 mm/s and 400 mm/s², respectively. Determine the corresponding values of the velocity and acceleration of point E.

Prob. IV.58

Prob. IV.59

IV.59 The flywheel has a constant clockwise angular velocity $\bar{\omega}$. Determine the velocity and acceleration of the midpoint E of bar BC.

IV.60 The angle of rotation of bar AB is given by $\varphi = (\pi/4)(1 - \cos 5\pi t)$, where the units of φ and t are radians and seconds, respectively. Determine the velocity and acceleration of point C when $\varphi = 0°$.

IV.61 In the position shown bar AB is rotating at a constant rate of 120 rev/min. Determine the velocity of the midpoint G of bar BC at this instant.

Prob. IV.60

Prob. IV.61

IV.62 Determine the acceleration of point G for the system of Problem IV.61.

IV.63 The diagram shows a possible mechanism for a log-cutting saw. The flywheel rotates at the constant rate of 60 rev/min. Determine the velocity and acceleration of the lowest point on the perimeter of the circular saw blade when $\theta = 0°$.

Prob. IV.63

IV.64 Arm AB in the escapement mechanism rotates clockwise at the constant rate of 6 rev/min. For the position shown, determine (a) the velocity and acceleration of the ratchet tooth E, and (b) the angular velocity and angular acceleration of the gear.

Prob. IV.64

IV.65 Two wheels which are interconnected by rod AC, roll along the ground without slipping. Wheel A has a constant velocity of 5 m/s to the right. Determine the velocity and acceleration of wheel B in the position where $\theta = 0$.

Prob. IV.65

IV.66 For the system of Problem IV.65, determine the velocity and acceleration of wheel B in the position where $\theta = 120°$.

IV.67   The machine shown in the sketch is used to test the wear of railroad wheels and tracks. The flywheel has a constant counterclockwise angular velocity of 600 rev/min. There is no slippage between the railroad wheel and the track. Determine (a) the angular velocity and (b) the angular acceleration of the wheel in the illustrated position. The diameter of the wheels is 1.2 m.

Prob. IV.67

IV.68   A truck is being driven up a 20° incline. The speedometer, which measures the speed of the rear wheel, indicates a constant 8 km/hr, and the wheels are not skidding. Determine the values of the angular velocities of the front and rear wheels and of the vehicle body for the position illustrated in the diagram. The diameter of the wheels is 0.9 m.

Prob. IV.68

IV.69   Determine the angular accelerations of the front and rear wheels and of the vehicle body for the truck in Problem IV.68.

IV.70   The slider has a constant velocity of 0.7 m/s to the right, thus causing gear A to move over the fixed gear B. Determine the angular velocity of gear A in this position.

IV.71   Determine the angular acceleration of gear A in Problem IV.70.

Prob. IV.70

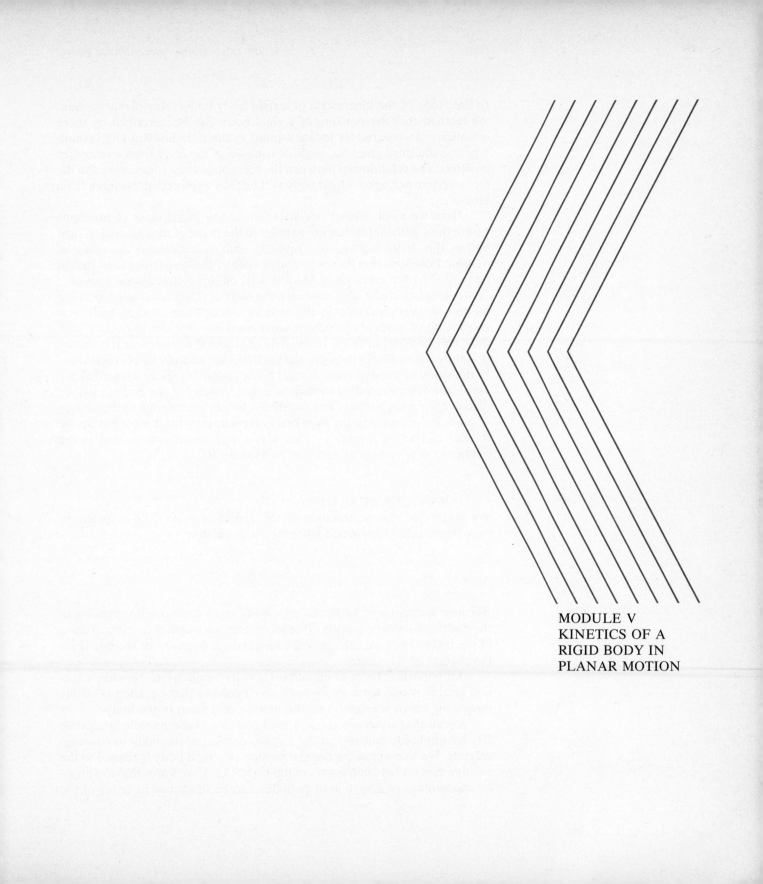

MODULE V
KINETICS OF A
RIGID BODY IN
PLANAR MOTION

In the study of the kinematics of a rigid body undergoing planar motion, we learned that the position of a rigid body can be described by three quantities: Two variables locate a point in the rigid body at any instant, whereas the third gives the angle of rotation of the body from a reference position. The relationship between the foregoing three parameters and the force system acting on a rigid body will be fully explored in this module on kinetics.

Here we shall restrict our attention to the broad class of problems where the significant forces are parallel to the plane of motion, and to rigid bodies that have a plane of symmetry that is parallel to the plane of motion. Problems that do not fit one or both of these specifications should be treated by the methods of Module VII, on three-dimensional motion.

A kinetics study must consider the inertial effect associated with the motion of every particle in the system. In essence, a rigid body is a continuum of particles of infinitesimal mass dm that are in contact with their neighboring particles. In the field of strength of materials it is shown that the contact forces between the particles are actually stress resultants. In the study of the dynamics of rigid bodies these forces are treated as the constraint forces required to maintain the particles of the body at a fixed distance from each other. The constraint forces are internal to the system of particles composing the rigid body. Hence, they have no effect on the overall motion of the body. This was a key result emphasized in the treatment of a system of particles in Module III.

A. EQUATIONS OF MOTION

We know that the acceleration of the center of mass G of a system of particles can be determined from the basic relation

$$\Sigma \bar{F} = m\bar{a}_G \tag{1}$$

Because the center of mass of a rigid body has a fixed position relative to the body, the use of equation (1) requires that we locate the center of mass of the body. Then we can apply the kinematical formulas of Module II for the motion of this point.

An important facet of equation (1) is its similarity to Newton's second law, $\bar{F} = m\bar{a}$, for a *single particle*. It follows that equation (1) is the means by which we determine the motion of a point in the body.

Recall that when a body was modeled as a single particle in Module III, we implicitly considered the angular motion of the body to be insignificant. We know that the angular motion of a rigid body is related to the relative motion between points in the body. Also, we know that the angular momentum of a system of particles can be described in terms of the relative velocity of the particles. Hence, we shall employ the angular momentum principle to study the rotation of a rigid body.

The simplest form for the basic relationship between the angular momentum of a system of particles and the resultant moment of the force system is

$$\Sigma \, \bar{M}_A = \dot{\bar{H}}_A \tag{2}$$

where point A is either (a) the center of mass, (b) a fixed point, or (c) a point accelerating directly toward or away from the center of mass. In treating conditions (b) and (c) we shall, for convenience, now impose the additional restriction that *point A must be fixed to the body.* This will enable us to use the relative motion concepts of the previous module.

Referring to the rigid body shown in Figure 1, where m_i denotes the element of mass located at point i, the angular momentum vector $\bar{H}_A$ in equation (2) is defined by

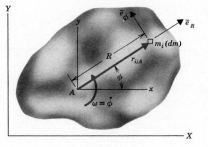

Figure 1

$$\bar{H}_A = \sum_{i=1}^{N} \bar{r}_{i/A} \times m_i \bar{v}_{i/A} \tag{3}$$

If we consider the rigid body of Figure 1 to be composed of an infinite number of infinitesimal masses $dm = m_i$, then the summation in equation (3) becomes an integral over all of the masses dm in the volume $\mathcal{V}$ of the body. Hence

$$\bar{H}_A = \int_{\mathcal{V}} (\bar{r}_{i/A} \times \bar{v}_{i/A}) \, dm \tag{4}$$

To describe the position and velocity of the differential element dm at point i relative to point A, we introduce a translating reference frame system xyz whose origin is placed coincident with point A. As postulated earlier, the body has a plane of symmetry parallel to the plane of motion. Let us choose the xy plane to coincide with this symmetry plane, so that the z axis is normal to the plane of motion in accordance with the right-hand rule. Then, by defining cylindrical coordinates with respect to this reference frame, as shown in Figure 1, we have

$$\bar{r}_{i/A} = R \bar{e}_R$$

To obtain the relative velocity we apply the velocity relationship for two points in a rigid body undergoing planar motion, which gives

$$\bar{v}_i = \bar{v}_A + \bar{\omega} \times \bar{r}_{i/A}$$

Thus

$$\bar{v}_{i/A} \equiv \bar{v}_i - \bar{v}_A = \bar{\omega} \times \bar{r}_{i/A}$$

Because $\bar{e}_h \equiv \bar{k}$ for the chosen reference frame,

$$\bar{\omega} = \omega \bar{e}_h$$

$$\bar{\omega} \times \bar{r}_{i/A} = \omega R \bar{e}_\phi$$

so that

$$\bar{r}_{i/A} \times \bar{v}_{i/A} = \omega R^2 \bar{e}_h = \omega R^2 \bar{k} = R^2 \bar{\omega}$$

We now substitute this result into equation (4) for the angular momentum, noting that the angular velocity $\bar{\omega}$ is a function of time only, so it may be removed from the integral over the volume of the body. Hence we have

$$\bar{H}_A = \int_{\mathscr{V}} \bar{r}_{i/A} \times \bar{v}_{i/A}\, dm = \bar{\omega} \int_{\mathscr{V}} R^2\, dm$$

$$\boxed{\bar{H}_A = I_{zz}\bar{\omega}} \tag{5}$$

where, in view of the basic relationship between the polar coordinate R and the Cartesian coordinates x and y (evident from Figure 1),

$$\boxed{I_{zz} = \int_{\mathscr{V}} R^2\, dm = \int_{\mathscr{V}} (x^2 + y^2)\, dm} \tag{6}$$

The quantity I_{zz} is called the *second moment of mass,* or alternatively, the *moment of inertia* of the body about the z axis. Remember that the z axis is perpendicular to the plane of motion, passing through the point A that was selected for the moment sum. A significant property of the moment of inertia I_{zz}, as derived above, is that it is a constant for the rigid body. This is so because we required point A to be a point that is fixed relative to the body, thus making the radial distance R to each mass element constant. We shall consider further properties of, and computational procedures for, moments of inertia in the next section. As an aside, the double subscript in I_{zz} has no significance for the analysis of problems in planar kinetics; it is introduced here in order to present a notation that is consistant with that used for studying the kinetics of rigid bodies in three-dimensional motion.

Returning now to the derivation of the equations of motion, by differentiating equation (5) with respect to time we may determine $\dot{\bar{H}}_A$. Thus, noting that the time derivative of the angular velocity $\bar{\omega}$ is the angular acceleration $\bar{\alpha}$, and that I_{zz} is a constant, differentiating equation (5) yields

$$\dot{\bar{H}}_A = I_{zz}\dot{\bar{\omega}} = I_{zz}\bar{\alpha}$$

Hence, equation (2) becomes

$$\boxed{\Sigma \, \bar{M}_A = I_{zz}\bar{\alpha}} \tag{7}$$

Equation (7) represents the third of the three scalar equations necessary to describe the three independent position variables of a rigid body in planar motion. Two equations are obtained from the planar components of the force equation (1), and the third equation is obtained by determining the moment of the planar force system about point A and applying equation (7).

An important feature of equation (7) is the similarity of its component form to the equation for rectilinear motion of a particle that we studied in Module Three. Thus, in situations where we wish to relate the angle of rotation ϕ, the angular speed ω, and time, we need only obtain a general expression for the moment of the force system and apply the identity $\alpha = \dot{\omega} = \ddot{\phi}$. This will allow us to form a differential equation of motion for the angle ϕ.

Before we can write the equations of motion for a specific problem, we must establish procedures for the determinations of the location of the center of mass and of the moment of inertia. These topics are the subjects of the next section. We shall close this section with a brief review regarding the computation of the moment for a planar force system.

Consider the force $\bar{F}$ shown in Figure 2. The moment $\bar{M}_A$ of the force $\bar{F}$ about point A may be determined by the vector operation

$$\bar{M}_A = \bar{r}_{P/A} \times \bar{F}$$

where point P is a point on the line of action of $\bar{F}$. In terms of the components of $\bar{r}_{P/A}$ illustrated in Figure 2, this moment is

$$\bar{M}_A = (x_P \, \bar{\imath} + y_P \, \bar{\jmath}) \times (F_x \, \bar{\imath} + F_y \, \bar{\jmath})$$

$$= (x_P F_y - y_P F_x)\bar{k}$$

Figure 2

This result can also be obtained directly by multiplying each force component by its lever arm about point A, the latter quantity being the perpendicular distance from point A to the line of action of the force component. To obtain the proper sign we note that, for the axes shown, the z axis is outward. Hence, in accord with the right-hand rule, counterclockwise moments are positive. For planar problems this approach usually proves as efficient as the cross product.

B. INERTIA PROPERTIES

Let us begin by considering the physical meaning of the center of mass and moment of inertia. The center of mass of a body may be determined experimentally by suspending it by a cable from two points. For consis-

Figure 3a

Figure 3b

Figure 4

tency with the concepts of planar motion, let the body be symmetric with respect to the xy plane, and let points A and B be two points that coincide with this plane, as shown in Figure 3a. When suspended from point A, the body assumes a specific orientation with respect to the direction of the earth's gravitational field, the vertical line AC being the line of action of the weight force. When the body is suspended from point B, it assumes a different orientation with respect to the direction of the earth's gravitational field. This line of action of the weight force is labeled BD in Figure 3b. The point of intersection of lines AC and BD is called the *mass center*, G, or synonymously, the *center of gravity*. If the body is suspended from still another point, the vertical line of action would also intersect point G.

From the foregoing experiment we deduce that we can locate the center of mass for a specific body by considering any two orientations of the body with respect to the earth's gravitational field. The orientations we shall employ, for computation purposes, will always be 90° apart.

A moment of inertia (I_{zz}) conveys information regarding the distribution of the mass of a body about the z axis. For instance, for the thin circular ring in Figure 4 each element of mass is essentially situated along the perimeter. Hence the distance from the z axis to each mass element is the radius r of the ring, so that

$$I_{zz} = \int_{\mathcal{V}}(x^2 + y^2)\, dm = r^2\int_{\mathcal{V}} dm = mr^2$$

Clearly, if two rings have the same mass, then the one with the larger radius has the larger moment of inertia.

Dividing the value of I_{zz} by the mass and then computing the square root of that quotient, we obtain the *radius of gyration* about the z axis,

$$k_z = \sqrt{\frac{I_{zz}}{m}} \tag{8}$$

Because $k_z = r$ for the ring in Figure 4, we can think of the radius of gyration of any body as the radius of a ring having the same inertia properties as the body under consideration. Note that the integrand in equation (6) is always positive, so that I_{zz} will always be positive and hence k_z is a real number.

An understandable confusion arises between the concepts of *mass moment of inertia* and *second moment of area*. The second moment of area plays an important role in fields such as strength of materials, because it arises in the derivation for the determination of the stress distribution within materials. In our study of inertia properties we are concerned with the *polar* second moment of area. To avoid confusion with

Figure 5

the symbols for mass properties, we shall denote this quantity by J_z. In what follows we shall compare I_{zz} and J_z.

For an area $\mathcal{A}$, shown in Figure 5, the definition of J_z is

$$J_z \equiv \int_{\mathcal{A}} (x^2 + y^2)\, d\mathcal{A} \tag{9}$$

The similarity of this quantity to the moment of inertia I_{zz} is obvious, because J_z measures the distribution of the area about the z axis.

Now let us consider a three-dimensional body whose outline in the xy plane is the area $\mathcal{A}$ of Figure 5. Let the thickness of the mass element dm in the z direction be h. In order to study the motion of this body by the methods of planar kinetics, we shall limit the analysis to cases where the (mass) density ρ depends only on the location of the differential mass elements with the xy plane. It then follows that

$$dm = \rho\, d\mathcal{V} = \rho h\, d\mathcal{A} \tag{10}$$

and hence

$$\boxed{m = \int_{\mathcal{V}} dm = \int_{\mathcal{A}} \rho h\, d\mathcal{A}} \tag{11}$$

$$\boxed{I_{zz} = \int_{\mathcal{V}} (x^2 + y^2)\, dm = \int_{\mathcal{A}} (x^2 + y^2)\, \rho h\, d\mathcal{A}} \tag{12}$$

Note that ρ and h were left in the integrand in the foregoing equations. If the body has an arbitrary shape and an arbitrary mass distribution, then ρ and h are not constant, and hence cannot be factored out of integrals that extend over the area $\mathcal{A}$. However, in the case when the body is *homogeneous,* so that ρ is constant, and when it has a constant thickness h, we then find that

$$m = \rho h \int_{\mathcal{A}} d\mathcal{A} = \rho h \mathcal{A} \tag{13}$$

$$I_{zz} = \rho h \int_{\mathcal{A}} (x^2 + y^2)\, d\mathcal{A} \equiv \rho h J_z \tag{14}$$

Equation (13) is the familiar result for the mass of a body with constant thickness, and equation (14) shows that in this special case the moment of inertia I_{zz} is proportional to the polar moment of area J_z. A more subtle point to note is that only in this special case of constant ρ and h does the center of mass of the body *always* coincide with the centroid of the area $\mathcal{A}$.

In general, unless we are given the value of the moment of inertia or the radius of gyration, the determination of I_{zz} becomes part of the task in the solution of problems. Also, with the exception of those cases where the geometry is such that we locate the center of mass by inspection, locating this point becomes part of our task. We shall now study some techniques for obtaining these quantities.

1 Inertia Properties by Integration

The fundamental method for determining the center of mass and moment of inertia is based on the definitions of these properties in terms of integrals.

To describe the position of the mass element we may use either rectangular Cartesian or polar coordinates. Remember that the two are interchangeable, because we can always express the transformation between the coordinate systems. For example, for the coordinate systems of Figure 5 we have

$$x = R \cos \phi \qquad y = R \sin \phi$$

Similarly, the cross-sectional area $d\mathcal{A}$ expressed in terms of Cartesian coordinates is

$$d\mathcal{A} = dx \, dy$$

whereas in terms of polar coordinates it is

$$d\mathcal{A} = R \, dR \, d\theta$$

We generally base the choice of coordinate system on whether the curve defining the perimeter of the area $\mathcal{A}$ is more easily expressed in terms of Cartesian or polar coordinates.

If the mass of the body is not given, this value may be obtained by evaluating the integral in equation (11). With the total mass determined, we can then locate the center of mass by using the result of the experiment we described in Figure 3 for determining its location. When the y axis in Figure 5 is vertical, the gravitational force on each element of mass is $(dm)g$ parallel to the y axis, so the moment about the z axis for each element is $x(dm)g$. Denoting the coordinates of the center of mass as (x_G, y_G), the total moment of the gravity force is $x_G mg$ because the resultant gravity force acts at this point. Equating this moment to the total moment of the force on each element, we obtain the *first moment of mass* with respect to the x coordinate,

$$mx_G = \int_{\mathcal{A}} x(\rho h \, d\mathcal{A}) \tag{15a}$$

Similarly, the first moment of mass with respect to the y coordinate is obtained if we imagine that gravity is aligned with the x axis. This yields

$$my_G = \int_{\mathcal{A}} y(\rho h \, d\mathcal{A})$$ (15b)

After evaluating the integrals in these expressions, we may solve for the coordinates (x_G, y_G). Conversely, equations (15) demonstrate that when the center of mass is situated on one of the coordinate axes, the first moment of mass with respect to the other coordinate is zero. This fact will prove useful for later derivations.

There is little theoretical discussion necessary to employ equation (12) to determine I_{zz}. Let us now study three examples that review the techniques of multiple integration.

EXAMPLE 1
A $\frac{3}{4}$-in. sheet of soft pine wood is cut into the triangular shape shown in the diagram. Using the methods of integration, locate the center of mass of the sheet and then determine the value of I_{zz} for the coordinate axes shown.

Solution
Organizing our thinking, we must first determine the mass of the system, then the coordinates of the mass center, and finally I_{zz}. Because the edges of this body are all straight, the decision to formulate the solution in terms of rectangular Cartesian coordinates is obvious. We begin by describing the boundaries of the body as functions of x and y, as is done in the sketch to the left. (We used the straight-line formula, $y = mx + b$, to describe the hypotenuse.)

As the sketch indicates, we shall first form a vertical differential strip by integrating in the y direction from $y = 0$ to $y = 1.5 - 3x/4$. Then all strips will be summed by integrating from $x = 0$ to $x = 2$ ft. The thickness of the sheet is constant at $h = 3/48$ ft, and the density of soft pine is $\rho = 0.933$ lb-s^2/ft^4. Thus the mass of the body is

$$m = \rho h \int_0^2 \int_0^{1.5-3x/4} dy \, dx = \rho h \int_0^2 \left(1.5 - \frac{3x}{4}\right) dx$$

$$= \rho h [1.5(2) - \tfrac{3}{8}(2)^2] = 0.08743 \text{ lb-s}^2\text{/ft}$$

To locate the center of mass, we use the first moment of mass, equations (15), which gives

$$mx_G = \rho \int_0^2 \int_0^{1.5-3x/4} x \, dy \, dx = \rho h \int_0^2 x\left(1.5 - \frac{3x}{4}\right) dx$$

$$= 0.05828 \text{ lb-s}^2$$

$$my_G = \rho h \int_0^2 \int_0^{1.5-3x/4} y \, dy \, dx = \rho h \int_0^2 \frac{1}{2}\left(1.5 - \frac{3x}{4}\right)^2 dx$$

$$= 0.04371 \text{ lb-s}^2$$

Solving for x_G and y_G, we obtain

$$(x_G, y_G) = (0.667, \ 0.500) \text{ ft}$$

Finally, using equation (12), the moment of inertia is

$$I_{zz} = \rho h \int_0^2 \int_0^{1.5-3x/4} (x^2 + y^2) \, dy \, dx$$

$$= \rho h \int_0^2 \left[x^2\left(1.5 - \frac{3x}{4}\right) + \frac{1}{3}\left(1.5 - \frac{3x}{4}\right)^3 \right] dx$$

$$= 0.0911 \text{ lb-s}^2\text{-ft}$$

EXAMPLE 2

A thin wire having a constant cross-sectional area A is bent into the form of a circular arc, as shown in the sketch. Determine (a) the center of mass and, (b) the value of I_{zz} for the coordinate axes shown.

Solution

Because the wire is thin, we can consider its mass to be distributed along the centerline of the cross section, so that the differential mass element dm is defined by the infinitesimal arc length ds. The circular arc is easily described in terms of the polar coordinates defined in the following sketch.

For this set of polar coordinates we have

$$x = r \sin \phi \qquad y = r - r \cos \phi \qquad ds = r \, d\phi$$

Then, because dm has cross-sectional area A and length ds, it follows that

$$dm = \rho A \, ds = \rho A r \, d\phi$$

In order to locate the center of mass, we first require the total mass, which is

$$m = \int dm = \rho A r \int_{-\theta/2}^{\theta/2} d\phi = \rho A r \theta$$

From the symmetry property of the system we note that the center of gravity is situated along the y axis, so $x_G = 0$. Therefore, we need only determine the first moment of mass with respect to the y coordinate. Using the polar coordinate expression for the value of y then gives

$$my_G = \int y \, dm = \rho A r \int_{-\theta/2}^{\theta/2} r(1 - \cos \phi) \, d\phi$$

$$= \rho A r^2 \left(\theta - 2 \sin \frac{\theta}{2} \right)$$

Dividing by m results in

$$y_G = \frac{\rho A r^2 (\theta - 2 \sin \theta/2)}{\rho A r \theta} = r\left(1 - \frac{2}{\theta} \sin \frac{\theta}{2} \right)$$

or

$$(x_G, y_G) = \left[0, r\left(1 - \frac{2}{\theta} \sin \frac{\theta}{2} \right) \right]$$

Lastly, we compute the moment of inertia as follows.

$$I_{zz} = \int (x^2 + y^2) \, dm = \rho A r \int_{-\theta/2}^{\theta/2} [r^2 \sin^2 \phi + r^2(1 - \cos \phi)^2] \, d\phi$$

$$= 2\rho A r^3 \left(\theta - 2 \sin \frac{\theta}{2} \right)$$

To eliminate the density ρ, we may solve for ρ in the result for the total mass, which gives

$$\rho = \frac{m}{A r \theta}$$

so that

$$I_{zz} = 2\left(\frac{m}{Ar\theta}\right)Ar^3\left(\theta - 2\sin\frac{\theta}{2}\right)$$

$$= 2mr^2\left(1 - \frac{2}{\theta}\sin\frac{\theta}{2}\right)$$

EXAMPLE 3

Determine, by integration, the moment of inertia of a homogeneous sphere of radius a about an axis passing through the center of the sphere.

Solution

The problem requires the determination of I_{zz}, where the z axis passes through the center of the sphere. Hence, we locate the origin of the xyz system at the center. This also allows us to utilize the symmetry properties of a sphere. The shape of a sphere makes it logical to use polar coordinates to locate the differential mass elements. The significant variables are shown in the diagram below.

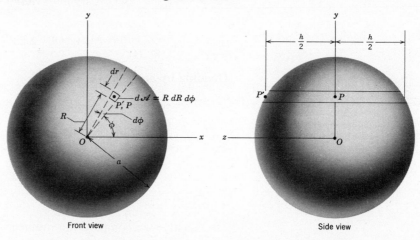

Front view   Side view

In this sketch the point P locates where the mass element intersects the xy plane and point P' locates where the mass element intersects the surface of the sphere. The length of the line OP is $R = \sqrt{x^2 + y^2}$, and the length of the line OP' is the radius a. Thus, because the mass element is perpendicular to the xy plane, the Pythagorean theorem gives

$$\frac{h}{2} = \sqrt{a^2 - R^2}$$

To obtain the contribution of all mass elements we first integrate over the range of ϕ from 0 to 2π, and then integrate over the range of R from 0 to a. Noting that ρ is a constant for a homogeneous body, we have

$$I_{zz} = \rho \int_0^a \int_0^{2\pi} R^2 h(R\ dR\ d\phi) = 2\rho \int_0^a \int_0^{2\pi} R^3 \sqrt{a^2 - R^2}\ d\phi\ dR$$

$$= 2\rho(2\pi) \int_0^a R^3 \sqrt{a^2 - R^2}\ dR$$

$$= \frac{8}{15}\ \pi\rho a^5$$

The last integral was evaluated with the aid of a table of integrals. In order to eliminate ρ, we note that the mass of a sphere is

$$m = \rho\mathscr{V} = \frac{4}{3}\ \pi\rho a^3$$

so that

$$\rho = \frac{3}{4\pi}\ \frac{m}{a^3}$$

Hence

$$I_{zz} = \frac{2}{5}\ ma^2$$

Each of the foregoing examples contained a body that can be termed an elementary shape. Any mechanical system can be approximated by a combination of elementary shapes. Some can be duplicated exactly. We shall concern ourselves with the latter. In such cases it is very useful to compute the centroidal and inertia properties of the shapes and compile them for future use. We present a typical standard table of properties in Appendix B.

In the next section we shall develop a formula that is of great value in inertia computations, especially when we wish to use the standard table of properties.

HOMEWORK PROBLEMS

By direct integration, determine the location of the center of mass and I_{zz} for each of the homogeneous bodies (of mass m) shown.

V.1   Slender Rod

$$I_{zz} = \tfrac{1}{3}ml^2$$

Prob. V.1

V.2 Rectangular plate

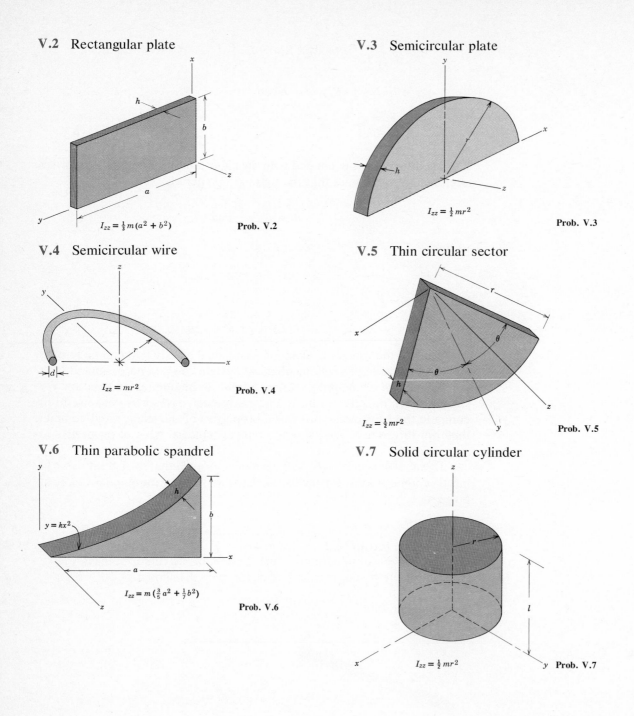

$$I_{zz} = \tfrac{1}{3} m (a^2 + b^2)$$

Prob. V.2

V.3 Semicircular plate

$$I_{zz} = \tfrac{1}{2} mr^2$$

Prob. V.3

V.4 Semicircular wire

$$I_{zz} = mr^2$$

Prob. V.4

V.5 Thin circular sector

$$I_{zz} = \tfrac{1}{2} mr^2$$

Prob. V.5

V.6 Thin parabolic spandrel

$y = kx^2$

$$I_{zz} = m \left(\tfrac{3}{5} a^2 + \tfrac{1}{7} b^2 \right)$$

Prob. V.6

V.7 Solid circular cylinder

$$I_{zz} = \tfrac{1}{2} mr^2$$

Prob. V.7

V.8   Solid cone

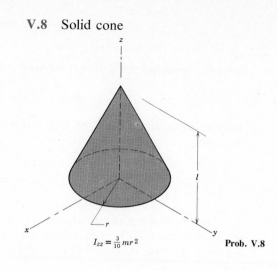

$I_{zz} = \frac{3}{10} mr^2$ **Prob. V.8**

2   The Parallel Axis Theorem

Standard tables of properties generally give the moment of inertia of a body about only one set of axes, usually axes that pass through the center of mass. Suppose that we know $I_{\hat{z}\hat{z}}$, where the $\hat{z}$ axis is perpendicular to the plane of motion of a rigid body and passes through its center of mass, as shown in Figure 6. It is our desire to establish the relation of this moment of inertia to I_{zz}, where the z axis is also perpendicular to the plane of motion (hence it is parallel to the $\hat{z}$ axis). Note that in Figure 6 we selected the two coordinate systems to contain mutually parallel axes. This proves to be an important convenience in the work to follow.

Letting the xy and $\hat{x}\hat{y}$ planes coincide, from Figure 6 we may write

Figure 6

$$x = \hat{x} - \hat{x}_A \qquad y = \hat{y} - \hat{y}_A$$

where $(\hat{x}_A, \hat{y}_A, 0)$ are the coordinates of the origin A of the xyz system with respect to the $\hat{x}\hat{y}\hat{z}$ system. Introducing this transformation into equation (6) gives

$$I_{zz} = \int_{\mathcal{V}} (x^2 + y^2)\, dm = \int_{\mathcal{V}} [\hat{x} - \hat{x}_A)^2 + (\hat{y} - \hat{y}_A)^2]\, dm$$

$$= \int_{\mathcal{V}} (\hat{x}^2 + \hat{y}^2)\, dm - 2\hat{x}_A \int_{\mathcal{V}} \hat{x}\, dm - 2\hat{y}_A \int_{\mathcal{V}} \hat{y}\, dm$$

$$+ [(\hat{x}_A)^2 + (\hat{y}_A)^2] \int_{\mathcal{V}} dm \tag{16}$$

As the origin of the $\hat{x}\hat{y}\hat{z}$ system is the center of mass, it follows that the first moments of mass with respect to this coordinate system are zero; that is,

$$\int_{\mathscr{V}} \hat{x} \, dm = \int_{\mathscr{V}} \hat{y} \, dm = 0$$

Further, denoting the distance between the z and $\hat{z}$ axes as d_z, we have

$$d_z = \sqrt{(\hat{x}_A)^2 + (\hat{y}_A)^2}$$

so that equation (16) gives the following form for the *parallel axis theorem:*

$$\boxed{I_{zz} = I_{\hat{z}\hat{z}} + md_z{}^2} \qquad\qquad (17)$$

The application of equation (17) is straightforward, provided that you remember that the $\hat{z}$ *axis must pass through the center of mass.* If it does not, then the first moments of mass do not vanish and equation (17) is not valid.

As indicated earlier, in applications it is more expedient to determine the inertia properties by using the table of properties in Appendix B rather than the methods of integration. The parallel axis theorem is a powerful tool in this regard because it allows us to determine the moment of inertia about an axis other than the one referred to in the tables.

HOMEWORK PROBLEMS

Using the parallel axis theorem, and the answers given to the problems in the previous section, find the moment of inertia, I_{zz}, about the mass center for each of the following problems.

V.9   Problem V.1

V.10   Problem V.2

V.11   Problem V.3

V.12   Problem V.4

V.13   Problem V.5

V.14   Problem V.6

For the following problems find I_{zz} about an axis parallel to the given z axis containing the point $y = r$.

V.15   Problem V.7

V.16   Problem V.8

3   Composite Bodies

As mentioned earlier, it is very useful to employ tabulated properties when we consider a body to be a composite shape, that is, a shape composed of several basic shapes. Such a body is shown in Figure 7 where the areas $\mathcal{A}_1$ and $\mathcal{A}_2$ represent two elementary geometric shapes forming the outline of the body of interest.

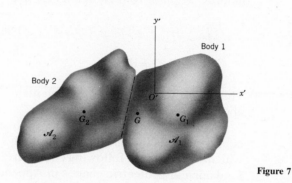

Figure 7

Any integral extending over the entire area $\mathcal{A}$ is the sum of the integrals over the areas $\mathcal{A}_1$ and $\mathcal{A}_2$. Thus, equation (11) for the mass becomes

$$m = \int_{\mathcal{A}_1} \rho h \, d\mathcal{A} + \int_{\mathcal{A}_2} \rho h \, d\mathcal{A} \quad m_1 + m_2 \tag{18}$$

This is the obvious result that the mass of the body is the sum of the mass of each component. Conversely the mass of a composite body, in the form of a body 1 from which a body 2 has been removed, can be found by *subtracting m_2 from m_2.*

We require an arbitrary reference frame to locate the center of mass G of the composite body. This is the primed frame shown in Figure 7; the prime emphasizes its arbitrary selection. In practice, the origin O' is selected as any convenient point for computing first moments of mass, such as the center of mass of either body 1 or 2. With the reference frame selected, the center of mass G is obtained by taking the first moment of mass with respect to both the x' and y' axes. Thus, considering the area $\mathcal{A}$ in terms of its component shapes, and using equation (15), we find

$$mx'_G \equiv \int_{\mathcal{A}} x'(\rho h \, d\mathcal{A}) = \int_{\mathcal{A}_1} x'(\rho h \, d\mathcal{A}) + \int_{\mathcal{A}_2} x'(\rho h \, d\mathcal{A})$$

$$= m_1 x'_{G_1} + m_2 x'_{G_2} \tag{19a}$$

Similarly,

$$my'_G = m_1 y'_{G_1} + m_2 y'_{G_2} \tag{19b}$$

where (x'_{G_1}, y'_{G_1}) and (x'_{G_2}, y'_{G_2}) are the $x'y'$ coordinates of the mass centers of bodies 1 and 2, respectively. Solution of equations (19) yields the values of x'_G and y'_G, locating the center of mass of the composite body.

The final task is the determination of the moment of inertia. In terms of *any* coordinate system xyz whose z axis is perpendicular to the plane of motion, we find from equation (12) that

$$I_{zz} = \int_{\mathcal{A}} (x^2 + y^2)\, \rho h\, d\mathcal{A} = \int_{\mathcal{A}_1} (x^2 + y^2)\, \rho h\, d\mathcal{A} + \int_{\mathcal{A}_2} (x^2 + y^2)\, \rho h\, d\mathcal{A}$$

$$= (I_{zz})_1 + (I_{zz})_2 \tag{20}$$

This result tells us that the moments of inertia of the component bodies about the *same* axis are additive.

When we are solving kinetics problems, the z axis we shall employ in equation (20) is the one passing through a particular point, the selection of which we discussed following equation (2). The moments of inertia for the fundamental geometric shapes in Appendix B may not be given about an axis that coincides with this z axis. In such cases it will be necessary first to apply the parallel axis theorem (17) to determine the value of I_{zz} for each component of the composite body before applying equation (20).

EXAMPLE 4
The aluminum body shown in the sketch has a thickness of 80 mm. Determine the mass of the body, the location of its center of mass, and its moment of inertia about an axis perpendicular to the plane of the diagram through the center of mass.

Solution
The body has a composite shape formed by removing a square parallelepiped, call it body 2, from a circular cylinder, call it body 1. Thus, because the density of aluminum is

$$\rho = 2.691(10^3)\ \text{kg/m}^3 = 2.691(10^{-6})\ \text{kg/mm}^3$$

we have

$$m_1 = \rho \mathcal{V}_1 = 2.691(10^{-6})\pi(400)^2(80) = 108.21\ \text{kg}$$

$$m_2 = \rho \mathcal{V}_2 = 2.691(10^{-6})(200)^2(80) = 8.61\ \text{kg}$$

$$m = m_1 - m_2 = 99.60\ \text{kg}$$

We next select a reference frame $x'y'z'$ for computing first moments. The center of the circle seems a convenient origin for this system, and it makes sense to orient the x' and y' axes to be parallel to the edges of the cut-out area, as shown in the sketch. (We shall generally find that selecting a coordinate system that is parallel to the one depicted in the tables of Appendix B will be most efficient for computations.) Because of the symmetry of the body, we can see that the center of mass G is situated along the 45° line connecting points G_1 and G_2, so it must be true that $x'_G = y'_G$. (This observation is merely for the purpose of eliminating some computations.)

Because the mass m_2 is subtracted from the mass m_1, the first moment of mass with respect to the x' coordinate is

$$mx'_G = m_1 x'_{G_1} - m_2 x'_{G_2}$$

$$= 108.21(0) - 8.61(100)$$

$$= -861 \text{ kg-mm}$$

Thus

$$x'_G = y'_G = \frac{-861}{99.60} = -8.64 \text{ mm} \qquad \triangleleft$$

The tabulated values of moment of inertia for both the cylinder and the parallelepiped are for axes $\hat{z}$ that pass through the respective centers of mass. We shall therefore need first to apply the parallel axis theorem to obtain the moments of inertia about an axis passing through point G. Knowing the location of point G, we now find the distances from points G_1 and G_2 to the center of mass G to be

$$(d_z)_1 = \sqrt{2}(8.64) = 12.22 \text{ mm}$$

$$(d_z)_2 = \sqrt{2}(100 + 8.64) = 153.64 \text{ mm}$$

Using the tabulated moments of inertia, we then find

$$(I_{zz})_1 = \tfrac{1}{2} m_1 r^2 + m_1 (d_z)_1{}^2$$

$$= 108.21[\tfrac{1}{2}(400)^2 + (12.22)^2]$$

$$= 8.673(10^6) \text{ kg-mm}^2 = 8.673 \text{ kg-m}^2$$

$$(I_{zz})_2 = \tfrac{1}{12} m_2 (a^2 + b^2) + m_2 (d_z)_2{}^2$$

$$= 8.61\{\tfrac{1}{12}[(200)^2 + (200)^2] + (153.64)^2\}$$

$$= 2.61(10^5) \text{ kg-mm}^2 = 0.261 \text{ kg-m}^2$$

$$I_{zz} = (I_{zz})_1 - (I_{zz})_2 = 8.41 \text{ kg-m}^2 \qquad \triangleleft$$

HOMEWORK PROBLEMS

In the following problems use the information in Appendix B to determine the location of the mass center of the composite body and the mass moment of inertia $I_{\hat{z}\hat{z}}$, where $\hat{z}$ is the axis passing through the mass center that is parallel to the z axis shown.

V.17 The steel plate shown below.

Prob. V.17

V.18 The aluminum plate shown below.

Prob. V.18

V.19 Solve Problem V.18 for the case where plate A is aluminum and plate B is steel.

V.20 The aluminum cylinder and cone shown to the left.

V.21 Solve Problem V.20 for the case where the cylinder is steel and the cone is aluminum.

Prob. V.20

V.22   The steel assembly shown below.

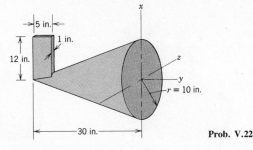

Prob. V.22

V.23   Solve Problem V.22 for the case where the cone is steel and the plate is aluminum.

V.24   The aluminum wires, 0.20 in.^{2} cross-sectional area, shown below.

Prob. V.24

C.   THREE TYPES OF RIGID BODY MOTION

Now that we have the ability to evaluate the inertial properties of a body, let us return to the consideration of the equations of motion (1) and (7). Our aim in this section is to gain further understanding of the significance of these equations. With this as our goal, let us consider an interpretation of equations (1) and (7) that builds on concepts developed in the study of the statics of rigid bodies.

According to the principle of equivalence of force systems, the effect on a rigid body of an arbitrary force system can be represented by an equivalent force-couple system acting at any point P. In this equivalent system, the force that is applied at point P is the resultant of the forces acting on the body, and the torque of the couple is the total moment of all the forces about point P.

Figure 8a, on the next page, shows a rigid body subjected to an arbitrary planar force system, where $\hat{x}\hat{y}\hat{z}$ is the coordinate system whose origin is the center of mass G. We use the center of mass G as the point P for the equivalent force-couple system because it is always a valid point for summing moments in the formulation of the dynamics problem. Thus,

Figure 8A and Figure 8b

equations (1) and (7) tell us that the force system in Figure 8a is equivalent to a force $m\bar{a}_G$ applied at point G and a couple $I_{zz}\bar{\alpha}$, as illustrated in Figure 8b. Using this concept of equivalence, let us examine each of the three types of rigid body motion: pure translation, pure rotation, and general motion.

1 Translational Motion

If a rigid body is translating, then the angular velocity $\bar{\omega}$ and angular acceleration $\bar{\alpha}$ are both zero. From our knowledge of kinematics it follows that every point in the rigid body has the same acceleration as the center of mass. In this case, the equivalent force system in Figure 8a reduces to that shown in Figure 9.

From Figure 9 we can deduce that if we had computed the moment of the force system about any point other than the center of mass, such as point P (in Figure 9), then, in general, we would have to equate the resulting moment to the moment about point P of the equivalent force $m\bar{a}_G$. An exception occurs if point P lies on the indicated dashed line, for then the vector $m\bar{a}_G$ passes through point P. However, there will be much less chance for error if we make it standard practice to sum moments about the center of mass if the body is translating. The equations of motion in this case reduce to

Figure 9

$$\Sigma \bar{F} = m\bar{a}_G \qquad \Sigma \bar{M}_G = \bar{0} \tag{21}$$

2 Rotational Motion

A body undergoing pure rotation in a plane has an axis perpendicular to the plane of motion that is fixed. It then follows that the center of mass is in circular motion about the axis of rotation. In terms of polar coordinates, defined with respect to point O on the fixed axis, we have

$$\bar{a}_G = -R\omega^2 \bar{e}_R + R\alpha \bar{e}_\phi$$

Figure 10

Using this expression for $\bar{a}_G$, the general equivalent force system in Figure 8b reduces to that shown in Figure 10.

From Figure 10 we note that the equivalent force $-mR\omega^2\bar{e}_R$ exerts no moment about point O. Therefore, when we compute $\Sigma \bar{M}_O$ for the actual force system and equate this to the sum of the moments about point O of the equivalent force system, we find that

$$\Sigma \bar{M}_O = I_{\hat{z}\hat{z}}\bar{\alpha} + mR\alpha(R)\bar{k} = (I_{\hat{z}\hat{z}} + mR^2)\alpha\bar{k}$$

In view of the parallel axis theorem, equation (17), this reduces to

$$\Sigma \bar{M}_O = I_{zz}\alpha\bar{K} = I_{zz}\bar{\alpha}$$

This, of course, is the same result as equation (7) for the case where we sum moments about a fixed point (thus demonstrating why a fixed point is an allowable point for summing moments). Notice that if we had computed the moment about any point other than the fixed point O or center of mass G, we would not have the simple relationship of equation (7).

The question of whether we should compute the moment of the force system about point O or point G is easily answered when we consider the forces acting on the system. In order to have a fixed axis of rotation, there must be constraint forces applied to this axis, such as the reaction forces produced by a pin joint. It follows then, that in order to eliminate these (unknown) constraint forces, we solve problems involving bodies executing pure rotational motion by summing moments about the fixed point O. The resulting equations of motion are therefore

$$\Sigma \bar{F} = m\bar{a}_G \qquad \Sigma \bar{M}_O = I_{zz}\bar{\alpha} \tag{22}$$

where the z axis is coincident with the fixed axis of rotation.

3 General Motion

For this type of motion, which is a superposition of translational and rotational effects, no point in the body is fixed. Also, whether or not there is any point in the body whose acceleration is directly toward or away from the center of mass depends on the geometry and constraints of the system. This can be illustrated by the simple problem of a circular wheel rolling without slipping down an incline.

If the wheel is balanced so that the center of mass G coincides with the geometric center O, then we have the kinematical situation illustrated in Figure 11a. For the no slip condition, point C of the body, which is in contact with the incline, must accelerate directly toward point O. Thus, if points O and G coincide, then point C is an allowable point for applying equation (17). In contrast, if the wheel is unbalanced, so that the center of mass is at a distance l from the geometric center O, we then have the

Figure 11a

Figure 11b

kinematical situation shown in Figure 11b. From this Figure we conclude that the contact point C does not accelerate directly toward point G, except for the special cases that arise at instants when points G, O, and C are collinear.

To have an approach that is valid in all situations, we therefore solve problems in general motion by computing moments about the center of mass G. Then, the general equations of motion are

$$\Sigma\,\bar{F} = m\bar{a}_G \qquad \Sigma\,\bar{M}_G = I_{zz}\bar{\alpha} \qquad (23)$$

where the z axis is perpendicular to the plane of motion, passing through point G.

D. PROBLEM SOLVING

We did not present examples using the equations of motion prior to this section for the reason that the procedures to use them are fairly straightforward. As we proceed, note that the steps we present are essentially a synthesis of the methods for treating force systems (developed in statics) and the methods of kinematics (developed in Module IV).

1   Determine the mass and location of the center of mass G of each body in the system for which such information is not given.

2   Draw a free body diagram of each individual body in the system, making use of Newton's law of action and reaction where appropriate. Remember that each constraint on the motion of a system can only be imposed by a corresponding force or moment. The free body diagram(s) should also show the location of the center of mass, even when the gravitational force is perpendicular to the plane of motion.

3   Select an xyz reference frame for each body. For bodies in translation or general motion, the origin A of this frame should be the center of mass of the body, whereas for bodies that are rotating about a fixed axis the origin should be situated on that axis. The x and y axes should be parallel to the plane of motion and their orientation should, where possible, fit the geometry of the physical system. In problems where there is more than one body, the xyz reference frame for each body should be parallel to each other, making it easier to equate various vectors.

4   Show all given values of forces in the free body diagram and write down any other given information in equation form, using the xyz frame to describe vector components.

5   Following the methods developed in Module IV for treating the kinematics of rigid bodies in planar motion, relate the angular acceleration and the x and y components of the acceleration of the center of mass. Be certain to satisfy all of the constraints imposed on the motion of the system.

6 Calculate I_{zz} for any body that has rotational motion.

7 Obtain the scalar equations of motion for each body by equating the x and y components of $\Sigma \bar{F} = m\bar{a}_G$ and the z component of $\Sigma \bar{M}_A = I_{zz}\bar{\alpha}$. In following this step, substitute the results of step 5.

8 Count the number of scalar equations and the number of scalar unknowns to be determined. If the two numbers do not match, you probably have either improperly employed the law of action and reaction, or else forgotten a constraint force or the corresponding kinematical constraint condition.

9 Solve the scalar equations of motion. Then check that the solution satisfies the initial assumptions. For example, if you have assumed that two surfaces do not slip relative to each other, check to be sure that the friction force $\bar{f}$ and normal force $\bar{N}$ satisfy the condition $\left|\bar{f}\right| < \mu_s \left|\bar{N}\right|$.

ILLUSTRATIVE PROBLEM 1

The automobile shown accelerates at a constant rate on a level road. It starts from rest and reaches the speed of 70 km/hr in a distance of 200 m. Determine the reactive force at the hitch of the 400-kg trailer during the motion. Neglect friction.

Solution

Step 1 The location of the center of mass G of the trailer is given.

Steps 2, 3, and 4 Because the reactive force at the hitch can be determined only when it appears as an external force, we isolate the trailer in a free body diagram. Note that we did not isolate the automobile, because we have no information about this body. It is reasonable to consider the hitch as a pin, hence it exerts horizontal and vertical reactions. The choice for the coordinate system is now obvious.

Step 5 The system is in pure translation, so $\omega = \alpha = 0$. There is no motion in the y direction, hence $v_y = a_y = 0$. For constant acceleration in the x direction we have

$$2a_x s_x = (v_x)_2{}^2 - (v_x)_1{}^2 = \left(70\frac{1000}{3600}\right)^2 - 0 = 378.1$$

$$a_x = \frac{378.1}{2(200)} = 0.9453 \text{ m/s}^2$$

Step 6   Because $\alpha \equiv 0$, I_{zz} is not needed.

Step 7

$$\Sigma F_x = A_x = m(a_G)_x = 400(0.9453)$$

$$\Sigma F_y = -400(9.806) + A_y + N = m(a_G)_y = 0$$

$$\Sigma M_{Gz} = (2A_y + 0.6A_x) = 0$$

Step 8   We have three scalar equations in the three scalar unknowns A_x, A_y, and N.

Step 9

$\bar{\imath}$ direction:	$A_x = 400(0.9453) = 378.1$ newtons
$\bar{\jmath}$ direction:	$A_y + N = 3922$
moment:	$A_y = -0.3A_x = -113.4$ newtons

Note that we only needed the first and third equations to solve for the components of the reactive force.

ILLUSTRATIVE PROBLEM 2

The 100-lb block is released from rest. Compute the angular velocity of the 60-lb uniform drum and the tension in the cable: (a) 5 s after release, (b) after the block has fallen 20 ft.

Solution

Step 1   We must cut the cable to determine its tension. Hence, we construct two free body diagrams. We can treat the block as a particle and assume that the center of mass of the drum coincides with the shaft supporting the drum.

Steps 2, 3, and 4   The shaft can apply horizontal and vertical reactions to support the drum, resulting in the free body diagram shown. Note that even if we were not interested in the tension force, we would have to treat the block and drum separately because jointly they do not form a rigid body.

Step 5   The acceleration of the 100-lb weight must be the same as the tangential acceleration of the point P on the circumference of the drum. Therefore,

$$-a_B \bar{\jmath} = -r\alpha_A \bar{\jmath} = -0.5\alpha_A \bar{\jmath} \text{ ft/s}^2$$

Step 6   From Appendix B, for body A we have

$$I_{zz} = \frac{1}{2} mr^2 = \frac{1}{2}\frac{60}{32.17}\left(\frac{1}{2}\right)^2 = 0.2331 \text{ lb-s}^2\text{-ft}$$

The moment of inertia of the block is not required because we are treating it as a particle.

Step 7 The force sum for the drum will involve the unknown reactions at point A. As we are not interested in these forces, we only sum moments for the drum. Hence, for body A we have

$$\Sigma\, M_{Gz} = -T(0.5) = 0.2331(-\alpha_A)$$

For body B we sum forces in the y direction and use the result of step 5. This gives

$$\Sigma\, F_y = T - 100 = \frac{100}{32.17}(-a_B) = -\frac{100}{32.17}(0.5\, \alpha_A)$$

Step 8 We have two scalar equations in the two scalar unknowns T and α_A.

Step 9 The solution of the equations of motion in step 7 is

$$\alpha_A = 49.49 \text{ rad/s}^2$$

$$T = 23.07 \text{ lb}$$

The value of T is constant. We obtain the desired angular velocities by applying the formulas for constant acceleration. Hence, when $t = 5\,\text{s}$ for part (a), we have

$$(\omega_A)_2 = (\omega_A)_1 + \alpha_A t$$

$$= 0 + 49.49(5)$$

or

$$\bar{\omega}_A = -247.5\bar{k} \text{ rad/s}$$

For part (b) we first determine the angle of rotation corresponding to a 20-ft drop of the block.

$$\theta \equiv \frac{20}{R} = \frac{20}{0.5} = 40 \text{ rad}$$

Then using

$$(\omega_A)_2{}^2 = (\omega_A)_1{}^2 + 2\alpha_A\theta$$

for $\theta = 40$ rad we find

$$(\omega_A)_2{}^2 = 0 + 2(49.49)40$$

or

$$\bar{\omega}_A = -62.92\bar{k} \text{ rad/s}$$

ILLUSTRATIVE PROBLEM 3

The uniform plate shown is released from its rest position in a vertical plane by cutting cable BC. Determine the forces in cables BD and AE at the instant of release.

Solution

Step 1 The center of mass of a triangular plate is at a distance from the base that is one-third of the height.

Step 2 Because we want the initial values of the cable forces, we draw the free body diagram for the instant of release.

Steps 3 and 4 Edge AB will always be horizontal, so the plate translates. Hence, the origin of the xyz coordinate system must be at the center of mass. The chosen orientation of xyz expedites the description of the force components and their points of application. As the weight of the plate is not given, call it W.

Step 5 For translational motion $\bar{\omega} = \bar{\alpha} = \bar{0}$. Because the plate is released from rest, $\bar{v}_G = \bar{0}$, but $\bar{a}_G \neq \bar{0}$. The value of $\bar{a}_G$ is not arbitrary, because the acceleration of all points in a translating body are identical and points A and B are constrained by the cables to follow a circular motion. It is convenient to use path variables to determine $\bar{a}_A$.

In terms of the path variable unit vectors with $v = 0$ we have

$$\bar{a}_G = \bar{a}_A = \dot{v}(\bar{e}_t)_A + \frac{v^2}{\rho}(\bar{e}_n)_A = \dot{v}(\bar{e}_t)_A$$

However, we shall need the x and y components of $\bar{a}_G$ for the force sum. Thus, transforming $(\bar{e}_t)_A$ yields

$$\bar{a}_G = \dot{v}(-\cos 20°\,\bar{i} - \sin 20°\,\bar{j})$$

Step 6 $\alpha \equiv 0$ so I_{zz} is not needed.

Step 7 Referring to the free body diagram in step 2, the equations of motion are

$$\Sigma F_x = -T_{AE} \sin 20° - T_{BD} \sin 20° = m(a_G)_x$$

$$= \frac{W}{g}(-\dot{v} \cos 20°)$$

$$\Sigma F_y = T_{AE} \cos 20° + T_{BD} \cos 20° - W = m(a_G)_y$$

$$= \frac{W}{g}(-\dot{v} \sin 20°)$$

$$\Sigma M_{Gz} = T_{AE} \sin 20° (0.15) - T_{AE} \cos 20° (0.60)$$

$$+ T_{BD} \sin 20° (0.15) + T_{BD} \cos 20° (0.60)$$

$$= I_{zz}\alpha = 0$$

Step 8 We have three scalar equations for the three scalar unknowns T_{AE}, T_{BD}, and $\dot{v}$.

Step 9 Solving the moment equation for T_{BD} and substituting this result into the two force equations (which are then solved simultaneously) yields

$$\dot{v} = 0.3214g$$

$$T_{AE} = 0.4816W \qquad T_{BD} = 0.4012W$$

ILLUSTRATIVE PROBLEM 4

A 30-kg uniform rod is sliding along the floor and wall. The coefficient of friction between all surfaces is 0.3. Determine the reactions at ends A and B for the instant shown, if end A has a velocity of 3 m/s to the left at this instant.

Solution

Step 1   The location of the center of mass G is known by symmetry.

Steps 2, 3, and 4   The friction forces are opposite the sliding motion. The rod is in general motion, so we place the origin of xyz at point G.

Step 5   Ends A and B are constrained to move horizontally and vertically, respectively. Hence, we can relate $\bar{a}_G$ to $\bar{\alpha}$ for the constrained motion. Following the methods established in Module Four, we first determine the angular velocity using the method of instant centers.

From the sketch, point C is the instant center, so

$$\omega = \frac{v_A}{2 \sin 45°} = \frac{3}{1.4142} = 2.121 \text{ rad/s} \qquad \text{clockwise}$$

$$\bar{\omega} = -2.121\bar{k}$$

For the acceleration we then have

$$\bar{a}_G = \bar{a}_A + \bar{\alpha} \times \bar{r}_{G/A} + \bar{\omega} \times (\bar{\omega} \times \bar{r}_{G/A})$$

$$= \bar{a}_B + \bar{\alpha} \times \bar{r}_{G/B} + \bar{\omega} \times (\bar{\omega} \times \bar{r}_{G/B})$$

Note that the given value for v_A is an instantaneous value and is not a constant. (An additional constraint force would be necessary to have no acceleration for this point.) The individual vectors are

$$\bar{a}_A = a_A \, \bar{i} \qquad \bar{a}_B = -a_B \, \bar{j} \qquad \bar{\alpha} = \alpha \bar{k}$$

$$\bar{r}_{G/A} = 0.7071(-\bar{i} + \bar{j}) \qquad \bar{r}_{G/B} = 0.7071(\bar{i} - \bar{j})$$

The acceleration equations then give

$$\bar{a}_G = [a_A - 0.7071\alpha + (2.121)^2(0.7071)]\bar{i}$$
$$+ [-0.7071\alpha - (2.121)^2(0.7071)] \, \bar{j}$$
$$= [0.7071\alpha - (2.121)^2(0.7071)]\bar{i}$$
$$+ [-a_B + 0.7071\alpha + (2.121)^2(0.7071)] \, \bar{j}$$

Equating corresponding components gives

$$(a_G)_x = a_A - 0.7071\alpha + 3.181 = 0.7071\alpha - 3.181$$
$$(a_G)_y = -0.7071\alpha - 3.181 = -a_B + 0.7071\alpha + 3.181$$

We are not interested in the values of a_A and a_B, so we write

$$(a_G)_x = 0.7071\alpha - 3.181 \qquad (a_G)_y = -0.7071\alpha - 3.181$$

Step 6 We obtain I_{zz} for the rod from Appendix B.

$$I_{zz} = \tfrac{1}{12}ml^2 = \tfrac{1}{12}(30)(2)^2 = 10 \text{ kg-m}^2$$

Step 7 From the free body diagram we have

$$\Sigma F_x = N_B - 0.3N_A = m(a_G)_x = 30(0.7071\alpha - 3.181)$$
$$\Sigma F_y = 0.3N_B + N_A - 30(9.806) = m(a_G)_y$$
$$= 30(-0.7071\alpha - 3.181)$$
$$\Sigma M_{Gz} = N_A(0.7071) - 0.3N_A \,(0.7071) - N_B(0.7071)$$
$$- 0.3N_B(0.7071) = I_{zz}\alpha = 10\alpha$$

Note that in the moment equation, clockwise moments are positive because the z axis is into the plane of the diagram.

Step 8 We have three equations for the three unknowns, N_A, N_B, and α.

Step 9 Solving the three equations simultaneously for N_A and N_B yields

$$N_A = 112.3 \text{ newtons} \qquad N_B = 19.03 \text{ newtons}$$

Hence, the reactions are

$$\bar{R}_A = -0.3N_A\bar{i} + N_A \, \bar{j} = 112.3\bar{i} - 33.69\bar{j} \text{ newtons}$$
$$\bar{R}_B = N_B\bar{i} + 0.3N_B \, \bar{j} = 19.03\bar{i} + 5.709\bar{j} \text{ newtons}$$

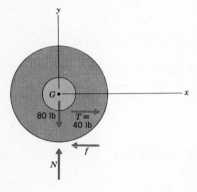

ILLUSTRATIVE PROBLEM 5

The diagram shows the cross section of an 80-lb spool of wire. The radius of gyration of the spool is 10 in. Knowing that the coefficients of static and kinetic friction are $\mu_s = 0.25$ and $\mu_k = 0.20$, respectively, determine the acceleration of the mass center G for a wire tension $T = 40$ lb.

Solution

Step 1   The location of the center of mass is known by symmetry.

Steps 2 and 3   We do not know whether there is slippage at the point of contact, so we *cannot* assume that the friction force $f = \mu N$. Hence, we consider f to be an unknown.

Regardless of whether or not there is slippage, the disk is in general motion, so the origin of the xyz axes is selected to coincide with the center of mass.

Step 4   The radius of gyration is $k_z = 10$ in. and the given value for T is shown in the free body diagram.

Step 5   The center of mass follows a horizontal path and its acceleration will probably be to the right, so

$$\bar{a}_G = a_G \bar{\imath}$$

For planar motion we know that $\bar{\alpha} = \alpha \bar{k}$, but we have two alternative kinematical hypotheses to consider. The constraint of no slippage gives $a_G = -R\alpha = -1.5\alpha$ (see Module Four), where the negative sign occurs because, in this case, negative α corresponds to clockwise angular acceleration and positive a_G. On the other hand, if there is slippage, then there is no constraint condition to relate a_G and α. However, for the latter case we know that $f = \mu_k N$, opposite the sliding motion.

Step 6   We use the given value of the radius of gyration to determine I_{zz}.

$$I_{zz} = mk_z{}^2 = \frac{80}{32.17}\left(\frac{10}{12}\right)^2 = 1.7269 \text{ lb-s}^2\text{-ft}$$

Step 7

$$\Sigma F_x = 40 - f = m(a_G)_x = \frac{80}{32.17}\, a_G$$

$$\Sigma F_y = N - 80 = m(a_G)_y = 0$$

$$\Sigma M_{Gz} = 40\left(\frac{6}{12}\right) - f\,\left(\frac{18}{12}\right) = I_{zz}\alpha = 1.7269\alpha$$

Steps 8 and 9   The three equations of motion have four unknowns: f, N, a_G, and α. For the fourth relation we assume one of the kinematical possibilities; if we are wrong in this assumption we shall find a contradiction. Let us assume that there is no slipping. Then $a_G = -1.5\alpha$ and the equations of motion reduce to

$$40 - f = 2.487 a_G$$

$$N - 80 = 0$$

$$20 - 1.5f = 1.7269\left(-\frac{a_G}{1.5}\right) = -1.151 a_G$$

The solution to these equations is

$$a_G = 8.194 \text{ ft/s}^2 \qquad f = 19.62 \text{ lb} \qquad N = 80 \text{ lb}$$

We now can check the assumption of no slippage, by ascertaining if $|\vec{f}| \leq \mu_s N$, where the equality sign is the maximum friction force that may be sustained without slipping. Using the given value for μ_s we have

$$\mu_s N = 0.25(80) = 20 > f$$

Hence the spool does not slip and the foregoing solution is correct. Thus

$$a_G = 8.194 \text{ ft/s}^2$$

Note the procedure we used to solve the problem. Whenever we are not sure whether surfaces slip over each other, we begin by assuming no slippage. If the required friction force exceeds $\mu_s N$, then we know that slipping will occur. In that case the friction force will be $\mu_k N$ in the same direction as the friction force obtained from the assumption of no slipping. Also, the values of a_G and α are then independent variables. The next problem features a situation where we know that slipping occurs.

ILLUSTRATIVE PROBLEM 6

The driver of an automobile causes the rear drive wheels to skid as the car accelerates. The coefficient of kinetic friction is 0.40. Assuming that the drive wheels have a constant angular velocity of 600 rev/min and that the automobile starts from rest as it moves to the left, determine the acceleration of the vehicle during the time interval when the rear wheels are skidding and also find the duration of this time interval. The mass of the vehicle is 1250 kg. The location of its center of mass C is shown in the diagram. The wheels have a mass of 15 kg, a radius of 400 mm, and a radius of gyration of 300 mm.

Solution

There are three types of bodies in this system: the chassis, the two rear drive wheels, and the two front wheels. If we wished to neglect the mass of the wheels, we could simply draw a free body diagram of the entire system, determine the acceleration of the vehicle, and then seek the condition where the no slip condition for a wheel, $v = \omega R$, is satisfied. However, as we are given the properties of the wheels, let us include the effect of their mass as we model the system.

Step 1   The given diagram locates the center of mass of all bodies.

Steps 2, 3, and 4   In order to accelerate to the left, the wheels must be turning counterclockwise. Hence, the friction force on the skidding drive wheels is to the left. The front wheels are not slipping, so this friction force is unknown, but it is probably to the right. We must not forget to include the counterclockwise couple exerted by the drive train on the rear wheels and to satisfy Newton's third law for any forces that correspond to interactions between the chassis and the wheels.

Note that the coordinate systems for each body are mutually parallel. The only given information not appearing in the free body diagram is that

$$\bar{\omega}_B = -600\left(\frac{2\pi}{60}\right)\bar{k} = -20\pi\bar{k} \text{ rad/s} \qquad \bar{\alpha}_B = \bar{0}$$

$$(k_z)_A = (k_z)_B = 300 \text{ mm}$$

Step 5   The centers of mass are all moving to the left at the same speed. Letting v denote that speed, we have

$$\bar{a}_A = \bar{a}_B = \bar{a}_G = \dot{v}\bar{\imath}$$

The chassis is translating, so

$$\bar{\alpha}_{chassis} \equiv \bar{0}$$

Finally, because the front wheels do not skid,

$$\bar{\alpha}_A = -\frac{\dot{v}}{R}\,\bar{k} = -\frac{\dot{v}}{0.4}\,\bar{k}$$

Step 6 The value of I_{zz} for the rotating wheels about their respective z axes is

$$I_{zz} = mk_z{}^2 = 15(0.3)^2 = 1.350 \text{ kg-m}^2$$

Step 7 Referring to the free body diagram for the forces, the equations of motion are

Chassis:

$$\Sigma F_x = -2A_x - 2B_x = m(a_G)_x = 1250\dot{v}$$

$$\Sigma F_y = 2A_y + 2B_y - 1250(9.806) = m(a_G)_y = 0$$

$$\Sigma M_{Cz} = 2A_y(1.0) - 2A_x(0.2) - 2B_y(1.2) - 2B_x(0.2) + 2M_B = 0$$

Wheel A:

$$\Sigma F_x = A_x - f_A = m_A(a_A)_x = 15\dot{v}$$

$$\Sigma F_y = N_A - A_y - 15(9.806) = m_A(a_A)_y = 0$$

$$\Sigma M_{Az} = -f_A(0.4) = I_{zz}\alpha_A = 1.350\left(-\frac{\dot{v}}{0.4}\right)$$

Wheel B:

$$\Sigma F_x = B_x + 0.4N_B = m_B(a_B)_x = 15\dot{v}$$

$$\Sigma F_y = N_B - B_y - 15(9.806) = m_B(a_B)_y = 0$$

$$\Sigma M_{Bz} = 0.4N_B(0.4) - M_B = I_{zz}\alpha_B = 0$$

Step 8 The nine equations in step 7 contain nine unknowns: A_x, A_y, B_x, B_y, M_B, f_A, N_A, N_B, and $\dot{v}$.

Step 9 The equations are not as complicated as they might appear. We can use the six equations governing the motion of the wheels to express N_A, f_A, N_B, M_B, A_x, and B_x in terms of A_y, B_y, and $\dot{v}$. Then subsituting these expressions into the equations for the chassis and solving gives

$$\dot{v} = 1.803 \text{ m/s}^2$$

The values of the forces are not pertinent to the problem, because we are interested only in the motion of the system. To determine the desired time interval, we examine the motion of the point of contact D between a rear wheel and the ground, as sketched to the left.

The kinematical formula for velocity gives

$$\bar{v}_D = -v_D\bar{i} = \bar{v}_B + \bar{\omega}_B \times \bar{r}_{D/B} = v\bar{i} - 0.4(20\pi)\bar{i}$$

$$v_D = 8\pi - v \text{ m/s}$$

Slipping ceases to occur when $v_D = 0$, for in that case there is no relative motion between the contact surfaces. Hence, for no slipping $v = 8\pi$ m/s. Then to determine time we use the result for $\dot{v}$ to obtain

$$v = 8\pi = \dot{v}t = 1.803t$$

$$t = 13.94 \text{ s}$$

Once the skidding ceases, it will not occur again as long as the controls of the vehicle are not changed.

HOMEWORK PROBLEMS

V.25 The truck is accelerating at 1 m/s² as the tailgate is held in the position shown by two cables (only one is shown in the side view). The tailgate has a mass of 15 kg. What is the tension force in each cable?

Prob. V.25

V.26 A person is trying to balance a stick in a tilted position. Determine the horizontal acceleration of this person's hand for which the angle θ is constant at 10°.

Prob. V.26

V.27 A homogeneous cylinder of mass m and radius r is pulled along a rough surface as shown by a truck whose acceleration is $g/10$. If the cylinder is sliding over the ground without rotating, determine the coefficient of friction and the force the truck applies to the two cables (only one is shown in the side view).

Prob. V.27

1.10 m

G

0.35 m   0.70 m

Prob. V.28

V.28 A cyclist moving at a speed of 20 km/hr applies the front wheel brakes only. The cyclist wishes to avoid being thrown over the handle bars. Determine the shortest safe stopping distance for which the rear wheel will not lose contact with the ground.

V.29 An automobile coasts freely down a 10% grade hill. Determine (a) the normal reactions between the road and tires, and (b) the acceleration of the 1500-kg vehicle. The mass of the wheels is negligible, as is rolling resistance.

G

600 mm

1.2 m   1.3 m

10

Prob. V.29

V.30 The automobile in Problem V.29 is accelerated on a level road where the coefficient of static friction between the tires and car is 0.65. Determine the maximum acceleration possible if (a) it has rear wheel drive, (b) it has front wheel drive, (c) it has four wheel drive.

V.31 A 2-ton coal car is pulled up a ramp. Neglecting the mass and friction of the wheels, determine the minimum cable tension at which one set of wheels loses contact with the ramp. Also determine the corresponding acceleration of the coal car.

60°

Hoist

2 ft

G

2.5 ft

2.5 ft

3 ft

Prob. V.31

G_c 500 mm

G_t 1 m

800 mm

1 m   1.5 m   1 m

Prob. V.32

V.32 The 1000-kg forklift truck is carrying an 800-kg crate. The brakes are applied at all four wheels in order to bring the truck to rest from an initial speed of 5 m/s to the right. Determine the shortest distance in which the truck may be safely stopped without causing the truck to tilt forward or causing the crate to slide. The coefficient of static friction between the fork and the crate is 0.4.

V.33 The 100-kg refrigerator is placed on casters and then pushed across a floor by the 1000-N force shown. Determine the limiting values for the

1000 N

1 m

G

h

800 mm

350 mm | 350 mm

Prob. V.33

height h at which this force can be applied without causing the refrigerator to tip over. Friction is negligible.

V.34 Solve Problem V.33 for the case where the casters are locked and the coefficient of kinetic friction is 0.30.

V.35 In Problem V.33, replace the 1000-N force by a horizontal force $\bar{F}$ and consider the case where the casters are locked. The coefficient of friction for sliding between each caster and the ground is μ. Determine the range of values of the height h for which tipping will not occur. Is there any value of h for which the refrigerator will not tip over regardless of the magnitude of $\bar{F}$?

V.36 A truck traveling at a speed of 45 miles/hr brakes to a stop. Determine the shortest stopping distance for the truck, such that the 200-lb crate will neither tilt nor slip. The coefficient of friction between the crate and the truck is 0.25.

18 in.

36 in.

Prob. V.36

V.37 Solve Problem V.36 when the truck is initially (a) headed up a road with a 10% grade, (b) headed down a road with a 10% grade.

V.38 A 40-kg solid homogeneous cylinder is connected to the rigid vertical support AB by two links of negligible weight. In the position shown, the cylinder is moving upward at a speed of 6 m/s and the hydraulic piston D is pulling point C downward with a force of 700 N. Determine the acceleration of the center of mass of the cylinder and the reaction force at pin A for this instant.

1 m

500 mm

200 mm

200 mm

C

A

E

1 m

1.5 m

Hydraulic piston

B

F

D

Prob. V.38

Prob. V.39

Prob. V.41

Prob. V.42

V.39   A 10-kg homogeneous plate is set into motion by severing cable *BE*. Determine the acceleration of point *D* and the tension in cables *AB* and *DC* at this instant.

V.40   The torque of an electric motor directly drives the platter of a record turntable. The platter weighs 6 lb and its radius of gyration is 10 in. What is the constant value of the torque *T* for which the platter attains a speed of $33\frac{1}{3}$ rev/min in 1 rev after starting from rest?

V.41   In Problem V.40, the torque of the electric motor varies with the angular speed ω according to the given graph. What is the value of the initial torque *T* for which the platter attains a speed of $33\frac{1}{3}$ rev/min in 1 rev after starting from rest?

V.42   The flywheel weighs 150 lb and has a radius of gyration of 10 in. It is rotating at 400 rev/min when the 100-lb beam is brought into contact with the rim. Determine how long it will take for the flywheel to come to rest, given that the coefficient of friction is 0.25.

V.43   The hydraulic cylinder *C* is used to brake the 800-kg flywheel whose radius of gyration is 160 mm. The coefficient of friction between the brake shoe *D* and the rim is 0.60. Determine the force that the hydraulic cylinder B must apply in order to bring the flywheel to rest in 25 rev from an initial angular velocity of 600 rev/min (a) clockwise, (b) counterclockwise.

Prob. V.43

V.44   A dynamometer is used to simulate the inertial resistance of an automobile. To do this the vehicle is held stationary and the drive wheels cause the drum of radius *R* to rotate clockwise. Letting m_a denote the mass of the automobile, derive an expression for the moment of inertia of the drum about its axis of rotation for which the tire traction required to impart an angular acceleration to the drum is the same as that required to actually accelerate the vehicle on level ground.

Prob. V.44

V.45   Determine the angular acceleration of the 100-kg pulley (a) if a 400-N force is applied to the end of the cable, (b) if a block whose weight is 400 N is suspended from the end of the cable.

Prob. V.45

V.46   The 20-kg double pulley shown has a 250-mm radius of gyration and is initially at rest. Determine (a) the angular acceleration of the pulley, (b) the angular velocity of the pulley when $t = 3$ s, and (c) the velocity of the 30-kg mass when it has moved 1 m.

Prob. V.47

Prob. V.46

V.47   It takes 5 s for gear A, which has an external torque of 20 lb-in. applied to it, to increase the angular speed of gear B from 100 to 300 rev/min. Gear B weighs 45 lb and has a radius of gyration of 7 in. Gear A weighs 10 lb. Determine the radius of gyration of gear A.

V.48   A hoist consists of a motor driven gear A which rotates the larger stepped gear B. The masses of the gears are $m_A = 1$ kg, $m_B = 8$ kg, and their radii of gyration are $k_A = 60$ mm, $k_B = 150$ mm. A torque $\bar{M}$ is applied to gear A, with the result that the 12-kg block C is accelerated to a speed of 4 m/s from rest after rising 1 m. What is the value of M?

Prob. V.48

V.49 Disks A and B have identical mass m. Initially, disk A is rotating clockwise at a rate of 300 rev/min and disk B is at rest. Disk B is then gently released and brought into contact with disk A. The coefficient of friction between the two disks is 0.25, and both disks rotate freely about their shafts. Determine the time required for the disks to attain the same angular speed and thus cease to slip over each other.

V.50 In Problem V.49, a clockwise couple $\bar{M}$ is applied to disk A to maintain its angular speed at 300 rev/min. Determine (a) the required value of M, (b) the time required for the disks to cease to slip over each other.

Prob. V.49

V.51 The homogeneous bar weighs 15 lb. The cord at end B breaks. Determine the angular acceleration of the bar and the reaction at the support A at this instant.

V.52 Solve Problem V.51 for the case where the support at end A exerts a constant frictional couple of 2 lb-ft opposing the motion of the bar.

Prob. V.51

V.53 A horizontal force $\bar{P}$ is applied at end B of the uniform rod of mass m. The bar is initially at rest in the position shown. If the distance $q = L/4$, determine (a) the angular acceleration of the rod in this position, (b) the components of the reaction force at pin C for this position.

V.54 A horizontal force $\bar{P}$ is applied at end B of the uniform rod of mass m in Problem V.53. The bar is at rest in the position shown. Determine the value of q for which the horizontal component of the reaction force at pin C is zero in this position, and also determine the corresponding angular acceleration.

Prob. V.53

V.55 The equilateral triangular plate of mass m pivots about pin A. (a) Derive a differential equation of motion for the angle θ for the triangular plate. (b) The angle ϕ for the simple pendulum formed from a cable and a particle of mass m shown obeys the same equation of motion as that found in part (a). What is the length l of the cable?

Prob. V.55

Prob. V.56

V.56 A 200-kg homogeneous block is suspended by cables *OA* and *OC* which are attached to the stationary crane at point *O*. In the position shown, the box is descending and the center of mass has a speed of 2 m/s. Determine the acceleration of the center of mass in this position and the tensile forces in the cables.

V.57 The acceleration of plate *A* is 12 ft/s² to the right and the acceleration of plate *B* is 4 ft/s² to the left. Assuming that the surfaces are not slipping over each other, determine the friction forces exerted by the plates on the 20-lb cylinder *C*.

Prob. V.57

V.58 Racks *A* and *B*, having mass of 30 kg and 60 kg, respectively, are constrained by their smooth guides to move horizontally. The stepped gear has a mass of 80 kg and a radius of gyration of 500 mm. A torque of 5 kN-m is applied clockwise to the gear. Determine the acceleration of the center of the gear and of the racks.

Prob. V.58

V.59 The 300-kg gear and drum assembly has a radius of gyration of 700 mm. The 1500-N force causes the gear to roll on the 100-kg rack, with negligible friction between the rack and the horizontal surface. Determine (a) the tension in the cable, (b) the length of cable wound on or off the drum in 5 s starting from rest.

Prob. V.59

V.60   The 4-lb stepped pulley in the sketch is pinned at its center O to the 8-lb sliding block. The tension forces in the cables are $T_A = 10$ lb, $T_B = 5$ lb. The radius of gyration of the pulley is 6 in., and friction is negligible. Determine the angular acceleration of the pulley and the acceleration of the block.

Prob. V.60

Prob. V.61

V.61   The 30-kg wheel A has a radius of gyration of 40 mm. Determine (a) the acceleration of the center of the wheel when the 20-kg counterweight B is released from rest, (b) the speed of the counterweight after it has fallen 2 m.

V.62   The stepped pulley B weighs 10 lb and its radius of gyration is 8 in., whereas both blocks weigh 20 lb. Determine the velocity of block A 2 s after the system is released from rest. The coefficient of friction between block C and the horizontal surface is 0.3.

Prob. V.62

V.63   The 60-kg drum and axle assembly has a 300-mm radius of gyration. The counterweight has a mass of 40 kg. Determine (a) the acceleration of the axle assembly and of the counterweight, (b) the velocity of the counterweight 2 s after it is released from rest.

Prob. V.63

Prob. V.65

Prob. V.66

Prob. V.67

V.64 Solve Problem V.63 if the counterweight has a mass of 10 kg and all other parameters are unchanged.

V.65 A uniform disk of mass m and radius r rolls without slipping down the incline as shown. Determine (a) the acceleration of the center of the disk, (b) the value of the coefficient of friction μ corresponding to the case of impending slippage.

V.66 A 1000-kg thin pipe rests on the flatbed of a truck in the position shown. The coefficient of static friction between the pipe and the horizontal surface is 0.30. The truck accelerates from rest. Determine (a) the maximum acceleration the truck can have for which the pipe will roll without slipping until it falls off the back of the truck, (b) for the acceleration found in part (a), the distance the truck will have traveled when the pipe falls off.

V.67 The 2.5-kg homogeneous cylinder rolls without slipping down the curved surface. In the position where $\theta = 36.87°$, the speed of the mass center is 4 m/s. Determine the acceleration of the cylinder in this position and the components of the reaction forces acting on the cylinder.

V.68 Derive the differential equation of motion for the angle θ for the 2.5 kg homogeneous cylinder in Problem V.67.

V.69 A 5-lb wheel of 8-in. radius was imperfectly manufactured, so its mass center G is 2 in. from its geometric center C, as shown. The radius of gyration with respect to an axis through the center C is 7 in. As a result of the imperfection, the velocity of point C varies when the wheel rolls. In the position where $\theta = 0$, point C is moving to the left at 2 in./s. Assuming that the wheel is rolling without slipping, determine the corresponding angular acceleration of the wheel and the acceleration of point C.

Prob. V.69

500 mm

G

200 N

Prob. V.71

V.70 Solve Problem V.69 if the given velocity for point C corresponds to the position where $\theta = 30°$.

V.71 The 100-kg roller is initially at rest and is acted upon by a 200-N force as shown. The body rolls without slipping. Determine (a) the velocity of the mass center after 6 s, (b) the friction force required to prevent slipping.

V.72 Solve Illustrative Problem 5 for a wire tension of 60 lbs.

200 mm

G

150 mm

40 N

θ

Prob. V.73

V.73 The stepped pulley has a mass of 10 kg and a radius of gyration of 180 mm, and the coefficients of friction between the pulley and the horizontal surface are $\mu_s = 0.40$, $\mu_k = 0.35$. The angle θ is held constant as the end of the cable is pulled by a 40-N force. Determine the acceleration of the center of mass G and the angular acceleration of the disk if $\theta = 90°$.

V.74 Solve Problem V.73 (a) if $\theta = 75°$, (b) if $\theta = 45°$.

V.75 When the thin hoop of radius R and weight W was placed on the floor, its center O was at rest and it was rotating at an angular velocity $\bar{\omega}_1$. Letting the coefficient of kinetic friction be μ_k, derive expressions for (a) the speed of point O at the instant when the hoop ceases to slip relative to the ground, (b) the time duration during which slipping occurs.

ω_1

R

O

O

25 ft/s

Prob. V.76

Prob. V.75

V.76 When projected onto the floor, the 16-lb bowling ball is moving horizontally at 25 ft/s and is not rotating. The coefficient of kinetic friction is 0.20. Determine (a) the speed of the ball when it begins to roll on the ground without slipping, (b) the distance traveled and the elapsed time for the slipping phase of the motion. Regard the bowling ball as a homogeneous 1-ft-diameter sphere.

V.77 A 12-kg uniform rod is released from rest in the position shown. Determine the required coefficient of static friction to prevent slipping at

Prob. V.77

Prob. V.79

Prob. V.80

Prob. V.83

Prob. V.84

point *A*. In which direction would point *A* slip if the actual coefficient of friction were less than this value?

V.78 The coefficients of static and kinetic friction between the 12-kg uniform bar in Problem V.77 and the ground are $\mu_s = 0.27$, $\mu_k = 0.25$. Determine the angular acceleration of the rod.

V.79 The semicylinder of mass *m* and radius *R* is released from rest in the position shown. Determine (a) the minimum value of the coefficient of friction for which the semicylinder will roll without slipping, (b) the corresponding initial angular acceleration.

V.80 Gear *A* is stationary. Gear *B* weighs 10 lb and has a radius of gyration of 5 in. The uniform bar *AB* weighs 3 lb. If the system is released from rest in the position shown, determine (a) the initial angular acceleration of the bar, and (b) the initial acceleration of the mass center of gear *B*.

V.81 The 250-kg marble slab is being moved by placing it on two cylindrical rollers and applying the 150-N towing force. Each cylinder has a mass of 25 kg. Determine the acceleration of the marble slab. Assume that there is no slippage.

Prob. V.81   **Prob. V.82**

V.82 The 250-kg marble slab is being moved by placing it on a lightweight frame that is supported by two 25-kg cylinders, as shown. A 150-N towing force is applied. Determine the acceleration of the slab.

V.83 A 20-ft beam weighing 2000 lb is lifted by two cables. The upward acceleration of cable *A* is 10 ft/s², whereas that of cable *B* is 4 ft/s². The beam is initially at rest in the position shown. Determine the tension in each cable at this instant.

V.84 The 5000-kg crate is at rest in the position shown when cables *A* and *B* are given upward accelerations of 4 m/s² and 6 m/s², respectively. Determine the tension in each cable. The crate may be regarded as a homogeneous body.

V.85 To correct the course of a missile, it is proposed to increase the thrust in engine A. The mass of the missile is (10^5) kg and each engine has a thrust of $2(10^3)$ kN when the increase is to occur. It is desired to rotate the missile by 1° in 1 s. Determine (a) the required increase in the thrust of engine A, (b) the acceleration of the center of mass G immediately after the thrust of engine A is increased. The missile may be approximated as a thin uniform bar 60 m long.

V.86 and V.87   The 4-kg homogeneous plate, which is suspended as shown, is initially at rest. Determine the initial acceleration of the mass center when the connection at point B breaks suddenly.

Prob. V.85

Prob. V.86

Prob. V.87

V.88   An 80-kg file cabinet is placed on the 80-kg scaffold. Cable B suddenly breaks when the system is at rest. Assuming that the cabinet does not move relative to the scaffold, determine the initial acceleration of the cabinet when it is (a) at the center of the scaffold, (b) at end B. Consider the cabinet to be a slender bar and the scaffold to be a uniform plate.

Prob. V.88

V.89   Point A on the 20-lb bracket has a constant speed of 10 in./s. For this position determine the tension in the cables and the couple $\bar{M}_G$ required for the motion.

Prob. V.89

Prob. V.90

V.90   The 20-lb homogeneous plate is released from rest in the position shown. Determine (a) the tension in the supporting cables AE and BF, (b) the acceleration of the center of mass.

V.91   Solve Problem V.90 if, in the position shown, the plate is moving to the left and the speed of the center of mass G is 5 ft/s.

V.92   The 500-kg box is released from rest in the position shown. The coefficient of friction between all surfaces is 0.20. Determine (a) the angular acceleration of the crate and (b) the reactions at A and B.

Prob. V.92

V.93   Solve Problem V.92 if in the position shown point B is moving to the right at 1 m/s.

V.94   The 100-g rigid link ABC is a portion of the control linkage of an automobile carburetor. At the instant shown, the control rod BE has a velocity of 20 mm/s and an acceleration of 10 mm/s², both to the left. Determine the value of the force F that is pushing control rod BE. The mass of both control rods is negligible and the linkage lies in the horizontal plane.

Prob. V.94

Prob. V.95

V.95   In the position shown, the power stroke of a one-cylinder engine applies a 5000-N force to the 1-kg piston head. The 2-kg connecting rod AB is pinned to crankshaft BC, which is rotating at a constant counterclockwise speed of 4000 rev/min. For the position where $\theta = 90°$, determine (a) the couple $\bar{M}$ that is required to hold the speed of the crankshaft constant for this position, (b) the reactions at pins A and B for this position. The linkage lies in the horizontal plane and the mass of the piston is negligible.

V.96   Solve Problem V.95 for $\theta = 45°$.

V.97   In the position shown, point A is moving to the left at 3 m/s and decelerating at 6 m/s². The mass of the wheel is 10 kg and its radius of gyration about point A is 80 mm. What is the value of the force F acting on the piston C if the mass of the piston and bar BC are negligible? Consider point B to be on the circumference of the wheel and assume that the wheel does not slip.

Prob. V.97

V.98   Solve Problem V.97 if the piston has a mass of 1 kg and the mass of bar BC is 3 kg.

V.99   The flywheel is rotated clockwise at a constant angular speed ω by a torque $\bar{M}$. The mass of each bar is m. Determine the required value of M for the instant shown.

V.100   Two identical 80-kg cylinders are interconnected by rod AC whose mass is 10 kg. The system is at rest in the position shown when the 400-N force is applied at point A. Determine the acceleration of each cylinder at this instant. Assume that there is no slipping.

Prob. V.99

V.101   Solve Problem V.100 if cylinder A is moving to the left at 5 m/s in the position shown when the 400-N force is applied.

Prob. V.100

V.102   The A-frame shown consists of two identical 100-lb bars that are 6 ft long. Cable AB suddenly breaks. Determine (a) the initial angular acceleration for each member and, (b) the initial reactions at points C and E.

Prob. V.102

Prob. V.103

V.103   Bars AB and BC have the same mass m and length l. Derive two differential equations of motion relating the value of θ_1 and θ_2 and their derivatives to the mass m and length l.

E. ENERGY AND MOMENTUM PRINCIPLES

Our study of energy and momentum principles for particle motion showed that these principles greatly facilitate problem solving for certain classes of problems, such as those involving conservative or impulsive forces. These principles are, in essence, special forms of the differential equations that arise when using Newton's second law.

In this section we shall adapt the energy and momentum principles for a system of particles, derived in Module Three, to the special case where the particles form a rigid body in planar motion. As in the previous sections, we shall restrict the study to the case of bodies that have a plane of symmetry parallel to the plane of motion, reserving the more general case for Module Seven on three-dimensional motion.

1 Work and Energy

Recalling the symbolism for the work-energy formulation for a system of particles, T is the total kinetic energy of the system, V is the potential energy of all conservative forces acting on the system, and $U_{1\to2}^{(nc)}$ is the total work done by the nonconservative forces acting on the system. In terms of these symbols, the work-energy equation is

$$T_2 + V_2 = T_1 + V_1 + U_{1\to2}^{(nc)}$$ (24)

Figure 12

Let us begin the study of this equation by formulating the kinetic energy for a rigid body. In Figure 12 a rigid body is modeled as a continuum of infinitesimal particles dm whose position with respect to the origin of a fixed reference frame is $\bar{r}_{i/O}$. Denoting the velocity of each particle as $\bar{v}_i \equiv \dot{\bar{r}}_{i/O}$, the kinetic energy of the body, which is the sum of the kinetic energy of all the particles in the system, may be written as an integral over the volume of the body.

$$T = \tfrac{1}{2} \int_{\mathcal{V}} \bar{v}_i \cdot \bar{v}_i \, dm$$ (25)

where the mass m is a function of the volume $\mathcal{V}$ occupied by the body. Our objective now is to place equation (25) in a more useful form for computational purposes. We begin by choosing a translating reference frame xyz whose origin coincides with an arbitrary point A in the body. Then, using the vectors shown in Figure 12, for a rigid body,

$$\bar{r}_{i/O} = \bar{r}_{A/O} + \bar{r}_{i/A}$$

$$\bar{v}_i = \bar{v}_A + \bar{\omega} \times \bar{r}_{i/A}$$

Hence, equation (25) becomes

$$T = \tfrac{1}{2} \int_{\mathcal{V}} [(\bar{v}_A + \bar{\omega} \times \bar{r}_{i/A}) \cdot (\bar{v}_A + \bar{\omega} \times \bar{r}_{i/A})] \, dm$$

$$= \tfrac{1}{2} \int_{\mathcal{V}} [\bar{v}_A \cdot \bar{v}_A + 2\bar{v}_A \cdot (\bar{\omega} \times \bar{r}_{i/A})$$

$$+ (\bar{\omega} \times \bar{r}_{i/A}) \cdot (\bar{\omega} \times \bar{r}_{i/A})] \, dm$$

$$= \tfrac{1}{2}(v_A)^2 \int_{\mathcal{V}} dm + \bar{v}_A \cdot [\bar{\omega} \times \int_{\mathcal{V}} \bar{r}_{i/A} \, dm]$$

$$+ \tfrac{1}{2} \int_{\mathcal{V}} [\bar{\omega} \times \bar{r}_{i/A}) \cdot (\bar{\omega} \times \bar{r}_{i/A})] \, dm \qquad (26)$$

The first integral in equation (26) gives the mass m of the rigid body. To investigate the meaning of the other two integrals, let us resolve $\bar{r}_{i/A}$ into its components. Referring once again to Figure 12,

$$\bar{r}_{i/A} = x\bar{i} + y\bar{j}$$

Also, for planar motion, $\bar{\omega} = \omega\bar{k}$, so the second and third integrals in equation (26) now become

$$\int_{\mathcal{V}} \bar{r}_{i/A} \, dm = \int_{\mathcal{V}} x \, dm \, \bar{i} + \int_{\mathcal{V}} y \, dm \, \bar{j}$$

$$\int_{\mathcal{V}} [(\bar{\omega} \times \bar{r}_{i/A}) \cdot (\bar{\omega} \times \bar{r}_{i/A})] \, dm$$

$$= \omega^2 \int_{\mathcal{V}} [(x\bar{j} - y\bar{i}) \cdot (x\bar{j} - y\bar{i})] \, dm$$

$$= \omega^2 \int_{\mathcal{V}} (x^2 + y^2) \, dm$$

The integrals occuring in the first equation are the first moments of mass. Hence, referring to equations (15), this term becomes

$$\int_{\mathcal{V}} \bar{r}_{i/A} \, dm = mx_G \, \bar{i} + my_G \, \bar{j} = m\bar{r}_{G/A}$$

Next, we note from equation (12) that the last integral from equation (26) is the moment of inertia. Hence

$$\int_{\mathcal{V}} [(\bar{\omega} \times \bar{r}_{i/A}) \cdot (\bar{\omega} \times \bar{r}_{i/A})] \, dm = \omega^2 I_{zz}$$

Substituting these results into equation (26) yields

$$T = \tfrac{1}{2}mv_A{}^2 + \bar{v}_A \cdot (\bar{\omega} \times m\bar{r}_{G/A}) + \tfrac{1}{2}I_{zz}\omega^2 \qquad (27)$$

This is the result when point A is arbitrarily chosen. The formula is considerably simplified when the *origin A is located at the center of mass G*, in which case $\bar{r}_{G/A} = \bar{0}$ and equation (27) reduces to

$$\boxed{T = \tfrac{1}{2}mv_G{}^2 + \tfrac{1}{2}I_{zz}\omega^2} \qquad (28a)$$

In effect, when using equation (28a), we are considering the motion of the body to be a superposition of a translation following the center of mass G plus a rotation about point G. In the language of dynamics, the kinetic energy of translation is $\tfrac{1}{2}mv_G{}^2$ and the kinetic energy of rotation is $\tfrac{1}{2}I_{zz}\omega^2$. Clearly, the latter term is zero for a translating body and for motion of a particle.

We shall use equation (28a) to describe rigid bodies that are in translation or in general motion. The third type of planar motion is a pure rotation about a fixed axis. For such cases we *locate the origin A at the fixed axis*. Then, setting $\bar{v}_A = 0$ in equation (27) yields

$$\boxed{T = \tfrac{1}{2}I_{zz}\omega^2} \qquad (28b)$$

This shows that the body has only rotational kinetic energy for a coordinate system whose origin coincides with the axis of rotation.

Having now derived the formula for the kinetic energy of a rigid body, let us examine the other terms of the work-energy principle, equation (24). The potential energy V is determined in the same manner as was done for particles. Each conservative force may be considered individually, as their potential energies are additive. We may also use this procedure for the nonconservative forces, tracing the path of the point of application of each force and computing the work of each force individually.

However, in some situations the points of application may not have simple motions. Let us derive an alternative method for computing the work term for such cases.

Consider the force $\bar{F}$ in Figure 13 which is applied at point P in the rigid body. The work done by the force $\bar{F}$ in changing the position of point P with respect to the fixed origin is

$$U_{1 \to 2} = \int_{(r_{P/O})_1}^{(r_{P/O})_2} \bar{F} \cdot d\bar{r}_{P/O}$$

where the limits in the integral denote the coordinates of the initial and final positions of point P.

We may now relate the motion of point P to that of the origin A.

Figure 13

Using the kinematic formula relating the velocity of two points in a rigid body gives

$$\bar{v}_P \equiv \frac{d\bar{r}_{P/O}}{dt} = \bar{v}_A + \bar{\omega} \times \bar{r}_{P/A} \equiv \frac{d\bar{r}_{A/O}}{dt} + \frac{d\bar{\theta}}{dt} \times \bar{r}_{P/A}$$

Upon multiplying both sides by dt, we obtain

$$d\bar{r}_{P/O} = d\bar{r}_{A/O} + d\bar{\theta} \times \bar{r}_{P/A} \tag{29}$$

This expression shows that an infinitesimal displacement of point A causes point P to displace in the same manner, whereas an infinitesimal rotation $d\bar{\theta}$ causes P to move perpendicular to the line from point A to point P, as indicated by the cross product term.

We now substitute equation (29) into the expression for the work to obtain

$$U_{1 \to 2} = \int_{(r_{P/O})_1}^{(r_{P/O})_2} \bar{F} \cdot (d\bar{r}_{A/O} + d\bar{\theta} \times \bar{r}_{P/A})$$

Employing the readily verified identity

$$\bar{F} \cdot (d\bar{\theta} \times \bar{r}_{P/A}) \equiv (\bar{r}_{P/A} \times \bar{F}) \cdot d\bar{\theta}$$

we break the expression for the work into two terms

$$\boxed{U_{1 \to 2} = \int_{(r_{A/O})_1}^{(r_{A/O})_2} \bar{F} \cdot d\bar{r}_{A/O} + \int_{\theta_1}^{\theta_2} (\bar{r}_{P/A} \times \bar{F}) \cdot d\bar{\theta}} \tag{30}$$

Recall from statics that the given force $\bar{F}$ acting at point P is equivalent to a force $\bar{F}$ acting at point A and a couple $\bar{M}_A$ that is the moment of the given force about point A; $\bar{M}_A = \bar{r}_{P/A} \times \bar{F}$. Equation (30) is a direct consequence of this fact; the first term gives the work done by the equivalent force in translating the body and the second term gives the work done by the equivalent couple in rotating the body. Therefore, if we are dealing with a system of nonconservative forces, we may resolve the forces into their equivalent force and couple, $\Sigma \bar{F}^{(nc)}$ and $\Sigma \bar{M}_A^{(nc)}$. The work done is then given by

$$U_{1 \to 2}^{(nc)} = \int_{(r_{A/O})_1}^{(r_{A/O})_2} \Sigma \bar{F}^{(nc)} \cdot d\bar{r}_{A/O} + \int_{\theta_1}^{\theta_2} \Sigma \bar{M}_A^{(nc)} \cdot d\bar{\theta} \tag{31}$$

An important consequence of equation (31) is that the forces internal to a rigid body do no work when the body moves. This is so because

Newton's third law tells us that all internal forces cancel each other in computing the total force and total moment.

In summary, we have found that the work-energy principle, equation (24), can be applied in conjunction with either of the kinetic energy expressions of equations (28) to relate the angular and linear velocity of a rigid body when it occupies different positions. The work of the nonconservative forces can be determined either by the same methods as that used in particle motion, or alternatively when it is difficult to follow the point of application of the force, by applying equation (30) or (31). Equally important, in those cases where there is *any* uncertainty as to whether a force is conservative or nonconservative, it is usually best to treat it as a nonconservative force, rather than pausing to determine whether it is indeed conservative. Remember that the concept of the potential energy of a conservative force was derived in Module Three merely as an aid in determining the work done by the force.

Finally, recall that it is sometimes impossible to evaluate the work done by a force until the motion of the system has been determined. In such situations we resort to formulating and solving the differential equations of motion.

EXAMPLE 5

A 200-N force is applied to the free end of cable wrapped around the drum of the stepped disk, as shown. The disk has a mass of 40 kg and a radius of gyration of 150 mm. Knowing that the disk starts from rest and rolls without slipping, determine its speed after moving 5 m up the incline.

Solution

The key to recognizing that this problem is ideally suited for the work-energy principle is that the speed is required at a known position. Because the ultimate success of this approach requires that we be able to evaluate the work term, we begin with a free body diagram to assist in this determination.

The origin of the *xyz* coordinate system is located at the mass center *G* because the disk is in general motion. The weight force is a conservative force, so we shall account for it in the potential energy term, although it would not be incorrect to ignore the conservative nature of this force. We shall determine $U_{1\to2}^{(nc)}$ by applying equation (31), because the remaining forces are applied to different points of the body at each instant; that is, different points of the disk come into contact with the ground and the 200-N force as the disk rolls. Hence, from the free body diagram we have

$$\Sigma \bar{F}^{(nc)} = (200 - f)\bar{i} + N\bar{j} \text{ N}$$

$$\Sigma \bar{M}_C^{(nc)} = [-f(0.25) - 200(0.10)]\bar{k} \text{ N-m}$$

The negative signs in the moment arise because clockwise moments are the direction of the negative *z* axis, according to the right-hand rule.

In order to compute the work, we next describe the differential displacement of point *G* and the differential rotation. For this system point *G* moves parallel to the incline in the *x* direction. Letting *ds* be the increment of distance traveled,

$$d\bar{r}_{G/O} = ds\,\bar{i}$$

The disk does not slip. Hence, this displacement will cause the disk to rotate clockwise by an amount $ds/R = ds/0.25$ rad, as sketched here. Hence

$$d\bar{\theta} = -\frac{ds}{0.25}\bar{k}$$

The initial value of *s* is zero and we want the velocity when *s* = 5 m. Therefore, equation (31) for the work gives

$$U_{1\to2}^{(nc)} = \int_{(r_{G/O})_1}^{(r_{G/O})_2} \Sigma \bar{F}^{(nc)} \cdot d\bar{r}_{G/O} + \int_{\theta_1}^{\theta_2} \Sigma \bar{M}_G^{(nc)} \cdot d\bar{\theta}$$

$$= \int_{s=0}^{s=5} [(200 - f)\bar{i} + N\bar{j}] \cdot ds\,\bar{i}$$

$$- \int_{s=0}^{s=5} (0.25f + 20)\bar{k} \cdot (-\frac{ds}{0.25}\bar{k})$$

$$= \int_0^5 (200 - f)\,ds + \int_0^5 (f + 80)\,ds$$

$$= (200 - f)5 + (f + 80)5 = 1400 \text{ J}$$

There are several noteworthy aspects to this result. That the normal force does no work is not surprising, because it is perpendicular to the

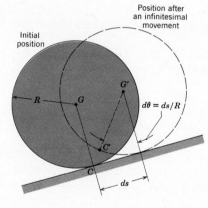

Initial position

Position after an infinitesimal movement

R G G' $d\theta = ds/R$

C C'

ds

direction of motion. However, the fact that the friction force does no work at first seems to conflict with the notion that friction retards motion. The friction force in this system does no work because the contact surfaces are not moving relative to each other. However, friction does slow the disk in that it tends to make the disk rotate as well as translate. Finally, note that the work done by the cable could have (alternatively) been obtained by multiplying the cable tension by the distance the end of the cable moves.

The solution can now be completed by writing the initial and final values of the kinetic and potential energies. The disk starts from rest, so choosing the initial elevation as the datum for gravitational potential energy gives

$$T_1 = 0 \qquad V_1 = 0$$

In the final position the center of mass has a speed of v and the no slip condition gives $\omega = v/R = v/0.25$. From equation (28a) and the given value for the radius of gyration, we write

$$T_2 = \tfrac{1}{2}mv^2 + \tfrac{1}{2}I_{zz}\omega^2 = \tfrac{1}{2}(40)v^2 + \tfrac{1}{2}(40)(0.15)^2\left(\frac{v}{0.25}\right)^2$$

$$= 27.20v^2$$

When the disk moves 5 m up the incline, its vertical elevation above the datum is 5 sin 10°, so

$$V_2 = mgh = 40(9.806)(5 \sin 10°)$$

$$= 340.6 \text{ J}$$

Hence, equation (24) gives

$$T_2 + V_2 = T_1 + V_1 + U^{(nc)}_{1\to2}$$

$$27.20v^2 + 340.6 = 0 + 0 + 1400$$

$$v = 6.24 \text{ m/s}$$

EXAMPLE 6
The system shown in the sketch is used to lower a 100-lb homogeneous package. The horizontal bar weighs 25 lb and the other bars weigh 10 lb. The spring has a stiffness of 50 lb/in. and an unstretched length of 9 in. When $\theta = 53.13°$ the package is descending with a speed of 10 ft/s. What constant horizontal force $\bar{F}$ is required to bring the system to rest at $\theta = 0°$?

Solution

We must find the value of F that produces a specific change in the velocity of the system between the initial and final positions. This is the key that suggests employing the work-energy principle. As an aid in understanding the role of the forces, we draw a free body diagram of each body, taking care that interaction forces between bodies satisfy Newton's third law. The bar EH and the package move in unison, and thus they may be treated as one body.

Note that we selected the coordinate systems for the bodies to be parallel to each other. The origins of the coordinate systems were chosen to coincide with the fixed axes of rotation for bars AB and CD, and with the center of mass of the combination of the package and bar EH.

The weight and spring forces are conservative. The reactions at points A and C are applied at fixed points, so they do no work. The work done by each interaction force in moving one body is canceled by the work done by its counterpart in moving the other body by the same amount. Hence, the only nonconservative force doing work in moving the system is $\bar{F}$. The work done by this force, and also the system's potential energy, is determined with the aid of sketches of the system in the initial and final positions.

The sketches show that bar EH moves to the left by 12 in. in the motion. Because $\bar{F} = F\bar{\imath}$, it follows that

$$U_{1\to2}^{(\text{nc})} = -F(12.0) \text{ in.-lb}$$

Let us choose the elevation of pins A and C as the datum for gravity. In the initial position the centers of mass of bars AB and CD are 4.5 in. below the datum, whereas the corresponding points for bar EH and the package are 9 in. and 6 in. below the datum, respectively. The length of the spring connecting points A and E is 9 in., so the spring is unstretched. Thus

$$V_1 = (V_{\text{gravity}} + V_{\text{spring}})_1 = 10(-4.5) + 10(-4.5)$$

$$+ 100(-6) + 25(-9) + 0 = -915.0 \text{ in.-lb}$$

In the final position the length of the spring is $\sqrt{(15)^2 + (12)^2} = 19.209$ in. Referring to the sketch for the locations of the center of mass, we have

$$V_2 = (V_{\text{gravity}} + V_{\text{spring}})_2 = 10(-7.5) + 10(-7.5)$$

$$+ 100(-12) + 25(-15) + \tfrac{1}{2}(50)(19.209 - 9.0)^2$$

$$= 880.6 \text{ in.-lb}$$

The last terms we require are those for the kinetic energy. We want the system to come to rest at $\theta = 0$. Hence, the value of T_2 is zero. Before we can determine T_1, the velocities of the bars must be related to the given initial speed for the package, for which a velocity sketch is now required. Remember that bar EH is translating.

From this sketch we see that

$$v_D = v_B = v_{\text{package}} = 120 \text{ in./s}$$

$$\omega_{AB} = \omega_{CD} = \frac{v_B}{15} = 8 \text{ rad/s}$$

The coordinate systems in the free body diagram require that we use equation (28a) for the kinetic energy of the translating body and equation (28b) for the rotating bars. With the assistance of Appendix B and the parallel axis theorem for the computation of I_{zz} with respect to points A and C, the total kinetic energy is found to be

$$T_1 = \tfrac{1}{2}(I_{zz})_{AB}\,\omega_{AB}{}^2 + \tfrac{1}{2}(I_{zz})_{CD}\,\omega_{CD}{}^2 + \tfrac{1}{2}(m_{EH} + m_{\text{package}})v_P{}^2$$

$$= \tfrac{1}{2}(2)\tfrac{1}{3}\left[\frac{10}{32.17(12)}(15)^2\right](8)^2 + \tfrac{1}{2}\left[\frac{125}{32.17(12)}\right](120)^2 = 2456 \text{ in.-lb}$$

Finally, we substitute all terms into the work-energy principle and solve for F.

$$T_2 + V_2 = T_1 + V_1 + U_{1\to2}^{(\text{nc})}$$

$$0 + 730.6 = 2456 - 1065.0 - 12F$$

$$F = 55.03 \text{ lb}$$

Prob. V.104

HOMEWORK PROBLEMS

V.104 The 50-kg cylinder is slipping as it rolls. It has a clockwise angular velocity of 120 rev/min and its center O has a speed of 54 km/hr. Determine the cylinder's kinetic energy when (a) the translational velocity is to the left, (b) the translational velocity is to the right.

Prob. V.105

Prob. V.107

Prob. V.109

V.105 The mass of each of the uniform bars is m and that of the flywheel disk is 8 m. In the position shown the disk is rotating at angular speed ω counterclockwise. Determine the kinetic energy of each bar and of the disk.

V.106 For the mechanism of Problem V.105, derive an expression for the couple $\bar{M}$ applied to the disk that will bring the system to rest in 1 rev of the disk.

V.107 In the position shown, the 5-kg uniform rod is rotating counterclockwise with an angular speed of 120 rev/min. Determine the kinetic energy of the bar (a) using equation (28a), (b) using equation (28b).

V.108 For the initial motion of the bar given in Problem V.107, determine the reactions at pin O when the bar reaches the vertical position.

V.109 The 50-lb pulley has a radius of gyration of 12 in. and is rotating at $\omega = 240$ rev/min clockwise. Determine the total kinetic energy of the system.

V.110 Initially ω for the system in Problem V.109 is zero and the system is in static equilibrium. While in this position a 10-lb mass is added to block B. Determine the velocity of each block when the pulley has made 2 rev.

V.111 A 2000-kg turbine engine has a radius of gyration of 1 m. When shut down while operating at 1500 rev/min, it requires 4000 rev to come to rest. Determine the average magnitude of the frictional torque caused by the bearings and air resistance.

V.112 The 10-kg cylinder B is placed in contact with a conveyor belt traveling at 40 m/s. The cylinder is initially at rest and the coefficient of friction between the cylinder and belt is 0.25. Determine the number of revolutions required for the cylinder to attain a constant angular velocity.

Prob. V.112

V.113 By pushing on point B with a 1000-N force, the hydraulic piston

C brakes the 800-kg flywheel whose radius of gyration is 150 mm. Initially the flywheel is rotating at 550 rev/min clockwise. The coefficient of friction between the rim and brake shoe *D* is 0.60. Determine the number of revolutions required for the flywheel to stop.

Prob. V.113

V.114   The identical inner gears *A* and *B* each have a mass of 2 kg and a radius of gyration of 100 mm. The radius of gyration of the 4-kg outer gear is 400 mm. A torque of 200 N-m is applied to gear *A*. Determine the angular velocity of gear *B* when gear *C* has made 1 rev starting from rest.

Prob. V.114

Prob. V.115

V.115   The 10-kg cylinder is released from rest. Determine the velocity of its mass center after it has made 2 rev.

Prob. V.117

V.116 For the system of Problem V.115, determine how far the cylinder must drop to attain a speed of 20 m/s.

V.117 The 5-kg wheel of radius of gyration 80 mm is hoisted by a constant force P-400 N. Assuming that the cord does not slip on the wheel, determine the distance the 10-kg mass A must be raised, starting from rest, to attain a speed of 15 m/s.

V.118 For the system of Problem V.117, determine the force $\bar{P}$ required for the mass A to have a speed of 10 m/s when it has fallen 5 m starting from rest.

V.119 The 10-lb force is applied to pin P on the 40-lb wheel when it is at rest in the position shown. The radius of gyration of the wheel is 0.9 ft. If the force remains horizontal as the wheel rolls without slipping, determine the velocity of the center of the wheel (a) after the wheel has undergone one-half a revolution, (b) after the center of the wheel has moved 4 ft.

Prob. V.119

V.120 The solid homogeneous semicylinder is released from rest in the position shown. Determine its angular speed when its mass center is at its lowest point. Assume that slipping does not occur.

V.121 A 15-lb sphere of 1-ft radius rolls without slipping inside a curved surface of 4-ft radius. Knowing that the sphere is released from rest at $\theta = 60°$, determine (a) the velocity of the center of the sphere as it passes $\theta = 0°$, (b) the magnitude of the normal reaction at that instant.

Prob. V.120

Prob. V.121

V.122 Solve Problem V.121 for the instant when the sphere is at $\theta = 30°$.

V.123 A solid homogeneous cylinder A of radius r rolls down a fixed cylindrical surface of radius R that is sufficiently rough to prevent slipping. If cylinder A starts from rest in the position of unstable equilibrium, Q, determine the angle θ where it will lose contact with the cylindrical surface.

Prob. V.123

V.124 Determine the minimum initial angular speed the 2-kg slender rod

Prob. V.124

must have when $\theta = 0$ to just attain a position of $\theta = 90°$. The stiffness of the spring is 20 N/m and the spring is unstretched in the position where $\theta = 0$.

V.125 The 20-lb rod AB is constrained to move in the vertical plane by block A, which follows the horizontal groove, and block B, which follows the vertical one. The unstretched length of the spring is 5 in., and $\theta = 0$ is the static equilibrium position. If the block is released from rest when $\theta = 45°$, determine the velocity of the slider block B when $\theta = 0°$, (a) if the slider blocks have negligible mass, (b) if the mass of each slider block is 2 lb.

Prob. V.125

Prob. V.126

V.126 The system shown in the diagram moves in the horizontal plane. The stiffness of the spring is k and the spring is unstretched when $\theta = 60°$. Each bar has a weight W, and the weight of the piston is negligible. Determine the constant force $\bar{F}$ that, starting the system from rest at $\theta = 60°$, gives the piston a speed v to the right at $\theta = 30°$.

V.127 A 60-N vertical force is applied at point B when the system is at rest in the position where $\theta = 60°$. The mass of the rods is 2 kg each and that of the disk is 4 kg. The spring has a stiffness of 50 N/m and an unstretched length of 1 m. Assuming that the disk rolls without slipping, determine the angular velocity of each rod when $\theta = 0$.

Prob. V.127

Prob. V.129

V.128   For the system of Problem V.127, determine the velocity of point C when $\theta = 30°$.

V.129   The mechanism shown lies in a horizontal plane. Bar AB weighs 10 lb and gear B weighs 5 lb and has a 3-in. radius of gyration. The spring has a stiffness of 20 lb/ft and an unstretched length of 12 in. If the system was at rest at $\theta = 0$ when gear B was given a very slight push to the right, determine (a) the maximum speed of point B and the corresponding value of x, (b) the maximum value of x attained by gear B in the motion. Neglect the mass of block A.

V.130   The 3-kg bar AB and the 4-kg bar BC form a linkage in the vertical plane. Bar AB is rotating clockwise at 180 rev/min in the position shown. What constant force $\bar{F}$ should be applied to the 1-kg collar C to bring the system to rest in the position where the bars are horizontal? Friction is negligible.

Prob. V.130

2   Impulse and Momentum

The linear impulse-momentum principle for a system of particles is the integral of Newton's second law for the total system. This principle (derived in Module Three) in equation form is

$$m(\bar{v}_G)_2 = m(\bar{v}_G)_1 + \int_{t_1}^{t_2} \Sigma\,\bar{F}\,dt \tag{32}$$

Equation (32) is equivalent to saying that the change in the linear momentum $m\bar{v}_G$ equals the total impulse $\int_{t_1}^{t_2} \Sigma\,\bar{F}\,dt$ of the forces acting on the system.

The principle of angular impulse and momentum has similar significance, in that it is the integral of the moment equation for a system of particles. Letting point A be a fixed point or the center of mass of the system, we have

$$(\bar{H}_A)_2 = (\bar{H}_A)_1 + \int_{t_1}^{t_2} \Sigma \, \bar{M}_A \, dt \tag{33}$$

where $\bar{H}_A$ is the total angular momentum with respect to point A. Equation (5) showed that for a rigid body in planar motion,

$$\bar{H}_A = I_{zz}\bar{\omega}$$

where the origin of the xyz coordinate system is located at point A and the z axis is normal to the plane of motion. Then $\bar{\omega} = \omega\bar{k}$ and equation (33) becomes

$$\boxed{I_{zz}\omega_2 = I_{zz}\omega_1 + \int_{t_1}^{t_2} \Sigma \, M_{Az} \, dt} \tag{34}$$

Clearly, equations (32) and (34) can be employed only if we can evaluate the impulse term; that is, we must know how $\Sigma \, \bar{F}$ and $\Sigma \, \bar{M}_{Az}$ vary with time. On the other hand, if these dependencies are known, it is a simple matter to integrate the basic differential equations of motion for a rigid body to relate its velocity at two instants of time, as we did in Illustrative Problem 2, rather than employing the momentum principles. However, equations (32) and (34) are of particular importance in the case of impulsive forces, that is, very large forces acting over a short interval of time. As we saw in treating forces of this type acting on a particle, the major effect of any impulsive force is described by its impulse, and not by its actual variation with time.

EXAMPLE 7

The 20-kg bar is the moving component of an impact machine used to test the strength of the object at point A. In a particular test the bar is released from rest in the horizontal position, strikes the object, and rebounds from the collision at a speed of 3 rad/s. The time of the collision was measured as 0.02 s. Determine the value of the distance h that minimizes the reactions at pin B during the collision, and also find the corresponding average value of the impact force.

Solution

In order to evaluate the impact, we must know the angular speed of the bar as it strikes the object. Because the given information defines the motion of the bar at another position, we can employ the work-energy principle to determine the angular velocity at impact. Hence, we must first solve a work-energy problem to obtain sufficient information to study the impact problem. A free body diagram of the falling bar is drawn below.

The reactions at point B act at a fixed point and therefore do no work. The only other force is gravity, which is conservative, so $U_{1 \to 2}^{(\text{nc})} = 0$. Letting the elevation of point B be the datum for gravity, in the initial position $\theta = 90°$, so $V_1 = 0$. In the final position $\theta = 0°$, so the center of mass is 0.5 m below the datum, and

$$V_2 = 20(9.806)(-0.50) = -98.06 \text{ J}$$

The origin of the xyz coordinate system was placed at point B because the bar is in pure rotation about the z axis through this point. Hence, with respect to the point B,

$$T = \tfrac{1}{2}I_{zz}\omega^2 = \tfrac{1}{2}(\tfrac{1}{3}ml^2)\omega^2 = \tfrac{1}{2}[\tfrac{1}{3}(20)\,(1.0)^2]\omega^2 = 3.333\omega^2$$

Because the bar is released from rest, $T_1 = 0$. Thus

$$T_2 + V_2 = T_1 + V_1 + U_{1 \to 2}^{(\text{nc})}$$

$$3.333\omega^2 - 98.06 = 0 + 0 + 0$$

$$\omega = 5.424 \text{ rad/s}$$

We can now study the impact problem. When treating an impulsive force, we can neglect any change in the position of the system. Hence, we draw a free body diagram of the bar in the vertical (impact) position.

The linear impulse-momentum relation is

$$m(\bar{v}_G)_2 = m(\bar{v}_G)_1 + \int_{t_1}^{t_2} [(B_x - F)\bar{\imath} + (B_y - 196.12)\bar{\jmath}]\, dt$$

For circular motion the speed of v_G is 0.5ω. Also, $\bar{v}_G$ is to the right initially and to the left at a known speed when the bar rebounds. Hence, using these values, the foregoing equation takes the form

$$20[-0.5(3.0)]\bar{\imath} = 20[0.5(5.424)]\bar{\imath} + [(B_x - F)\bar{\imath} + (B_y - 196.12)\bar{\jmath}](0.02) \text{ kg-m/s}$$

Equating the components of this equation yields

$$B_x - F = -\frac{1}{0.02}[30.0 + 54.24] = -4212 \text{ N}$$

and

$$(B_y - 196.12)0.02 = 0$$

from which $B_y = 196.12$ N.

We are left with one equation in two unknowns. However, the angular impulse-momentum principle may also be applied. We do so with respect to the fixed point B in order to eliminate the reactions from the moment term. The value of I_{zz} was determined in the first portion of the problem. Because the z axis is outward, counterclockwise rotations and moments are positive according to the right-hand rule. Referring to the free body diagram, we see that the only nonzero moment is that of the impact force $\bar{F}$. Hence

$$I_{zz}\omega_2 = I_{zz}\omega_1 + \int_{t_1}^{t_2} \Sigma M_{Bz}\, dt$$

$$6.667(-3.0) = 6.667(5.424) - Fh(0.02)$$

This gives $Fh = 2808$ kg-m^2/s.

We now have two equations for the values of B_x and F in terms of h. Solving them yields

$$F = \frac{2808}{h} \qquad B_x = \frac{2808}{h} - 4212$$

The reaction at point B is minimized when the magnitude of B_x has its smallest value. Setting $B_x = 0$ yields

$$h = \frac{2808}{4212} = 0.667 \text{ m} \qquad \triangle$$

The other component of the reaction, B_y, is, of course, the weight force. The corresponding value of F is

$$F = 4212 \text{ N} \qquad \triangle$$

The location on a rotating body where the reaction to an impulsive force is minimized is called the *center of percussion*.

Prob. V.131

HOMEWORK PROBLEMS

V.131   The pendulum consists of a 2-kg rod rigidly attached to a 4-kg sphere. At the instant shown it is rotating at 240 rev/min clockwise. Determine (a) the linear momentum, and (b) the angular momentum of the pendulum.

V.132   The 600-kg turbine of a jet engine is rotating at 4000 rev/min when

Prob. V.134

Prob. V.135

Prob. V.136

the engine is shut down. The radius of gyration of the turbine is 700 mm. Friction in the bearings produces a torque of 500 N-m. Determine the time required for the engine to come to rest.

V.133 Solve Problem V.42 using the momentum principles.

V.134 The 50-lb stepped pulley whose radius of gyration is 27 in. is released from rest. Use the linear and angular impulse-momentum principles to determine the velocity of its mass center 10 s later.

V.135 A homogeneous cylinder and a homogeneous sphere having the same radius roll without slipping down an incline starting from rest. Use the linear and angular impulse-momentum principles to derive expressions for the velocity of the center of each body after t s. Which body rolls faster?

V.136 Two cables are wrapped in opposite directions around the drum of the wire spool. The radius of gyration of the spool is 500 mm. If a constant tensile force of 40 N is applied to one of the cables as shown, determine the velocity of the disk's mass center 5 s after starting from rest. Assume that the drum does not slip on the ground. The mass of the spool is 10 kg.

V.137 The 20-kg flywheel A, whose radius of gyration is 250 mm, and the 40-kg disk B are connected by means of a belt that does not slip over the contacting surfaces. The flywheel is driven by a torque $M = 100t^{1/2}$ where M is in newton-meters and t is in seconds. Determine the time required for the disk to attain an angular velocity of 100 rad/s.

Prob. V.137

V.138 A 120-lb ice skater spins with knees bent and arms outstretched at an angular speed of 12 rev/min. In this position the skater has an 18-in. radius of gyration about the axis of rotation. When standing the radius of gyration is 6 in. and the mass center rises by 1 ft. Neglecting friction from the ice, determine (a) the angular velocity of the skater in the upright position, (b) the work done by the skater's muscles in going from the kneeling to the upright position.

V.139 A 50-kg crate is hoisted up the incline as shown. The torque applied by the 5-kg hoisting drum is $M = 800t^{1/3}$, where M is in newton-meters and t is in seconds. The radius of the drum is 600 mm and the radius of gyration is 500 mm. The coefficient of friction between the crate and the plane is 0.20. Knowing that the system was at rest when $t = 0$, determine the velocity of the block when $t = 4$ s.

Prob. V.139

V.140 A 100-g cylindrical block slides inside the 1-kg, slender cylindrical tube, which rotates freely about the vertical shaft. Friction is negligible. Initially, the system is rotating at 20 rev/min, as shown, and the block is stationary with respect to the tube at $R = 250$ mm. The block is then released and slides outward. Determine the angular speed of the tube when the block reaches $R = 400$ mm.

V.141 For the system in Problem V.140 determine (a) the angular speed of the tube as the block emerges from the open end B, (b) the velocity of the block as it emerges.

Prob. V.140

V.142 The 2-kg bar is initially translating at 10 m/s, as shown, when it is subjected to the impulsive force $\bar{F}$, whose average magnitude over a 0.10-s interval is 500 N. If $h = 250$ mm, determine the velocity of the center of mass and the angular velocity of the bar immediately after the impact.

V.143 Immediately after the application of the impulsive force $\bar{F}$ in Problem V.142, point A is observed to be the instantaneous axis of rotation. What are the corresponding values of h and of the angular velocity after the impact?

Prob. V.142

2000 km/hr

Prob. V.144

V.144 A 50-g bullet traveling at 2000 km/hr in the direction shown hits the 10-kg square plate. The bullet becomes embedded, coming to rest within the plate 0.002 s after the initial contact. Determine (a) the angular velocity of the plate immediately after the impact, (b) the impulsive reaction at pin A.

V.145 The 20-lb wooden block is initially at rest when it is struck by a 2-oz bullet that is traveling at 1500 ft/s in the direction shown. The bullet becomes embedded, coming to rest with respect to the block 0.004 s after the initial contact. Determine (a) the velocity of the center of mass of the block immediately after the impact, (b) the average tensile force within each cable during the impact.

Prob. V.145

V.146 The 300-g billiard ball of 20-mm radius is struck by a cue stick that exerts an average force of 600 N horizontally over a 0.005-s interval. If $h = 10$ mm, determine the velocity of the center of the ball and the angular velocity of the ball immediately after the impact.

Prob. V.146

V.147 Immediately after being hit by the cue stick, the billiard ball in Problem V.146 is rolling without slipping. Determine (a) the height h for the cue stick, (b) the velocity of the ball after the impact.

3   Eccentric Impact

Recall that Example 7 considered a collision between bodies in which the normal to the plane of contact did not coincide with the line connecting the mass centers of the two bodies. This is called an *eccentric impact*. In that example the motion of the system after the collision was given. Our aim here is to learn how to determine the motion of a system when it rebounds from a collision.

When we studied the collision of particles in Module III (for which the line of impact coincides with the line connecting the mass centers), we were able to solve problems by using the concept of conservation of linear momentum in conjunction with the coefficient of restitution, ϵ, which was a measure of the loss of energy during the collision. In

Immediately
before impact

Impact

Immediately
after impact

Figure 14

essence, we recognized that particles are not really rigid and that energy can be lost due to deformation during impact. The problem for rigid bodies is quite similar, but it is complicated slightly by the fact that the velocity at the point of impact is not the same as the velocity of the mass center.

Consider the general type of collision depicted in Figure 14. For convenience we employ an xyz coordinate system having an origin at the point of contact with the y axis tangent to the plane of contact and the x axis along the line of impact. In the figure points A and B are the centers of mass of the bodies, and points C and D are the points on each body that come into contact. In the following we shall establish procedures for determining the values of the kinematical variables $\bar{v}_A$, $\bar{\omega}_A$, $\bar{v}_B$, and $\bar{\omega}_B$ after the impact in terms of their values before the collision.

The actual collision process is quite complicated. Fortunately, we have the experimental observation that the component of the relative velocity of points C and D parallel to the line of impact is related to the coefficient of restitution, according to

$$\epsilon = -\left[\frac{(v_{Dx})_2 - (v_{Cx})_2}{(v_{Dx})_1 - (v_{Cx})_1}\right] \tag{35}$$

As an aside, this relationship applies to collisions where friction is negligible, and is accurate only for a limited range of relative velocities.

Equation (35) relates the velocities of the points of contact on each body, but we are interested in the velocities of the centers of mass. We therefore employ the relative velocity relations for each of the rigid bodies:

$$\bar{v}_C = \bar{v}_A + \bar{\omega}_A \times \bar{r}_{C/A} \qquad \bar{v}_D = \bar{v}_B + \bar{\omega}_B \times \bar{r}_{D/B} \tag{36}$$

When we use equations (36) to evaluate the x components of $\bar{v}_A$ and $\bar{v}_C$, equation (35) then represents one relation for the unknowns. Other relationships are obtained from the linear and/or angular impulse-momentum principles, which, of course, involves the impulse of the contact force $\bar{F}$ exerted between the two bodies. As shown in Figure 14, this force is directed along the line of impact.

To see how these equations are used, suppose that the bodies in Figure 14 are unconstrained in their motion, so that the contact force $\bar{F}$ is the only impulsive force acting on the body. In such a situation, we apply the linear and angular momentum principles for each body, the latter with respect to the centers of mass. There are x and y components of linear momentum and a z component of angular momentum. This gives three scalar equations of momentum for each body. Thus, after using equation (35), we have a total of seven scalar equations. The seven scalar unknowns are the x and y components of $\bar{v}_A$ and $\bar{v}_B$, the z components of $\bar{\omega}_A$ and $\bar{\omega}_B$, and the impulse of the contact force $\bar{F}$. Hence, we have a well-posed problem.

In contrast, when a body rotates about a fixed axis, there are impulsive reaction forces, as was seen in Example 7. To eliminate these reactions in the impact problem, we formulate the angular impulse-momentum relation with respect to the point on the fixed axis. The linear impulse-momentum relation for the body is then discarded, as it merely defines the impulse of the reaction.

EXAMPLE 8

A 1000-kg crate is dropped by parachute, hitting the ground in the position shown. At the instant of impact the crate is in rectilinear translation, downward, with a speed of 20 m/s. Knowing that the coefficient of restitution is 0.20, determine the motion of the crate at the end of the impact. The radius of gyration of the crate about its center of mass G is 0.80 m.

Solution

The forces exerted by the parachute and the weight force are negligible compared to the impulsive force generated by the collision. Choosing the x axis normal to the plane of contact, we sketch a free body diagram for the crate.

The earth is massive enough to be considered immovable. From the given information we know that

$$(\bar{v}_G)_1 = -20\bar{i} \text{ m/s} \qquad (\bar{\omega})_1 = \bar{0}$$

In order to use the equation for the coefficient of restitution, we shall need the general expression for v_{Cx} in terms of $\bar{v}_G$ and $\bar{\omega}$. Hence

$$\bar{v}_C = \bar{v}_G + \bar{\omega} \times \bar{r}_{C/G}$$

Because we want the x component of this equation, $\bar{r}_{C/G}$ must be expressed in terms of its x and y components. To do this we refer to the free body diagram, where we see that

$$\theta = \tan^{-1} \frac{0.5}{1.0} = 26.57°$$

$$\left|\bar{r}_{C/G}\right| = \sqrt{(1.0)^2 + (0.5)^2} = 1.1180 \text{ m}$$

from which it follows that

$$\bar{r}_{C/G} = \left|\bar{r}_{C/G}\right|[-\sin(\theta + 15°)\bar{i} + \cos(\theta + 15°)\bar{j}]$$

$$= -0.7418\bar{i} + 0.8365\bar{j} \text{ m}$$

Hence, because $\bar{\omega} = \omega\bar{k}$ for planar motion,

$$\bar{v}_C = \bar{v}_G + \omega\bar{k} \times (-0.7418\bar{i} + 0.8365\bar{j})$$

$$v_{Cx} = v_{Gx} - 0.8365\omega$$

Thus, setting the velocity of the earth to zero, the coefficient of restitution gives

$$\epsilon = 0.2 = -\frac{(v_{Cx})_2}{(v_{Cx})_1} = -\frac{(v_{Gx})_2 - 0.8365\omega_2}{-20 - 0.8365(0)}$$

so

$$(v_{Gx})_2 - 0.8365\omega_2 = 4.0$$

The linear impulse-momentum equation for the 1000-kg crate is

$$1000(\bar{v}_G)_2 = 1000(\bar{v}_G)_1 + \int_{t_1}^{t_2} (F\bar{i}) \, dt$$

Hence, equating components and lettering $L \equiv \int_{t_1}^{t_2} F \, dt$, we obtain

$\bar{i}$ direction: $1000(v_{Gx})_2 = 1000(-20) + L$

$\bar{j}$ direction: $1000(v_{Gy})_2 = 1000(0)$

The crate is in general motion, so we formulate the angular momentum principle with respect to a $\hat{x}\hat{y}\hat{z}$ coordinate system, parallel to xyz, with origin at the center of mass G. From the given radius of gyration we have

$$I_{\hat{z}\hat{z}} = 1000(0.8)^2 = 640 \text{ kg-m}^2$$

The moment of the force $\bar{F}$ about point G is

$$\Sigma \, \bar{M}_G = \bar{r}_{C/G} \times \bar{F} = (-0.7418\bar{i} + 0.8365\bar{j}) \times F\bar{i}$$

$$= -0.8365F\bar{k}$$

Hence the resulting angular impulse-momentum equation is

$$I_{\hat{z}\hat{z}}\omega_2 = I_{\hat{z}\hat{z}}\omega_1 + \int_{t_1}^{t_2} \Sigma \, M_{Gz} \, dt$$

$$640\omega_2 = 640(0) - 0.8365 \int_{t_1}^{t_2} F \, dt$$

$$= -0.8365L$$

We now have four equations (one from ϵ, two from linear momentum, and one from angular momentum) for the four scalar unknowns $(v_{Gx})_2$, $(v_{Gy})_2$, ω_2, and L. The solution of these equations is

$$(v_{Gx})_2 = -8.53 \text{ m/s} \qquad (v_{Gy})_2 = 0$$

$$\omega_2 = -14.99 \text{ rad/s}$$

Hence, after the collision the center of mass moves downward at a slower rate and the crate is rotating counterclockwise.

HOMEWORK PROBLEMS

V.148 The 2-kg slender rod is hit by a 150-g ball traveling at 20 m/s in the direction shown. The coefficient of restitution is 0.80. Determine the angular velocity of the bar and the velocity of the ball immediately after the collision.

Prob. V.148   Prob. V.149

V.149 A 5.50-oz baseball travelling horizontally at 30 miles/hr hits the 4-lb wooden bar. The coefficient of restitution for the collision is 0.50. Determine the velocity of both ends of the bar immediately after the collision.

Prob. V.150

Prob. V.152

Prob. V.154

Prob. V.156

V.150 A 10-kg package on an assembly line slides along the curved chute in the vertical plane until it hits a small cleat at point A. If the speed of the package just before it hits the cleat is $v = 5$ m/s, determine the angular velocity of the package immediately after the collision. Assume that the impact is perfectly plastic.

V.151 What is the maximum speed v at which the package in Problem V.150 can hit the cleat without tipping over onto edge AD?

V.152 The 100-lb slender beam is initially at rest in the vertical position when the lower end B is given a very slight push to the left, causing the bar to fall. The floor and wall are both smooth. Determine the impulse of the reaction forces at both ends of the bar as end A hits the ground. The impact is perfectly plastic. Confine the analysis to the first impact.

V.153 In Problem V.152, the coefficient of restitution is 0.40. Determine the velocity of both ends of the beam immediately after the first impact.

V.154 The 1-kg homogeneous sphere is rolling without slipping down the incline at 10 m/s when it hits the change in grade. Assuming that the collision with the horizontal surface is perfectly plastic, determine the velocity of the center of the sphere and the angular velocity immediately after the impact. Does the sphere roll without slipping immediately after the impact?

V.155 Solve Problem V.154 if the coefficient of restitution for the collision with the horizontal surface is 0.90.

V.156 A 30-g ball bearing is rolling without slipping at 20 m/s on the horizontal surface when it hits the 5-mm ledge. The corner of the ledge is slightly rounded, so the ball bearing may slip relative to this edge. The collision is perfectly plastic. What is the velocity of the center of mass and the angular velocity of the ball bearing immediately after the collision?

Solve Problem V.156 if the coefficient of restitution for the collision is 0.80.

V.158 In Problem V.156, the corner of the ledge is sharp, so the velocity of the point of contact immediately after the collision is zero. Determine the velocity of the center of the ball bearing after the collision. *Hint:* In order to bring the point of contact to rest, an impulsive force tangent to the contact plane at the corner is required.

MODULE VI
KINEMATICS IN A
MOVING REFERENCE
FRAME

Picture the following situation. You are in a spaceship and you observe an interesting phenomenon whose location you wish to describe via radio to people on the earth. Considering the fact that your spaceship is translating and rotating with respect to the earth, what information should you convey to the people at the earth station? If you stop to think about it, the earth is also rotating and translating. To solve the dilemma let us pick a "fixed" point O in space and radio back the position of the phenomenon with respect to the fixed point. This situation is shown in Figure 1, where $\bar{r}_{P/O'}$ represents the position of the phenomenon at point P with respect to our moving spaceship at point O', $\bar{r}_{O'/O}$ represents the position of the spaceship with respect to a fixed point (a distant star), and $\bar{r}_{P/O}$ represents the position of P with respect to the "fixed" star.

Figure 1

From this figure we see that we can describe the vector $\bar{r}_{P/O}$ by giving the position of the spaceship with respect to the fixed star ($\bar{r}_{O'/O}$) and the position of the phenomenon with respect to the spaceship ($\bar{r}_{P/O'}$). This relationship can be written in vector form as

$$\bar{r}_{P/O} = \bar{r}_{O'/O} + \bar{r}_{P/O'} \tag{1}$$

Furthermore, because you are in the spaceship you will probably choose to use the spaceship as your frame of reference for describing $\bar{r}_{P/O'}$, for instance by giving the distances that the phenomenon is ahead of, to the right of, and above the vehicle.

Whereas the preceding argument discussed a system of advanced technology, we could have made the same arguments with respect to the moving parts of a machine. In either case the new task before us is clear: to describe the motion of a particle with respect to a moving reference frame. The kinematical tools that will emerge from the development to follow are essential to the solution of three-dimensional dynamics problems. Equally important, when applied to two-dimensional problems, in many cases they can make a previously complex analysis quite trivial.

A.   REFERENCE FRAMES AND ANGULAR VELOCITY

Consider the rectangular coordinate frame xyz shown in Figure 2, which is free to translate and rotate with respect to the fixed Cartesian coordinate system XYZ. The vector $\bar{r}_{O'/O}$ locates the origin of the moving system. Clearly the components of $\bar{r}_{O'/O}$ may be written as functions of either the unit vectors of the fixed system $(\bar{I}, \bar{J}, \bar{K})$ or the moving system $(\bar{\imath}, \bar{\jmath}, \bar{k})$. Thus describing the motion of the origin O' of the moving system will not present a significant problem, for we are already familiar with the kinematics of the absolute motion of a particle (Module II).

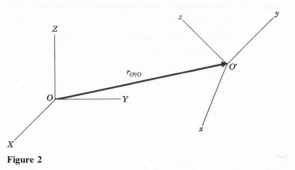

Figure 2

Let us now consider a particle whose position vector $\bar{r}_{P/O'}$ is given with respect to the xyz system, as seen in Figure 3. Writing this given vector in terms of the moving reference frame, we have

$$\bar{r}_{P/O'} = x\bar{\imath} + y\bar{\jmath} + z\bar{k} \tag{2}$$

Because the frame of reference we shall use is rotating, we need to describe its rotational characteristics. To do this, let us consider an infinitesimal rotation that the system undergoes during the infinitesimal time interval from t to $t + dt$. The resulting rotation may be described in terms of the infinitesimal rotations $d\theta_x$, $d\theta_y$, and $d\theta_z$, where $d\theta_x$ is a rotation about an axis that is parallel to the x axis at time t, and so on. (By an infinitesimal rotation we mean that the magnitude of the angle of rotation in radians squared is negligible when compared to the angle itself.)

A rotation about an axis can be described in vector form by the right-hand rule. Our intuition tell us that the total (infinitesimal) rotation vector can be obtained by vector addition of the individual rotations, in which case

Figure 3

$$d\bar{\theta} = d\theta_x\,\bar{i} + d\theta_y\,\bar{j} + d\theta_z\,\bar{k} \qquad (3)$$

The basic definition of angular velocity is that $\bar{\omega}$ is the rate of change of the angle of rotation directed in accordance with the right-hand rule. Therefore

$$\bar{\omega} \equiv \frac{d\bar{\theta}}{dt} = \frac{d\theta_x}{dt}\,\bar{i} + \frac{d\theta_y}{dt}\,\bar{j} + \frac{d\theta_z}{dt}\,\bar{k} = \omega_x\,\bar{i} + \omega_y\,\bar{j} + \omega_z\,\bar{k} \qquad (4)$$

The direction of the vector $\bar{\omega}$ describes the orientation of the axis of rotation of the reference frame xyz at the instant t. The magnitude of $\bar{\omega}$ is the instantaneous rate of rotation of the xyz frame about this axis.

The implication of equation (3) is that after the xyz frame undergoes a set of infinitesimal rotations, its resulting orientation is independent of the order of the rotations. Knowing that, it follows that the angular velocity $\bar{\omega}$ of the moving reference system is obtained by vector addition of the rates or rotation about the various axes, as stated by equation (4). This result is proven rigorously in Section A.2. In Section A.1 we demonstrate that in the case of finite (large) rotations, the final orientation *does* depend on the order of rotation.

Sections A.1 and A.2 are presented for those who desire a rigorous treatment. If you choose, without loss of continuity, you can accept equations (3) and (4) on the basis of the foregoing discussion and proceed directly to Section B.

1   Finite Rotations

It is a simple matter to demonstrate by example that, when we deal with finite rotations, we must know the order in which the rotations are imposed. Figures 4a and 4b illustrate what happens when rectangle $OABC$, which initially lies in the xy plane, is rotated first about the x axis by $-90°$ (Figure 4a), and then about the z axis by $90°$ (Figure 4b). The result of the two rotations is that the rectangle lies in the yz plane.

Figure 4a   **Figure 4b**

Alternatively, the rotation about the z axis could have been imposed first, followed by the rotation about the x axis. This is illustrated in Figures 5a and 5b. In this case the position of the rectangle is the xz plane. Therefore we can conclude that, in general, the order of rotation is important when considering finite rotations.

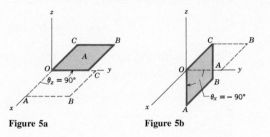

Figure 5a   Figure 5b

2 Infinitesimal Rotations and Unit Vectors

Let us now consider the case of infinitesimal rotations by considering how an arbitrary unit vector

$$\bar{e} = e_x \bar{\imath} + e_y \bar{\jmath} + e_z \bar{k}$$

is changed by successive rotations $d\theta_x$, $d\theta_y$, and $d\theta_z$. Figure 6 is a view looking down the x axis at the result of the rotation $d\theta_x$. In this rotation the vector $\bar{e}$ changes by the amount $d\bar{e}_1$ to become the vector $\bar{e}_1$. The vectors $(\bar{e})_\perp$ and $(\bar{e}_1)_\perp$ in Figure 6 are the components of $\bar{e}$ and $\bar{e}_1$, respectively, which are perpendicular to the x axis, and hence tangent to the yz plane. Thus

Figure 6

$$(\bar{e})_\perp = e_y \bar{\jmath} + e_z \bar{k} = \bar{e} - e_x \bar{\imath} \tag{5}$$

In the limit as the value of $d\theta_x$ goes to zero, we see from Figure 6 that $d\bar{e}_1$, which is parallel to the yz plane, becomes perpendicular to $(\bar{e})_\perp$. Furthermore, because the chord of an infinitesimal circular sector is equal to the arc length, we have

$$\left| d\bar{e}_1 \right| = \left| (\bar{e})_\perp \right| d\theta_x$$

Thus, in vector form,

$$d\bar{e}_1 = (d\theta_x \bar{\imath}) \times (\bar{e})_\perp$$

Now, in view of equation (5), we have

$$d\bar{e}_1 = (d\theta_x \bar{\imath}) \times (\bar{e} - e_x \bar{\imath}) \equiv (d\theta_x \bar{\imath}) \times \bar{e}$$

Therefore, we may find $\bar{e}_1$ by adding $d\bar{e}_1$ to $\bar{e}$, yielding

$$\bar{e}_1 = \bar{e} + d\bar{e}_1 = \bar{e} + (d\theta_x \, \bar{\imath}) \times \bar{e} \tag{6}$$

We shall now operate on $\bar{e}_1$ by performing the rotation $d\theta_y$, which results in $\bar{e}_1$ changing by $d\bar{e}_2$ to become $\bar{e}_2$. By following similar steps to those used in obtaining equation (6), it is easy to show that

$$\bar{e}_2 = \bar{e}_1 + d\bar{e}_2 = \bar{e}_1 + (d\theta_y \, \bar{\jmath}) \times \bar{e}_1 \tag{7}$$

Finally, with the rotation $d\theta_z$ the vector $\bar{e}_2$ changes by $d\bar{e}_3$ to become the vector $\bar{e}_3$, according to

$$\bar{e}_3 = \bar{e}_2 + d\bar{e}_3 = \bar{e}_2 + (d\theta_z \, \bar{k}) \times \bar{e}_2 \tag{8}$$

Equations (6)–(8) detail how a unit vector $\bar{e}$ becomes a vector $\bar{e}_3$. If we substitute equation (6) into equation (7) and then substitute equation (7) into equation (8), we shall obtain an expression for $\bar{e}_3$ in terms of $\bar{e}$. The first of these two steps gives

$$\bar{e}_2 = \bar{e} + (d\theta_x \, \bar{\imath}) \times \bar{e} + (d\theta_y \, \bar{\jmath}) \times [\bar{e} + (d\theta_x \, \bar{\imath}) \times \bar{e}]$$

and the second yields

$$\begin{aligned}
\bar{e}_3 = \bar{e} &+ (d\theta_x \, \bar{\imath}) \times \bar{e} + (d\theta_y \, \bar{\jmath}) \times \bar{e} \\
&+ (d\theta_y \, \bar{\jmath}) \times [(d\theta_x \, \bar{\imath}) \times \bar{e})] + (d\theta_z \, \bar{k}) \times \{\bar{e} \\
&+ (d\theta_x \, \bar{\imath}) \times \bar{e} + (d\theta_y \, \bar{\jmath}) \times \bar{e} \\
&+ d\theta_y \, \bar{\jmath} \times [(d\theta_x \, \bar{\imath}) \times \bar{e}]\}
\end{aligned}$$

Recall that the concept of infinitesimal quantities means that their products are negligible in comparison with the quantities themselves. Therefore, when we neglect products of $d\theta_x$, $d\theta_y$, and $d\theta_z$, we find that

$$\begin{aligned}
\bar{e}_3 &= \bar{e} + (d\theta_x \, \bar{\imath}) \times \bar{e} + (d\theta_y \, \bar{\jmath}) \times \bar{e} + (d\theta_z \, \bar{k}) \times \bar{e} \\
&= \bar{e} + (d\theta_x \, \bar{\imath} + d\theta_y \, \bar{\jmath} + d\theta_z \, \bar{k}) \times \bar{e} \\
&= \bar{e} + d\bar{\theta} \times \bar{e}
\end{aligned}$$

This equation tells us that the total infinitesimal change in the unit vector $\bar{e}$ is

$$d\bar{e} = \bar{e}_3 - \bar{e} = d\bar{\theta} \times \bar{e} \tag{9}$$

From this we can conclude that only the resultant rotation vector $d\bar{\theta}$ is significant to infinitesimal rotations.

B. RELATIVE VELOCITY—DIFFERENTIATION IN A MOVING REFERENCE FRAME

In Section A.2 we proved that the change of a unit vector $\bar{e}$ due to an arbitrary infinitesimal rotation $d\bar{\theta}$ is given by

$$d\bar{e} = d\bar{\theta} \times \bar{e} \tag{10}$$

Equation (10) simply states that in an infinitesimal rotation $d\bar{\theta}$ the unit vector $\bar{e}$ rotates by an angle $d\bar{\theta}$ about an instantaneous axis of rotation that is parallel to the direction of $d\bar{\theta}$.

It follows from equation (10) that, as a unit vector rotates, its rate of change is given by

$$\frac{d\bar{e}}{dt} \equiv \dot{\bar{e}} = \frac{d\bar{\theta}}{dt} \times \bar{e}$$

$$\boxed{\dot{\bar{e}} = \bar{\omega} \times \bar{e}} \tag{11}$$

Equation (11) provides a method for describing the rotational characteristics of a moving reference frame. We shall now utilize this result to study the motion of a point P. In Figure 3 we selected an xyz system that is rotating and translating. The origin of the xyz system is at point O', and the vector $\bar{r}_{P/O'}$ locates the position of point P within this system. The velocity $\bar{v}_{P/O'}$ of point P with respect to origin O' accounts for the total rate of change of the position vector, according to

$$\bar{v}_{P/O'} \equiv \frac{d}{dt}(\bar{r}_{P/O'})$$

Differentiating the component form of $\bar{r}_{P/O'}$, equation (2), we find that

$$\bar{v}_{P/O'} = \dot{x}\bar{\imath} + x\dot{\bar{\imath}} + \dot{y}\bar{\jmath} + y\dot{\bar{\jmath}} + \dot{z}\bar{k} + z\dot{\bar{k}} \tag{12}$$

To evaluate the derivatives of the unit vectors that occur in this equation, *let $\bar{\omega}$ be the angular velocity of the xyz frame* (and hence of the unit vectors $\bar{\imath}, \bar{\jmath}, \bar{k}$). Thus, using equation (10) we may write

$$\dot{\bar{\imath}} = \bar{\omega} \times \bar{\imath} \qquad \dot{\bar{\jmath}} = \bar{\omega} \times \bar{\jmath} \qquad \dot{\bar{k}} = \bar{\omega} \times \bar{k}$$

With these results equation (11) becomes

$$\bar{v}_{P/O'} = \frac{d}{dt}(\bar{r}_{P/O'}) = \dot{x}\bar{\imath} + \dot{y}\bar{\jmath} + \dot{z}\bar{k} + x(\bar{\omega} \times \bar{\imath}) + y(\bar{\omega} \times \bar{\jmath}) + z(\bar{\omega} \times \bar{k})$$

$$= \dot{x}\bar{\imath} + \dot{y}\bar{\jmath} + \dot{z}\bar{k} + \bar{\omega} \times (x\bar{\imath} + y\bar{\jmath} + z\bar{k})$$

$$= (\dot{x}\bar{\imath} + \dot{y}\bar{\jmath} + \dot{z}\bar{k}) + \bar{\omega} \times \bar{r}_{P/O'}$$

We shall now define a new process of differentiation of a vector using the symbol $\delta(\ )/\delta t$. This differentiation process will account only for the rate of change of the scalar components of a vector. It ignores the fact that the unit vectors for the components are not constant. In terms of this definition,

$$\frac{\delta}{\delta t}\left(\bar{r}_{P/O'}\right) = \dot{x}\bar{\imath} + \dot{y}\bar{\imath} + \dot{z}\bar{k} \tag{13}$$

Therefore, $\delta(\bar{r}_{P/O'})/\delta t$ is the rate of change of $\bar{r}_{P/O'}$ that someone would observe if that person did not realize that the reference frame was rotating. In other words, $\delta(\bar{r}_{P/O'})/\delta t$ *is the velocity of point P relative to the moving reference frame.* For notational convenience let us denote this term by

$$\frac{\delta}{\delta t}\left(\bar{r}_{P/O'}\right) \equiv \left(\bar{v}_{P/O'}\right)_{\text{rel}}$$

This allows us to write the velocity of P with respect to O' as

$$\bar{v}_{P/O'} = \frac{d}{dt}\left(\bar{r}_{P/O'}\right) = \left(\bar{v}_{P/O'}\right)_{\text{rel}} + \bar{\omega} \times \bar{r}_{P/O'} \tag{14}$$

In words, this equation states that the velocity of point P with respect to the origin of the moving reference frame xyz (point O') is the velocity of point P relative to xyz, $(\bar{v}_{P/O'})_{\text{rel}}$, plus an additional term $\bar{\omega} \times \bar{r}_{P/O'}$ that accounts for the velocity caused by the angular motion $\bar{\omega}$ of the moving reference system.

Equation (14) describes the rate of change of the vector $\bar{r}_{P/O'}$. It follows then that if we consider any vector, for instance $\bar{A}$, we may write

$$\boxed{\frac{d\bar{A}}{dt} = \frac{\delta\bar{A}}{\delta t} + \bar{\omega} \times \bar{A}} \tag{15}$$

C.   ABSOLUTE VELOCITY

Let us return to the example of Figure 1, where the information we wished to transmit was the position of P with respect to the fixed star at location O. Thus, repeating equation (1), we have

$$\bar{r}_{P/O} = \bar{r}_{O'/O} + \bar{r}_{P/O'} \tag{1}$$

so that the position of the phenomenon in space ($\bar{r}_{P/O}$) may be described by knowing the position of the spaceship ($\bar{r}_{O'/O}$) and the position of the phenomenon with respect to the spaceship ($\bar{r}_{P/O'}$).

In order to describe this phenomenon to the people on the earth, we shall need to describe its velocity with respect to the fixed star. We shall refer to this quantity as the *absolute velocity* $\bar{v}_P$. In view of the basic definition of velocity, we have

$$\bar{v}_P = \frac{d}{dt}(\bar{r}_{P/O})$$

so that differentiating equation (1) gives

$$\bar{v}_P = \frac{d}{dt}(\bar{r}_{O'/O}) + \frac{d}{dt}(\bar{r}_{P/O'}) = \bar{v}_{O'} + \bar{v}_{P/O'}$$

By using equation (14) we obtain the final result,

$$\boxed{\bar{v}_P = \bar{v}_{O'} + (\bar{v}_{P/O'})_{\text{rel}} + \bar{\omega} \times \bar{r}_{P/O'}} \qquad (16)$$

From equation (16) we see that the absolute velocity of point P equals the absolute velocity of the origin of the moving reference frame plus the velocity of point P relative to the moving reference frame plus the cross product of the angular velocity $\bar{\omega}$ of the moving reference frame and the relative position vector $\bar{r}_{P/O'}$. In the simple case of a reference frame that is only translating, $\bar{\omega} \equiv \bar{0}$ and $\bar{v}_{P/O'} = (\bar{v}_{P/O'})_{\text{rel}}$. Equation (16) then becomes the relative motion equation that we studied in Module II, specifically,

$$\bar{v}_P = \bar{v}_{O'} + \bar{v}_{P/O'}$$

On the other hand, if we attach an *xyz* frame to a rigid body, then the coordinates of any point P of the body with respect to the *xyz* frame are constant, so that $(\bar{v}_{P/O'})_{\text{rel}} \equiv \bar{0}$. For this case equation (16) becomes

$$\bar{v}_P = \bar{v}_{O'} + \bar{\omega}_{\text{body}} \times \bar{r}_{P/O'} \qquad (17)$$

This result, of course, is the same as the velocity formula that we obtained for rigid bodies in planar motion (Module Four). The new feature is that we have now shown the result to be true for three dimensional motion, where the axis of rotation of the body can have a variable direction.

Before considering some examples that utilize equation (16), let us discuss criteria to be employed in the selection of a moving reference frame. Basically, we have to make three decisions regarding the *xyz* frame:

1   Where should its origin O' be located?
2   What angular velocity should it have?
3   What orientation should it have at the instant of interest?

These questions can be answered by studying equation (16). First, we note that the absolute velocity of point O' is required; hence the origin should have a fairly simple type of motion. For example, a fixed point or a point following a planar path would be ideal. The decision regarding the angular velocity $\bar{\omega}$ to be ascribed to the moving frame is based on the fact that another term we need to compute is the velocity of point P relative to the moving reference frame. Hence we look for a reference frame whose rotation makes it comparatively easy to determine $(\bar{v}_{P/O'})_{\text{rel}}$.

Finally, to select the instantaneous orientation of the xyz frame, we note that we shall require the components of various vectors. These components must *all* be with respect to the same coordinate system in order to perform the operations of vector addition and multiplication. Obviously, we can use components with respect to either the fixed unit vectors $\bar{I}$, $\bar{J}$, $\bar{K}$, or the moving unit vectors $\bar{i}$, $\bar{j}$, $\bar{k}$. As a general rule, dynamicists tend to use $\bar{i}$, $\bar{j}$, $\bar{k}$ components. We shall therefore select the orientation of the xyz frame to be such that the components of the vectors are easy to describe.

EXAMPLE 1

Turbine rotors and impeller pumps, which have fluid flowing over the vanes or blades, lead to important and interesting kinematical studies. Let us idealize such a problem. A disk of radius R rotates with constant angular velocity $\bar{\omega}$ about a fixed axis. A straight vane of length l is welded rigidly to the disk, and the angle between the vane and a radial line is θ, as shown in the sketch. If a particle of fluid slides outward along the vane at a speed u relative to the vane, determine the velocity of the particle as it flies off the end of the vane.

Solution

Because the axis of the disk is fixed, point A, which is on this axis of rotation, can serve as the fixed point of reference. The axes XYZ can be oriented with the Z axis parallel to the axis of rotation. We must now choose the moving reference frame. There are several possibilities for its origin: (a) on the axis of rotation, that is, point A, (b) at the junction of the disk and the vane, that is, point B, (c) at the end of the vane, that is, point C on the vane, (d) traveling with the particle P, that is, at point C at this instant, but moving with the particle. Let us study each of these possibilities.

Approach a

Select xyz to have origin O' at point A and let xyz rotate with the disk. We select the orientation of xyz to facilitate the writing of the expressions for the relative position and velocity vectors. This situation is depicted in the sketch.

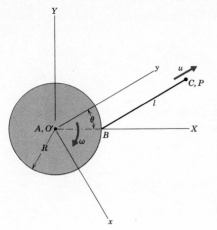

Because the origin of xyz is the fixed point A we have, at this instant,

$$\bar{r}_{O'/A} = \bar{0}$$

$$\bar{r}_{P/O'} \equiv \bar{r}_{P/A} = (R \sin \theta)\bar{i} + (R \cos \theta + l)\bar{j}$$

To determine the velocity we first obtain the individual terms in the velocity equation (16) and then perform the vector algebra. Noting that the origin O' is fixed, we have

$$\bar{v}_{O'} = \bar{0}$$

With respect to the rotating reference frame, the particle seems to be only sliding along the vane at speed u. Thus, the velocity of the particle relative to the moving reference frame is rectilinear motion parallel to the vane, so that

$$(\bar{v}_{P/O'})_{\text{rel}} = u\,\bar{j}$$

Because xyz is attached to the disk, the angular velocity according to the right-hand rule is

$$\bar{\omega} = \omega(-\bar{K}) = -\omega\bar{k}$$

Note that we have expressed $\bar{\omega}$ in terms of both XYZ and xyz components. This is done in order to anticipate a step in the acceleration problem, which we shall consider in the next section. However, when we substitute $\bar{\omega}$ and the other vectors into the velocity equation, we must use only one coordinate system for components. Thus

$$\bar{v}_P = \bar{v}_{O'} + (\bar{v}_{P/O'})_{\text{rel}} + \bar{\omega} \times \bar{r}_{P/O'}$$

$$= \bar{0} + u\,\bar{j} + (-\omega\bar{k}) \times [R \sin \theta\,\bar{i} + (R \cos \theta + l)\,\bar{j}]$$

$$= \omega(R \cos \theta + l)\bar{i} + (u - \omega R \sin \theta)\,\bar{j} \qquad \triangleleft$$

Approach b

Let the origin O' be point B, and let xyz rotate with the disk. Again, we select the orientation of xyz to make it as easy as possible to describe the vector components, as shown in the sketch.

For this coordinate system,

$$\bar{r}_{P/O'} = l\,\bar{j}$$

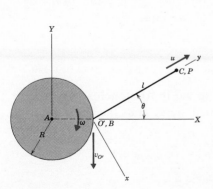

In this approach origin O' is moving. Its velocity may be derived in one of two ways. We can either note that point O' is following a circular path centered at point A, so that its velocity is ωR tangent to the path, or alternatively that point O' is a point on the rigid disk, in which case we can employ the rigid body equation (17). Let us follow the latter approach. Then

$$\bar{v}_{O'} = \bar{v}_A + \bar{\omega}_{\text{disk}} \times \bar{r}_{O'/A} = \bar{0} + (-\omega\bar{k}) \times [R(\cos\theta\,\bar{j} + \sin\theta\,\bar{i})]$$

$$= \omega(\cos\bar{i} - \sin\theta\,\bar{j})$$

We next determine the relative velocity by noting that with respect to xyz the motion is parallel to the y axis at speed u, so that

$$(\bar{v}_{P/O'})_{\text{rel}} = u\,\bar{j}$$

Finally, because xyz rotates with the disk, its angular velocity is

$$\bar{\omega} = -\omega\bar{K} = -\omega\bar{k}$$

Thus

$$\bar{v}_P = \bar{v}_{O'} + (\bar{v}_{P/O'})_{\text{rel}} + \bar{\omega} \times \bar{r}_{P/O'}$$

$$= \omega R(\cos\theta\,\bar{i} - \sin\theta\,\bar{j}) + u\,\bar{j} + (-\omega\bar{k}) \times (l\,\bar{j})$$

$$= \omega(R\cos\theta + l)\bar{i} + (u - \omega R\sin\theta)\bar{j} \qquad \triangleleft$$

Because the xyz system we have utilized in this approach is parallel to the one we used in Approach a, the components of $\bar{v}_P$ should be identical for each approach. A comparison shows that this is true.

Approach c

Let the origin O' be at the tip of the vane, so that the particle is at point O' at this instant. Let xyz rotate with the vane and orient xyz parallel to the reference frames used in the previous approaches. It follows, then, that, with respect to the xyz axes, the particle is moving at speed u parallel to the vane.

The tip of the vane is a point in the rigid body (consisting of the disk and the vane). Therefore, from the sketch we find

$$\bar{r}_{P/O'} = \bar{0}$$

$$\bar{v}_{O'} = \bar{v}_A + \bar{\omega}_{\text{disk}} \times \bar{r}_{A/O'}$$

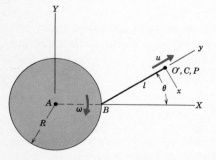

Because xyz in this approach is parallel to xyz in Approach a, we can use the components of $\bar{r}_{O'/A}$ that we determined there. Thus the velocity of the origin of the moving system is

$$\bar{v}_{O'} = \bar{0} + (-\omega\bar{k}) \times [R\sin\theta\,\bar{i} + (R\cos\theta + l)\bar{j}]$$

$$= \omega(R\cos\theta + l)\bar{i} - \omega R\sin\theta\,\bar{j}$$

Also, we know that

$$(\bar{v}_{P/O'})_{\text{rel}} = u\,\bar{j}$$

and

$$\bar{\omega} = -\omega\bar{K} = -\omega\bar{k}$$

so that

$$\bar{v}_P = \bar{v}_{O'} + (\bar{v}_{P/O'})_{\text{rel}} + \bar{\omega} \times \bar{r}_{P/O'}$$

$$= \omega(R\cos\theta + l)\bar{i} - \omega R\sin\theta\,\bar{j} + u\,\bar{j} + \bar{0}$$

$$= \omega(R\cos\theta + l)\bar{i} + (u - \omega R\sin\theta)\bar{j}$$

Approach d

Let the origin of the moving reference frame follow the motion of the particle, so that the condition $\bar{r}_{P/O'} = \bar{0}$ always holds. It then follows that the relative velocity $(\bar{v}_{P/O'})_{\text{rel}} = \bar{0}$. Thus, regardless of the angular velocity of xyz, the velocity equation reduces to

$$\bar{v}_P = \bar{v}_{O'} + (\bar{v}_{P/O'})_{\text{rel}} + \bar{\omega} \times \bar{r}_{P/O'}$$

$$= \bar{v}_{O'} + \bar{0} + \bar{0}$$

In other words, "The answer is the answer." In this case the use of a moving reference frame does not aid the solution for $\bar{v}_P$.

Let us summarize the foregoing by commenting on the advantages and disadvantages of each approach.

Clearly, the last approach is of no use. We should *never* let the origin of the moving reference frame follow the point in which we are interested. In Approach c the method we used for determining the velocity of the origin may not have been obvious to you, and the alternative of using the concepts of circular motion leads to some complications due to the geometry of the system.

The solution was obtained fairly directly in both Approaches a and b. However, of the two, in Approach b the origin O' was a moving point whose velocity was easy to compute, and each term in the velocity equation had components that were very simple to write. This turns out to be true in general. That is, if there is a point intermediate to the moving point P and the fixed point O, and if the kinematical description of this point is fairly direct, we shall use such a point as the origin of the moving reference system.

EXAMPLE 2

A centrifuge is being used to study the tolerance of human beings to motion. It consists of a main arm that rotates about the A–A axis at an angular speed ω_1, as shown in the sketch, and a cockpit that may rotate with respect to the main arm at an angular speed ω_2 about the C–C axis. The individual system rotations are computer controlled, and the head of the person being tested is stationary with respect to the cockpit seat.

Consider a test where $\omega_1 = 60$ rad/min, $\omega_2 = 90$ rad/min, the rider is 40 ft from the A–A axis, and the rider's eyes are 3 ft from the C–C axis. Determine the absolute velocity of the rider's eyes at the instant when the rider is in the horizontal position shown below.

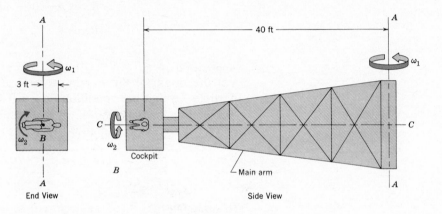

End View   Side View

Solution

In the previous example we discussed the matter of the choice of the origin of the moving reference frame. From the experience we gained there we see that for this problem, either the moving point B (which is on the rotating axis C–C and therefore follows a circular path) or a fixed point on the A–A axis would make a reasonable choice for the origin O'. Further, in Example 1 we saw that the moving point is preferable. Therefore, we shall let the origin O' coincide with B, so that for circular motion $v_{O'} = \omega_1(40)$ directed outward from the plane of the diagram. Note that for this step we could have used equation (17) because point O' is a point on the rigid main arm.

Now let us turn our attention to the question of what angular velocity $\bar{\omega}$ we should ascribe to reference frame xyz. The reference frame can be attached either to the main arm or to the cockpit. In what follows we shall work both approaches. Regardless of which approach you select, as part of your solution you *must state* the body to which you have attached your reference frame.

Approach a

Let xyz be attached to the main arm. In order to facilitate expressing vector components at the instant when the rider is horizontal, orient the *xyz* axes so that one axis is parallel to the line from point *B* to the rider's eyes and one axis is parallel to axis *C–C* along the arm. The resulting fixed and moving reference systems are shown in the sketch.

As we can see from the sketch, at this instant

$$\bar{r}_{P/O'} = -3\bar{k} \text{ ft}$$

Also, because $\bar{v}_{O'}$ is outward when viewed from the side, we have

$$\bar{v}_{O'} = 40\omega_1(-\bar{k}) = -40(60 \text{ rad/min})(1 \text{ min/60 s})\bar{k}$$

$$= -40\bar{k} \text{ ft/s}$$

To determine the velocity of the rider's eyes relative to the moving reference frame, we note that the cockpit, and hence the rider, is in pure rotation about axis *C–C* relative to the main arm. The *xyz* frame is fixed to the main arm, so point *P* is following a circular path. Therefore, $(\bar{v}_{P/O'})_{\text{rel}}$ is $-3\omega_2 \bar{j}$ at this instant, so that

$$(\bar{v}_{P/O'})_{\text{rel}} = -3(90 \text{ rad/min})(1 \text{ min/60 s})\bar{j}$$

$$= -4.5\bar{j} \text{ ft/s}$$

The next item to be considered is the matter of the angular velocity of *xyz*. Although we have not yet discussed the problem of accelerations, it will be seen later that there are two separate uses for $\bar{\omega}$. One use, which fits the aim of this problem, is to substitute the $\bar{\omega}$ vector into the velocity equation in terms of its components with respect to the *xyz* axes, in order to perform the necessary computations. A further use is to have an expression for the $\bar{\omega}$ vector that is valid at any arbitrary instant. That expression can then be differentiated with respect to time in order to determine the angular acceleration $\bar{\alpha}$.

Thus, in order to ensure that the procedures we develop for the velocity problem lead directly into the acceleration problem, we shall first express $\bar{\omega}$ in component form according to a description that is always true. Then $\bar{\omega}$ will be evaluated for the desired instant of time for the purpose of substitution into the relative motion formulas. In this example xyz is attached to the arm that is rotating about axis $A\text{--}A$ at the rate ω_1. Therefore we have, in general,

$$\bar{\omega} = \omega_1 \bar{J} = \omega_1 \bar{j}$$

At this instant, and also at all instants if ω_1 is constant, this expression becomes

$$\bar{\omega} = 1\bar{j} \text{ rad/s}$$

As a last step we substitute each term into equation (16).

$$\bar{v}_P = \bar{v}_{O'} + (\bar{v}_{P/O'})_{\text{rel}} + \bar{\omega} \times \bar{r}_{P/O'}$$
$$= -40\bar{k} - 4.5\bar{j} + \bar{j} \times (-3\bar{k})$$
$$= -3\bar{i} - 4.5\bar{j} - 40\bar{k} \text{ ft/s}$$

Approach b

In this approach we attach the xyz axes to the cockpit. Again, for ease in describing vector components when the rider is horizontal, we shall orient the xyz axes in the same manner as that shown in Approach a. The relative position vector is therefore

$$\bar{r}_{P/O'} = -3\bar{k} \text{ ft}$$

and the velocity of the origin is again

$$\bar{v}_{O'} = -40\bar{k} \text{ ft/s}$$

With regard to the motion of the rider's eyes with respect to the xyz axes, there is no motion with respect to the cockpit. Therefore, because the xyz axes are fixed to the cockpit, we have

$$(\bar{v}_{P/O'}) = \bar{0}$$

To compute $\bar{\omega}$ of the xyz axes, we note that the cockpit and the axes are rotating with respect to the main arm at ω_1 as the main arm rotates about the fixed axis A–A at ω_2. To determine $\bar{\omega}$ we add the *vector* rates of rotations. Thus, it is generally true that

$$\bar{\omega} = \omega_1 \bar{J} - \omega_2 \bar{I}$$

This expression is always valid because the A–A axis and the Y axis are always parallel. (This form is the one we would use for the determination of angular acceleration.) On the other hand, for the velocity computation we must express $\bar{\omega}$ in terms of $(\bar{i}, \bar{j}, \bar{k})$ components only. At this instant the rider is horizontal, so the y axis is parallel to the Y axis, in which case

$$\bar{\omega} = \omega_1 \bar{j} - \omega_2 \bar{i} = 1\bar{j} - 1.5\bar{i} \text{ rad/s}$$

Finally, the velocity equation (16) gives

$$\begin{aligned}
\bar{v}_P &= \bar{v}_{O'} + (\bar{v}_{P/O'})_{\text{rel}} + \bar{\omega} \times \bar{r}_{P/O'} \\
&= -40\bar{k} + \bar{0} + (\bar{j} - 1.5\bar{i}) \times (-3\bar{k}) \\
&= -3\bar{i} - 4.5\bar{j} - 40\bar{k} \text{ ft/s}
\end{aligned}$$

Because the xyz axes in this approach are parallel at this instant to the xyz axes in Approach a, the components of $\bar{v}_P$ in both approaches should be identical, as they are.

Let us assess the comparative advantages of the two approaches. In Approach a the angular velocity of the reference frame (one component) was easier to describe and understand than the angular velocity of the xyz axes in Approach b, which had two components. However, in Approach b the motion relative to the moving reference system was much simpler to formulate. You should be familiar with both approaches, for when we treat the kinetics of rigid bodies in three-dimensional motion it will be necessary to describe the angular motion of bodies that rotate about more than one axis.

Prob. VI.1

HOMEWORK PROBLEMS

VI.1 Rod ABC rotates clockwise at 10 rad/s as the slider moves outward at 1 m/s. Determine the velocity of the slider when it is at (a) position B, (b) position C.

Probs. VI.2 and VI.3

VI.2 Sliders A and B move within their grooves at the same constant speed $u = 100$ mm/s relative to the disk, which is rotating at the constant rate $\omega = 4$ rad/s clockwise about an axis passing through point O. Determine the absolute velocity of each slider in the position shown.

VI.3 Sliders A and B move within their grooves at the same constant speed $u = 100$ mm/s relative to the disk, which is rotating at the constant rate $\omega = 4$ rad/s clockwise about an axis passing through point C. Determine the absolute velocity of each slider in the position shown.

VI.4 A passenger in the front seat of automobile A follows automobile B by observing the separation distance R and the angle ϕ, which is measured with respect to the longitudinal axis of automobile A. In the position shown, automobile A has a speed of 45 miles/hr, $\phi = 45°$, and $\dot{\phi} = -0.2$ rad/s. Determine the speed of automobile B.

VI.5 Water in the radial flow turbine shown in the sketch enters with a velocity of 30 m/s at an angle of 75° from the radial line. At the inlet A the tangent to the blade forms an angle of 30°. (a) Determine the angular speed ω of the turbine for which the water enters with a relative velocity that is tangent to the blade. (b) Assume that the velocity of the water relative to the blade has constant magnitude and that it is always tangent to the blade. For the conditions of part (a), determine the exit angle θ for which the water leaving the outlet B has only a radial velocity component.

Prob. VI.4

Prob. VI.5

VI.6 The slider oscillates within the groove of the disk, such that the distance $s = 20 \sin 30t$, where s is measured in millimeters and t is in seconds. The disk is fixed to the horizontal axis, which is rotating at 200 rad/s. Determine the velocity of the slider as it passes $s = 10$ mm, moving outward in the groove.

Prob. VI.6

VI.7 The circular hoop is rotating about the vertical axis at 20 rad/s as the block moves relative to the hoop at a constant speed of 4 m/s. Determine the velocity of the slider (a) when $\theta = 0°$, (b) when $\theta = 90°$, (c) when $\theta = 60°$.

Prob. VI.7

VI.8 Bar BC is pinned at end B to bar AB, which rotates about the vertical axis at 240 rev/min. At the instant when $\theta = 45°$, this angle is decreasing at the rate of 5 rad/s. Determine the velocity of end C at this instant.

Prob. VI.8

VI.9 The radar antenna rotates about the fixed vertical axis at 2 rev/min as the angle θ oscillates at $\theta = \pi/6 + (\pi/3) \sin \pi t$, where θ is in units of radians and t is in seconds. Determine the velocity of the probe P when $t = 0.25s$, using (a) a moving reference system that is attached to the vertical shaft, (b) a reference system that is attached to the antenna.

VI.10 A small electric motor pivots about the horizontal pivot A. The disk that is attached to the shaft has an angular speed of 900 rev/min. In the position where $\theta = 30°$, the motor is rotating downward at 5 rad/s.

Prob. VI.9

Determine (a) the angular velocity of the disk, (b) the velocity of point C on the disk. The perpendicular distance from the pivot axis through A to the disk is L.

Prob. VI.10

D.   ABSOLUTE ACCELERATION

We shall now obtain a formula, comparable to equation (16), for describing the absolute acceleration, $\bar{a}_P$, of a point P. By definition, acceleration is the rate of change of velocity. When we differentiate equation (16) we find that

$$\bar{a}_P = \dot{\bar{v}}_P = \frac{d}{dt}(\bar{v}_{O'}) + \frac{d}{dt}[(\bar{v}_{P/O'})_{\text{rel}}] + \frac{d}{dt}(\bar{\omega} \times \bar{r}_{P/O'})$$

The term $d(\bar{v}_{O'})/dt$ is the acceleration of the origin O', which in our notation is $\bar{a}_{O'}$. Then, by applying the law for the derivative of the cross product of two vectors, we transform this equation to

$$\bar{a}_P = \bar{a}_{O'} + \frac{d}{dt}[(\bar{v}_{P/O'})_{\text{rel}}] + \dot{\bar{\omega}} \times \bar{r}_{P/O'} + \bar{\omega} \times \left[\frac{d}{dt}(\bar{r}_{P/O'})\right] \tag{18}$$

The vector $\dot{\bar{\omega}}$ is the time rate of change of the angular velocity of xyz. In other words, $\dot{\bar{\omega}}$ is the angular acceleration $\bar{\alpha}$ of the xyz system.

$$\boxed{\bar{\alpha} = \dot{\bar{\omega}}} \tag{19}$$

One of the terms in equation (18) is $d(\bar{r}_{P/O'})/dt$. We have encountered this derivative previously; it is given by equation (14). Differentiating $(\bar{v}_{P/O'})_{\text{rel}}$ in equation (18) presents us with a similar problem, because both vectors describe what an observer in the moving reference system, such as the spaceship pilot of Figure 1 is witnessing. This suggests that we again employ the concept of a derivative with respect to the moving reference system, as defined by equation (15). This step gives

$$\frac{d}{dt}[(\bar{v}_{P/O'})_{\text{rel}}] = \frac{\delta}{\delta t}[(\bar{v}_{P/O'})_{\text{rel}}] + \bar{\omega} \times (\bar{v}_{P/O'})_{\text{rel}} \tag{20}$$

The first term in the right-hand side of equation (20) expresses how an observer in the moving reference system would see the relative velocity change. In other words, it is the relative acceleration,

$$(\bar{a}_{P/O'})_{\text{rel}} \equiv \frac{\delta}{\delta t}[(\bar{v}_{P/O'})_{\text{rel}}] \equiv \frac{\delta^2}{\delta t^2}(\bar{r}_{P/O'})_{\text{rel}} \tag{21}$$

Substitution of equations (14), (19), (20), and (21) into equation (18) yields the final form of the acceleration equation in relative motion,

$$\boxed{\bar{a}_P = \bar{a}_{O'} + (\bar{a}_{P/O'})_{\text{rel}} + 2\bar{\omega} \times (\bar{v}_{P/O'})_{\text{rel}} + \bar{\alpha} \times \bar{r}_{P/O'} \times \bar{\omega} \times (\bar{\omega} \times \bar{r}_{P/O'})} \tag{22}$$

A study of the terms in this equation will help in understanding the significance of the result. Suppose that the xyz frame is in pure translation, in which case $\bar{\omega}$ and $\bar{\alpha}$ are both zero. Then equation (22) gives the same result as that of Module II for translating reference frames,

$$\bar{a}_P = \bar{a}_{O'} + (\bar{a}_{P/O'})_{\text{rel}} \qquad (\bar{a}_{P/O'})_{\text{rel}} \equiv \bar{a}_{P/O'}$$

Another special case to consider is the situation when point P is a point in a rigid body and xyz is fixed to the body. In this case there is no motion relative to xyz, so equation (22) reduces to

$$\bar{a}_P = \bar{a}_{O'} + \bar{\alpha}_{\text{body}} \times \bar{r}_{P/O'} + \bar{\omega}_{\text{body}} \times (\bar{\omega}_{\text{body}} \times \bar{r}_{P/O'}) \tag{23}$$

This formula is identical to the one derived in Module IV to describe the kinematics of rigid bodies in planar motion.

When we refer to Figure 7, we can see that the last term in equation (22) is once again the centripetal acceleration directed toward the axis of rotation. The term $\bar{\alpha} \times \bar{r}_{P/O'}$ represents the effect of the angular acceleration. The term $2\bar{\omega} \times (\bar{v}_{P/O'})_{\text{rel}}$ did not occur in our previous investigations of translating reference frame and of rigid body motion. If we had naively tried to construct the general equation (22) on the basis of the foregoing special cases, we surely would have omitted this important effect, which represents the interaction of the motion relative to the xyz frame and the rotation of xyz. We refer to $2\bar{\omega} \times (\bar{v}_{P/O'})_{\text{rel}}$ as the Coriolis acceleration. Note its similarity to the transverse acceleration term $2\dot{R}\dot{\theta}$ in cylindrical coordinates (Module II).

A primary new feature of three-dimensional kinematics is that the axis of rotation need not have a fixed orientation. Hence $\bar{\alpha}$ will generally not be parallel to $\bar{\omega}$, and $\bar{\alpha} \times \bar{r}_{P/O'}$ will not represent solely a transverse acceleration.

ω

Centripetal acceleration

P

$\omega \times (\omega \times r_{P/O'})$

$r_{P/O'}$

$\omega \times r_{P/O'}$

Instantaneous orientation of the axis of rotation

O'

Figure 7

Note that equation (17) for the velocity of a rigid body and equation (23) for the acceleration demonstrate that the superposition principle for the motion of a rigid body is valid for an arbitrary three-dimensional motion. Specifically, this principle states that the general motion of a rigid body can be pictured as the combination of a translatory motion following an arbitrary point O', and a rotation having angular velocity $\bar{\omega}$ and angular acceleration $\bar{\alpha}$ about an instantaneous axis passing through point O'.

The conclusions we arrived at regarding the selection of the moving reference frame for velocity considerations also apply for the determination of acceleration. The only quantity that we have not determined in previous examples is the angular acceleration $\bar{\alpha}$.

Our approach for angular velocity and angular acceleration consists of *first* expressing $\bar{\omega}$ in general terms. This is achieved by first using the unit vectors of either the fixed or moving coordinate system to describe the various axes of rotation, whichever gives a constant description. Thus, if an axis of rotation is fixed in space, we describe it in terms of the fixed $\bar{I}\,\bar{J}\,\bar{K}$, whereas if an axis of rotation has varying orientation, we describe it in terms of the moving $\bar{i}\,\bar{j}\,\bar{k}$. Equally important, we use algebraic variables to denote the rates or rotation. *Second,* we differentiate $\bar{\omega}$ to obtain a general expression for $\bar{\alpha}$, using $\dot{\bar{e}} = \bar{\omega} \times \bar{e}$ to evaluate the derivatives of $\bar{i}, \bar{j},$ and $\bar{k}$. *Finally,* the instantaneous values of $\bar{\omega}$ and $\bar{\alpha}$ are determined by substituting the instantaneous values of the rates of rotation, and expressing the unit vectors of XYZ by their xyz equivalents.

These three steps will be a key feature of the second of the following two examples.

EXAMPLE 3
The turbine rotor of Example 1 is shown again in the accompanying sketch. Find the acceleration of a particle of fluid immediately prior to the instant when it flies off the end of the vane if the rotation rate ω and relative speed u are both constant.

Solution
We shall proceed as we did in Example 1, except that our efforts will be confined to Approaches a and b, in that the other procedures were shown to be inefficient.

Approach a: xyz fixed to the disk with origin at A.
Because we have selected the origin of xyz to coincide with the fixed point A, we have $\bar{v}_{O'} = \bar{a}_{O'} = \bar{O}$. Also, following our method for describing the angular motion, the general expressions for $\bar{\omega}$ and $\bar{\alpha}$ are

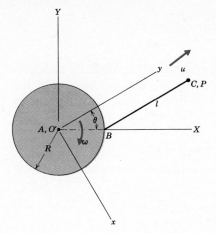

$$\bar{\omega} = \omega(-\bar{K})$$

$$\bar{\alpha} = \dot{\omega}(-\bar{K}) + \omega(-\dot{\bar{K}}) = -\dot{\omega}\bar{K}$$

Because the rate ω is constant and $\bar{K} \equiv \bar{k}$, we have

$$\bar{\omega} = -\omega\bar{k} \qquad \bar{\alpha} = \bar{0}$$

Also, as was determined in Example 1,

$$\bar{r}_{P/O'} = (r \sin \theta)\bar{\imath} + (R \cos \theta + l)\bar{\jmath}$$

Before we can substitute into the acceleration equation (22), we must determine the relative acceleration. As we saw in Example 1, the vane is fixed to xyz, so the motion of the particle relative to xyz is a simple rectilinear motion tangent to the vane with constant speed. Thus

$$(\bar{v}_{P/O'})_{\text{rel}} = u\,\bar{\jmath}$$

$$(\bar{a}_{P/O'})_{\text{rel}} = \bar{0}$$

Incidentally, you probably realize that the relative acceleration would not have been zero if the speed had not been constant, but do you know what the effect would have been if the vane had a curved shape, such as the arc of a circle? In such a case we could use path variables to describe the relative acceleration. The relative motion would then consist of a velocity $\bar{u}$ tangent to the curved vane, so that

$$(\bar{v}_{P/O'})_{\text{rel}} = u\bar{e}_t$$

$$(\bar{a}_{P/O'})_{\text{rel}} = \dot{u}\bar{e}_t + \frac{u^2}{\rho}\,\bar{e}_n$$

Before these expressions could be used in the computation of acceleration, the unit vector $\bar{e}_t$ (tangent to the vane) and $\bar{e}_n$ (normal to the vane toward the center of curvature of the vane) would have to be expressed in terms of $(\bar{\imath}, \bar{\jmath}, \bar{k})$ components.

The last step in the solution of the problem at hand is the substitution of the individual terms into equation (22). This gives

$$\bar{a}_P = \bar{0} + \bar{0} + 2(-\omega\bar{k}) \times u\,\bar{\jmath} + (-\omega\bar{k}) \times \{(-\omega\bar{k})$$
$$\times [(R \sin \theta)\bar{\imath} + (R \cos \theta + l)\bar{\jmath}]\}$$
$$= (2\omega u - R\omega^2 \sin \theta)\bar{\imath} - \omega^2(R \cos \theta + l)\bar{\jmath}$$

Approach b: xyz fixed to the disk with origin at B

Because xyz is rotating with the disk, we have

$$\bar{\omega} = -\omega\bar{K} = -\omega\bar{k}$$

$$\bar{\alpha} = \bar{0}$$

In this approach the origin O' is on the perimeter of the disk. Its acceleration can be determined either by describing the circular motion of the point or else by using the rigid body motion formula (23). The latter approach gives

$$\bar{a}_{O'} = \bar{a}_A + \bar{\alpha}_{\text{disk}} \times \bar{r}_{A/O'} + \bar{\omega}_{\text{disk}} \times (\bar{\omega}_{\text{disk}} \times \bar{r}_{A/O'})$$

To compute the $(\bar{\imath}, \bar{\jmath}, \bar{k})$ components of $\bar{a}_{O'}$, we first determine the position vector.

$$\bar{r}_{A/O'} = R\bar{I} = R(\sin\theta\,\bar{\imath} + \cos\theta\,\bar{\jmath})$$

which gives

$$\bar{a}_{O'} = \bar{0} + \bar{0} + (-\omega\bar{k}) \times [(-\omega\bar{k}) \times R(\sin\theta\,\bar{\imath} + \cos\theta\,\bar{\jmath})]$$

$$= -\omega^2 R(\sin\theta\,\bar{\imath} + \cos\theta\,\bar{\jmath})$$

The sketch shows that

$$\bar{r}_{P/O'} = l\,\bar{\jmath}$$

Also, we see that the motion of the fluid particle relative to xyz is rectilinear motion along the vane at constant speed. Therefore

$$(\bar{v}_{P/O'})_{\text{rel}} = u\,\bar{\jmath}$$

$$(\bar{a}_{P/O'})_{\text{rel}} = \bar{0}$$

Substituting the individual terms into equation (22) then gives

$$\bar{a}_P = -\omega^2 R\,(\sin\theta\,\bar{\imath} + \cos\theta\,\bar{\jmath}) + \bar{0} + \bar{0}$$

$$+ 2(-\omega\bar{k}) \times (u\,\bar{\jmath}) + (-\omega\bar{k}) \times [(-\omega\bar{k}) \times (l\,\bar{\jmath})]$$

$$= (2\omega u - R\omega^2 \sin\theta)\bar{\imath} - \omega^2(R\cos\theta + l)\,\bar{\jmath}$$

The moving reference frame in this approach is parallel to the one in Approach a, so the expressions for $\bar{a}_P$ are identical, as they should be.

EXAMPLE 4
In Example 2 we determined the velocity of the eyes of a test subject in a centrifuge for a given set of rotation rates at a specific instant. In this example we shall determine the acceleration of the rider's eyes at an

arbitrary instant when the seat is oriented at an angle θ from the vertical, as shown in the diagram. We shall also allow the angular rates ω_1 and ω_2 to be variables.

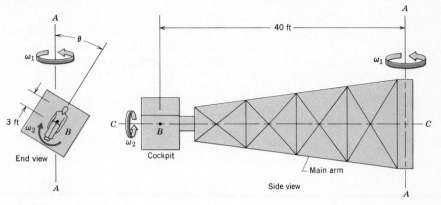

SOLUTION

Without loss of generality, we can select a fixed reference frame by letting the Y axis coincide with the fixed axis A–A, and by letting the X axis coincide with the instantaneous orientation of axis C–C. In both of the methods used to solve Example 2, point B was selected as the origin for the xyz frame. The difference in the methods was that in Approach a, xyz rotated only with the main arm, whereas in Approach b, xyz rotated with the cockpit. We shall again show both of these methods.

We must first decide on the orientation of xyz at the general instant of interest. If we choose the axes of xyz to coincide with the horizontal and vertical directions, the result will be an expression for $\bar{a}_P$ in terms of horizontal and vertical directions. If, on the other hand, we align the axes with the instantaneous orientation of the rider's body, the result will be an expression for $\bar{a}_P$ in terms of what the rider is experiencing. Either choice is correct; we shall employ the latter. Thus for both approaches, the orientations of the reference frames are as shown in the sketch.

For both methods we shall employ the basic acceleration equation (22), which becomes

$$\bar{a}_P = \bar{a}_B + (\bar{a}_{P/B})_{\text{rel}} + \bar{\alpha} \times \bar{r}_{P/B} + 2\bar{\omega} \times (\bar{v}_{P/B})_{\text{rel}} + \bar{\omega} \times (\bar{\omega} \times \bar{r}_{P/B})$$

As usual we have the choice of describing $\bar{a}_B$ by using the fact that point B is following a circular path centered at point O at angular speed ω_1, or by using the fact that point B is on the main arm. The former approach gives

$$\bar{a}_B = -40(\omega_1)^2(\bar{e}_R)_B + 40\dot{\omega}_1(\bar{e}_\theta)_B$$

where, for this instant, the radial direction coincides with the X axis and the transverse direction is parallel to the negative Z axis. Hence,

$$\bar{a}_B = -40(\omega_1)^2(\bar{e}_R)_B + 40\dot{\omega}_1(\bar{e}_\phi)_B$$

However, we shall need the $(\bar{\imath}, \bar{\jmath}, \bar{k})$ components of $\bar{a}_B$, which we can obtain by using the sketch to transform the components of the unit vectors. The sketch shows that $\bar{I} = \bar{\imath}$, and that $\bar{K}$ is in the yz plane, so

$$\bar{a}_B = -40(\omega_1)^2\bar{\imath} - 40\dot{\omega}_1(\cos\theta\,\bar{k} - \sin\theta\,\bar{\jmath})$$

$$= -40(\omega_1)^2\bar{\imath} + 40\dot{\omega}_1\sin\theta\,\bar{\jmath} - 40\dot{\omega}_1\cos\theta\,\bar{k}$$

We shall now consider how the two approaches for describing the angular motion of xyz affect the determination of $\bar{a}_P$.

Approach a: *xyz* fixed to the main arm
In this approach the xyz system rotates only about the fixed Y axis at ω_1. Thus, following the method for angular motion detailed on page 344, the general expressions for the rotation of xyz are

$$\bar{\omega} = \omega_1\bar{J} \qquad \text{always}$$

$$\bar{\alpha} = \dot{\omega}_1\bar{J} + \omega_1\dot{\bar{J}} = \dot{\omega}_1\bar{J} \qquad \text{always}$$

The $(\bar{\imath}, \bar{\jmath}, \bar{k})$ components of $\bar{\alpha}$ and $\bar{\omega}$ at this instant are found by transforming the unit vectors. Hence, using the end view in our sketch to express the components of the unit vector $\bar{J}$, we find that

$$\bar{\omega} = \omega_1(\cos\theta\,\bar{\jmath} + \sin\theta\,\bar{k}) \qquad \text{instantaneously}$$

$$\bar{\alpha} = \dot{\omega}_1(\cos\theta\,\bar{\jmath} + \sin\theta\,\bar{k}) \qquad \text{instantaneously}$$

Now we consider the relative motion of the rider's eyes. As we can see from the sketch,

$$\bar{r}_{P/B} = 3\bar{\jmath} \text{ ft}$$

The motion of the cockpit, and thus the rider, relative to xyz (i.e., relative to the main arm) consists of a rotation about the x axis at angular speed ω_2. The rider's eyes are 3 ft from the x axis. Hence, we have

$$(\bar{v}_{P/B})_{\text{rel}} = 3\omega_2(\bar{e}_\theta)_P$$

$$(\bar{a}_{P/B})_{\text{rel}} = -3(\omega_2)^2(\bar{e}_R)_P + 3\dot{\omega}_2(\bar{e}_\theta)_P$$

For this relative motion the radial direction at point P is along the y axis, and the transverse direction in the sense of increasing θ is parallel to the negative z axis. Thus

$$(\bar{v}_{P/B})_{\text{rel}} = 3\omega_2(-\bar{k})$$

$$(\bar{a}_{P/B})_{\text{rel}} = -3(\omega_2)^2\bar{j} + 3\dot{\omega}_2(-\bar{k})$$

To conclude the solution, we substitute the individual terms into equation (22) for $\bar{a}_P$. This gives

$$
\begin{aligned}
\bar{a}_P &= [-40(\omega_1)^2\bar{i} + 40\dot{\omega}_1 \sin\theta\,\bar{j} - 40\dot{\omega}_1 \cos\theta\,\bar{k}] \\
&\quad + [-3(\omega_2)^2\bar{j} - 3\dot{\omega}_2\bar{k}] + [\dot{\omega}_1(\cos\theta\,\bar{j} \\
&\quad + \sin\theta\,\bar{k}) \times 3\bar{j}] + 2[\omega_1(\cos\theta\,\bar{j} \\
&\quad + \sin\theta\,\bar{k}) \times (-3\omega_2\bar{k})] + \omega_1(\cos\theta\,\bar{j} \\
&\quad + \sin\theta\,\bar{k}) \times [\omega_1(\cos\theta\,\bar{j} + \sin\theta\,\bar{k}) \times 3\bar{j}] \\
&= -[40(\omega_1)^2 + 3\dot{\omega}_1 \sin\theta + 6\omega_1\,\omega_2 \cos\theta]\bar{i} \\
&\quad + [40\dot{\omega}_1 \sin\theta - 3(\omega_2)^2 - 3\omega_1^2 \sin^2\theta]\bar{j} \\
&\quad + [-40\dot{\omega}_1 \cos\theta - 3\dot{\omega}_2 + 3(\omega_1)^2 \sin\theta \cos\theta]\bar{k}
\end{aligned}
$$

The unit of length used in this problem was feet, so if the rotation rates are radians per second, the solution for $\bar{a}_P$ has the units of feet per second squared.

Approach b: *xyz* fixed to the cockpit

In this approach xyz has two rotational effects: a rotation with the main arm about the fixed Y axis at the rate ω_1 and a rotation relative to the main arm about the axis $C\text{--}C$ at the rate ω_2. Note that the axis $C\text{--}C$ is always the same as the x axis in this approach, and that the rotation about that axis is negative according to the right hand rule. The general expressions for the rotation of xyz are therefore

$$\bar{\omega} = \omega_1\bar{J} + \omega_2(-\bar{i})$$

$$\bar{\alpha} = \dot{\omega}_1\bar{J} + \omega_1\dot{\bar{J}} - \dot{\omega}_2\bar{i} - \omega_2\dot{\bar{i}}$$

The Y axis is fixed, so $\dot{\bar{J}} = 0$. Because the unit vector $\bar{\imath}$ has the same rotation as the xyz frame, we have $\dot{\bar{\imath}} = \bar{\omega} \times \bar{\imath}$. Thus

$$\bar{\omega} = \omega_1 \bar{J} - \omega_2 \bar{\imath} \qquad \text{always}$$

$$\bar{\alpha} = \dot{\omega}_1 \bar{J} - \dot{\omega}_2 \bar{\imath} - \omega_2(\bar{\omega} \times \bar{\imath}) \qquad \text{always}$$

We find the instantaneous values of $\bar{\omega}$ and $\bar{\alpha}$ in terms of their $(\bar{\imath}, \bar{\jmath}, \bar{k})$ components by transforming all unit vectors. Thus, using the sketch to express the components of $\bar{J}$, we obtain

$$\bar{\omega} = \omega_1(\cos\theta\,\bar{\jmath} + \sin\theta\,\bar{k}) - \omega_2 \bar{\imath} \qquad \text{instantaneously}$$

$$\bar{\alpha} = \dot{\omega}_1(\cos\theta\,\bar{\jmath} + \sin\theta\,\bar{k}) - \dot{\omega}_2 \bar{\imath}$$

$$- \omega_2[\omega_1(\cos\theta\,\bar{\jmath} + \sin\theta\,\bar{k}) + \omega_2\,\bar{\imath}] \times \bar{\imath}$$

$$= -\dot{\omega}_2 \bar{\imath} + (\dot{\omega}_1 \cos\theta - \omega_1\omega_2 \sin\theta)\,\bar{\jmath}$$

$$+ (\dot{\omega}_1 \sin\theta + \omega_1\omega_2 \cos\theta)\bar{k} \qquad \text{instantaneously}$$

Before continuing, let us consider the meaning of the terms containing the product $\omega_1\omega_2$. The vector representing the rotation of xyz relative to the main arm is $-\omega_2\bar{\imath}$. This vector rotates with the main arm about the Y axis. Therefore at the instant we are considering, the tip of the unit vector $\bar{\imath}$ is moving out of the plane of the sketch, that is, in the negative z direction. The terms containing $\omega_1\omega_2$ represent this effect in terms of the $(\bar{\imath}, \bar{\jmath}, \bar{k})$ components.

The relative motion in this approach is very simple. From the diagram we see that

$$\bar{r}_{P/B} = 3\bar{\jmath} \text{ ft}$$

As long as the rider's head is held stationary relative to the cockpit, the person's eyes have no movement relative to xyz. Thus

$$(\bar{v}_{P/B})_{\text{rel}} = \bar{0}$$

$$(\bar{a}_{P/B})_{\text{rel}} = \bar{0}$$

Finally, we substitute into equation (22) for $\bar{a}_P$ to obtain

$$\bar{a}_P = -40(\omega_1)^2\bar{\imath} + 40\dot{\omega}_1 \sin\theta\,\bar{\jmath} - 40\dot{\omega}_1 \cos\theta\bar{k}$$

$$+ \bar{0} + [-\dot{\omega}_2\bar{\imath} + (\dot{\omega}_1 \cos\theta - \omega_1\omega_2 \sin\theta)\bar{\jmath}$$

$$+ (\dot{\omega}_1 \sin\theta + \omega_1\omega_2 \cos\theta)\bar{k}] \times 3\bar{\jmath}$$

$$+ \bar{0} + (-\omega_2\bar{\imath} + \omega_1 \cos\theta\,\bar{\jmath} + \omega_1 \sin\theta\,\bar{k})$$

$$\times [(-\omega_2\bar{\imath} + \omega_1 \cos\theta\,\bar{\jmath} + \omega_1 \sin\theta\,\bar{k}) \times 3\bar{\jmath}]$$

$$= -[40(\omega_1)^2 + 3\dot{\omega}_1 \sin\theta + 6\omega_1\omega_2 \cos\theta]\bar{i}$$

$$+[40\dot{\omega}_1 \sin\theta - 3(\omega_2)^2 - 3(\omega_1)^2 \sin^2\theta]\bar{j}$$

$$+[-40\dot{\omega}_1 \cos\theta - 3\dot{\omega}_2 + 3(\omega_1)^2 \sin\theta \cos\theta]\bar{k}$$

Note that the answers from Approaches a and b are identical. A comparison of the two approaches leads to the same thinking as that discussed at the conclusion of Example 2. Basically, you should be familiar with both approaches.

Perhaps the most impressive feature of Example 4 is that we were able to solve the problem at all. Such a problem is beyond the scope of the simple kinematical methods of Module II. The two approaches we employed in utilizing the relative motion formulas are not the only ones possible, but they can be regarded as the most direct approaches.

HOMEWORK PROBLEMS

VI.11 In Problem VI.2, determine the absolute acceleration of each slider in the position shown.

VI.12 The pin P slides within the circular groove of the rectangular plate at a constant relative speed of 50 in./s. In the position shown, the plate has an angular velocity of 5 rad/s and an angular acceleration of 1 rad/s, both clockwise. Determine the acceleration of the pin if at this instant it is located at (a) point A, (b) point B.

Prob. VI.12

VI.13 In Problem VI.3, determine the absolute acceleration of each slider in the position shown.

VI.14 The radar equipment on board aircraft A follows aircraft B by measuring the distance R and angle ϕ with respect to the aircraft's longitudinal axis. Both aircraft are in level flight at the same altitude. Aircraft

A is in a tight horizontal turn having a radius of 1000 m, and its speed is constant at 400 km/hr. At the instant shown, the radar indicates that $R = 500$ m, $\phi = 75°$, $\dot{\phi} = -0.1$ rad/s, $\dot{R} = 80$ m/s, $\ddot{R} = 60$ m/s², $\ddot{\phi} = 0.04$ rad/s². Determine the absolute velocity and absolute acceleration of aircraft *B*. Show the results in a sketch.

Prob. VI.14

Prob. VI.15

VI.15 At the instant shown, the circular hoop is rotating about the vertical axis at $\Omega = 20$ rad/s, and $\dot{\Omega} = 5$ rad/s². The block is moving at a constant speed of 4 m/s relative to the hoop. Determine the acceleration of the slider at the instant when (a) $\theta = 0°$, (b) $\theta = 90°$, (c) $\theta = 60°$.

VI.16 Determine the acceleration of the slider in Problem VI.6 as a function of time.

VI.17 Determine the acceleration of the probe of the antenna in Problem VI.9 when $t = 0.25$ s using the two methods outlined there.

VI.18 The propeller of an airplane is rotating at 1000 rev/min as the airplane follows a 1000-m vertical circle at 360 km/hr. When the airplane is at the top of the circle, determine the angular acceleration of the propeller.

Prob. VI.18

VI.19 In Problem VI.10, the motor is rotating downward at 5 rad/s and this rate is increasing at 10 rad/s². The disk is rotating about the axis of the motor shaft at the constant rate of 900 rev/min. For the instant when $\theta = 30°$ determine (a) the angular acceleration of the disk, (b) the acceleration of point *C* on the disk.

Prob. VI.20

VI.20 The motion of rod AB is constrained by a collar at each end. Collar B moves outward along rod CD at the constant rate of 30 mm/s. Point C is fixed. Determine the velocity and acceleration of each collar for the position shown when $\bar{\omega}$ of the axes is the constant (a) $5\bar{i}$ rad/s, (b) $5\bar{i} + 5\bar{j}$ rad/s.

E. THE EARTH AS A MOVING REFERENCE FRAME

The principles we have developed in this and previous modules have all depended on the existence of a fixed point that could serve as a point of reference. However, according to Einstein's theory of relativity, no point in the universe is truly fixed. Nevertheless, in most engineering problems concerning phenomena close to the earth's surface, we can neglect the motion of the center of the earth in its orbit about the sun, thereby making the earth's center a fixed point of reference.

In this section we shall study the effect of the fact that the earth is rotating about an axis that contains this fixed point of reference and passes through the north and south poles. Specifically, let us determine the effect of the earth's rotation on the motion of the particle P shown in Figure 8.

We shall consider an elementary model for the earth. In this model the earth is regarded as a homogeneous, perfect sphere of radius $R_e = 6370$ km rotating about its fixed polar axis in a counter-clockwise direction as seen by a fixed observer looking down at the North Pole. This rate of rotation is 1 rev/day, so that

$$\omega_e = \frac{2\pi \text{ rad}}{23.934 \text{ hr}} \times \frac{1 \text{ hr}}{3600 \text{ s}} = 7.292 \times (10^{-5}) \text{ rad/s}$$

Therefore, our model ignores such effects as deviation from a spherical shape, local variation in the mass density of the earth, and the very slight wobble (nutation) of the polar axis.

Let the resultant force on the particle of Figure 8 be $\bar{F}$. From Newton's second law we know that the absolute acceleration of the particle is

$$\bar{a}_P = \frac{\bar{F}}{m}$$

To describe the acceleration we use an xyz reference frame which is attached to the earth with origin O' at the earth's surface. The relative motion equation (22) then gives

$$\bar{a}_P = \frac{\bar{F}}{m} = \bar{a}_{O'} + (\bar{a}_{P/O'})_{\text{rel}} + \bar{\alpha} \times \bar{r}_{P/O'} + 2\bar{\omega} \times (\bar{v}_{P/O'})_{\text{rel}} + \bar{\omega} \times (\bar{\omega} \times \bar{r}_{P/O'}) \quad (24)$$

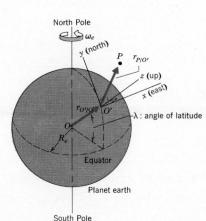

Figure 8

Using our knowledge of the earth's motion, let us study the individual terms in this equation.

The angular velocity of the earth is parallel to the polar axis. The *xyz* frame is attached to the surface of the earth, so the angular velocity vector is

$$\bar{\omega} = \omega_e(\cos \lambda \, \bar{j} + \sin \lambda \, \bar{k}) \tag{25}$$

Further, because we regard both the polar axis and the rate of rotation to be invariant, it follows that

$$\bar{\alpha} = \bar{0}$$

To determine $\bar{a}_{O'}$ we note that point O' is a point on the earth's surface. It therefore follows a circular path in a plane perpendicular to the polar axis. To obtain an explicit expression for $\bar{a}_{O'}$ we use equation (23) for rigid body motion, which yields

$$\bar{a}_{O'} = \bar{a}_O + \bar{\alpha} \times \bar{r}_{O'/O} + \bar{\omega} \times (\bar{\omega} \times \bar{r}_{O'/O})$$

$$= \bar{\omega} \times (\bar{\omega} \times \bar{r}_{O'/O})$$

Comparing this result to the last term in equation (24), $\bar{\omega} \times (\bar{\omega} \times \bar{r}_{P/O'})$, we can conclude that for a particle near the earth's surface,

$$\left| \bar{r}_{P/O'} \right| \ll \left| \bar{r}_{O'/O} \right| = R_e = 6370 \text{ km} \tag{26}$$

Hence the term $\bar{\omega} \times (\bar{\omega} \times \bar{r}_{P/O'})$ is negligible compared to $\bar{a}_{O'}$.

With the foregoing in mind, equation (24) becomes

$$\frac{\bar{F}}{m} = (\bar{a}_{P/O'})_{\text{rel}} + 2\bar{\omega} \times (\bar{v}_{P/O'})_{\text{rel}} + \bar{\omega} \times (\bar{\omega} \times \bar{r}_{O'/O}) \tag{27}$$

where the kinematical quantities for an observer on the earth at point O' are

$$\bar{r}_{P/O'} = x\bar{i} + y\bar{j} + z\bar{k} \tag{28}$$

$$(\bar{v}_{P/O'})_{\text{rel}} = \dot{x}\bar{i} + \dot{y}\bar{j} + \dot{z}\bar{k}$$

$$(\bar{a}_{P/O'})_{\text{rel}} = \ddot{x}\bar{i} + \ddot{y}\bar{j} + \ddot{z}\bar{k}$$

In the examples that follow we shall consider two cases where equation (27) has special significance.

EXAMPLE 5

A ball is suspended near the earth's surface from a cord and is brought to rest relative to the earth. Determine the force in the cord and the direction in which the cord is oriented.

Solution

We first draw a free body diagram of the ball that also shows the *xyz* frame fixed to the surface of the earth.

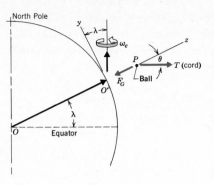

Note that we have given the cord an arbitrary orientation, rather than assuming it to be oriented along the radial line from the center of the earth. Also, we do not assume that the gravitational force $\bar{F}_G$ is the weight. The reasons for doing this will be obvious in the work that follows.

The magnitude of $\bar{F}_G$ is given by the universal law of gravitation, which is given by

$$\left|\bar{F}_G\right| = \frac{GM_e m}{R_e^2}$$

Therefore, from the free body diagram we see that the force sum is

$$\bar{F} = \left(T \cos \theta - \frac{GM_e m}{R_e^2}\right)\bar{k} - T \sin \theta\, \bar{j}$$

The ball is constrained to have no motion relative to the earth, so equation (27) reduces to

$$\frac{\bar{F}}{m} = \bar{\omega} \times (\bar{\omega} \times \bar{r}_{O'/O})$$

As we have seen in the general discussion, $\bar{\omega} = \omega_e(\cos \lambda\, \bar{j} + \sin \lambda\, \bar{k})$ and $\bar{r}_{O'/O} = R_e\bar{k}$. The centripetal acceleration is therefore

$$\bar{\omega} \times (\bar{\omega} \times \bar{r}_{O'/O}) = R_e\omega_e^2 \cos \lambda(\sin \lambda\, \bar{j} - \cos \lambda\, \bar{k})$$

Notice that the computation shows that the centripetal acceleration is perpendicular to the polar axis, as it should be.

By equating this term to $\bar{F}/m$, we have the two scalar equations

$$\frac{T}{m} \cos \theta - \frac{GM_e}{R_e^2} = -R_e\omega_e^2 \cos^2 \lambda$$

$$-\frac{T}{m} \sin \theta = R_e\omega_e^2 \sin \lambda \cos \lambda$$

Solving for T and θ, we find that

$$T = m\left[\left(\frac{GM_e}{R_e^2}\right)^2 - R_e\omega_e^2\left(2\frac{GM_e}{R_e^2} - R_e\omega_e^2\right)\cos^2 \lambda\right]^{1/2}$$

$$\tan \theta = -\frac{R_e\omega_e^2 \sin \lambda \cos \lambda}{(GM_e/R_e^2) - R_e\omega_e^2 \cos^2 \lambda}$$

Let us now interpret this solution. In the sense of a statics problem, we would say that the tension T equals the weight mg, where g is the *local* free fall acceleration. We would also say that the cord is oriented vertically. From the results we see that the apparent weight of the body is not solely the force of gravity and that the vertical direction is not truly along the radial line from the center of the earth. However, from the solution for θ we can see that $\theta \equiv 0$ at $\lambda = 0, \pm 90°$, (i.e., at the equator and the north and south poles). Furthermore, because

$$\frac{GM}{R_e{}^2} = \frac{6.670(10^{-11})(5.976)(10^{24})}{(6.370)^2(10^6)^2} \text{ m/s}^2$$

$$= 9.820 \text{ m/s}^2$$

and

$$R_e\omega_e{}^2 = (6.371)(10^6)(7.292)^2(10^{-5})^2$$

$$= 3.388(10^{-2}) \text{ m/s}^2$$

we can show that the maximum value $|\theta| = 0.099°$ occurs at $\lambda = \pm 44.90°$. This value of θ is sufficiently small to consider it negligible in engineering problems. Similarly, if we let $T = mg$, we can see that the value of g increases steadily from $g = 9.786$ m/s² at the equator to $g = 9.820$ m/s² at the poles. This fluctuation is not significant to the usual accuracy required for engineering problems, and hence it has become common practice to use $g = 9.806$ m/s² as an average. However, for completeness, we note that there is an *international gravity formula* that accounts for both the rotation of the earth and the deviation of the earth from a spherical shape. It is

$$g = 9.780(1 + 0.005288 \sin^2 \lambda - 0.000006 \sin^2 2\lambda) \text{ m/s}^2$$

EXAMPLE 6
A low-pressure area forms within a fluid, such as an open drain in a tank of water, or a low-pressure atmospheric weather system. Assuming that the fluid is constrained to move solely in a horizontal plane, describe the resulting fluid motion.

Solution
In the previous example we saw that the term $\bar{a}_{O'} = \bar{\omega} \times (\bar{\omega} \times \bar{r}_{O'/O})$ is essentially a small correction to the weight force. We can therefore neglect this term when we apply equation (27). Hence, we have the following expression for the relative acceleration:

$$(\bar{a}_{P/O'})_{\text{rel}} = \frac{\bar{F}}{m} - 2\bar{\omega} \times (\bar{v}_{P/O'})_{\text{rel}}$$

We know that the relative velocity and acceleration are both parallel to the horizontal (xy) plane, because the fluid particles are constrained to move in this plane. We therefore need consider only the horizontal component of the Coriolis acceleration, which depends only on the vertical component of $\bar{\omega}$,

$$(\bar{a}_{\text{cor}})_{\text{hor}} = 2(\omega_z \bar{k}) \times (\bar{v}_{P/O'})_{\text{rel}}$$

Also, from equation (25) we know that

$$\omega_z = \omega_e \sin \lambda$$

The horizontal component of the force $\bar{F}$ is due to the pressure difference, and is therefore directed toward the center of the low-pressure area.

We now introduce some qualitative reasoning. As a particle of fluid starts from rest relative to the earth, the pressure difference causes the particle to accelerate toward the center of the low pressure area. Therefore, $(\bar{v}_{P/O'})_{\text{rel}}$ initially is also toward the center of the low-pressure area. The initial relative acceleration is the vector sum of the terms $\bar{F}/m$ and $-(\bar{a}_{\text{cor}})_{\text{hor}} = -2(\omega_z \bar{k}) \times (\bar{v}_{P/O'})_{\text{rel}}$; the latter term, of course, is perpendicular to $(\bar{v}_{P/O'})_{\text{rel}}$. This vector sum is illustrated in the sketch, which shows a number of fluid particles around a low-pressure region in the Northern Hemisphere.

The dashed lines are a trace of the path that the particles follow. We see that the Coriolis effect causes a relative acceleration that is to the right of the initial direction of motion, thus causing the particle to assume a counterclockwise motion about the low pressure region. In the Southern Hemisphere the value of ω_z is negative, thus giving a Coriolis term that is opposite in sign from the one shown in the sketch.

We conclude that the fluid motion will consist of an inward movement, plus a counterclockwise rotation in the Northern Hemisphere or a clockwise rotation in the Southern Hemisphere.

Knowledge of this effect is essential to the understanding of meteorological and oceanographic phenomena, such as hurricanes, whirlpools, and ocean currents. However, the preceding discussion does not provide a full description because we have not considered the forces that are generated within the fluid by its motion.

HOMEWORK PROBLEMS

VI.21 A 1500-kg automobile is traveling north at 90 km/hr along a straight road. It is in the Northern Hemisphere, at 45° latitude. Determine the reaction force exerted by the ground on the vehicle in terms of north, east, and vertical components.

VI.22 The automobile in Problem VI.21 is moving (a) south, (b) west. Determine the corresponding reaction forces.

VI.23 A ball bearing rolls on a frictionless horizontal plane located at latitude 60° north. It was given an initial velocity of 20 ft/s east. Determine

Prob. VI.21

the acceleration of the ball bearing and the radius of curvature of its path, both with respect to the earth, at the instant it was released.

Prob. VI.23   Prob. VI.24

VI.24   An object is falling vertically at 25 m/s in the Northern Hemisphere, at 30° latitude. Neglecting air resistance, determine the acceleration of the object and the radius of curvature of its path, both with respect to the earth.

VI.25   A block of mass m, released from rest, slides down a vertical pole without opposition from friction. Derive expressions for the horizontal components of the contact force exerted by the pole on the block in terms of the time t that the block has been falling and the angle of latitude λ.

VI.26   A baseball is thrown south with a velocity of 90 miles/hr at an angle of elevation of 30° above the horizontal. The location is at latitude 45° north. Neglecting air resistance, determine the initial acceleration of the baseball with respect to the earth.

Prob. VI.26

F.   PROBLEM SOLVING

Following the approach of previous modules, we now present a systematic procedure for the determination of expressions for linear and angular velocity and acceleration. These expressions may be the only results required in a problem, or they may represent the kinematical input needed for the study of the kinetics of the system.

1   Sketch the system, showing the axes of rotation and, symbolically, the angular speeds about these axes. This sketch should also show all significant points, including points whose motion is desired and all points that are constrained.

2   Show your choice for the moving reference frame xyz in the sketch. The general criteria for this step are summarized as follows:

a   Let the origin O' of xyz be a point in the system that is executing a simple motion, such as a point following a known planar path.

b   Let *xyz* have as many rotational velocities as you can easily comprehend, in order to expedite the description of $(\bar{v}_{P/O'})_{rel}$ and $(\bar{a}_{P/O'})_{rel}$. In the situation where it is desired to study the angular motion of a specific body, attach *xyz* to that body.

c   Choose the instantaneous orientation of the *x, y,* and *z* axes to give a simple description for the geometry of the system, in particular, the description of the axes of rotation and of the relative position vectors.

When some aspect of the rotation of the system is described with respect to a fixed observer, select a fixed reference frame *XYZ* such that its origin *O* is a convenient fixed point and one of the axes is parallel to the fixed axis of rotation. (For an example of a situation where the fixed *XYZ* system is not required, see Illustrative Problem 3.)

3   Following the method described on page 344, obtain a general description for the angular velocity $\bar{\omega}$ of the *xyz* frame. If the problem involves acceleration, differentiate this expression to obtain a general expression for $\bar{\alpha}\ (=\dot{\bar{\omega}})$. The instantaneous expressions for $\bar{\omega}$ and $\bar{\alpha}$ are then obtained by replacing the components of the $(\bar{I}, \bar{J}, \bar{K})$ unit vectors by the $(\bar{i}, \bar{j}, \bar{k})$ representation of the *xyz* frame. In problems where constraint conditions must be satisfied, it will prove useful to leave all velocity and acceleration rates in algebraic (symbolic) form.

Velocity Problem

4   If any angular speeds were not given, relate the velocities of all constrained points to the rates of rotation using the relative velocity equation (16). The rates of rotation are obtained by satisfying the constraints on the velocities of these points. Then substitute all rotation rates into the expressions for $\bar{\omega}$ and $\bar{\alpha}$.

5   If the problem requires the velocity of any unconstrained points, use equation (16) for these determinations.

Acceleration Problem

6 and 7   Using the relative acceleration equation (22), repeat the procedures of steps 4 and 5 in terms of the acceleration variables and the restrictions on the accelerations of the constrained points.

ILLUSTRATIVE PROBLEM 1

$\omega_{CD} = 2$ rad/s
$\alpha_{CD} = 1$ rad/s²

The linkage arm *CD* rotates clockwise about the pivot *D* with an angular velocity of 2 rad/s and an angular acceleration of 1 rad/s². The collar at *C* is pinned to rod *CD* and slides along rod *AB*. For the position shown, determine (a) the angular velocity and acceleration of rod *AB,* (b) the velocity and acceleration of the collar relative to rod *AB*.

Solution

This problem is similar to the planar linkage problems of Module IV where all members were pinned together. The new feature of this problem is that the collar allows for motion of point C *relative* to bar AB. To describe this motion we shall employ a moving reference frame. On the other hand, the collar is pinned to rod CD, so the rigid body motion formulas will be applied to describe the motion of this rod.

Step 1

Step 2 The fixed reference frame is selected such that the XY plane is the plane of the problem, and its origin is placed at point A because we shall be treating motion relative to bar AB. Its orientation is not crucial to this problem, but is chosen so that the X axis coincides with the fixed line AD. To describe the motion relative to rod AB, we attach the moving xyz frame to this rod with its origin also at the fixed point A. At the instant of interest the x axis is also selected to coincide with line AD; this expedites the writing of the position vectors. Both the fixed and moving reference frames are shown in the sketch in step 1.

Step 3 Because the xyz frame is attached to rod AB, and because this is a planar motion problem, the rotation of this reference frame is given by

$$\bar{\omega} = \omega_{AB}\,(-\bar{K}) = -\omega_{AB}\bar{k}$$

$$\bar{\alpha} = \dot{\omega}_{AB}(-\bar{K}) = -\alpha_{AB}\bar{k}$$

The magnitudes ω_{AB} and $\dot{\omega}_{AB}$ are two variables that we seek as part of the solution.

Step 4 The constrained points in this problem are the fixed points A and D, and point C on the collar. We can relate points A and C using the relative motion equation (16). Thus

$$\bar{v}_C = \bar{v}_A + (\bar{v}_{C/A})_{\text{rel}} + \bar{\omega} \times \bar{r}_{C/A}$$

However, point C is also a point on rod CD. Its velocity is therefore given by the

rigid body equation (17), which becomes

$$\bar{v}_C = \bar{v}_D + \bar{\omega}_{CD} \times \bar{r}_{C/D}$$

Because points A and D are fixed, $\bar{v}_A = \bar{v}_D = \bar{0}$. Also, we know from the given information that

$$\bar{\omega}_{CD} = 2(-\bar{k}) \text{ rad/s}$$

The two expressions for $\bar{v}_C$ will be equated, but we must first describe the individual terms in these expressions. Toward this end we draw a sketch that illustrates the instantaneous geometry of the system.

From this sketch we see that

$$\bar{r}_{C/A} = -(0.60 \cos 40° - 0.40)\bar{i} + 0.60 \sin 40° \,\bar{j}$$

$$= -0.0596\bar{i} + 0.3857\bar{j} \text{ m}$$

$$\bar{r}_{C/D} = -0.60 \cos 40° \,\bar{i} + 0.60 \sin 40° \,\bar{j}$$

$$= -0.4596\bar{i} + 0.3857\bar{j} \text{ m}$$

$$\beta = \tan^{-1}\left(\frac{0.3857}{0.0596}\right) = 81.21°$$

The angle β is needed to describe $(\bar{v}_{C/A})_{\text{rel}}$. Noting that the collar is free to slide along rod AB, this relative velocity must be parallel to the rod. The magnitude of this term, denoted by v_{rel}, is one of the unknown quantities that is being sought. Assuming the relative velocity to be outward, we therefore have

$$(\bar{v}_{C/A})_{\text{rel}} = v_{\text{rel}}(-\cos \beta \, \bar{i} + \sin \beta \, \bar{j})$$

$$= v_{\text{rel}}(-0.1528\bar{i} + 0.9883\bar{j})$$

Substituting the appropriate terms into the first expression for $\bar{v}_C$, we find that

$$\bar{v}_C = \bar{0} + v_{\text{rel}}(-0.1528\bar{i} + 0.9883\bar{j}) + (-\omega_{AB}\bar{k}) \times (-0.0596\bar{i} + 0.3857\bar{j})$$

$$= (-0.1528v_{\text{rel}} + 0.3857\omega_{AB})\bar{i} + (0.9883v_{\text{rel}} + 0.0596\omega_{AB})\bar{j} \text{ m/s}$$

Similarly, the second expression for $\bar{v}_C$ gives

$$\bar{v}_C = \bar{0} + (-2\bar{k}) \times (-0.4596\bar{i} + 0.3857\bar{j})$$

$$= 0.7714\bar{i} + 0.9192\bar{j} \text{ m/s}$$

By equating the individual components of $\bar{v}_C$, we obtain the following two simultaneous equations:

$$\bar{\imath} \text{ component:} \qquad (v_C)_x = -0.1528v_{rel} + 0.3857\omega_{AB} = 0.7714$$

$$\bar{\jmath} \text{ component:} \qquad (v_C)_y = 0.9883v_{\text{rel}} + 0.0596\omega_{AB} = 0.9192$$

The solution of these equations is

$$\omega_{AB} = 2.313 \qquad \bar{\omega}_{AB} = \bar{\omega} = -2.313\bar{k} \text{ rad/s}$$

$$v_{\text{rel}} = 0.7901 \qquad (\bar{v}_{C/A})_{\text{rel}} = -0.1207\bar{\imath} + 0.7809\bar{\jmath} \text{ m/s}$$

Step 5 Not applicable.

Step 6 We repeat the procedures of step 4 using the acceleration variables. Thus,

$$\bar{a}_C = \bar{a}_A + (\bar{a}_{C/A})_{\text{rel}} + 2\bar{\omega} \times (\bar{v}_{C/A})_{\text{rel}} + \bar{\alpha} \times \bar{r}_{C/A} + \bar{\omega} \times (\bar{\omega} \times \bar{r}_{C/A})$$

$$= \bar{a}_D + \bar{\alpha}_{CD} \times \bar{r}_{C/D} + \bar{\omega} \times (\bar{\omega} \times \bar{r}_{C/D})$$

Noting that points A and D are fixed, $\bar{a}_A = \bar{a}_D = \bar{0}$. From the given information we have

$$\bar{\alpha}_{CD} = 1(-\bar{k}) \text{ rad/s}^2$$

All of the other terms, with the exception of the relative acceleration, are defined in steps 3 and 4. For the relative acceleration we note that because the xyz frame is attached to rod AB, $(\bar{a}_{C/A})_{\text{rel}}$ must account for the sliding motion of the collar along the rod. Thus, letting $\left|(\bar{a}_{C/A})_{\text{rel}}\right| \equiv a_{\text{rel}}$, we have

$$(\bar{a}_{C/A})_{\text{rel}} = a_{\text{rel}}(-\cos\beta\,\bar{\imath} + \sin\beta\,\bar{\jmath})$$

$$= a_{\text{rel}}(-0.1528\bar{\imath} + 0.9883\bar{\jmath})$$

When we substitute the individual terms into the first expression for $\bar{a}_C$, we obtain

$$\bar{a}_C = \bar{0} + a_{\text{rel}}(-0.1528\bar{\imath} + 0.9883\bar{\jmath})$$

$$+ 2(-2.313\bar{k}) \times (-0.1207\bar{\imath} + 0.7809\bar{\jmath})$$

$$+ (-\alpha_{AB}\bar{k}) \times (-0.0596\bar{\imath} + 0.3857\bar{\jmath})$$

$$+ (-2.313\bar{k}) \times [(-2.313\bar{k}) \times (-0.0596\bar{\imath} + 0.3875\bar{\jmath})]$$

$$= (-0.1528a_{\text{rel}} + 3.612 + 0.3857\alpha_{AB} + 0.3189)\bar{\imath}$$

$$+ (0.9883a_{\text{rel}} + 0.5584 + 0.0596\alpha_{AB} + 2.064)\bar{\jmath} \text{ m/s}^2$$

The second expression for $\bar{a}_C$ gives

$$\bar{a}_C = \bar{0} + (-\bar{k}) \times (-0.4596\bar{\imath} + 0.3875\bar{\jmath})$$

$$+ (-2\bar{k}) \times [(-2\bar{k}) \times (-0.4596\bar{i} + 0.3875\bar{j})]$$

$$= 2.226\bar{i} - 1.090\bar{j} \text{ m/s}^2$$

The two scalar equations obtained by equating the two expressions for $\bar{a}_C$ are

$\bar{i}$ component: $\qquad (a_C)_x = -0.1528a_{\text{rel}} + 3.931 + 0.3875\dot{\omega}_{AB}$

$$= 2.226$$

$\bar{j}$ component: $\qquad (a_C)_y = 0.9883a_{\text{rel}} + 2.622 + 0.0596\dot{\omega}_{AB}$

$$= -1.090$$

The solution is

$$\alpha_{AB} = -5.774 \qquad \bar{\alpha}_{AB} = \bar{\alpha} = -5.744\bar{k} \text{ rad/s}$$

$$a_{\text{rel}} = -3.408 \qquad (\bar{a}_{C/A})_{\text{rel}} = 0.5207\bar{i} - 3.368\bar{j} \text{ m/s}^2$$

In closing, let us note the physical significance of the method of the solution. The given rotation values for rod CD define the absolute velocity and acceleration of the slider. The relative motion equations then use the component of these kinematic values perpendicular to rod AB to define the angular velocity and angular acceleration of that rod, whereas the components parallel to rod AB are used to define the velocity and acceleration of the slider relative to rod AB.

ILLUSTRATIVE PROBLEM 2

Cone A is fixed in space and cone B rolls without slipping over the surface of cone A, making five revolutions about the fixed vertical axis each minute. Determine the angular velocity and angular acceleration of cone B.

Solution

The motion of cone B consists of two rotations: the rotation ω_1 about the vertical axis of cone A, which requires 1 minute for 5 rev, plus a rotation of the cone about its own axis at an angular speed ω_2. The value of ω_2 will be determined by satisfying the condition that there is no slipping in the rolling motion.

Step 1 The line of contact CD between the two cones is illustrated by a side view. The angles β and γ defined in the sketch are obtained by solving the appropriate right triangles, specifically,

$$\beta = \tan^{-1}\left(\frac{15.6}{20.8}\right) = \tan^{-1}\left(\frac{3}{4}\right) = 36.87°$$

$$\gamma = \sin^{-1}\left(\frac{10}{\sqrt{(15.6)^2 + (20.8)^2}}\right) = \sin^{-1}\left(\frac{5}{13}\right) = 22.62°$$

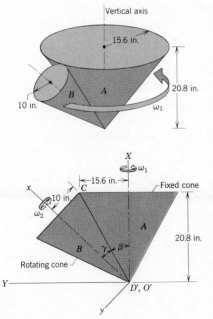

Step 2   Point D, which is the apex of cone B, is the only point on this cone that is fixed. Therefore, choose this point as the origin of the fixed XYZ system, letting the X axis coincide with the vertical axis and the Y axis be parallel to the plane of the sketch in step 1. Now attach the moving xyz frame to cone B so that the angular motion of the reference frame will be the same as that of the cone. It is convenient to select point D to be the origin of this frame also, aligning the x axis along the axis of cone B and placing the y axis in the plane of the sketch at this instant. This orientation suits the description of the axes of rotation and the position vectors of points on cone B. Note that the sense of the rotation ω_2 shown in the sketch is in accordance with our physical intuition. If our intuition should be wrong, the value we obtain for ω_2 will be negative.

Step 3   The total rotation of cone B is the sum of the rotations about the fixed X axis and moving x axis. The general expressions for angular velocity and acceleration are therefore

$$\bar{\omega} = \omega_1 \bar{I} + \omega_2 \bar{\imath} \qquad \text{always}$$

$$\bar{\alpha} = \dot{\omega}_1 \bar{I} + \dot{\omega}_2 \bar{\imath} + \omega_2 \dot{\bar{\imath}}$$

$$= \dot{\omega}_1 \bar{I} + \dot{\omega}_2 \bar{\imath} + \omega_2 (\bar{\omega} \times \bar{\imath}) \qquad \text{always}$$

The information given in the problem is that

$$\omega_1 = 10\,\pi \text{ rad/60 s} = 0.5236 \text{ rad/s}$$

and because ω_1 is constant, $\dot{\omega}_1 = 0$. From the sketch in step 1,

$$\bar{I} = \cos(\beta + \gamma)\,\bar{\imath} - \sin(\beta + \gamma)\,\bar{\jmath}$$

$$= 0.5077\bar{\imath} - 0.8615\bar{\jmath}$$

Hence, with the rotation rates in algebraic form, we have

$$\bar{\omega} = \omega_1 (0.5077\bar{\imath} - 0.8615\bar{\jmath}) + \omega_2 \bar{\imath}$$

$$= (0.5077\omega_1 + \omega_2)\bar{\imath} - 0.8615\omega_1 \bar{\jmath} \qquad \text{instantaneously}$$

$$\bar{\alpha} = \dot{\omega}_2 \bar{\imath} + \omega_2 [(0.5077\omega_1 + \omega_2)\bar{\imath} - 0.8615\omega_1 \bar{\jmath}] \times \bar{\imath}$$

$$= \dot{\omega}_2 \bar{\imath} + 0.8615\omega_1 \omega_2 \bar{k} \qquad \text{instantaneously}$$

Step 4   We now satisfy the constraint condition that there is rolling without slipping. Physically this requirement means that all points on the line CD of cone B that are in contact with line CD of fixed cone A must have zero velocity. To satisfy this condition, we relate the velocities of points C and D on cone B by using equation (17). Thus

$$\bar{v}_C = \bar{v}_D + \bar{\omega}_{\text{cone}} \times \bar{r}_{C/D} = \bar{0} + \bar{\omega} \times \bar{r}_{C/D} \equiv \bar{0}$$

Interpreting this result, recall that $\bar{\omega}$ is parallel to the instantaneous axis of rotation. Hence, it follows that the condition $\bar{\omega} \times \bar{r}_{C/D} = \bar{0}$ means that the instantaneous axis of rotation must be parallel to line CD.

From the sketch we have

$$r_{C/D} = \sqrt{(15.6)^2 + (20.8)^2} \, (\cos \gamma \, \bar{i} - \sin \gamma \, \bar{j}$$

$$= 24\bar{i} - 10\bar{j} \text{ in.}$$

so that

$$\bar{\omega} \times \bar{r}_{C/D} = [(0.5077\omega_1 + \omega_2)\bar{i} - 0.8615\omega_1 \, \bar{j}] \times (24\bar{i} - 10\bar{j})$$

$$= (-5.08\omega_1 - 10\omega_2 + 20.68\omega_1)\bar{k} = \bar{0}$$

As a result, we obtain

$$\omega_2 = 1.560\omega_1$$

Recall that the sketch represented an *arbitrary* instant. Therefore, this relationship between ω_1 and ω_2 is valid for all time t, thus enabling us to differentiate it to obtain an expression for $\dot{\omega}_2$.

$$\dot{\omega}_2 = 1.560\dot{\omega}_1$$

Then, for the numerical values appearing in step 3 we have

$$\omega_2 = 1.560(0.5236) = 0.8168 \text{ rad/s}$$

$$\dot{\omega}_2 = 0$$

Placing these results into the general form for $\bar{\omega}$ and $\bar{\alpha}$ yields

$$\bar{\omega} = [0.5077(0.5236) + 0.8168]\bar{i} - 0.8615(0.5236)\bar{j}$$

$$= 1.083\bar{i} - 0.451\bar{j} \text{ rad/s}$$

$$\bar{\alpha} = 0.8615(0.5236)(0.8168)\bar{k}$$

$$= 0.3684\bar{k} \text{ rad/s}$$

The solution is now complete; the remaining steps are unnecessary.

ILLUSTRATIVE PROBLEM 3

The rotations of an airplane about a set of xyz axes attached to the airplane are termed the roll rate ω_x, the pitching rate ω_y, and the yaw rate ω_z. A passenger on a jumbo jet is 30 m behind and 2 m above the center of mass G of the airplane and at that instant is walking aft at the constant rate of 1 m/s. Determine the difference between the acceleration that this passenger experiences and the acceleration that the center of mass of the airplane undergoes, if the roll rate is 0.5 rad/s, the pitching rate is -0.3 rad/s, and the yaw rate is 0.1 rad/s. All rotation rates are constant.

Solution

Step 1 The diagram accompanying the statement of the problem will suffice for the solution.

Step 2 We shall use the moving xyz frame given in the problem, which is attached to the aircraft. Because all rotation rates are referred to the axes of this coordinate system, and because we are not concerned with the absolute acceleration of the origin G, we do not need a fixed reference frame XYZ.

Step 3 The roll, pitch, and yaw rates are with respect to the x, y, and z axes, respectively. Hence

$$\bar{\omega} = \omega_x \, \bar{\imath} + \omega_y \, \bar{\jmath} + \omega_z \, \bar{K} \qquad \text{always and instantaneously}$$

$$\bar{\alpha} = \dot{\omega}_x \, \bar{\imath} + \omega_x \, \dot{\bar{\imath}} + \dot{\omega}_y \, \bar{\jmath} + \omega_y \, \dot{\bar{\jmath}} + \dot{\omega}_z \, \bar{k} + \omega_z \, \dot{\bar{k}}$$

$$= \dot{\omega}_x \, \bar{\imath} + \dot{\omega}_y \, \bar{\jmath} \, \dot{\omega}_z \, \bar{K} + \omega_x(\bar{\omega} \times \bar{\imath}) + \omega_y(\bar{\omega} \times \bar{\jmath}) + \omega_z(\bar{\omega} \times \bar{k})$$

$$= \dot{\omega}_x \, \bar{\imath} + \dot{\omega}_y \, \bar{\jmath} + \dot{\omega}_z \, \bar{k} + \bar{\omega} \times (\omega_x \, \bar{\imath} + \omega_y \, \bar{\jmath} + \omega_z \, \bar{k})$$

$$= \dot{\omega}_x \, \bar{\imath} + \dot{\omega}_y \, \bar{\jmath} + \dot{\omega}_z \, \bar{k} \qquad \text{always and instantaneously}$$

This result for angular acceleration is interesting, because it states that the components of angular acceleration of the airplane are the derivatives of the components of the angular velocity. The fact that this is always true when we fix the moving reference system to the body will have great significance for our treatment of the equations of motion of a rigid body in the next module.

We may now substitute all rotation rates, for all values are given. Thus

$$\bar{\omega} = 0.5\bar{\imath} - 0.3\bar{\jmath} + 0.1\bar{k} \text{ rad/s}$$

The rates of rotation are constant, so $\bar{\alpha} = \bar{0}$.

Steps 4-6 These steps are not applicable, for there are no constraint conditions to satisfy and we do not seek the absolute velocity of any points.

Step 7 We are interested in the value of $(\bar{a}_P - \bar{a}_G)$, where point P is the position of the passenger. This is obtained from equation (22), namely,

$$\bar{a}_P - \bar{a}_G = (\bar{a}_{P/G})_{\text{rel}} + 2\bar{\omega} \times (\bar{v}_{P/G})_{\text{rel}} + \bar{\alpha} \times \bar{r}_{P/G} + \bar{\omega} \times (\bar{\omega} \times \bar{r}_{P/G})$$

From the information given in the problem and the choice of coordinate axes in the sketch, we may write,

$$\bar{r}_{P/G} = -30\bar{\imath} + 2\bar{k} \text{ m}$$

Also, from the given information we know that the passenger has a constant speed of 1 m/s in the negative x direction relative to the airplane. Thus

$$(\bar{v}_{P/G})_{\text{rel}} = -1.0\bar{\imath} \text{ m/s}$$

$$(\bar{a}_{P/G})_{\text{rel}} = \bar{0}$$

The solution is then found by substituting the values of $\bar{\omega}$ and $\bar{\alpha}$ in step 4, and the relative motion variables determined above into the acceleration equation. This step gives

$$\begin{aligned}
\bar{a}_P - \bar{a}_G &= \bar{0} + 2(0.5\bar{\imath} - 0.3\bar{\jmath} + 0.1\bar{k}) \times (-1.0\bar{\imath}) \\
&\quad + \bar{0} + (0.5\bar{\imath} - 0.3\bar{\jmath} + 0.1\bar{k}) \times [(0.5\bar{\imath} - 0.3\bar{\jmath} \\
&\quad + 0.1\bar{k}) \times (-30\bar{\imath} + 2\bar{k})] \\
&= 3.10\bar{\imath} + 4.24\bar{\jmath} - 2.78\bar{k} \text{ m/s}^2
\end{aligned}$$

The magnitude of this result is $0.61g$, which exceeds the value desired in normal operation of passenger aircraft. It should be noted, however, that atmospheric disturbances can cause rotation rates comparable to those considered in this problem.

ILLUSTRATIVE PROBLEM 4

The sketch shows a universal joint of the type commonly employed in the drive train of automobiles. Considering ω_1 as a known value, determine the corresponding value of ω_2 and the angular velocity of the crossbar in the position where arm CD of the crossbar is horizontal and arm AB coincides with the vertical plane, as shown.

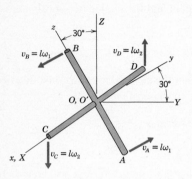

Solution

Step 1 The angular speeds ω_1 and ω_2 are related to each other by the fact that the shafts are connected by a crossbar. For this reason we focus attention on the motion of the crossbar.

Step 2 We place the origin of xyz at the fixed midpoint of the crossbar and attach xyz to that body, so as to eliminate the relative velocity term when we express the velocities of the endpoints. The x and z axes are oriented along the arms in order to expedite writing the position vectors. The fixed XYZ system is oriented horizontally and vertically. This is illustrated in the sketch.

Step 3 Being unsure of the exact nature of the angular motion of the crossbar, we let the angular velocity be an unknown general vector

$$\bar{\omega} = \omega_x\,\bar{\imath} + \omega_y\,\bar{\jmath} + \omega_z\,\bar{k}$$

Note that we were asked only to relate angular speeds, hence we are not concerned here with accelerations.

Step 4 The constrained points of the crossbar are points B and C (the motion of the other two end points differ only in sign). Point B, being a point that is also on the angled shaft, follows a circular path whose plane is perpendicular to this shaft. Because arm AB is oriented in the vertical plane at this instant, the velocity of point B is $l\omega_1$ horizontally. By similar reasoning the velocity of point C at this instant is $l\omega_2$ vertically. Both velocities are shown in the sketch; in component form they are

$$\bar{v}_B = l\omega_1\bar{\imath}$$

$$\bar{v}_C = -l\omega_2\bar{K} = -l\omega_2(\sin 30°\,\bar{\jmath} + \cos 30°\,\bar{k})$$

These are the (instantaneous) constraint conditions on the velocity. We also know that points B and C are both common to the cross bar. Applying the formula for the velocity of points in a rigid body and using the angular velocity $\bar{\omega}$ from step 3, we then have

$$\bar{v}_C = \bar{v}_B + \bar{\omega} \times \bar{r}_{C/B}$$

$$-l\omega_2(\sin 30°\,\bar{\jmath} + \cos 30°\,\bar{k}) = l\omega_1\bar{\imath} + (\omega_x\,\bar{\imath} + \omega_y\,\bar{\jmath} + \omega_z\,\bar{k}) \times (-l\bar{k} + l\bar{\imath})$$

$$= (l\omega_1 - l\omega_y)\bar{\imath} + (l\omega_x + l\omega_z)\bar{\jmath} - l\omega_y\bar{k}$$

Equating components yields

$\bar{\imath}$ component: $0 = l\omega_1 - l\omega_y$

$\bar{\jmath}$ component: $-l\omega_2 \sin 30° = l\omega_x + l\omega_z$

$\bar{k}$ component: $-l\omega_2 \cos 30° = -l\omega_y$

Thus

$$\omega_1 = \omega_y = \omega_2 \cos 30°$$

$$\omega_x + \omega_z = -\omega_2 \sin 30°$$

From this we obtain the solution to the problem,

$$\omega_2 = \frac{\omega_1}{\cos 30°} = 1.1547\omega_1$$

Notice that the foregoing equations are not sufficient to define fully the angular motion of the corssbar. A constraint condition we have not employed is that $\bar{v}_0 = \bar{0}$. Relating $\bar{v}_0$ and $\bar{v}_B$, we have

$$\bar{v}_B = \bar{v}_0 + \bar{\omega} \times \bar{r}_{B/0} = l\omega_1\bar{\imath}$$

$$= \bar{0} + (\omega_x\,\bar{\imath} + \omega_y\,\bar{\jmath} + \omega_z\,\bar{k}) \times l\bar{k} = l\omega_y\,\bar{\imath} - l\omega_x\,\bar{\jmath}$$

Matching components then yields

$$\bar{\imath} \text{ component:} \qquad \omega_1 = \omega_y$$

$$\bar{\jmath} \text{ component:} \qquad 0 = \omega_x$$

We thus find from the earlier equations that

$$\omega_z = -\omega_2 \sin 30° = -\left(\frac{\omega_1}{\cos 30°}\right)\sin 30° = -\omega_1 \tan 30°$$

$$\bar{\omega} = \omega_x\,\bar{\imath} + \omega_y\,\bar{\jmath} + \omega_z\,\bar{k} = \omega_1(\bar{\jmath} - \tan 30° \, \bar{k})$$

This vector is sketched below. An interesting aspect of this result is that

$$\left|\bar{\omega}\right| = \omega_1\sqrt{1 + \tan^2 30°} = \frac{\omega_1}{\cos 30°} \equiv \omega_2$$

Further, note that $\bar{\omega}$ is oriented horizontally.

ILLUSTRATIVE PROBLEM 5 (Optional)

In order to define the orientation of a spinning top, a set of *Eulerian angles* is defined. The *precession angle* ϕ describes the angle of rotation about the fixed Z axis of the centerline OA of the top, and the *nutation angle* θ describes the angle between the Z axis and the centerline. Then the *spin angle* ψ defines the rotation of the top about its centerline. Treating these angles as arbitrary functions of time, obtain expressions for the angular velocity and angular acceleration of the top.

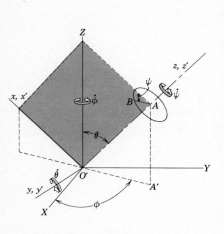

Solution

As we shall see, the new feature of this problem is that the three angles give rise to rotations about three axes, which will cause us to employ two moving reference systems.

Step 1 To simplify the sketch we draw only the centerline OA and the line AB, which are sufficient for illustrating the Eulerian angles.

Steps 2 and 3 The fixed reference frame XYZ is given. The origin O' of the moving reference frame xyz is located at the fixed point O. Further, xyz is attached to the top because we want to describe the angular motion of this body. Let the z axis always coincide with the centerline OA. For convenience we select the instantaneous orientations of the other axes such that the y axis coincides with the XY plane and the x axis coincides with the shaded plane formed by the fixed Z axis and the z axis.

By this choice of axes we are able to say that the precession rate $\dot{\phi}$ is always about the fixed Z axis and the spin rate $\dot{\psi}$ is always about the moving z axis. What about the nutation rate $\dot{\theta}$? The corresponding axis of rotation is perpendicular to the plane formed by the z and Z axes. At this instant the y axis is perpendicular to this plane. Hence, instantaneously the nutation rate is about this axis. However, when the spin angle ψ changes, the y axis will rotate away from the XY plane and will not longer be perpendicular to the Zz plane. For this reason we introduce another reference frame $x'y'z'$ that does not undergo any spin motion. As shown in the sketch in step 1, the xyz and $x'y'z'$ axes are coincident at the instant under consideration, but only the z and z' axes are coincident at all instants. Then according to the right-hand rule, the nutation rate $\dot{\theta}$ is always about the negative y' axis.

Let us denote the angular velocity of xyz as $\bar{\omega}$ and the angular velocity of $x'y'z'$ as $\bar{\Omega}$. The general expressions are

$$\bar{\omega} = \dot{\phi}\bar{K} + \dot{\theta}(-\bar{j}') + \dot{\psi}\bar{k} \qquad \text{always}$$

$$\bar{\Omega} = \dot{\phi}\bar{K} + \dot{\theta}(-\bar{j}') \qquad \text{always}$$

As previously mentioned, $\bar{\omega}$ is the angular velocity of the top. To find the angular acceleration we compute $\bar{\alpha} \equiv \dot{\bar{\omega}}$, as follows:

$$\bar{\alpha} = \dot{\bar{\omega}} = \ddot{\phi}\bar{K} - \ddot{\theta}\bar{j}' - \dot{\theta}\,\dot{\bar{j}}' + \ddot{\psi}\bar{k} + \dot{\psi}\dot{\bar{k}} \qquad \text{always}$$

The unit vector $\bar{j}'$ is fixed to the $x'y'z'$ system which is rotating at the angular speed $\bar{\Omega}$, whereas the unit vector $\bar{k}$ is fixed to the xyz system which is rotating at the angular speed $\bar{\omega}$. Hence

$$\bar{\alpha} = \ddot{\phi}\bar{K} - \ddot{\theta}\bar{j}' - \dot{\theta}(\bar{\Omega} \times \bar{j}') + \ddot{\psi}\bar{k} + \dot{\psi}(\bar{\omega} \times \bar{k}) \qquad \text{always}$$

For the last step we obtain the instantaneous angular velocity and angular acceleration vectors by expressing the unit vectors in terms of their xyz components. Referring to the sketch in step 1, we see that the Z axis coincides with the *xz* plane and that θ is the angle between the Z and z axes, so

$$\bar{K} = \sin \theta \, \bar{\imath} + \cos \theta \, \bar{k}$$

Also, the y and y' axes are coincident at this instant, so $\bar{\jmath}' = \bar{\jmath}$. Thus, the general expressions become

$$\bar{\omega} = \dot{\phi}(\sin \theta \, \bar{\imath} + \cos \theta \, \bar{k}) - \dot{\theta}\bar{\jmath} + \dot{\psi}\bar{k}$$

$$= \dot{\phi} \sin \theta \, \bar{\imath} - \dot{\theta} \, \bar{\jmath} + (\dot{\phi} \cos \theta + \dot{\psi})\bar{k}$$

$$\bar{\Omega} = \dot{\phi} \sin \theta \, \bar{\imath} - \dot{\theta}\bar{\jmath} + \dot{\phi} \cos \theta \, \bar{k}$$

$$\bar{\alpha} = \ddot{\phi}(\sin \theta \, \bar{\imath} + \cos \theta \, \bar{k}) - \ddot{\theta}\bar{\jmath} - \dot{\theta}(\dot{\phi} \sin \theta \, \bar{\imath} - \dot{\theta} \, \bar{\jmath} + \dot{\phi} \cos \theta \, \bar{k}) \times \bar{\jmath}$$

$$+ \ddot{\psi}\bar{k} + \dot{\psi}[\dot{\phi} \sin \theta \, \bar{\imath} - \dot{\theta}\bar{\jmath} + (\dot{\phi} \cos \theta + \dot{\psi})\bar{k}] \times \bar{k}$$

$$= (\ddot{\phi} \sin \theta + \dot{\phi} \, \dot{\theta} \cos \theta - \dot{\theta} \, \dot{\psi})\bar{\imath} - (\ddot{\theta} + \dot{\phi}\dot{\psi} \sin \theta) \, \bar{\jmath}$$

$$+ (\ddot{\psi} + \ddot{\phi} \cos \theta - \dot{\phi} \, \dot{\theta} \sin \theta) \, \bar{k}$$

The foregoing expressions are always valid, for the expressions were evaluated at a general instant (the angles are arbitrary). Equations such as these are widely employed for advanced studies of the motion of rigid bodies, especially for computerized studies.

HOMEWORK PROBLEMS

VI.27 Two automobiles are traveling in opposite directions along a curve in a divided highway. Both vehicles are traveling at a constant speed of 55 miles/hr. For the position where $\theta = 0$, determine (a) the absolute velocity and acceleration of car B relative to car A, (b) the velocity and acceleration of car B relative to a passenger in car A who is unaware of the motion of that vehicle.

Prob. VI.27

VI.28 Solve Problem VI.27 in the position where $\theta = 20°$.

VI.29 Airplane A is in straight level flight at an altitude of 15 km and airplane B is following a horizontal curve of 1-km radius at an altitude of

12 km. Both aircraft have a constant speed of 750 km/hr. Determine the velocity and acceleration of aircraft A as viewed by the pilot of aircraft B. The aircraft forms the frame of reference for the pilot.

Prob. VI.29

VI.30 A person who is stationed at point B on the perimeter of the rotating platform is observing a baseball in free flight. At the instant shown, the ball is at point C and is moving horizontally with a speed of 40 ft/s. The angular velocity of the platform is 2 rad/s clockwise. For this instant, determine (a) the velocity of the baseball as seen by this person, (b) the acceleration of the baseball as seen by this person, (c) the radius of curvature in the *horizontal plane* of the path of the ball with respect to the person. Draw a sketch of the results.

Prob. VI.30

VI.31 Water flows through the arms of the garden sprinkler at the constant rate of 15 m/s as the sprinkler rotates at 20 rad/s. At the nozzle the radius of curvature of the sprinkler arm is 200 m and the tangent to an arm is at a 30° angle from the radial direction. (a) What is the acceleration of the water immediately before it emerges from an arm? (b) What is the actual velocity of the fluid that emerges? Give the magnitude and the angle with respect to the radial line.

30°

$v_{rel} = 15$ m/s

200 mm

$\omega = 20$ rad/s

Prob. VI.31

VI.32 The horizontal arm of the amusement park ride rotates about the vertical axis at $\Omega = 6$ rev/min and the cars rotate at the angular speed $\omega_{rel} = 9$ rev/min *relative to that arm*. Determine the velocity and acceleration of the rider's head for the position shown.

12 m

6 m

$\omega_{rel} = 9$ rev/min

$\Omega = 6$ rev/min

45°

Prob. VI.32

VI.33 Collar *B* is pinned to bar *AB* and slides over bar *CD*. In the position shown, bar *CD* is rotating clockwise at 50 rad/s and is slowing down at the rate of 20 rad/s². Determine the angular velocity and angular acceleration of bar *AB* at this instant.

Prob. VI.33

Both bars slide through holes in block *E*, with the result that the angle between the bars is constant at 45°. Bar *AB* has a constant counterclockwise angular velocity of 10 rad/s. Determine the corresponding velocity of block *E* in the position shown.

Prob. VI.34

VI.35 Determine the acceleration of block *E* in Problem VI.34 and the radius of curvature of its path in the position shown.

VI.36 Slider *A* is moving downward at the constant rate of 600 mm/s. Collar *C* is pinned to bar *CD* and slides over bar *AB*. Determine the angular velocities of both bars in the position shown.

Prob. VI.36

VI.37 Determine the angular acceleration of both bars in Problem VI.36.

VI.38 The bead slides over the bent rod with a constant relative acceleration $\dot{u}$, starting from rest at point A at $t = 0$. The rod is rotating about the vertical axis at a constant angular velocity $\bar{\omega}$. Determine the velocity and acceleration of the bead as a function of t.

VI.39 An insect crawls along the exterior surface of the cylindrical hoop shown at a constant rate of 8 in./s as the hoop rotates about the fixed horizontal axis at the constant rate of 0.25 rad/s. Derive expressions for the components of the acceleration tangent and normal to the *surface* of the hoop in terms of the value of θ. What are the maximum values of each component?

Prob. VI.38

Prob. VI.39

VI.40 Sliders A and B are pinned to bar AB. Slider A has a constant downward velocity of 1.2 m/s as the T-bar assembly rotates about the vertical axis at 3 rad/s. Determine (a) the velocity of slider B, (b) the acceleration of slider B for the position shown.

Prob. VI.40

VI.41 For the system in Problem VI.40, determine (a) the angular velocity of bar AB, (b) the angular acceleration of bar AB.

VI.42   Solve Illustrative Problem 4 for the position where arm AB is horizontal.

VI.43 and VI.44   Sliders A and B are connected to bar AB by ball joints. Slider B has a speed of 300 mm/s to the right. Determine the velocity of slider A in the position shown. *Hint:* Considering $\bar{\omega}_{AB}$ as a general vector that is to be determined results in three algebraic equations for four unknowns. Nevertheless, $\bar{v}_A$ may be solved for by a process of elimination. This situation occurs because the ball joints allow the bar to have an arbitrary angular velocity about its own axis.

Prob. VI.44

Prob. VI.43

VI.45   The speed of slider B in problem VI.43 is constant. What is the corresponding acceleration of slider A?

VI.46   Slider C and bars AB and BC are interconnected by ball joints. Bar AB is rotating about the fixed point A at the constant rate of 5 rad/s. Determine the velocity of slider C for the position shown. *Hint:* See the comment regarding Problem VI.43.

VI.47   Determine the acceleration of slider C in Problem VI.46.

Prob. VI.46

VI.48   The disk has a constant angular velocity of 80 rad/s about its axis as the fork that supports the disk rotates about the horizontal axis at 10 rad/s. Determine the velocity and acceleration (a) of point B, (b) of point C.

Prob. VI.48

VI.49   The disk is rotating about its axis at a constant rate of 1000 rev/min as arm AB pivots about a horizontal axis. In the position shown, arm

AB is inclined 30° from the vertical and it has an angular speed of 10 rad/s which is increasing at the rate of 2 rad/s². Determine the angular velocity and angular acceleration of the disk.

Prob. VI.49

VI.50 Determine the velocity and acceleration of point *C* of the disk in Problem VI.49.

VI.51 The three-bladed propeller of an airplane is turning at 1200 rev/min as the airplane follows a horizontal circle of 1000-ft radius at a speed of 180 miles/hr. Determine the angular velocity and angular acceleration of the propeller blades.

Prob. VI.51

VI.52 The blades of the airplane in Problem VI.51 are 2 ft long. At the instant when blade *A* is oriented vertically determine the velocity and acceleration (a) of the tip of blade *A* and (b) the tip of blade *C*.

VI.53 Gear *B* rotates freely about its shaft, which has an angular acceleration of 20 rad/s² and an instantaneous angular speed of 10 rad/s about the vertical axis. Gear *A* is stationary. Determine (a) the angular velocity and angular acceleration of gear *B*, (b) the velocity and acceleration of point *C*.

Prob. VI.53

$\omega = 300$ rev/min

43°

A

C

B

50 mm

250 mm

Prob. VI.54

VI.54   The shaft of gear A rotates about the fixed vertical axis at a constant rate of 300 rev/min. This gear rotates freely about its shaft as it meshes with the stationary gear B. Determine (a) the angular velocity and angular acceleration of gear A, (b) the velocity and acceleration of point C on gear A.

VI.55   A motorcyclist banks the vehicle into a turn at an angle of 30° from the vertical and follows a horizontal 30-m-radius curve at 50 km/hr. Each wheel has a diameter of 800 mm. Determine the angular velocity and angular acceleration of a wheel.

$v = 50$ km/hr

800 mm

30°

Prob. VI.55

VI.56   In the position shown, the turret on the tank is rotating about the vertical axis of 0.4 rad/s and the barrel is being raised at 0.6 rad/s; both rotation rates are constant. The tank has a constant forward velocity of 60 km/hr. If a cannon shell is fired at this instant with a muzzle speed of 500 m/s relative to the barrel, determine the velocity and acceleration of the shell immediately before it leaves the barrel.

0.4 rad/s

$v_{\text{muzzle}} = 500$ m/s

36.87°

0.6 rad/s

60 km/hr

Prob. VI.56

VI.57   The left tread of the crane is moving faster than the right one, with the result that the chassis is following a circular path. The radius of the path of point A is 40 ft and the speed of that point is a constant 5 ft/s. When $\theta = 53.13°$, the 100-ft-long boom is being lowered such that $\dot{\theta} = -4$ rad/min and $\ddot{\theta} = 0.5$ rad/s². Determine the angular velocity and angular acceleration of the boom.

900 rev/min

30°

ω

Prob. VI.59

Prob. VI.57

VI.58   For the parameters in Problem VI.57, determine the velocity and acceleration of end B of the boom. The boom is 60 feet long.

VI.59   An oscillating electric fan is aimed at 30° above the horizontal. The rotation of the fan about the vertical axis is defined by $\omega = \sin \pi t/5$, where ω is radians per second and t is in seconds. The fan blades are spinning at 900 rev/min. Determine the angular velocity and angular acceleration of the fan blades as a function of time.

VI.60   The motor is mounted on a bracket that is rotating about the vertical shaft at 10 rev/min. The angular speed of the disk about the axis of the motor shaft is 300 rev/min, and the centerline of the shaft is inclined at a constant 36.87° angle from the horizontal. Determine (a) the angular velocity and the angular acceleration of the disk, (b) the velocity and acceleration of point C on the disk.

Prob. VI.60

VI.61 Using the moving xyz reference frame shown, derive expressions for the position, velocity, and acceleration in terms of the cylindrical coordinates (R, ϕ, h) and the corresponding unit vectors $(\bar{e}_R, \bar{e}_\phi, \bar{e}_h)$.

VI.62 Spherical coordinates are another useful set of extrinsic coordinates. In this system the location of a point is defined by the distance r from the origin, the angle θ from the Z axis, and the angle ϕ from the X axis. Use the moving xyz reference frame shown to derive expressions for the position, velocity, and acceleration in terms of the spherical coordinates (r, ϕ, θ) and the corresponding orthogonal unit vectors $(\bar{e}_r, \bar{e}_\phi, \bar{e}_\theta)$.

Prob. VI.61

Prob. VI.62

VI.63 In the position shown, the aircraft is executing a diving turn to the right by pitching its nose down at $\omega_P = 0.5$ rad/s and rolling at $\omega_r = 0.3$ rad/s. The roll rate is decreasing at 0.6 rad/s². Determine the angular velocity and angular acceleration of the aircraft.

Prob. VI.63

VI.64 In Problem VI.63, determine the velocity and acceleration of point A relative to the center of mass G.

VI.65 The center of mass G of a large ship has a constant forward velocity of 6 m/s. Due to the action of the ocean waves, however, it is pitching, rolling, and yawing about the x, y, and z axes, respectively, which are attached to the ship. At this instant these axes have the horizontal and vertical orientations they would have in a calm sea but $\omega_x = 0.08$ rad/s, $\omega_y = 0.04$ rad/s, $\omega_z = 0.02$ rad/s. Also, all rotation rates have their maximum value at this instant. Determine the absolute velocity and acceleration of the bow of the ship, which is 120 m forward and is 15 m above point G.

Prob. VI.65

VI.66 In Problem VI.65, an elevator on the ship is 40 m behind, 10 m above, and 5 m to the right (looking forward) of point G. The elevator is moving upward at the constant rate of 4 m/s relative to the ship. Determine the absolute velocity and acceleration of the elevator.

VI.67 The disk is rotating about bar AB at 100 rad/s as the fixed vertical axis rotates at 10 rad/s. In the position shown, $\theta = 90°$, $\dot{\theta} = 2$ rad/s, $\ddot{\theta} = 0$. Determine the velocity and acceleration of (a) point C, (b) point D. *Hint:* Follow the development in Illustrative Problem 5, where two moving reference frames are employed.

Prob. VI.67

VI.68 The disk of the gyroscope rotates about its own axis at 600 rev/min, and the gimbal support is rotating about the vertical axis at the constant rate of 1 rad/s. In the position where $\theta = 90°$ it is given that $\dot{\theta} =$

10 rad/s and $\ddot{\theta} = 0$. Determine the angular velocity and angular acceleration of the disk (a) using two moving reference frames as in Illustrative Problem 5, (b) using the formulas for Eulerian angles derived in Illustrative Problem 5.

VI.69   Given that $\dot{\theta} = 5$ rad/s and $\ddot{\theta} = 10$ rad/s^2 when $\theta = 60°$, solve Problem VI.68.

Prob. VI.68

MODULE VII
KINETICS OF A
RIGID BODY IN
THREE-DIMENSIONAL
MOTION

Our purpose in this module is to relate the three-dimensional motion of a rigid body to the forces that cause this motion. In the task confronting us we shall consider a rigid body of known dimensions and mass distribution being acted upon by a general system of forces. Similar to the kinetics studies of particles and rigid bodies in planar motion, this study of the laws of kinetics will consist of investigations of the equations of motion, the work-energy principle, and the linear and angular impulse-momentum principles. The new feature here is that we shall be dealing with the most general case of motion of a rigid system.

When we studied three-dimensional rigid body kinematics in the previous module, we saw that the motion we seek is fully determined once we know both the motion of a point in the body and the angular motion of the body, just as it was in the case of a rigid body in planar motion. However, in contrast to what we observed in the study of planar motion, when a rigid body executes an arbitrary motion in three dimensions its angular velocity can have a variable direction as well as a variable magnitude. We shall see that this variable characteristic produces some fundamental phenomena that are unique to three-dimensional motion.

A.   FUNDAMENTAL PRINCIPLES

Let us pause to recall the principles of statics, where it was shown that there are six scalar equations of equilibrium when a rigid body is subjected to a three-dimensional force system. These equations represent the components of the vector equations obtained by requiring that the resultant of the force system satisfy $\Sigma \bar{F} = \bar{0}$, and that the resultant moment of the force system about an arbitrary point A satisfy $\Sigma \bar{M}_A = \bar{0}$.

We shall find in this module that, correspondingly, there are six scalar equations of motion governing a rigid body undergoing an arbitrary three-dimensional motion. For this study we shall employ the principles in Module III governing a system of particles. Three of the scalar equations we seek are obtained from

$$\boxed{\Sigma \bar{F} = m\bar{a}_G} \tag{1}$$

The mass m of the rigid body is considered to be the collection of an infinite number of infinitesimal masses dm filling the volume $\mathcal{V}$ occupied by the body. Therefore

$$m = \int_{\mathcal{V}} dm$$

Equation (1) serves to describe the motion of one point in the body, the center of mass G. Therefore, the major task confronting us is the determination of the equations of motion governing the rotation of the body. These equations are most conveniently written in terms of the moment equation for a system of particles. Specifically, the vector equation is

$$\Sigma \, \bar{M}_A = \frac{d}{dt} \, \bar{H}_A \equiv \dot{\bar{H}}_A \tag{2}$$

where

$$\bar{H}_A = \sum_{i=1}^{N} (\bar{r}_{i/A} \times m_i \bar{v}_{i/A}) \tag{3}$$

Remember that in order for equation (2) to be valid, point A must be either a fixed point, the center of mass of the system, or a point accelerating toward the center of mass.

B.   ANGULAR MOMENTUM OF A RIGID BODY

Equation (2) illustrates the importance of the vector $\bar{H}_A$: the angular momentum of a rigid body relative to point A. Hence, our objective in this section is to develop a general formula for the angular momentum of a rigid body undergoing three-dimensional motion.

Figure 1 shows an infinitesimal particle dm within an arbitrary rigid body. In addition to the requirement that point A be one of the three special points mentioned in Section A, we further restrict the choice of this point to be a point that is fixed to the body. Then, because the body is rigid, we know that $\bar{r}_{i/A}$ is a vector with constant magnitude that rotates at the angular velocity $\bar{\omega}$ of the body. Thus, from our study of rigid body kinematics in Module VI, we have

$$\bar{v}_{i/A} = \bar{\omega} \times \bar{r}_{i/A}$$

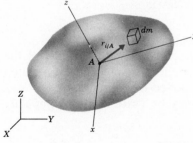

Figure 1

Furthermore, the contribution of each differential element dm to the value of $\bar{H}_A$ must be evaluated by an integral, rather than by a conventional summation process. Equation (3) therefore becomes

$$\bar{H}_A = \int_{\mathscr{V}} [\bar{r}_{i/A} \times (\bar{\omega} \times \bar{r}_{i/A})] dm \tag{4}$$

In order to proceed further, we must choose a coordinate system for describing the vector components and the boundary of the body. For this we could choose the fixed reference system XYZ. Such a choice creates the awkward situation where the boundary of the body, in terms of the (X, Y, Z) coordinates, will change as a function of time. This would cause the computation of the time variation of $\bar{H}_A$ to be difficult. To avoid this problem we attach a moving reference frame xyz to the body, so that all points in the body are fixed with respect to the *xyz* frame. There are some special geometries, such as bodies of revolution, where we could allow the body to have some rotational motion relative to the reference frame. However, attaching the *xyz* frame to the body ensures that our formulation will be valid in all cases.

For simplicity, let the origin of the *xyz* frame coincide with point *A*. Such a coordinate system is shown in Figure 1. (The question of a suitable choice for the orientation of the *xyz* frame will be discussed later.) The position vector of any mass element *dm* at point *i* can then be written as

$$\bar{r}_{i/A} = x\,\bar{\imath} + y\,\bar{\jmath} + z\,\bar{k}$$

To be consistent, we express the angular velocity of the body in terms of its *x*, *y*, and *z* components. Thus

$$\bar{\omega} = \omega_x\,\bar{\imath} + \omega_y\,\bar{\jmath} + \omega_z\,\bar{k}$$

Now, substituting $\bar{r}_{i/A}$ and $\bar{\omega}$ into equation (4) and calculating the cross products, we get

$$\bar{H}_A = \int_{\mathscr{V}}\{[\omega_x(y^2 + z^2) - \omega_y xy - \omega_z xz]\,\bar{\imath}$$

$$+ [\omega_y(z^2 + x^2) - \omega_z yz - \omega_x xy]\,\bar{\jmath}$$

$$+ [\omega_z(x^2 + y^2) - \omega_x xz - \omega_z yz]\,\bar{k}\}\,dm$$

Recall that the angular velocity is a property of the motion of the body. Hence, the components of $\bar{\omega}$ do not depend on a particular location within the body, and may therefore be removed from the integrand. As a result, the angular momentum equation may be written more compactly as

$$\boxed{\begin{aligned}\bar{H}_A &= (I_{xx}\omega_x - I_{xy}\omega_y - I_{xz}\omega_z)\bar{\imath}\\ &\quad + (I_{yy}\omega_y - I_{xy}\omega_x - I_{yz}\omega_z)\bar{\jmath}\\ &\quad + (I_{zz}\omega_z - I_{xz}\omega_x - I_{yz}\omega_y)\bar{k}\end{aligned}} \tag{5}$$

where

$$\boxed{\begin{aligned}I_{xx} &= \int_{\mathscr{V}}(y^2 + z^2)\,dm\\ I_{yy} &= \int_{\mathscr{V}}(x^2 + z^2)\,dm\\ I_{zz} &= \int_{\mathscr{V}}(x^2 + y^2)\,dm\end{aligned}}$$

$$I_{xy} = \int_{\mathcal{V}} xy \; dm$$

$$I_{xz} = \int_{\mathcal{V}} xz \; dm$$

$$I_{yz} = \int_{\mathcal{V}} yz \; dm \qquad (6)$$

The quantities I_{xx}, I_{yy}, and I_{zz} are called the mass *moments of inertia*. The quantities I_{xy}, I_{xz}, and I_{yz} are called the mass *products of inertia*. In the next section we shall treat these inertia properties in detail. For now we shall only make some simple observations regarding them.

Foremost, we can now see the significance of selecting a moving reference frame *xyz* for which the boundaries of the body are constant. Such a selection means that the inertia terms defined by equations (6) will be constant properties of the body. Equally important, we shall find that a proper selection for the orientation of the *xyz* frame with respect to the body can greatly facilitate the calculation of the inertia properties. For instance, we shall see that in some cases the choice for the *xyz* frame will result in $I_{xy} = I_{yz} = I_{xz} = 0$. In such cases we shall refer to *xyz* as *principal axes*. Also, by properly orienting *xyz*, we shall frequently be able to make use of tables that list the inertia properties of basic geometric shapes.

To demonstrate how the angular momentum of a body is determined, we shall now consider two examples of the rotation of a thin disk. For the inertia properties, recall that we defined principal axes as those for which $I_{xy} = I_{yz} = I_{xz} = 0$. It is for such axes that inertia properties of standard geometric shapes are most often tabulated (as they are in Appendix B). Figure 2 shows such information for a thin disk. In the problems that follow, note how we select the axes to make use of this information.

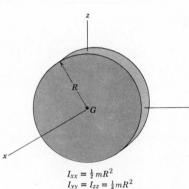

$I_{xx} = \frac{1}{2}mR^2$
$I_{yy} = I_{zz} = \frac{1}{4}mR^2$
$I_{xy} = I_{yz} = I_{xz} = 0$

Figure 2

EXAMPLE 1

A thin disk of radius r is mounted obliquely on a rigid shaft, with the result that the center of mass G is on the axis of the shaft, but the axis of symmetry of the disk is not parallel to the shaft. If the shaft is rotating at the constant rate ω, what is the angular momentum of the disk about its center of mass?

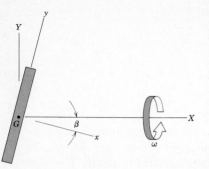

Solution

In order to use the angular momentum equation (5), we must first select an xyz frame. We want the angular momentum about the center of mass. Hence, we place the origin at that point and fix the axes to the disk. Further, orienting the reference frame as shown in the figure allows us to make direct use of the given inertia properties. For the fixed reference frame XYZ, let us choose the X axis to be parallel to the shaft, as shown in the sketch.

Note that because xyz is fixed to the disk, the z axis is not always parallel to the Z axis, but the xy plane always contains the X axis.

The inertia properties are known from Figure 2. Thus, we need only determine the components of angular velocity prior to evaluating the angular momentum. In terms of the chosen coordinate axes, we have

$$\bar{\omega} = \omega \bar{I} \quad \text{always}$$

$$= \omega(\cos \beta \, \bar{i} + \sin \beta \, \bar{j}) \quad \text{instantaneously}$$

The second expression is also always valid, because β is a constant angle between $\bar{i}$ and $\bar{I}$. Thus the components of $\bar{\omega}$ are constant.

$$\omega_x = \omega \cos \beta \qquad \omega_y = \omega \sin \beta \qquad \omega_z = 0$$

Upon substituting the moments of inertia and components of $\bar{\omega}$ into equation (5), we get

$$\bar{H}_G = \tfrac{1}{4}mr^2\omega(2 \cos \beta \, \bar{i} + \sin \beta \, \bar{j})$$

The resulting angular velocity and angular momentum vectors are shown below. As can be seen, the angular momentum vector is not parallel to the angular velocity vector.

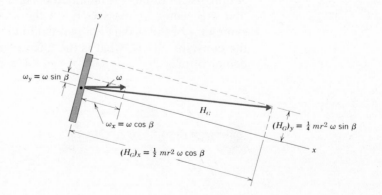

In this problem, the $(\bar{\imath}, \bar{\jmath}, \bar{k})$ components of $\bar{H}_G$ are constant because ω_x, ω_y and ω_z are constant values. Therefore $\bar{H}_G$ is fixed relative to xyz, with the result that $\bar{H}_G$ rotates about the fixed shaft at angular velocity $\bar{\omega}$.

EXAMPLE 2

An electric motor is mounted horizontally on a turntable that rotates about the vertical axis at 300 rev/min as shown. The flywheel of the motor is concentrically mounted on the shaft of the motor, whose rate of rotation is 1500 rev/min. The center of the mass of the flywheel is 0.5 m from the axis of the turntable. Approximate the flywheel as a thin disk of radius R to determine its angular momentum relative to its center of mass.

Solution

The relative angular momentum depends only on the angular motion of the body and the location of the origin. As always, we attach the xyz frame to the body with its origin coincident with the point of reference for the angular momentum, in this case the center of mass. We orient the xyz frame relative to the body for greatest convenience in describing the moments and products of inertia. Hence we let the x axis be the axis of the disk, so that we can use the given inertia properties directly. This still leaves the choice of the y and z axes undefined. We make the choice for the instantaneous orientation based on the geometry of the system. Here, we shall let the y axis be vertical at the instant to be considered, as is illustrated in the sketch.

The inertia properties for a thin disk are given in Figure 2. Following the kinematical method developed in the previous module, the general expression for the angular velocity of the disk is

$$\bar{\omega} = \omega_1 \bar{J} + \omega_2 \bar{\imath} \qquad \text{always}$$

Noting that $\bar{J} = \bar{\jmath}$ at the instant illustrated in the sketch, the instantaneous expression for $\bar{\omega}$ is

$$\bar{\omega} = \omega_1 \bar{\jmath} + \omega_2 \bar{\imath} \qquad \text{instantaneously}$$

Converting the given information to the proper units for angular rotation (radians), we have

$$\omega_x = \omega_2 = 1500\left(\frac{2\pi}{60}\right) = 50\pi \text{ rad/s}$$

$$\omega_y = \omega_1 = 300\left(\frac{2\pi}{60}\right) = 10\pi \text{ rad/s}$$

Then, from equation (5) we find that

$$\bar{H}_G = \tfrac{1}{2}mR^2(50\pi)\bar{i} + \tfrac{1}{4}mR^2(10\pi)\bar{j}$$

$$= 2.5\pi mR^2(10\bar{i} + \bar{j})$$

The resulting vectors $\bar{\omega}$ and $\bar{H}_G$ are shown below.

Prob. VII.1

Prob. VII.2

Prob. VII.3

HOMEWORK PROBLEMS

VII.1   A thin disk of mass m and radius r is mounted on a shaft whose axis intersects the center of the disk. The shaft is driven at the constant rate ω. Determine the angular momentum of the disk about its mass center. Sketch the resulting $\bar{H}_G$ and $\bar{\omega}$ vectors.

VII.2   During a maneuver a space capsule is spinning at the angular velocity $\bar{\omega} = 0.2\bar{i} + 0.8\bar{j} + 0.1\bar{k}$ rad/s. The xyz axes are principal axes for which the moments of inertia are $I_{xx} = I_{zz} = 8000$ lb-s²-ft and $I_{yy} = 6000$ lb-s²-ft. Determine the angular momentum $\bar{H}_G$ of the space capsule about its mass center G. Sketch the resulting $\bar{H}_G$ and $\bar{\omega}$ vectors.

VII.3   The thin disk of mass 8 kg and radius $r = 500$ mm rotates at constant angular speed $\omega_1 = 25$ rad/s while the forked rod is driven at constant angular speed $\omega_2 = 10$ rev/min. Determine the angular momentum of the disk about its mass center. Sketch the resulting $\bar{H}_G$ and $\bar{\omega}$ vectors for the disk.

VII.4   Consider a circular disk of mass m and radius r rolling on the ground in a situation where the plane of the disk is oriented vertically. The center of the disk has speed v_0 and there is no slipping. Determine the angular momentum of the disk about its mass center G when this point is

Prob. VII.4

following (a) a straight path, (b) a circular path of radius R. Why are $\bar{H}_G$ and $\bar{\omega}$ parallel in the first case, but not in the second?

VII.5 A 6-kg thin disk of 160 mm radius is welded on a shaft mounted in a rotating frame. The disk rotates at a constant angular speed $\omega_1 = 25$ rev/min about the axis of the shaft, as the frame rotates at a constant angular speed $\omega_2 = 40$ rad/s. Determine the angular momentum of the disk about its mass center. Sketch the resulting $\bar{H}_G$ and $\bar{\omega}$ vectors.

VII.6 A disk of mass m and radius r is mounted on the rigid bar AB. The bar is welded to the vertical shaft, which is rotating at the angular speed ω_1 about the vertical axis. The angular speed of the disk with respect to bar AB is ω_2, as shown. Determine (a) $\bar{H}_G$ for the disk, (b) the angle β between $\bar{H}_G$ and the vertical direction.

Prob. VII.5

Prob. VII.6

Prob. VII.7

VII.7 The system shown is similar to the one in the preceding problem, except that the disk is mounted obliquely on its shaft. Determine $\bar{H}_G$ for the disk when it is in the position shown.

VII.8 Determine the angular momentum of the disk in Problem VII.7 when it has rotated 180° about bar AB from the position shown.

C. MOMENTS AND PRODUCTS OF INERTIA

From the discussion and examples in the previous section, it is obvious that the evaluation of the inertia properties plays a crucial role in the determination of the angular momentum vector. A question that arises is: What is the physical significance of the inertia properties?

Let us first consider one of the moments of inertia, for instance,

$$I_{zz} = \int_{\mathscr{V}}(x^2 + y^2)\, dm$$

Looking down the z axis at an arbitrary body, such as that in Figure 3, we see that the definition of I_{zz} is the same quantity as the moment of inertia determined for a planar body in Module V. One new feature of the

Figure 3

kinetics of rigid bodies in three-dimensional motion is that the moments of inertia about all three coordinate axes are important to the motion. However, the meaning of each term is unchanged. They describe how the mass is distributed about the corresponding axis. For instance, if two bodies have the same mass, the one whose mass is distributed further from the x axis has the greater value for I_{xx}.

One common way of describing a moment of inertia about some axis, such as the z axis, is to give the *radius of gyration* k_z. This quantity is the radius of a fictional circular ring having the same mass and moment of inertia about its centerline as the actual body it stimulates. For such a ring, the mass is concentrated at the perimeter, so that

$$I_{zz} = mk_z{}^2 \tag{7}$$

where the z axis is normal to the plane of the ring, passing through the center.

Now let us turn our attention to the products of inertia. Considering I_{xy} for the body of Figure 3, by definition,

$$I_{xy} = \int_{\mathscr{V}} xy \, dm$$

In other words, I_{xy} is the sum of the products of the x and y coordinates of a differential element and the mass of that element. Clearly, if the mass of a body is predominantly in one octant, then I_{xy} will have the sign associated with the product of the x and y values in that octant. For example, if a body is entirely situated in an octant where $x < 0$, $y > 0$, then I_{xy} will be negative. Therefore, we can conclude that the products of inertia describe how the mass of the body is distributed within the various octants of the xyz coordinate system.

An important case arises when we deal with bodies having a plane of symmetry. In Figure 4 we have drawn the cross section of a body that is symmetric about the xz plane. For this cross section in the yz plane at some value of x we show a typical element dm_r in the right half of the body and its mirror image mass element dm_l in the left half of the body. The property of symmetry means that there is an element dm_l corresponding to every dm_r. Thus, as is shown in Figure 4, if the coordinates of dm_r are (x, y, z), then the coordinates of dm_l must be $(x, -y, z)$. Hence the product $xy \, dm_l$ for the left element is the negative of the value $xy \, dm_r$ for the right element, and the two terms cancel each other in the integral over the volume $\mathscr{V}$. As a result we find that $I_{xy} = 0$. Similar reasoning shows that $I_{yz} = 0$ also. From this we obtain the following important fact:

Figure 4

Cross section at x

If one of the coordinate axes is normal to the plane of symmetry of a body, with the origin of the coordinate system coincident with the plane of symmetry, the two products of inertia having this axis in the subscript are zero.

An important corollary of this statement follows for the case of a body having two perpendicular planes of symmetry, such as a body of revolution.

> A coordinate system whose axes form two perpendicular planes of symmetry constitutes a principal set of axes, for which all products of inertia are zero.

Placing the discussion in this section in perspective, it is sufficient to note that the mass moments and products of inertia are terms that appear in the derivation of the angular momentum formula. Hence, it is necessary to be able to evaluate them. It then becomes expedient to compute the inertia quantities for common geometric shapes and tabulate them for easy reference. A typical set of tables may be found in Appendix B. For uncommon geometric shapes, which do not resemble the tabulated ones, we compute the inertia quantities from the basic definitions. This is the topic of the next section.

HOMEWORK PROBLEMS

VII.9   Select body-fixed axes having an origin at the mass center G that are principal axes of inertia.

Prob. **VII.9**

VII.10   For the shapes in Problem VII.9, select body-fixed axes having an origin at point *A* that are principal axes of inertia.

VII.11   The disk is welded obliquely to the vertical shaft. Sketch the principal axes having an origin at the center of mass *G* (a) for the disk, (b) for the shaft. (c) Are the principal axes for the system of the shaft and the disk obvious from symmetry? Explain your answer.

Prob. VII.11

1   Inertia Properties by Integration

Our purpose here is to review the basic calculus necessary to set up and calculate the moments and products of inertia of a rigid body, starting from the definitions. We shall proceed by solving two examples that demonstrate how the inertia properties given in standard tables are obtained, and also how the theorems pertaining to planes of symmetry are utilized.

EXAMPLE 3

Determine the inertia properties of the rectangular parallelepiped relative to the reference frame shown.

Solution

For this body and the selected axes, the *yz* plane is a plane of symmetry. Therefore, because the *x* axis is normal to the plane of symmetry and the origin is coincident with this plane, we find that $I_{xy} = I_{xz} = 0$. To determine the other inertia properties, we shall set up and evaluate the appropriate integrals. Because the surface of the body coincides with planes of constant values for the *x, y,* and *z* coordinates, we shall formulate the integral in terms of Cartesian coordinates.

Let ρ denote the constant mass density of the body. Then

$$dm = \rho \, dx \, dy \, dz$$

To evaulate the integrals we shall consider a differential slice of the body at some distance *z* from the origin and integrate first with respect to $x \, (-a/2 \leq x \leq a/2)$ and then with respect to $y \, (0 \leq y \leq b)$. Then the integration with respect to $z \, (0 \leq z \leq c)$ will represent the sum of the contributions of all the cross-sectional slices. Thus

$$I_{xx} = \rho \int_0^c \int_0^b \int_{-a/2}^{a/2} (y^2 + z^2) \, dx \, dy \, dz$$

$$= \rho \int_0^c \int_0^b a(y^2 + z^2) \, dy \, dz$$

$$= \rho a \int_0^c \left(\frac{b^3}{3} + bz^2 \right) dz = \rho ab \left(\frac{b^2 c}{3} + \frac{c^3}{3} \right)$$

$$= \tfrac{1}{3}\rho abc(b^2 + c^2)$$

$$I_{yy} = \rho \int_0^c \int_0^b \int_{-a/2}^{a/2} (x^2 + z^2)\ dx\ dy\ dz$$

$$= \tfrac{1}{12}\rho abc(a^2 + 4c^2)$$

$$I_{zz} = \rho \int_0^c \int_0^b \int_{-a/2}^{a/2} (x^2 + y^2)\ dx\ dy\ dz$$

$$= \tfrac{1}{12}\rho abc(a^2 + 4b^2)$$

$$I_{yz} = \rho \int_0^c \int_0^b \int_{-a/2}^{a/2} yz\ dx\ dy\ dx = \tfrac{1}{4}\rho ab^2c^2$$

These results can be simplified by noting that the volume of the body is $\mathcal{V} = abc$, so that the total mass is

$$m = \rho abc \quad \text{or} \quad \rho = \frac{m}{abc}$$

Thus we can eliminate ρ in the foregoing results to get

$$I_{xx} = \tfrac{1}{3}m(b^2 + c^2) \qquad I_{yy} = \tfrac{1}{12}m(a^2 + 4c^2)$$

$$I_{zz} = \tfrac{1}{12}m(a^2 + 4b^2) \qquad I_{yz} = \tfrac{1}{4}mbc$$

$$I_{xy} = I_{xz} = 0$$

EXAMPLE 4
Determine the inertia properties of a homogeneous right circular cone for the coordinate axes shown.

Solution
The cone is symmetric with respect to both the xy and xz planes, so that xyz are principal axes; that is, $I_{xy} = I_{yz} = I_{xz} = 0$. Because the cone is a body of revolution, we shall formulate the integrals in terms of cylindrical coordinates, with the axial direction coincident with the x axis.

Let us consider a typical slice of the cone at a distance x from the apex. In terms of the cylindrical coordinates (R, θ, x), a differential volume is $R\ dR\ d\theta\ dx$, so that we have

$$y = R \sin \theta \qquad z = R \cos \theta \qquad dm = \rho R\ dR\ d\theta\ dx$$

We can integrate over the cross section by integrating first with respect to θ $(0 \le \theta \le 2\pi)$ and then with respect to R $(0 \le R \le R_0)$. Noting that the radius R_0 of the cross section is proportional to the distance from the apex, we can write

$$R_0 = \frac{x}{h} r$$

Cross section at x

Thus, the moment of inertia about the x axis becomes

$$I_{xx} = \rho \int_0^h \int_0^{R_0} \int_0^{2\pi} (R^2 \sin^2 \theta + R^2 \cos^2 \theta)R \, d\theta \, dR \, dx$$

But $\sin^2 \theta + \cos^2 \theta \equiv \; =$, so that

$$I_{xx} = 2\pi\rho \int_0^h \int_0^{xr/h} R^3 \, dR \, dx = 2\pi\rho \int_0^h \tfrac{1}{4}\left(\frac{xr}{h}\right)^4 dx = \tfrac{1}{10}\pi\rho \, r^4 h$$

The volume of a cone is $(\pi/3)r^2h$, so we know that the mass of the cone is

$$m = \frac{\pi}{3}\rho r^2 h$$

Upon substituting for ρ we get

$$I_{xx} = \tfrac{3}{10}mr^2 \qquad \triangleleft$$

To complete the solution we need only determine I_{yy}, because $I_{zz} \equiv I_{yy}$ as a result of the symmetry of the body.

$$I_{yy} = \rho \int_0^h \int_0^{R_0} \int_0^{2\pi} (x^2 + R^2 \cos^2 \theta)R \, d\theta \, dR \, dx$$

$$= \rho \int_0^h \int_0^{xr/h} (2\pi x^2 + \pi R^2)R \, dr \, dx$$

$$= \pi\rho r^2 h\left(\frac{h^2}{5} + \frac{r^2}{20}\right)$$

Then substituting for ρ gives

$$I_{yy} = \tfrac{3}{20}m(4h^2 + r^2) = I_{zz} \qquad \triangleleft$$

HOMEWORK PROBLEMS

VII.12 Verify by integration that the moment of inertia of the uniform right circular cone of mass m about the y axis shown is $I_{yy} = m(3r^2 + 2h^2)/20$.

VII.13 Determine by integration the moments of inertia of the triangular prism of mass m about the axes shown.

VII.14 Determine by integration the products of inertia of the triangular prism of mass m about the axes shown in Problem VII.13.

Prob. VII.15

Prob. VII.16

VII.15 Determine by integration the moments of inertia of the paraboloid of revolution of mass m about the axes shown.

VII.16 Determine by integration the mass moments of inertia of the rectangular tetrahedron of mass m about the given axes. *Hint:* Determine the formula for one moment of inertia and derive the others by induction.

VII.17 Determine by integration the mass products of inertia of the rectangular tetrahedron of mass m about the given axes in Problem VII.16. See the hint in Problem VII.16.

VII.18 Determine by integration the moments of inertia of the circular cylinder of mass m about the axes shown. Then, by taking the limit as the appropriate dimension goes to zero, determine the corresponding formulas for (a) a slender bar, and (b) a thin disk.

Prob. VII.18

Prob. VII.19

VII.19 The density of the material of the truncated cone is given by $\rho = cz^2$, where z is the distance of the mass element from the base and c is a constant. The mass of this body is 50 kg. Determine (a) the value of c, (b) the location of the mass center G, and (c) the mass moments of inertia about a set of principal axes whose origin is point G.

2 Parallel Axis Theorems

All of the systems we shall study will be composed of bodies whose inertia properties have been compiled in standard tables, such as those in Appendix B. In this way we shall reduce the amount of computations necessary to solve problems, in that it is obviously preferable to use a tabulated value rather than evaluating an integral. This restriction on the types of bodies to be considered will not affect the generality of the presentation of three-dimensional kinetics.

The use of the tabulated information will sometimes require that we transform the inertia properties from the coordinate system given in the tables to a more appropriate set of coordinates for the problem at hand.

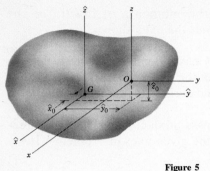

Figure 5

We shall now develop the formulas necessary to perform such transformations.

Figure 5 illustrates an arbitrary body. We consider the moment and products of inertia to be known quantities with respect to the coordinate system $\hat{x}\hat{y}\hat{z}$ whose origin is the center of mass G.

The problem we wish to consider is the determination of the inertia properties of this body with respect to the coordinate system xyz whose axes are parallel to the $\hat{x}$, $\hat{y}$, and $\hat{z}$ axes, respectively. We denote the $(\hat{x}, \hat{y}, \hat{z})$ coordinates of the origin O of the xyz coordinate system by the values $(\hat{x}_0, \hat{y}_0, \hat{z}_0)$. Hence, as can be seen from Figure 5, the (x, y, z) coordinates of any point are related to the $(\hat{x}, \hat{y}, \hat{z})$ coordinates by

$$x = \hat{x} - \hat{x}_0 \qquad y = \hat{y} - \hat{y}_0 \qquad z = \hat{z} - \hat{z}_0 \tag{8}$$

Let us substitute these expressions into the moment and product of inertia equations (6). For I_{xx} we get

$$I_{xx} = \int_{\mathscr{P}} (y^2 + z^2)\, dm$$

$$= \int_{\mathscr{P}} [(\hat{y} - y_0)^2 + (\hat{z} - \hat{z}_0)^2]\, dm$$

$$= \int_{\mathscr{P}} (\hat{y}^2 + \hat{z}^2)\, dm - 2\hat{y}_0 \int_{\mathscr{P}} \hat{y}\, dm$$

$$\quad - 2\hat{z}_0 \int_{\mathscr{P}} \hat{z}\, dm + (\hat{y}_0 + \hat{z}_0{}^2) \int_{\mathscr{P}} dm$$

The first integral is, by definition, the moment of inertia $I_{\hat{x}\hat{x}}$. The second and third integrals are the first moments of mass. These integrals are zero, because the origin of the xyz frame is the center of mass, for which the first moment of mass are

$$\int_{\mathscr{P}} \bar{r}_{i/G}\, dm = \int_{\mathscr{P}} (\hat{x}\bar{i} + \hat{y}\bar{j} + \hat{z}\bar{k})\, dm = m\bar{r}_{G/G} \equiv \bar{0}$$

In addition, the last integral is obviously m, the total mass of the body. Following similar steps for the other moments of inertia, we obtain the final form of the *parallel axis theorems for moments of inertia:*

$$\boxed{\begin{aligned} I_{xx} &= I_{\hat{x}\hat{x}} + m(\hat{y}_0{}^2 + \hat{z}_0{}^2) \\ I_{yy} &= I_{\hat{y}\hat{y}} + m(\hat{x}_0{}^2 + \hat{z}_0{}^2) \\ I_{zz} &= I_{\hat{z}\hat{z}} + m(\hat{x}_0{}^2 + \hat{y}_0{}^2) \end{aligned}} \tag{9}$$

This result takes on a simple meaning once we note that Figure 5 indicates that the distances between the parallel pairs of axes are

$$d_x = (y_0{}^2 + z_0{}^2)^{1/2}$$

$$d_y = (x_0{}^2 + z_0{}^2)^{1/2}$$

$$d_z = (x_0{}^2 + y_0{}^2)^{1/2}$$

Thus equations (9) are the extension to three dimensions of the parallel axis theorem for planar bodies. Specifically, they state that the moment of inertia about any axis equals the moment of inertia about a parallel axis that *passes through the center of mass* plus the product of the mass and the square of the distance between the axes.

Turning now to the determination of the products of inertia with respect to the xyz coordinate system, we proceed in the same manner as that employed for the moments of inertia. From equations (6) for I_{xy}, we write

$$I_{xy} = \int_{\mathscr{V}} xy \, dm = \int_{\mathscr{V}} (x - \hat{x}_0)(y - \hat{y}_0) \, dm$$

$$= \int_{\mathscr{V}} \hat{x}\hat{y} \, dm - \hat{x}_0 \int_{\mathscr{V}} \hat{y} \, dm - \hat{y}_0 \int_{\mathscr{V}} \hat{x} \, dm + \hat{x}_0\hat{y}_0 \int_{\mathscr{V}} dm$$

Using the fact that the origin of $\hat{x}\hat{y}\hat{z}$ is at the center of mass this, and similarly the other two products of inertia, reduce to the form we call the *parallel axis theorems for products of inertia*:

$$\boxed{\begin{aligned} I_{xy} &= I_{\hat{x}\hat{y}} + m\hat{x}_0\hat{y}_0 \\ I_{xz} &= I_{\hat{x}\hat{z}} + m\hat{x}_0\hat{z}_0 \\ I_{yz} &= I_{\hat{y}\hat{z}} + m\hat{y}_0\hat{z}_0 \end{aligned}} \tag{10}$$

Again, it is important to note that in applying both equations (9) and (10), the origin of the $\hat{x}\hat{y}\hat{z}$ reference frame must be the center of mass of the body, and that $(\hat{x}_0, \hat{y}_0, \hat{z}_0)$ represent the coordinates of the origin O of xyz with respect to $\hat{x}\hat{y}\hat{z}$.

3 Composite Bodies

Before we consider an example that illustrates the application of the parallel axis theorems, let us consider one more facet of the determination of the inertia properties, specifically, the matter of composite bodies. By a composite body we mean an arbitrary body whose form is composed of several basic shapes. An example of a composite body is shown in Figure 6, where the cylinder 1 and the rectangular parallelepiped 2 are the components of the composite body.

The basic mass properties of any body are its total mass, the location of the center of mass, and the moments and products of inertia with

Figure 6

respect to a given coordinate system xyz (whose origin is not necessarily the center of mass). The method for locating the center of mass of an arbitrary composite body is a straightforward extension of the method in Module Five for a planar composite body. Although some of you may have studied this topic as part of statics, we shall review it here for completeness.

The total mass m of the body is obviously the sum of the masses of the components, so that

$$m = \sum_{i=1}^{N} m_i \tag{11}$$

To describe the location of the center of mass G of the composite body, we select any convenient $x'y'z'$ coordinate system; usually the origin of $x'y'z'$ coincides with the center of mass of one of the components. The location of point G is obtained from the first moment of mass. Let (x'_G, y'_G, z'_G) and $(x'_{Gi}, y'_{Gi}, z'_{Gi})$ denote the coordinates relative to the $x'y'z'$ system of point G and of the center of mass of each component G_i, respectively. Then, to obtain the x' coordinate of point G, we have

$$mx'_G = \sum_{i=1}^{N} m_i x'_{Gi} \tag{12}$$

The other coordinates of point G are obtained by permuting the symbol x' in equation (12) to y' or z'.

The next step in the description of the mass characteristics of the composite body is the determination of the moments and products of inertia with respect to the coordinate system xyz whose origin is point A; generally, point A will be the point that was selected for the moment sum in equation (2). To obtain the inertia properties, equations (6) may be decomposed into integrals over the volumes $\mathcal{V}_i$ of the basic components. It follows that, when *evaluated for a given set of axes,* the inertia properties are additive, in which case

$$I_{xx} = \sum_{i=1}^{N} (I_{xx})_i \qquad I_{yy} = \sum_{i=1}^{N} (I_{yy})_i \qquad I_{zz} = \sum_{i=1}^{N} (I_{zz})_i$$

$$I_{xy} = \sum_{i=1}^{N} (I_{xy})_i \qquad I_{xz} = \sum_{i=1}^{N} (I_{xz})_i \qquad I_{yz} = \sum_{i=1}^{N} (I_{yz})_i \tag{13}$$

To use equations (13) properly, it is important to observe that when using tabulated values for the inertia properties of basic shapes, it will probably be necessary to use the parallel axis theorems for each shape in order to transform the basic properties to the desired xyz coordinate system.

All the aforementioned features of the parallel axis theorems and the methods for handling composite bodies are incorporated in the following example.

EXAMPLE 5

Without performing integrations, locate the center of mass of the steel body shown and determine its inertia quantities I_{xx} and I_{xz} with respect to the xyz axes shown in the diagram.

Solution

The composite body can most easily be depicted as a steel block, call it body 1, of dimensions 600 mm × 200 mm × 400 mm, from which a smaller block, call it body 2, of dimensions 300 mm × 100 mm × 200 mm, has been cut away. Hence, for this system we shall subtract the properties of body 2 with respect to the coordinate axes through O from those of body 1. In general, to determine the inertia properties of composite bodies we shall need coordinate axes $\hat{x}_i, \hat{y}_i, \hat{z}_i$ for each component shape i. The origin of $\hat{x}_i, \hat{y}_i, \hat{z}_i$ is the center of mass G_i of that shape, and each $\hat{x}_i, \hat{y}_i, \hat{z}_i$ system should be parallel to the xyz coordinate system of interest. In this problem the center of mass G_1 is coincident with point O, as illustrated in the sketch.

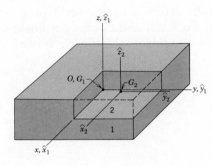

We must first determine the mass of the body. The mass density of steel is $\rho = 7833$ kg/m³, hence in length units of meters,

$$m_1 = \rho \mathcal{V}_1 = 7833(0.6)(0.2)(0.4) = 376.0 \text{ kg}$$

$$m_2 = \rho \mathcal{V}_2 = 7833(0.3)(0.1)(0.2) = 47.0 \text{ kg}$$

$$m = m_1 - m_2 = 329.0 \text{ kg}$$

To obtain the coordinates of the center of mass (x_G, y_G, z_G) with respect to the xyz coordinate system, we calculate the first moment of mass with respect to each coordinate axis. These are

$$mx_G = m_1 x_{G1} - m_2 x_{G2}$$

$$my_G = m_1 y_{G1} - m_2 y_{G2}$$

$$mz_G = m_1 z_{G1} - m_2 z_{G2}$$

where (x_{Gi}, y_{Gi}, z_{Gi}) are the coordinates of point G_i. Substituting the dimensions of the body into these equations yields

$$x_G = \frac{376(0) - 47.0(0.1)}{329} = -0.01429 \text{ meters}$$

$$y_G = \frac{376(0) - 47.0(0.15)}{329} = -0.2143 \text{ meters}$$

$$z_G = \frac{376(0) - 47.0(0.05)}{329} = -0.0714 \text{ meters}$$

Hence, with respect to the xyz system, the coordinates of the center of mass are $(-0.0143, -0.0214, -0.0714)$ meters.

We now turn our attention to the inertia terms. The tables in Appendix B detail the inertia values of a rectangular parallelepiped for a coordinate system with origin at the center of mass, as shown in the sketch.

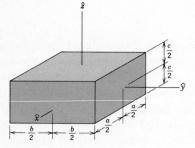

Also, because the $\hat{x}\hat{y}$, $\hat{x}\hat{z}$, and $\hat{y}\hat{z}$ planes are all planes of symmetry for the parallelepiped, we know that $\hat{x}\hat{y}\hat{z}$ are the principal axes for this shape, so that the products of inertia are zero.

From these tabulated values we start by determining the inertia properties of blocks 1 and 2 with respect to the $\hat{x}_1\hat{y}_1\hat{z}_1$ and $\hat{x}_2\hat{y}_2\hat{z}_2$ coordinate systems, respectively. The $\hat{x}_1\hat{y}_1\hat{z}_1$ frame is coincident with the xyz frame. Hence we have

$$(I_{xx})_1 \equiv (I_{\hat{x}_1\hat{x}_1})_1 = \tfrac{1}{12}m_1(0.6^2 + 0.2^2)$$

$$= 12.533 \text{ kg-m}^2$$

$$(I_{xz})_1 \equiv (I_{\hat{x}_1\hat{z}_1})_1 = 0$$

For body 2 the tabulated formulas give

$$(I_{\hat{x}_2\hat{x}_2})_2 = \tfrac{1}{12}m_2(0.3^2 + 0.1^2)$$

$$= 0.392 \text{ kg-m}^2$$

$$(I_{\hat{x}_2\hat{z}_2})_2 = 0$$

To combine the inertia terms of the individual components we must transform the properties of block 2 to the xyz system, using the parallel axis theorems. With respect to $\hat{x}_2\hat{y}_2\hat{z}_2$, the coordinates of the origin O of the xyz system are $(-0.1, -0.15, -0.05)$ meters. Therefore,

$$(I_{xx})_2 = (I_{\hat{x}_2\hat{x}_2})_2 + m_2(\hat{y}_o{}^2 + \hat{z}_o{}^2)$$

$$= 0.392 + 47.0[(-0.15)^2 + (-0.05)^2]$$

$$= 1.567 \text{ kg-m}^2$$

$$(I_{xz})_2 = (I_{\hat{x}_2\hat{z}_2})_2 + m_2\hat{x}_o\hat{z}_o$$

$$= 0 + 47.0(-0.1)(-0.05)$$

$$= 0.235 \text{ kg-m}^2$$

As the last step, we obtain the properties of the composite body by combining the results for the two components.

$$I_{xx} = (I_{xx})_1 - (I_{xx})_2 = 12.533 - 1.567$$

$$= 10.966 \text{ kg-m}^2$$

$$I_{xy} = (I_{xz})_1 - (I_{xz})_2 = 0 - 0.235$$

$$= -0.235 \text{ kg-m}^2$$

HOMEWORK PROBLEMS

VII.20 Using the formula for I_{yy} given in Problem VII.12, verify the formula in Appendix B for the moment of inertia of a cone about the $\hat{y}$ axis passing through the center of mass G.

Prob. VII.20

VII.21 Using the formula for $I_{\hat{y}\hat{y}}$ given in Appendix B, verify the formula derived in Example 4 for the moment of inertia of the cone in Problem VII.20 about the y' axis passing through the apex A.

VII.22 Use the tables in Appendix B and the parallel axis theorems to determine the moments and products of inertia of the circular cylinder with respect to the xyz coordinate system shown. The cylinder has a mass of 10 kg.

Prob. VII.22   **Prob. VII.23**

VII.23 Use the inertia properties for a hemisphere given in Appendix B to derive formulas for the moments of inertia of the hollow hemisphere with respect to the coordinate system shown, expressing the results in terms of the mass m, radius r, and thickness h. Then, by taking the limit as $h \to 0$, derive the corresponding properties for a thin hemispherical shell.

VII.24 Use the inertia properties for a complete circular cylinder to derive corresponding formulas for a semicircular cylinder. Then use those results to determine the mass moments of inertia of the hollow semicircular cylinder shown with respect to the xyz coordinate system.

Prob. VII.24

VII.25 and VII.26 The diagrams show aluminum castings. Determine the principal mass moments of inertia for each body for axes whose origin is at the center of mass. How would the results change if the casting was made of some other material?

Prob. VII.25

Prob. VII.26

VII.27 Steel spheres are welded to each end of the circular steel bar. Determine the value of I_{zz} (a) considering the spheres and the bar to be three-dimensional bodies, (b) neglecting the mass of the bar and considering the spheres to be particles of the same mass as the actual spheres. What conclusions can you draw from these results?

Prob. VII.27

VII.28 The built-up section shown consists of aluminum components. Determine the location of its mass center and the principal moments of inertia about a set of axes whose origin is at the mass center.

VII.29 Solve Problem VII.28 when the cylinder is steel and the other components are aluminum.

Prob. VII.28

VII.30 Determine the location of the mass center of the shelf and bracket shown and also determine the moments and products of inertia with respect to the xyz coordinate system shown. The two pieces are made of wood having a density of 750 kg/m³.

Prob. VII.30

VII.31 Determine the inertia properties with respect to the xyz coordinate system of the crankshaft shown, which is formed from a slender wire whose mass per unit length is m/l.

Prob. VII.31

Prob. VII.32

VII.32 The 60-lb thin disk is welded to the 15-lb slender shaft AC. Determine the location of the center of mass and the principal moments of inertia of the system for a set of axes having their origin at point A.

4 Rotation Transformation of Inertia Properties

The parallel axis theorems detail the transformation of the inertia properties for two coordinate systems having different origins but identical orientations. In some situations it may be necessary to transform the inertia properties between two coordinate systems having the same origin but different orientations. Let us develop such a situation by considering the propeller shown in Figure 7.

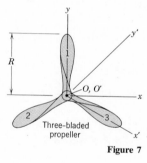

Three-bladed propeller

Figure 7

Suppose that we want the inertia properties with respect to the xyz coordinate system shown. Appendix B and the parallel axis theorems can be used to find the properties of each blade with respect to a coordinate system that has an axis parallel to the blade, such as the $x'y'z'$ system shown in Figure 7 for blade 3. Our task is to obtain the inertia properties of all three blades with respect to the xyz system. This determination involves a rotational transformation between the coordinate systems. For example, for the propeller blade 3, if we were to rotate the $x'y'z'$ system by 30° counterclockwise about the z' axis, we would obtain the xyz system. Therefore, we shall develop general relationships for two coordinate systems xyz and $x'y'z'$ whose origins are coincident, under the hypothesis that the inertia properties are known with respect to the $x'y'z'$ system. To do this we use a set of direction cosines.

Let $l_{pq'}(\equiv l_{q'p})$ be the cosine of the angle between axis p of the xyz system and axis q' of the $x'y'z'$ system. For example, in Figure 8, shown on the following page, we show an arbitrary orientation of x with respect to the $x'y'z'$ frame, for which we have $l_{xx'} = \cos \alpha$, $l_{xy'} = \cos \beta$, and $l_{xz'} = \cos \gamma$.

As can be seen in Figure 8, the unit vector $\bar{i}$ can be written in terms of its components in the $x'y'z'$ frame as

Figure 8

$$\bar{i} = (\cos \alpha)\bar{i}', + (\cos \beta)\bar{j}' + (\cos \gamma)\bar{k}'$$

$$\equiv l_{xx'}\bar{i}' + l_{xy'}\bar{j}' + l_{xz'}\bar{k}' \tag{14a}$$

If we had completed the x, y, z triad in Figure 8, the other unit vectors would be

$$\bar{j} = l_{yx'}\bar{i}' + l_{yy'}\bar{j}' + l_{yz'}\bar{k}' \tag{14b}$$

$$\bar{k} = l_{zx'}\bar{i}' + l_{zy'}\bar{j}' + l_{zz'}\bar{k}' \tag{14c}$$

A rectangular Cartesian coordinate system has the properties that

$$\bar{i} \cdot \bar{i} = \bar{j} \cdot \bar{j} = \bar{k} \cdot \bar{k} = 1$$

$$\bar{i} \cdot \bar{j} = \bar{j} \cdot \bar{k} = \bar{i} \cdot \bar{k} = 0$$

Then in view of equations (14) and the fact that $x'y'z'$ is also a rectangular Cartesian system, the foregoing relations require that

$$\begin{aligned} (l_{px'})^2 + (l_{py'})^2 + (l_{pz'})^2 &= 1 & p &= x, y, \text{ or } z \\ I_{px'}l_{qx'} + l_{py'}l_{qy'} + l_{pz'}l_{qz'} &= 0 & p, q &= x, y, \text{ or } z, \quad p \neq q \end{aligned} \tag{15}$$

As an aside, note that equations (15) represent six relationships that the nine direction cosines must satisfy, so that only three of the direction cosines, such as $l_{xx'}$, $l_{yy'}$, and $l_{zz'}$, are independent quantities.

Let us now consider the matter of the computation of the moment of inertia I_{xx}. By definition,

$$I_{xx} = \int_{\mathcal{P}} (y^2 + z^2) \, dm$$

But $(y^2 + z^2)^{1/2} = R$ is the distance from the differential element at point i to the x axis. In other words, it is the component of $\bar{r}_{i/o}$ perpendicular to the x axis, as illustrated in Figure 9 on the next page. From the vector algebra we know that the component of a vector perpendicular to an axis

Figure 9

can be obtained from a cross product. Specifically, for our problem, this step gives

$$R = \left| \bar{r}_{i/O} \times \bar{\imath} \right|$$

so that

$$R^2 \equiv (y^2 + z^2) = (\bar{r}_{i/O} \times \bar{\imath}) \cdot (\bar{r}_{i/O} \times \bar{\imath})$$

We shall refer all components to the $x'y'z'$ reference frame, because we are considering the inertia properties to be known with respect to this coordinate system. Hence, using equation (14a),

$$\bar{r}_{i/O} = x'\bar{\imath}' + y'\bar{\jmath}' + z'\bar{k}'$$

$$\bar{\imath} = l_{xx'}\bar{\imath}' + l_{xy'}\bar{\jmath}' + l_{xz'}\bar{k}'$$

it follows that

$$\bar{r}_{i/O} \times \bar{\imath} = (l_{xz'}y' - l_{xy'}z')\bar{\imath}' + (l_{xx'}z' - l_{xz'}x')\bar{\jmath}' + (l_{xy'}x' - l_{xx'}y')\bar{k}'$$

From this we find

$$R^2 \equiv (\bar{r}_{i/O} \times \bar{\imath}) \cdot (\bar{r}_{i/O} \times \bar{\imath})$$

$$= (l_{xz'}y' - l_{xy'}z')^2 + (l_{xx'}z' - l_{xz'}x')^2 + (l_{xy'}x' - l_{xx'}y')^2$$

Upon collecting terms in this equation for R^2, we can form

$$I_{xx} = \int_{\mathscr{V}} R^2 \, dm = \int_{\mathscr{V}} \{ l_{xx'}^2[(y')^2 + (z')^2] + l_{xy'}^2[(x')^2 + (z')^2]$$

$$+ l_{xz'}^2[(x')^2 + (y')^2] - 2l_{xx'}l_{xy'}x'y' - 2l_{xy'}l_{xz'}y'z' - 2l_{xx'}l_{xz'}x'z' \} \, dm$$

Comparable formulas for I_{yy} and I_{zz} can be obtained by suitable permutation of the symbols in this equation. Letting p represent either x, y, or z, the general formula for the moments of inertia with respect to the xyz system is then

$$
\begin{aligned}
I_{pp} = {}& l^2_{px'}I_{x'x'} + l^2_{py'}I_{y'y'} + l^2_{pz'}I_{z'z'} \\
& - 2l_{px'}l_{py'}I_{x'y'} - 2l_{py'}l_{pz'}I_{y'z'} \\
& - 2l_{px'}l_{pz'}I_{x'z'} \qquad p = x,\, y,\, \text{or } z
\end{aligned}
\tag{16}
$$

We shall now consider how to obtain a product of inertia, such as I_{xy}, from the inertia properties that are known with respect to the $x'y'z'$ coordinate system. We know that

$$
I_{xy} = \int_{\mathcal{V}} xy\, dm
$$

where x and y are the components of $\bar{r}_{i/O}$ in the $\bar{i}$ and $\bar{j}$ directions, respectively. Thus, in view of equations (14a) and (14b),

$$
x = \left| \bar{r}_{i/O} \cdot \bar{i} \right| = l_{xx'}x' + l_{xy'}y' + l_{xz'}z'
$$

$$
y = \left| \bar{r}_{i/O} \cdot \bar{j} \right| = l_{yx'}x' + l_{yy'}y' + l_{yz'}z'
$$

By forming the product of this expression and making use of identities (15), we find

$$
x'y' = -\,(l_{xy'}l_{yy'} + l_{xz'}l_{yz'})(x')^2 - (l_{xx'}l_{yx'} + l_{xz'}l_{yz'})(y')^2 - (l_{xx'}l_{yx'} + l_{xy'}l_{yy'})(z')^2
$$
$$
+ (l_{xx'}l_{yy'} + l_{xy'}l_{yx'})x'y' + (l_{xy'}l_{yz'} + l_{xz'}l_{yy'})y'z' + (l_{xz'}l_{yx'} + l_{xx'}l_{yz'})x'z'
$$

We now regroup terms and form the product of inertia.

$$
\begin{aligned}
I_{x'y'} = \int_{\mathcal{V}} \big\{ & -l_{xy'}l_{yy'}[(x')^2 + (z')^2] - l_{xz'}l_{yz'}[(x')^2 + (y')^2] \\
& - l_{xx'}l_{yx'}[(y')^2 + (z')^2] + (l_{xx'}l_{yy'} + l_{xy'}l_{yx'})x'y' \\
& + (l_{xy'}l_{yz'} + l_{xz'}l_{yy'})y'z' + (l_{xz'}l_{yx'} + l_{xx'}l_{yz'})x'z' \big\}\, dm
\end{aligned}
$$

Once again, the other products of inertia can be obtained by suitable permutation of the symbols in the foregoing equation. Letting p and q represent x, y, or z, the general formula for the product of inertia with respect to the xyz system is

$$I_{pq} = (l_{px'}l_{qy'} + l_{py'}l_{qx'})I_{x'y'} + (l_{py'}l_{qz'} + l_{pz'}l_{qy'})I_{y'z'} + (l_{px'}l_{qz'} + l_{pz'}l_{qx'})I_{x'z'} \quad (17)$$
$$- l_{px'}l_{qx'}I_{x'x'} - l_{py'}l_{qy'}I_{y'y'} - l_{pz'}l_{qz'}I_{z'z'} \qquad p, q = x, y, \text{ or } z, \text{ but } p \neq q$$

An example of the application of equation (16) follows. The method for equation (17) is essentially the same, so it will not be a feature of the example.

EXAMPLE 6

Consider the propeller of Figure 8. Approximating each propeller blade as a slender bar of mass m, determine the inertia properties of this body with respect to the coordinate system shown in that figure.

Solution

We are told to consider the propeller to be a planar composite body consisting of three uniform bars of mass m. As is illustrated in the sketch, this planar body is symmetric about both the xy and yz planes, so that xyz are principal axes, and

$$I_{xy} = I_{yz} = I_{xz} = 0$$

In order to use the tabulated values for moments of inertia, we treat each blade individually. For a slender uniform bar, Appendix B gives

$$I_{\hat{x}\hat{x}} = 0$$

$$I_{\hat{y}\hat{y}} = I_{\hat{z}\hat{z}} = \tfrac{1}{12}ml^2$$

For blade 1, therefore, we need only change the symbols for the axes from those given and apply the parallel axis theorems, equations (9), to find

$$(I_{yy})_1 \equiv (I_{\hat{y}\hat{y}})_1 = 0$$

$$(I_{xx})_1 \equiv (I_{zz})_1 = (I_{\hat{x}\hat{x}})_1 + m\left(\frac{R}{2}\right)^2 = \tfrac{1}{3}mR^2$$

From the symmetry of the system we see that blades 2 and 3 have identical moments of inertia with respect to xyz, so we need only consider blade 2. Our procedure will be first to use the parallel axis theorems to transfer to a coordinate system $x'y'z'$ that has the same origin as that of xyz, and then to use the rotation transformation formulas to obtain the properties with respect to xyz.

The values for blade 2 with respect to $x'y'z'$ are the same as those of blade 1 with respect to $xyz;$ that is,

$$(I_{x'x'})_2 \equiv (I_{z'z'})_2 = \tfrac{1}{3}mR^2 \qquad (I_{y'y'})_2 = 0$$

By symmetry, $x'y'z'$ are principal axes for blade 2.

In order to transform from $x'y'z'$ to xyz, we need the direction cosines of xyz with respect to $x'y'z'$. These are indicated in the table.

p \ q'	x'	y'	z'
x	$\cos 120° = -1/2$	$\cos 150° = -\sqrt{3/2}$	$\cos 90° = 0$
y	$\cos 30° = \sqrt{3/2}$	$\cos 120° = -1/2$	$\cos 90° = 0$
z	$\cos 90° = 0$	$\cos 90° = 0$	$\cos 0 = 1$

The transformation equation (16) may now be applied, using the fact that $x'y'z'$ are principal axes. Thus

$$(I_{xx})_2 = l_{xx'}{}^2 (I_{x'x'})_2 + l_{xy'}^2 (I_{y'y'})_2 + l_{xz'}^2 (I_{z'z'})_2$$

$$= \left(-\frac{1}{2} \right)^2 \left(\frac{1}{3}\,mR^2 \right) + \left(-\frac{\sqrt{3}}{2} \right)^2 (0) + (0)^2 \left(\frac{1}{3}\,mR^2 \right)$$

$$= \tfrac{1}{12}mR^2$$

$$(I_{yy})_2 = l_{yx'}^2 (I_{x'x'})_2 + l_{yy'}^2 (I_{y'y'})_2 + l_{yz'}^2 (I_{z'z'})_2$$

$$= \left(\frac{\sqrt{3}}{2} \right)^2 \left(\frac{1}{3}\,mR^2 \right) + 0 + 0 = \tfrac{1}{4}\,mR^2$$

$$(I_{zz})_2 \equiv (I_{z'z'})_2 = \tfrac{1}{3}mR^2$$

Then, because blades 2 and 3 have the same properties, for the system we have

$$I_{xx} = (I_{xx})_1 + (I_{xx})_2 + (I_{xx})_3$$

$$= \tfrac{1}{3}mR^2 + 2(\tfrac{1}{12}mR^2) = \tfrac{1}{2}mR^2$$

$$I_{yy} = (I_{yy})_1 + (I_{yy})_2 + (I_{yy})_3$$

$$= 0 + 2(\tfrac{1}{4}mR^2) = \tfrac{1}{2}mR^2$$

$$I_{zz} = (I_{zz})_1 + (I_{zz})_2 + (I_{zz})_3$$

$$= 3(\tfrac{1}{3}mR^2) = mR^2$$

In the preceding example the tabulated inertia properties were fairly simple because of the symmetry of a slender bar. Also, because of the nature of the problem, it was not necessary to perform a rotation transformation for all of the inertia values. As a result, the calculations for the solution were not unduly complicated. On the other hand, it is not difficult to see that a rotation transformation for a more general problem can be quite awkward. To assist in computations, the transformations defined by equations (16) and (17) can be written in either matrix or tensor form. Matrices and tensors are not presented in this textbook. However, with the aid of either matrices or tensors, we could then determine the principal axes of any body, not just symmetric ones. Principal axes have obvious usefulness, because of the simplifications that arise in the angular momentum equation (5). The student who is interested in these topics is referred to any higher-level textbook in dynamics.

Here, we shall find that elementary problems in kinetics are more easily solved by using the most convenient set of axes, regardless of whether or not they are principal ones. Such an approach avoids the extra step of determining principal axes. In actual engineering practice, the decision to use principal axes should be carefully made, especially when formulating a problem for solution by electronic computer. The computational savings associated with the simplifications in the expression for the angular momentum resulting from principal axes must be weighed against the computations necessary to determine such axes.

Prob. VII.33

Prob. VII.36

HOMEWORK PROBLEMS

VII.33 Determine I_{xx} for the circular disk of mass m.

VII.34 Determine I_{yy} for the circular disk of mass m in Problem VII.33.

VII.35 Determine I_{xy} for the circular disk of mass m in Problem VII.33.

VII.36 Determine I_{xx} for the 5-kg right circular cone.

VII.37 Determine I_{yy} for the 5 kg right circular cone in Problem VII.36.

VII.38 Determine I_{xy} for the 5-kg right circular cone in Problem VII.36.

D. EQUATIONS OF MOTION

Now that we have studied how the inertia properties of a body are obtained, we can continue with the development of the relationship between angular momentum and the moment exerted by the force system, as given in equation (2).

$$\Sigma \bar{M}_A = \dot{\bar{H}}_A \tag{2}$$

Because they cannot be emphasized enough, we begin by repeating the steps necessary to utilize the derived expression for the angular momentum of a rigid body, equation (5).

a   Restrict the choice of point A to be either a fixed point in space that is also fixed relative to the body, or the center of mass of the body, or else a point in the body that is accelerating directly toward or away from the center of mass, thus assuring the validity of the moment equation (2).
b   Select a moving reference frame xyz that is attached to the body with origin at point A. This choice for xyz ensures that the moments and products of inertia with respect to the reference frame are constants.
c   Orient the reference frame relative to the body in a manner that facilitates the determination of the moments and products of inertia.
d   Substitute these inertia properties and the $(\bar{\imath}, \bar{\jmath}, \bar{k})$ components of the angular velocity $\bar{\omega}$ of the body into equation (5), thus obtaining an expression for the components of $\bar{H}_A$ relative to xyz.

In the case where a body undergoes a translational motion, the only allowable point for summing moments is the center of mass G. Then, because $\bar{\omega} \equiv \bar{0}$ for translation, the moment equation (2) reduces to $\Sigma \bar{M}_G = \bar{0}$.

On the other hand, when a body has rotational motion, the procedure described above yields the $(\bar{\imath}, \bar{\jmath}, \bar{k})$ components of $\bar{H}_A$. Therefore, the evaluation of $\dot{\bar{H}}_A$ requires that we employ the concept of differentiation with respect to a moving reference frame as derived in Module VI. Recall that, for an arbitrary vector $\bar{A}$, the general formula for this procedure is

$$\frac{d\bar{A}}{dt} = \frac{\delta \bar{A}}{\delta t} + \bar{\omega} \times \bar{A} \tag{18}$$

Hence, operating on $\bar{H}_A$, we have

$$\boxed{\Sigma \bar{M}_A = \dot{\bar{H}}_A = \frac{\delta \bar{H}_A}{\delta t} + \bar{\omega} \times \bar{H}_A} \tag{19}$$

where, by definition, the components of $\delta \bar{H}_A / \delta t$ are the derivatives of the components of $\bar{H}_A$. Then, in view of the fact that the moments and products of inertia are constants, differentiating the components of $\bar{H}_A$ in equation (5) gives

$$
\begin{aligned}
\frac{\delta \bar{H}_A}{\delta t} = & (I_{xx}\dot{\omega}_x - I_{xy}\dot{\omega}_y - I_{xz}\dot{\omega}_z)\bar{\imath} \\
& + (I_{yy}\dot{\omega}_y - I_{xy}\dot{\omega}_x - I_{yz}\dot{\omega}_z)\bar{\jmath} \\
& + (I_{zz}\dot{\omega}_z - I_{xz}\dot{\omega}_x - I_{yz}\dot{\omega}_y)\bar{k}
\end{aligned} \tag{20}
$$

The interpretation of this equation involves an important subtlety. In general, it is a difficult matter to differentiate the components of $\bar{\omega}$ because such an evaluation would require expressions for ω_x, ω_y, and ω_z that are valid for all time instants. However, in the approach we have developed, the xyz reference frame is attached to the body and therefore has the same angular velocity as the body. As a result, equation (18) gives

$$\frac{d\bar{\omega}}{dt} \equiv \bar{\alpha} = \frac{\delta\omega}{\delta t} + \bar{\omega} \times \bar{\omega} \equiv \frac{\delta\bar{\omega}}{\delta t}$$

or, in component form,

$$\alpha_x = \dot{\omega}_x \qquad \alpha_y = \dot{\omega}_y \qquad \alpha_z = \dot{\omega}_z$$

In other words, because the xyz frame is attached to the body, the components of the angular acceleration of the body are identical to the derivatives of the angular velocity components. This allows us to follow the procedures of Module VI to describe angular velocity and angular acceleration, and then simply substitute the results into the equations of kinetics.

As a result, we now write equation (20) in the more explicit form

$$\boxed{\begin{aligned}\frac{\delta \bar{H}_A}{\delta t} &= (I_{xx}\alpha_x - I_{xy}\alpha_y - I_{xz}\alpha_z)\bar{i} \\ &+ (I_{yy}\alpha_y - I_{xy}\alpha_x - I_{yz}\alpha_z)\bar{j} \\ &+ (I_{zz}\alpha_z - I_{xz}\alpha_x - I_{yz}\alpha_y)\bar{k}\end{aligned}} \qquad (21)$$

Equations (5), (19), and (21) form the basic relationships between the resultant moment applied to a rigid body and the angular motion of the body. Clearly, the moment equation has three components, so that in conjunction with equation (1) governing the effect of the resultant force,

$$\boxed{\Sigma \bar{F} = m\bar{a}_G} \qquad (1)$$

we have the six scalar equations of motion alluded to earlier.

For a different perspective of the preceding development, let us see what conditions must exist for equation (19) to reduce to the now familiar $\Sigma \bar{M} = I\bar{\alpha}$ equation for planar motion. For an arbitrary body moving parallel to the fixed XY plane, we select the z axis to be perpendicular to the plane of motion. Then, for planar motion we know that

$$\bar{\omega} = \omega\bar{k} \qquad \bar{\alpha} = \alpha\bar{k}$$

Equations (5) and (20) then give

$$\bar{H}_A = -I_{xz}\omega\bar{\imath} - I_{yz}\omega\bar{\jmath} + I_{zz}\omega\bar{k}$$

$$\frac{\delta\bar{H}_A}{\delta t} = -I_{xz}\alpha\bar{\imath} - I_{yz}\alpha\bar{\jmath} + I_{zz}\alpha\bar{K}$$

By substituting the foregoing expressions, the moment equation (19) is specialized to the case of planar motion, becoming

$$\begin{aligned}
\Sigma\,\bar{M}_A &= \alpha(-I_{xz}\,\bar{\imath} - I_{yz}\,\bar{\jmath} + I_{zz}\,\bar{k}) + \omega\bar{k} \times \omega(-I_{xz}\,\bar{\imath} \\
&\quad - I_{yz}\,\bar{\jmath} + I_{zz}\,\bar{k}) \\
&= -(I_{xz}\alpha - I_{xz}\omega^2)\bar{\imath} - (I_{yz}\alpha + I_{xz}\omega^2)\,\bar{\jmath} + I_{zz}\alpha\bar{k}
\end{aligned} \tag{22}$$

Notice that the $\bar{\imath}$ and $\bar{\jmath}$ components in this equation cause it to be different from the moment equation $\Sigma\,\bar{M}_A = I_{zz}\alpha\bar{k}$ employed in Module V. The $\bar{\imath}$ and $\bar{\jmath}$ components in equation (22) describe the resultant moments of the force system about the x and y axes necessary to constrain an arbitrary body to planar motion. This is the situation that arises when a body that is rotating about a fixed axis is dynamically unbalanced (as will be illustrated in Example 7 to follow). Only in situations where $I_{xz} = I_{yz} = 0$, that is, where the z axis is a principal axis, will these contraint moments be unnecessary to maintain a body in planar motion. This condition was assured in Module V, because there we considered only bodies that are symmetric about the xy plane.

Let us now return to the study of the equations of motion for a body in three-dimensional motion. In those cases where the chosen xyz reference system does not form principal axes, the most direct way of treating the problem is to evaluate $\bar{H}_A$ and $\delta\bar{H}_A/\delta t$ for the axes chosen and to then substitute these expressions into equation (19) for $\Sigma\,\bar{M}_A$.

When the xyz frame forms principal axes, the expressions for $\bar{H}_A$ and $\delta\bar{H}_A/\delta t$ simplify considerably, as does equation (19). The result is called *Euler's equations of motion*, named after Leonhard Euler (1707–1783). Thus, *when the xyz coordinate system is a set of principal axes of the body*, the vector equation for the rotational motion of the body is

$$\begin{aligned}
\Sigma\,\bar{M}_A &= [I_{xx}\alpha_x - (I_{yy} - I_{zz})\omega_y\omega_z]\bar{\imath} \\
&\quad + [I_{yy}\alpha_y - (I_{zz} - I_{xx})\omega_x\omega_z]\,\bar{\jmath} \\
&\quad + [I_{zz}\alpha_z - (I_{xx} - I_{yy})\omega_x\omega_y]\bar{k}
\end{aligned} \tag{23}$$

It should be evident that for problems where the xyz system forms a principal set of axes, equation (23) involves fewer computations than equations (5), (19), and (21). However, the separate computation of $\bar{H}_A$ and $\delta\bar{H}_A/\delta t$ has the advantage of enabling us to gain insight about the underlying physics of the system being studied. Also, this approach is the one that must be employed if the xyz system is a nonprincipal set of axes. To gain some appreciation of the application of both approaches, let us consider the following two examples.

EXAMPLE 7

The obliquely mounted disk of Example 1 is illustrated again here. The rigid shaft, which is welded to the disk, is rotating at the constant angular speed ω. The bearings are frictionless and can only apply transverse shear forces to restrain the transverse motion of the shaft. Additionally, bearing A can exert a thrust force to restrain axial motion of the shaft. Determine the reaction forces exerted by the bearings due to the motion of the system.

Solution

We begin with a free body diagram of the system that also shows the orientation of xyz relative to the system.

We have not included the gravity force because it is a static effect; if we sought the total reactive forces we would include it. The xyz reference frame is the same as that used in Example 1, so that we may use the results for the vectors $\bar{\omega}$ and $\bar{H}_G$ derived there.

$$\bar{\omega} = \omega \bar{I} \qquad \text{always}$$

$$= \omega(\cos \beta \, \bar{i} + \sin \beta \, \bar{j}) \qquad \text{instantaneously and always}$$

$$\bar{H}_G = \tfrac{1}{4} mr^2 \omega (2 \cos \beta \, \bar{i} + \sin \beta \, \bar{j})$$

Both the rate of rotation ω and the unit vector $\bar{I}$ are constants, so we find $\bar{\alpha} \equiv \dot{\bar{\omega}} = \bar{0}$. This means that $\delta \bar{H}_G / \delta t = \bar{0}$, and

$$\dot{\bar{H}}_G \equiv \frac{\delta \bar{H}_G}{\delta t} + \bar{\omega} \times \bar{H}_G = -\tfrac{1}{4} mr^2 \omega^2 \sin \beta \cos \beta \, \bar{k}$$

$$= - \tfrac{1}{8} mr^2 \omega^2 \sin 2\beta \, \bar{k}$$

We now consider the external force system. Because the center of mass G is on the fixed axis of rotation, it has no acceleration, and

$$\Sigma \bar{F} = m \bar{a}_G = \bar{0}$$

This equation can be resolved into components with respect to any coordinate system. In view of the directions of the forces in the free body diagram, we shall use the fixed coordinate system for this step. Thus

$$\bar{I} \text{ direction:} \qquad A_X = 0$$

$$\bar{J} \text{ direction:} \qquad A_Y + B_Y = 0$$

$$\bar{K} \text{ direction:} \qquad A_Z + B_Z = 0$$

We next compute the moment sum, as follows:

$$\Sigma \bar{M}_G = \bar{r}_{A/G} \times (A_X \bar{I} + A_Y \bar{J} + A_Z \bar{K})$$

$$+ \bar{r}_{B/G} \times (B_Y \bar{J} + B_Z \bar{K})$$

$$= -\frac{l}{2} \bar{I} \times (A_X \bar{I} + A_Y \bar{J} + A_Z \bar{K})$$

$$+ \frac{l}{2} \bar{I} \times (B_Y \bar{J} + B_Z \bar{K})$$

$$= \frac{l}{2} (A_Z - B_Z) \bar{J} - \frac{l}{2} (A_Y - B_Y) \bar{K}$$

However, the formulation of the equation for moment equilibrium requires that we sum moments about the axes of the moving system xyz and equate each component to the corresponding component of $\dot{\bar{H}}_G$. To do this we transform the unit vectors, referring to the free body diagram for the orientations of the axes. Thus, using

$$\bar{I} = \cos \beta \, \bar{i} + \sin \beta \, \bar{j}$$

$$\bar{J} = - \sin \beta \, \bar{i} + \cos \beta \, \bar{j}$$

$$\bar{K} = \bar{k}$$

the moment equation becomes

$$\Sigma \bar{M}_G = \frac{l}{2}(A_Z - B_Z)(-\sin \beta \, \bar{i} + \cos \beta \, \bar{j}) - \frac{l}{2}(A_Y - B_Y)\bar{k}$$

$$= \dot{H}_G = -\tfrac{1}{8}mr^2\omega^2 \sin 2\beta \, \bar{k}$$

Matching components yields

$\bar{i}$ and $\bar{j}$ components: $A_Z = B_Z = 0$

$\bar{k}$ component: $-\dfrac{l}{2}(A_Y - B_Y) = -\tfrac{1}{8}mr^2\omega^2 \sin 2\beta$

Simultaneous solution of the force and moment equations yields

$$A_X = A_Z = B_Z = 0$$

$$A_Y = -B_Y = \tfrac{1}{16}m\,\frac{r^2}{l}\,\omega^2 \sin 2\beta$$

The force components A_Y and B_Y are a consequence of the fact that $\bar{H}_G$ does not have a fixed direction. Example 1 showed that $\bar{H}_G$ rotates about the shaft at angular velocity $\bar{\omega}$ in the manner illustrated in the sketch.

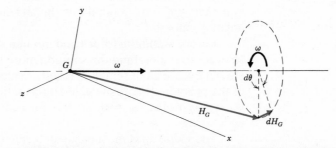

As the system rotates an infinitesimal amount $d\theta$ from its instantaneous position, the angular momentum changes by the amount $d\bar{H}_G$, as shown in the sketch. This demonstrates why $\dot{H}_G$ is in the negative $\bar{k}$ direction.

It is common to refer to the component of $\dot{H}_G$ that is perpendicular to $\bar{H}_G$ (and therefore a consequence of the variable direction of the angular momentum) as the *gyroscopic moment*. The reaction forces A_Y and B_Y in this example exert the gyroscopic moment on the system.

We selected the orientation of xyz such that the yz plane was the plane of the disk at an arbitrary instant. For this problem the gyroscopic moment is always in the negative $\bar{k}$ direction, so that the reaction forces A_Y and B_Y must always lie in the xy plane. This leads to the conclusion that the reaction forces rotate with the motion of the disk.

It should surprise you that the system is balanced in the static sense (the center of mass is on the axis of rotation), and yet reaction forces are generated by the rotation of the system. Such a system is said to be *dynamically unbalanced*. If we had solved this problem for an arbitrary body, rather than a disk, we could have shown that a system is dynamically balanced only when it is statically balanced *and* the fixed axis of rotation is a principal axis.

EXAMPLE 8

Let us now reconsider the system of Example 2 in which a disk is welded onto the shaft of a motor fixed to a turntable, as shown in the sketch. Initially, the turntable is rotating at the rate $\omega_1 = 10\pi$ rad/s and the motor is rotating at $\omega_2 = 50\pi$ rad/s. If the turntable is suddenly decelerated at the rate of 5π rad/s², what force and couple will the motor shaft exert on the disk?

Solution

The shaft is fixed to the disk, so it restricts the motion of the point of connection between the bodies in three dimensions and also restricts the rotation of the disk about three coordinate axes. Therefore, the shaft exerts an unknown three-dimensional force $\bar{F}_s$ on the disk and also an unknown three-dimensional couple $\bar{M}_s$. This is illustrated in the free body diagram. In this sketch we also show the xyz reference frame.

The instantaneous orientation of xyz has been chosen for the convenience it affords in obtaining the inertia properties and in describing the angular velocity.

Before summing forces and moments, we must determine $\bar{a}_G$ and $\dot{\bar{H}}_G$. In order to understand the solution more fully, and also for computational efficiency, we shall retain all results in algebraic form until the completion of the problem. From the given information we know that $\omega_1 = 10\pi$ rad/s, $\omega_2 = 50\pi$ rad/s, $\dot{\omega}_2 = 0$, $\dot{\omega}_1 = -5\pi$ rad/s² ($\dot{\omega}_1$ is negative because ω_1 is decreasing).

The value $\bar{a}_G$ is most easily determined by noting that point G follows a circular path of radius 0.5 m with angular speed ω_1. Furthermore, at the instant shown in the free body diagram, point G is moving into the plane of the diagram, that is, in the negative z direction. Therefore, using cylindrical coordinates with $\bar{e}_R = \bar{\imath}$ and $\bar{e}_\phi = -\bar{k}$, we find that

$$\bar{a}_G = -0.5\omega_1^2\bar{e}_R + 0.5\dot{\omega}_1\bar{e}_\phi$$

$$= -0.5\omega_1^2\bar{\imath} - 0.5\dot{\omega}_1\bar{k}$$

The angular velocity of the disk was obtained in Example 2. For this problem we shall also require the angular acceleration. The general expressions for these quantities are

$$\bar{\omega} = \omega_1 \bar{J} + \omega_2 \bar{i} \qquad \text{always}$$

$$\bar{\alpha} \equiv \dot{\omega}_1 \bar{J} + \dot{\omega}_2 \bar{i} + \omega_2 \dot{\bar{i}}$$

$$= \dot{\omega}_1 \bar{J} + \bar{0} + \omega_2 (\bar{\omega} \times \bar{i}) \qquad \text{always}$$

The instantaneous expressions are obtained by noting that at the instant illustrated in the free body diagram $\bar{J} = \bar{j}$, so that

$$\bar{\omega} = \omega_1 \bar{j} + \omega_2 \bar{i} \qquad \text{instantaneously}$$

$$\bar{\alpha} = \dot{\omega}_1 \bar{j} + \omega_2 (\omega_1 \bar{j} + \omega_2 \bar{i}) \times \bar{i}$$

$$= \dot{\omega}_1 \bar{j} - \omega_1 \omega_2 \bar{k} \qquad \text{instantaneously}$$

The inertia properties of the disk were obtained in Example 2 directly from the tables in Appendix B. By symmetry, xyz are principal axes, and

$$I_{xx} = \tfrac{1}{2} mR^2 \qquad I_{yy} = I_{zz} = \tfrac{1}{4} mR^2$$

We next form the equations of motion. From the free body diagram and the results for $\bar{a}_G$, we have

$$\Sigma F_x = (F_s)_x = m(a_G)_x = -0.5 m \omega_1{}^2$$

$$\Sigma F_y = (F_s)_y - mg = m(a_G)_y = 0$$

$$\Sigma F_z = (F_s)_z = m(a_G)_z = -0.5 m \dot{\omega}_1$$

Thus

$$\bar{F}_s = -0.5 m \omega_1{}^2 \bar{i} + mg \, \bar{j} - 0.5 m \dot{\omega}_1 \bar{k}$$

In this example we shall demonstrate the application of Euler's equation (23). Thus, by summing moments about point G and considering the disk to be sufficiently thin to allow us to consider the force $\bar{F}_s$ to be applied at point G, we find that

$$\Sigma M_{Gx} = (M_s)_x = I_{xx} \alpha_x - (I_{yy} - I_{zz}) \omega_y \omega_z$$

$$= \tfrac{1}{2} mR^2 (0) - (0)(\omega_1)(0) = 0$$

$$\Sigma M_{Gy} = (M_s)_y = I_{yy} \alpha_y - (I_{zz} - I_{xx}) \omega_z \omega_x$$

$$= \tfrac{1}{4} mR^2 \dot{\omega}_1 - (\tfrac{1}{4} - \tfrac{1}{2}) mR^2 (0)(\omega_2) = \tfrac{1}{4} mR^2 \dot{\omega}_1$$

$$\Sigma M_{Gz} = (M_s)_z = I_{zz} \alpha_z - (I_{xx} - I_{yy}) \omega_x \omega_y$$

$$= \tfrac{1}{4} mR^2 (-\omega_1 \omega_2) - (\tfrac{1}{2} - \tfrac{1}{4}) mR^2 (\omega_2)(\omega_1)$$

$$= -\tfrac{1}{2} mR^2 \omega_1 \omega_2$$

Thus

$$\bar{M}_s = mR^2(\tfrac{1}{4}\dot{\omega}_1\,\bar{j} - \tfrac{1}{2}\omega_1\omega_2\bar{k})$$

Now, substituting the given values for ω_1, $\dot{\omega}_1$, and ω_2 gives

$$\bar{F}_s = m[-493.5\bar{i} + 9.806\bar{j} - 7.854\bar{k}]\ \text{N}$$

$$\bar{M}_s = mR^2[-3.927\bar{j} - 2467\bar{k}]\ \text{N-m}$$

These answers, including the large component of $\bar{M}_s$ in the z direction, have simple explanations. The shaft force $\bar{F}_s$ causes the center of mass to follow a circular path. Thus, for this problem $(F_s)_x$ imposes a normal acceleration, $(F_s)_y$ counterbalances gravity to produce planar motion, and $(F_s)_z$ imposes a tangential acceleration. The shaft couple $\bar{M}_s$ causes the disk to assume the proper angular motion. The value of $(M_s)_y$ is independent of the speed of the motor ω_2. It represents the couple that must be applied to the disk if the turntable is to slow down. The large value of $(M_s)_z$ is the gyroscopic moment that must be applied to the disk if its angular motion is to be maintained. This is best illustrated by considering the result for $\bar{H}_G$ from Example 2, which is sketched again here.

Because of the rotation ω_1 about the Y axis, the tip of $\bar{H}_G$ must move into the plane of the sketch, thus requiring that the force system apply a gyroscopic moment in the negative z direction. Once again, if we had relied on planar kinetics for intuition, we would have completely ignored the gyroscopic moment, thus missing a major effect in the system.

HOMEWORK PROBLEMS

VII.39   Determine $\dot{H}_G$ for the system in Problem VII.1.

VII.40   Determine $\dot{H}_G$ for the system in Problem VII.3.

VII.41   Determine $\dot{H}_G$ for the system in Problem VII.5.

VII.42   The slender bar of mass m and length l is welded to the horizontal shaft, which is rotating at a constant rate ω. In terms of the constant angle θ, determine (a) $\bar{H}_G$ and $\dot{H}_G$, (b) $\bar{H}_A$ and $\dot{H}_A$. Explain why $\dot{H}_G = \dot{H}_A = 0$ for $\theta = 0$ and $\theta = 90°$.

Prob. VII.42

VII.43   For the system in Problem VII.42, determine the total force-couple reaction at point A that the horizontal shaft exerts on the bar when the bar coincides (a) with the horizontal plane, (b) with the vertical plane.

Prob. VII.44

VII.44   The 10-kg thin rectangular plate shown is welded diagonally onto the horizontal shaft AB. This shaft is rotating at $\omega = 240$ rev/min when a couple M is applied, resulting in a deceleration $-\dot{\omega}$. Determine the value of the deceleration rate for which the component of $\dot{H}_G$ associated with $\dot{\omega}$ will have the same magnitude as the gyroscopic component of $\dot{H}_G$.

VII.45   For the system in Problem VII.44, the couple $M = 20$ N-m. Determine the value of $\dot{\omega}$ and the bearing reactions at points A and B (a) when $\omega = 240$ rev/min, (b) when $\omega = 120$ rev/min.

VII.46   Two identical 40-kg disks are welded concentrically to a horizontal shaft that rotates at the angular rate ω. Identical small masses are attached to the perimeter of the disks. Is the system unbalanced, statically balanced only, or dynamically balanced when masses are added (a) nowhere, (b) at point A_1, (c) at points A_1 and A_2, (d) at points A_1 and C_2, (e) at points A_1, B_1, C_2, and D_2, (f) at points A_1, C_1, B_2, and D_2?

Prob. VII.46

Prob. VII.47

Prob. VII.49

VII.47   A 6-in.-radius sphere weighing 80 lb rotates about bar AB at 3000 rev/min as the bar pivots about pin A. The bar weighs 20 lb. When $\theta = 45°$, $\dot{\theta} = 5$ rad/s and $\ddot{\theta} = -10$ rad/s². Determine (a) $\bar{H}_A$ and $\dot{H}_A$ for the sphere, (b) $\bar{H}_A$ and $\dot{H}_A$ for bar AB.

VII.48   For the motion of the system defined in Problem VII.47, determine the couple that the pin at point A applies to the bar (a) if the bar is massless, (b) if the sphere is massless, and (c) if neither body is massless. Compare the results.

VII.49   The system of Problem VII.6 is presented again in the diagram. Determine $\dot{H}_G$ for the disk of mass m, considering ω_1 and ω_2 to be constant.

VII.50   For the system of Problem VII.49, determine $\bar{H}_A$ and $\dot{H}_A$ for the disk of mass m, using a coordinate system parallel to the one shown. Consider ω_1 and ω_2 to be constant. Draw a sketch showing $\bar{H}_A$.

VII.51 For the system of Problem VII.49, at the instant shown $\omega_1 = 5$ rad/s, $\omega_2 = 100$ rad/s, and ω_2 is increasing at 2 rad/s². The mass of the disk is m. Determine $\bar{H}_A$ and $\dot{H}_A$, using a coordinate system parallel to the one shown.

VII.52 For the system of Problem VII.49, the mass of shaft AB is one-half that of the disk. For the case where ω_1 and ω_2 are constant, compare (a) $\dot{H}_A$ for the disk to $\dot{H}_A$ for the shaft, (b) the force-couple reaction exerted upon the shaft at point A when the mass of the shaft is considered to the reaction obtained when the mass of the shaft is neglected.

VII.53 The two identical spheres of mass m and radius r rotate independently at rates ω_1 and ω_2 about shaft AB, as shown. This shaft is supported by a pin joint at its midpoint C and rotates about the vertical axis at the rate Ω. All rates of rotation are constant, and the mass of shaft AB is negligible. Determine (a) $\bar{H}_C$ and $\dot{H}_C$ for each sphere, (b) the couple that must be applied at the pin joint to sustain this motion. Why is this couple zero when $\omega_1 = \omega_2$?

Prob. VII.53

E. PROBLEM SOLVING

The steps for solving problems involving the equations of three-dimensional motion of rigid bodies may already be apparent to you from the examples. They are as follows:

1 Draw a complete free body diagram of the system. In doing this, to be sure that you define the problem fully, remember that if the motion of a point is restricted in a certain direction, then there is a constraint force in that direction. Similarly, if the rotation of a body about a certain axis is restricted, then there is a constraint moment about that axis.

2 Write down, in equation form, all given information that does not appear in the free body diagram. List all quantities desired in the solution.

3 Choose a point A for summing moments. This point should be either the center of mass or a fixed point in space that is also fixed relative to the body. The third type of point is rather special and can generally be ignored.

4 Attach xyz axes to the body with origin at point A and orient xyz relative to the body so as to minimize any computations for inertia properties. If possible, orient xyz so as to make it possible to use the inertia properties in Appendix B and the parallel axis theorems. Select XYZ to fit best the fixed axes of rotation of the body. Show these reference frames in the free body diagram of step 1.

5 Determine the inertia properties. Use integration methods and/or rotation transformations only as a last resort.

6 Determine the angular velocity $\bar{\omega}$, the angular acceleration $\bar{\alpha}$, and the acceleration of the center of mass $\bar{a}_G$. In doing this, be sure to satisfy

any kinematical constraints on the motion of the system. (Here it sometimes proves useful to leave the rotation rates and their derivatives as
algebraic symbols.)

7   Compute each component $\Sigma\ \bar{M}_A$ and equate them to the corresponding components of $\dot{H}_A$. Substitute all known quantities into these
equations. In general, use equations (5), (19), and (21) to perform this
step, but in the special case where the xyz axes are principal axes you can
use equation (23) instead.

8   Equate each component of $\Sigma\ \bar{F}$ to the corresponding component of
$m\bar{a}_G$.

9   Solve the equations.

As an aside, note that steps 7 and 8 include all the kinematics and kinematical constraints that you determined in step 6. Hence the number
of equations and the number of unknowns at this point should balance. If
not, the two most common conditions that may have been overlooked are

a   a kinematical constraint condition, or
b   a friction force relationship.

ILLUSTRATIVE PROBLEM 1

The crane shown in the sketch is initially at rest when the vertical support post is
given a constant angular acceleration $\dot{\Omega} = 0.5\ \text{rad/s}^2$. The boom, which has a mass
of 600 kg, is 10 m long, and the cable maintains the angle β at a constant value.
Determine the tension in the cable as a function of time.

Solution

Step 1   The cable tension appears as an external force in a free body diagram of
the boom. To construct this diagram we note that the pin connection at point A
constrains all motions of the boom except for rotations through the angle β. It
therefore follows that the force-couple system representing the reaction at this
joint consists of a three-dimensional force $\bar{F}_A$ and a moment $\bar{M}_A$ that has no
component perpendicular to the plane formed by the post and the boom. The free
body diagram of the boom is as shown here.

Step 2   The given information tells us that $\Omega = 0$ initially and that $\dot{\Omega} = 0.5$ rad/s², so that $\Omega = 0.5t$ rad/s. We wish to find the force T as a function of time t.

Step 3   We select the fixed point A for the origin of the moving reference system, thereby eliminating the reaction force $\bar{F}_A$ from the moment equation.

Step 4   Considering the boom to be a slender rod and selecting the x axis to coincide with the axis of the boom allows us to obtain the inertia properties directly from the tables in Appendix B. For convenience in describing the angular motion, we select the fixed Y axis to coincide with the vertical post and let the y axis lie in the XY plane, as shown in the free body diagram.

Step 5   The inertia properties for a slender rod appearing in Appendix B are repeated here.

$$I_{\hat{x}\hat{x}} = 0 \quad I_{\hat{y}\hat{y}} = I_{\hat{z}\hat{z}} = \tfrac{1}{12}mL^2$$

To determine the inertia properties with respect to the xyz coordinate system we use the parallel axis theorems, which yields

$$I_{xx} = I_{\hat{x}\hat{x}} = 0$$

$$I_{yy} = I_{zz} = I_{\hat{y}\hat{y}} + m\left(\frac{L}{2}\right)^4 = \tfrac{1}{3}mL^2$$

$$= \tfrac{1}{3}(600)(10)^2 = 2(10)^4 \text{ kg-m}^2$$

$$I_{xy} = I_{yz} = I_{xz} = 0$$

Note that the xyz system forms principal axes because the mass of a slender rod is considered to be situated entirely along the axis of the rod.

Step 6   The boom is rotating about the vertical axis, so that the general expressions for $\bar{\omega}$ and $\bar{\alpha}$ are

$$\bar{\omega} = \Omega\bar{J}, \quad \bar{\alpha} = \dot{\Omega}\bar{J} \qquad \text{always}$$

Resolving these expressions into $(\bar{i}, \bar{j}, \bar{k})$ components gives

$$\bar{\omega} = \Omega(\cos 60°\,\bar{i} + \sin 60°\,\bar{j}) = \Omega(0.5\bar{i} + 0.866\bar{j}) \qquad \text{instantaneously}$$

$$\bar{\alpha} = \dot{\Omega}(\cos 60°\,\bar{i} + \sin 60°\,\bar{j}) = \dot{\Omega}(0.5\bar{i} + 0.866\bar{j}) \qquad \text{instantaneously}$$

We omit calculating $\bar{a}_G$ because the force equation will only yield the reaction force $\bar{F}_A$.

Step 7   Because the xyz system is a principal set of axes, we can use Euler's equation (23). Hence, using the inertia values found in step 5 and the values of $\bar{\omega}$ and $\bar{\alpha}$ found in step 6, we find that

$$\Sigma \bar{M}_A = [0 - 0]\bar{\imath} + [2(10^4)(0.5\dot{\Omega}) - 0]\bar{\jmath}$$

$$+ \{0 - [0 - 2(10^4)](0.5\Omega)(0.866\Omega)\}\bar{k}$$

$$= 10^4\dot{\Omega}\bar{\jmath} + 0.866(10^4)\Omega^2\bar{k}$$

We now use the free body diagram to compute the moment of the force system. The reaction couple $\bar{M}_A$ has no component about the z axis, so we find that

$$\Sigma \bar{M}_A = (M_A)_x\,\bar{\imath} + (M_A)_y\,\bar{\jmath}$$

$$+ [T(5) - 600(9.806)(5 \sin 60°)]\bar{k}$$

$$= (M_A)_x\,\bar{\imath} + (M_A)_y\,\bar{\jmath} + (5T - 25480)\bar{k}$$

Step 8 We omit this step because we are not interested in the reaction force $\bar{F}_A$.

Step 9 To obtain the value of T we use the $\bar{k}$ component of the moment equation. Thus

$$(\Sigma \bar{M}_A) \cdot \bar{k} = (\dot{H}_A) \cdot \bar{k}$$

$$(5T - 25480) = 0.8660(10^4)\Omega^2$$

Then solving for T and using the expression for Ω in step 2, we finally obtain

$$T = 5095 + \frac{0.8660}{5}\,(10^4)(0.5t)^2$$

$$= 5095 + 433t^2$$

The first term in this result is the tension within the cable caused by gravity, which is a static effect. The second term represents the inertial effect of the rotation.

ILLUSTRATIVE PROBLEM 2

A 70-kg person is standing motionless with respect to the 140-kg wooden raft illustrated in the sketch. At the instant when the raft is horizontal, the centroid C of the raft is moving forward with a constant velocity of 2 m/s, while the raft's rider observes the raft to be pitching about the centerline CA at $\omega_1 = 0.2$ rad/s and rolling about the centerline CB at $\omega_2 = 0.5$ rad/s. The derivatives of these rates are zero at this instant. Determine the force and moment about the centroid C that the water must exert on the raft in order to produce this motion. For the purpose of computing inertial properties, the raft may be treated as a thin plate and the rider may be treated as a slender bar 1.80 m tall.

Solution

Step 1 Consider the rider and the raft to form a composite body. Let $\bar{F}$ and $\bar{M}_C$ be the force-couple resultant at point C.

Step 2 Given $\omega_1 = 0.2$ rad/s, $\omega_2 = 0.5$ rad/s, $\dot{\omega}_1 = 0$, $\dot{\omega}_2 = 0$. Find $\bar{F}$ and $\bar{M}_C$.

Step 3 The only allowable point for summing moments is the center of mass G of the *composite* body.

Step 4 Let *xyz* denote the reference frame whose origin is point G. Attach this reference frame to the body and orient the axes to be parallel to the axes of symmetry of the raft, as shown in the free body diagram. We shall need to determine the inertia properties of the composite body, so we also show the *xyz* system and the parallel centroidal axes for each component in a separate sketch.

Step 5 Locate point G by computing the first moments of mass about point C. (Note that using point C minimizes the number of terms in the calculations.) With respect to the $\hat{x}_1\hat{y}_1\hat{z}_1$ system, the coordinates of point G_2 are $(-1.2, 0, 0.90)$. Denote the coordinates of point G with respect to the $\hat{x}_1\hat{y}_1\hat{z}_1$ coordinate system as $(\hat{x}, \hat{y}, \hat{z})$. Because the total mass is 210 kg, we find that

$$\hat{x} = \frac{70(-1.2)}{210} = -0.40 \text{ m} \qquad \hat{y} = \frac{70(0)}{210} = 0 \qquad \hat{z} = \frac{70(0.9)}{210} = 0.30 \text{ m}$$

We can now determine the inertia properties of each component with respect to the *xyz* axes. For the raft the tabulated properties of a thin rectangular parallelepiped and the symmetry principles give

$$(I_{\hat{x}_1\hat{x}_1})_1 = \tfrac{1}{12}(140)(2^2 + 0^2) = 46.67 \text{ kg-m}^2$$

$$(I_{\hat{y}_1\hat{y}_1})_1 = \tfrac{1}{12}(140)(4^2 + 0^2) = 186.67 \text{ kg-m}^2$$

$$(I_{\hat{z}_1\hat{z}_1})_1 = \tfrac{1}{12}(140)(2^2 + 4^2) = 233.33 \text{ kg-m}^2$$

$$(I_{\hat{x}_1\hat{y}_1})_1 = (I_{\hat{x}_1\hat{z}_1})_1 = (I_{\hat{y}_1\hat{z}_1})_1 = 0$$

Then, because the $\hat{x}_1\hat{y}_1\hat{z}_1$ coordinates of point G are $(-0.40, 0, 0.30)$, the parallel axis theorems give

$$(I_{xx})_1 = (I_{\hat{x}_1\hat{x}_1})_1 + (140)(0^2 + 0.3^2) = 59.27 \text{ kg-m}^2$$

$$(I_{yy})_1 = (I_{\hat{y}_1\hat{y}_1})_1 + (140)(0.4^2 + 0.3^2) = 221.67 \text{ kg-m}^2$$

$$(I_{zz})_1 = (I_{\hat{z}_1\hat{z}_1})_1 + (140)(0.4^2 + 0^2) = 255.73 \text{ kg-m}^2$$

$$(I_{xy})_1 = (I_{\hat{x}_1\hat{y}_1})_1 + (140)(-0.4)(0) = 0$$

$$(I_{xz})_1 = (I_{\hat{x}_1\hat{z}_1})_1 + (140)(-0.4)(0.3) = -16.80 \text{ kg-m}^2$$

$$(I_{yz})_1 = (I_{\hat{y}_1\hat{z}_1})_1 + (140)(0)(0.3) = 0$$

Treating the rider as a slender bar, the tabulated properties yield

$$(I_{\hat{x}_1\hat{x}_2})_2 = (I_{\hat{y}_2\hat{y}_2})_2 = \tfrac{1}{12}(70)(1.8)^2 = 18.90 \text{ kg-m}^2$$

$$(I_{\hat{z}_2\hat{z}_2})_2 = (I_{\hat{x}_2\hat{y}_2})_2 = (I_{\hat{x}_2\hat{z}_2})_2 = (I_{\hat{y}_2\hat{z}_2})_2 = 0$$

The $\hat{x}_2\hat{y}_2\hat{z}_2$ coordinates of point G are $(0.8, 0, -0.6)$, so the parallel axis theorems give

$$(I_{xx})_2 = (I_{\hat{x}_2\hat{x}_2})_2 + (70)(0^2 + 0.6^2) = 44.10 \text{ kg-m}^2$$

$$(I_{yy})_2 = (I_{\hat{y}_2\hat{y}_2})_2 + (70)(0.8^2 + 0.6^2) = 88.90 \text{ kg-m}^2$$

$$(I_{zz})_2 = (I_{\hat{z}_2\hat{z}_2})_2 + (70)(0.8^2 + 0^2) = 44.80 \text{ kg-m}^2$$

$$(I_{xy})_2 = (I_{yz})_2 = (I_{\hat{y}_2\hat{z}_2})_2 = 0$$

$$(I_{xz})_2 = (I_{\hat{x}_2\hat{z}_2})_2 + 70(0.8)(-0.6) = -33.60 \text{ kg-m}^2$$

We now add the properties of each component to obtain the inertia properties of the system about its mass center G.

$$I_{xx} = (I_{xx})_1 + (I_{xx})_2 = 103.4 \text{ kg-m}^2$$

$$I_{yy} = (I_{yy})_1 + (I_{yy})_2 = 310.6 \text{ kg-m}^2$$

$$I_{zz} = (I_{zz})_1 + (I_{zz})_2 = 300.5 \text{ kg-m}^2$$

$$I_{xy} = I_{yz} = 0$$

$$I_{xz} = (I_{xz})_1 + (I_{xz})_2 = -50.4 \text{ kg-m}^2$$

Step 6   The given angular speeds ω_1 and ω_2 describe the rotation of the body about axes that are fixed to the body. Then, in view of the sense of the rotations shown in the given diagram, it is generally true that

$$\bar{\omega} = \omega_2(-\bar{\imath}) + \omega_1\bar{\jmath} \qquad \text{always}$$

$$\bar{\alpha} = -\dot{\omega}_2\bar{\imath} - \omega_2\dot{\bar{\imath}} + \dot{\omega}_1\bar{\jmath} + \omega_1\dot{\bar{\jmath}}$$

$$= -\dot{\omega}_2\bar{\imath} - \omega_2(\bar{\omega} \times \bar{\imath}) + \dot{\omega}_1\bar{\jmath} + \omega_1(\bar{\omega} \times \bar{\jmath}) \qquad \text{always}$$

Thus, substituting the given rotation rates, we find that

$$\bar{\omega} = -0.5\bar{\imath} + 0.2\bar{\jmath} \text{ rad/s}$$

$$\bar{\alpha} = \bar{0} - 0.5(-0.5\bar{\imath} + 0.2\bar{\jmath}) \times \bar{\imath} + \bar{0}$$

$$+ 0.2(-0.5\bar{\imath} + 0.2\bar{\jmath}) \times \bar{\jmath} = \bar{0}$$

(As an alternative procedure, we could also have found $\bar{\alpha} = \bar{0}$ by using the fact that $\bar{\alpha} \equiv \delta\bar{\omega}/\delta t$ when the xyz frame is attached to the body.)

To describe the acceleration of the center of mass G, we use the rigid body motion formula, relating point G to point C, which is moving with constant velocity so that $\bar{a}_C = \bar{0}$. This gives

$$\bar{a}_G = \bar{0} + \bar{\alpha} \times \bar{r}_{G/C} + \bar{\omega} \times (\bar{\omega} \times \bar{r}_{G/C})$$

Substitution of the location of point G and the values of $\bar{\omega}$ and $\bar{\alpha}$ gives

$$\bar{a}_G = \bar{0} + \bar{0} + (-0.5\bar{\imath} + 0.2\bar{\jmath}) \times [(-0.5\bar{\imath}$$

$$+ 0.2\bar{\jmath}) \times (-0.4\bar{\imath} + 0.3\bar{k})]$$

$$= 0.0160\bar{\imath} + 0.040\bar{\jmath} - 0.0870\bar{k} \text{ m/s}^2$$

Step 7   We now compute $\dot{\bar{H}}_G$. With respect to the nonprincipal axes xyz, equation (5) and the expression for $\bar{\omega}$ found in step 6 give

$$\bar{H}_G = [(103.4)(-0.5) - 0 - 0]\bar{\imath}$$

$$+ [(310.6)(0.2) - 0 - 0]\bar{\jmath}$$

$$+ [0 - (-50.4)(-0.5) - 0]\bar{k}$$

$$= -51.7\bar{\imath} + 62.1\bar{\jmath} - 25.2\bar{k} \text{ kg-m}^2\text{/s}$$

Because $\bar{\alpha} = \bar{0}$, equation (21) gives $\delta\bar{H}_G/\delta t = \bar{0}$. Therefore

$$\dot{\bar{H}}_G = \frac{\delta\bar{H}_G}{\delta t} + \bar{\omega} \times \bar{H}_G$$

$$= (-0.5\bar{\imath} + 0.2\bar{\jmath}) \times (-51.7\bar{\imath} + 62.1\bar{\jmath} - 25.2\bar{k})$$

$$= -5.04\bar{\imath} - 12.60\bar{\jmath} - 20.71\bar{k} \text{ N-m}$$

Using the free body diagram of step 1, we compute $\Sigma \bar{M}_G$. The resultant gravitational force acts through the center of mass, so that

$$\Sigma \bar{M}_G = \bar{r}_{C/G} \times \bar{F} + \bar{M}_C$$

$$= (0.4\bar{i} - 0.3\bar{k}) \times (F_x\,\bar{i} + F_y\,\bar{j} + F_z\,\bar{k}) + \bar{M}_C$$

$$= [0.3F_y + (M_C)_x]\bar{i} + [-0.3F_x - 0.4F_z$$

$$+ (M_C)_y]\bar{j} + [0.4F_y + (M_C)_z]\bar{j}$$

Equating the scalar components of $\Sigma\,\bar{M}_G$ to the scalar components of $\dot{H}_G$ yields the three scalar equations for moment equilibrium

$\bar{i}$ component: $\qquad 0.3F_y + (M_C)_x = -5.04$

$\bar{j}$ component: $\qquad -0.3F_x - 0.4F_z + (M_C)_y = -12.60$

$\bar{k}$ component: $\qquad 0.4F_y + (M_C)_z = -20.71$

Step 8 Using the free body diagram to sum forces, we find that

$$\Sigma\,\bar{F} = F_x\,\bar{i} + F_y\,\bar{j} + (F_z - 686.4 - 1373.8)\bar{k}$$

$$= 210(\bar{a}_G) = 3.360\bar{i} + 8.400\bar{j} - 18.270\bar{k} \text{ newtons}$$

The scalar equations are therefore

$\bar{i}$ component: $\qquad F_x = 3.36$

$\bar{j}$ component: $\qquad F_y = 8.40$

$\bar{k}$ component: $\qquad F_z - 2059 = -18.27$

Step 9 We first solve the equations of step 8 for the force components.

$$F_x = 3.36 \text{ N} \qquad F_y = 8.40 \text{ N} \qquad F_z = 2041 \text{ N} \qquad \triangleleft$$

Now, solving the equations of step 8 for the moment components gives

$$(M_C)_x = -7.6 \text{ N-m} \qquad (M_C)_y = 804.8 \text{ N-m} \qquad (M_C)_z = 24.1 \text{ N-m} \qquad \triangleleft$$

In conclusion, we should note that these computations are similar to those encountered in studying the motion of aircraft and ships when the rotations are described in terms of a set of body-fixed axes.

ILLUSTRATIVE PROBLEM 3

Gear A, which meshes with the fixed horizontal gear B, freely rotates about shaft OA. This shaft is attached to the vertical shaft by a clevis joint. The vertical shaft is initially at rest when it is given a constant angular acceleration $\dot{\Omega}$. Obtain an expression for the normal force exerted between the two gears as a function of time. (In this particular design of the gears, the desired force is perpendicular to shaft OA.) The weight of gear A is 15 lb, and the weight of the shaft is negligible.

Solution

Step 1 For the free body diagram we isolate gear A and shaft OA in order to utilize the special restraints of a clevis joint. A side view of the system is most suitable.

Acting at the clevis joint is an arbitrary (three-dimensional) force $\bar{F}_O$ that prevents point O from moving. The axis of the pin is perpendicular to the plane of the free body diagram, so there is no couple reaction about the z axis. Also because there is no friction between gear A and shaft OA, there is no tendency for this shaft to twist about the x axis. Hence, the only couple reaction is $(\bar{M}_O)_y$.

Step 2 Given that $\dot{\Omega}$ is constant and $\Omega = 0$ initially, we conclude that $\Omega = \dot{\Omega}t$. The problem is to find the value of N as a function of time.

Step 3 Point O is a fixed point in space that is also fixed relative to gear A. We select this point for summing moments, rather than the center of mass A, in order to eliminate the force $\bar{F}_O$ from the moment equations.

Step 4 In order to utilize the tabulated inertia properties most directly, we attach the xyz frame to gear A, which is approximated as a thin disk, and align one axis to coincide with shaft OA. One of the fixed XYZ axes is chosen to coincide with the fixed vertical axis. The resulting coordinate systems are shown in the free body diagram.

Step 5 For the determination of the inertia properties, we note that the disk is symmetric with respect to both the xy and xz planes, so that the xyz axes are principal axes. To determine the moments of inertia, we use the tabulated properties and the parallel axis theorems. The coordinates of point O with respect to a reference $\hat{x}\hat{y}\hat{z}$ frame (not shown) whose origin is the center of mass A and whose axes are parallel to the xyz system are $(-16 \text{ in.}, 0, 0)$. Therefore, using inches for the length unit,

$$I_{xx} = I_{\hat{x}\hat{x}} = \frac{1}{2} mR^2 = \frac{1}{2}\left(\frac{15}{386.4}\right)8^2 = 1.242 \text{ in.-lb-s}^2$$

$$I_{yy} = I_{zz} = I_{\hat{y}\hat{y}} + m(16)^2 = m\left(\frac{1}{4} R^2 + 16^2\right)$$

$$= 10.559 \text{ in.-lb-s}^2$$

$$I_{xy} = I_{yz} = I_{xz} = 0$$

Step 6 The study of the kinematics of the system is begun by drawing a sketch that shows the angular speed ω_1 of gear A about shaft OA. Note that the contact point C has no velocity (it is an instant center) because gear B is fixed. Also note that point O is fixed.

The general expressions for the angular velocity and angular acceleration are

$$\bar{\omega} = -\Omega\bar{J} + \omega_1\bar{\imath} \qquad \text{always}$$

$$\bar{\alpha} = -\dot{\omega}\bar{J} + \dot{\omega}_1\bar{\imath} + \omega_1(\bar{\omega} \times \bar{\imath}) \qquad \text{always}$$

At the instant of the sketch we have

$$-\bar{J} = \cos\theta\,\bar{\imath} - \sin\theta\,\bar{\jmath}$$

Thus, the instantaneous values of $\bar{\omega}$ and $\bar{\alpha}$ are

$$\bar{\omega} = (\omega_1 + \Omega\cos\theta)\bar{\imath} - \Omega\sin\theta\,\bar{\jmath} \qquad \text{instantaneously}$$

$$\bar{\alpha} = (\dot{\omega}_1 + \dot{\Omega}\cos\theta)\bar{\imath} - \dot{\Omega}\sin\theta\,\bar{\jmath} + \omega_1\Omega\sin\theta\,\bar{k} \qquad \text{instantaneously}$$

We shall now determine ω_1 in terms of Ω by satisfying the constraint condition that $\bar{v}_C = \bar{0}$. Point O and point C on gear A are both fixed relative to that gear, so that $(\bar{v}_{C/O})_{\text{rel}} = \bar{0}$, and

$$\bar{v}_C = \bar{v}_O + \bar{\omega} \times \bar{r}_{C/O}, \quad \text{i.e.,} \quad \bar{0} = \bar{\omega} \times \bar{r}_{C/O}$$

This means that the instantaneous axis of rotation of gear A is parallel to line OC. From the sketch we have

$$\bar{r}_{C/O} = 16\bar{\imath} - 8\bar{\jmath}$$

so that using the derived expression for $\bar{\omega}$ gives

$$\bar{\omega} \times \bar{r}_{C/O} = [-8(\omega_1 + \Omega\cos\theta) + 16\,\Omega\sin\theta]\bar{k} = \bar{0}$$

Solving for ω_1, we find that

$$\omega_1 = \Omega(2\sin\theta - \cos\theta)$$

Now, because the sketch describes the system for a general position, this relationship can be differentiated. Thus

$$\dot{\omega}_1 = \dot{\Omega}(2\sin\theta - \cos\theta)$$

Substituting these results into the expressions for $\bar{\omega}$ and $\bar{\alpha}$ gives

$$\bar{\omega} = \Omega\sin\theta(2\bar{\imath} - \bar{\jmath})$$

$$\bar{\alpha} = \dot{\Omega}\sin\theta(2\bar{\imath} - \bar{\jmath})$$

$$+ \Omega^2\sin\theta(2\sin\theta - \cos\theta)\bar{k}$$

The determination of $\bar{a}_G$ is omitted because the equations of motion obtained from summing forces only define the force $\bar{F}_O$. We shall therefore also omit step 8.

Step 7 Substitute the inertia properties from step 5 and the expressions for $\bar{\omega}$ and $\bar{\alpha}$ from step 6 into the equations of kinetics. For principal axes these equations are

$$\bar{H}_O = I_{xx}\omega_x\,\bar{\imath} + I_{yy}\omega_y\,\bar{\jmath} + I_{zz}\omega_z\,\bar{k}$$

$$= (1.242)(2\Omega \sin \theta)\bar{i} + (10.559)(-\Omega \sin \theta)\bar{j}$$

$$= 2.484\Omega \sin \theta \, \bar{i} - 10.559\Omega \sin \theta \, \bar{j} \text{ in.-lb-s}$$

$$\frac{\delta \bar{H}_O}{\delta t} = I_{xx}\alpha_x\bar{i} + I_{yy}\alpha_y\bar{j} + I_{zz}\alpha_z\bar{k}$$

$$= (1.242)(2\dot{\Omega} \sin \theta)\bar{i} + (10.559)(-\dot{\Omega} \sin \theta)\bar{j}$$

$$+ (10.559)[\Omega^2 \sin \theta(2 \sin \theta - \cos \theta)]\bar{k}$$

$$= \sin \theta[2.484\dot{\Omega}\bar{i} - 10.559\dot{\Omega}\bar{j}$$

$$+ 10.559\Omega^2(2 \sin \theta - \cos \theta)]\bar{k} \text{ in.-lb}$$

$$\dot{H}_O = \frac{\delta \bar{H}_O}{\delta t} + \bar{\omega} \times \bar{H}_O$$

$$= \sin \theta[2.484\dot{\Omega}\bar{i} - 10.559\dot{\Omega}\bar{j}$$

$$+ \Omega^2(2.484 \sin \theta - 10.559 \cos \theta)\bar{k}]$$

We now use the free body diagram to evaluate the sum of the moments about the origin O.

$$\Sigma \bar{M}_O = (M_O)_y\bar{j} + \bar{r}_{A/O} \times 15(\cos \theta \, \bar{i} - \sin \theta \, \bar{j})$$

$$+ \bar{r}_{C/O} \times (N\bar{j} + f\bar{k})$$

$$= -8f\bar{i}$$

$$+ [(M_O)_y - 16f]\bar{j}$$

$$+ (-240 \sin \theta + 16N)\bar{k}$$

Step 8   Omit.

Step 9   The value of N is found from the $\bar{k}$ component of the moment equation. Thus

$$(\Sigma \bar{M}_O) \cdot \bar{k} = (\dot{H}_O) \cdot \bar{k}$$

$$-240 \sin \theta + 16N = \Omega^2 \sin \theta(2.484 \sin \theta - 10.559 \cos \theta)$$

We now solve for N and substitute $\Omega = \dot{\Omega}t$ from step 2. Hence

$$N = 15 \sin \theta + \tfrac{1}{16}(\dot{\Omega}t)^2 \sin \theta(2.484 \sin \theta - 10.559 \cos \theta) \text{ lb} \qquad \triangleleft$$

The first term on the right side of this result represents the static gravitational effect, whereas the second term represents two opposing inertial effects. If gear A did not mesh with the fixed gear, then the angular speed ω_1 would be zero and the centripetal acceleration of point A would have the effect of decreasing the value of N. However, due to the rotation ω_1, a gyroscopic moment is created that has the effect of forcing the gear downward. If $\theta = \tan^{-1}(10.559/2.484) = 76.76°$, then the two dynamic effects counterbalance each other and N reduces to its static value $N = 15 \sin \theta$.

ILLUSTRATIVE PROBLEM 4

The body shown is in free flight with negligible air resistance. Its properties are arbitrary, with the exception that it is rotationally symmetric about the z axis. From the fact that the gravity force acts at the center of mass, it follows that $\Sigma \bar{M}_G = \dot{\bar{H}}_G = \bar{0}$, so the angular momentum $\bar{H}_G$ is constant. Letting Z be a fixed axis of arbitrary orientation, it is desired to investigate if the motion of the body consists of a constant spinning rotation ω_1 about the z axis, as the z axis itself rotates at the constant angle θ and constant rate ω_2 about the fixed Z axis. Verify that this assumed motion is admissible by showing that, for the foregoing angular motion, the equations of motion require that $\bar{H}_G$ have a constant magnitude, and that it be parallel to the fixed Z axis. Further, determine ω_2 in terms of ω_1 and θ, and derive an expression for the angle β between the angular velocity $\bar{\omega}$ and the z axis in terms of θ.

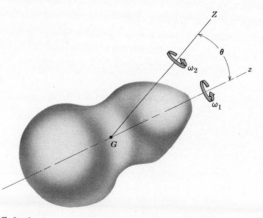

Solution

Steps 1–4   From the given information we know that the only force is gravity. The free body diagram shown here also shows the given information about the rotations. The body is in general motion, so the origin of the xyz axes is located at the center of mass. For convenience, the x axis is chosen such that the Z axis lies in the xz plane.

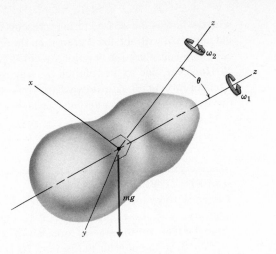

Step 5   For rotational symmetry, $I_{zz} = I_1$, $I_{xx} = I_{yy} = I_2$, where I_1, I_2, and the mass m are arbitrary constants. Also, xyz are principal axes.

Step 6   In the assumed motion, ω_1, ω_2, and θ are constant, so the general expressions for $\bar{\omega}$ and $\bar{\alpha}$ are

$$\bar{\omega} = \omega_2 \bar{K} + \omega_1 \bar{k} \qquad \text{always}$$

$$\bar{\alpha} = \omega_1 (\bar{\omega} \times \bar{k}) \qquad \text{always}$$

For the instant depicted in the sketch in step 1,

$$\bar{K} = \sin \theta \, \bar{\imath} + \cos \theta \, \bar{k}$$

so the instantaneous expressions are

$$\bar{\omega} = \omega_2 \sin \theta \, \bar{\imath} + (\omega_1 + \omega_2 \cos \theta)\bar{k} \qquad \text{instantaneously}$$

$$\bar{\alpha} = \omega_1 [\omega_2 \sin \theta \, \bar{\imath} + (\omega_1 + \omega_2 \cos \theta)\bar{k}] \times \bar{k}$$

$$= -\omega_1 \omega_2 \sin \theta \, \bar{\jmath} \qquad \text{instantaneously}$$

We omit considering $\bar{a}_G$, for we are not interested in it.

Step 7   Because xyz are principal axes, we may use Euler's equation (23). Substituting the inertia properties from step 5 and the values of $\bar{\omega}$ and $\bar{\alpha}$ found in step 6, we find that

$$\Sigma \bar{M}_A \equiv \bar{0} = [I_2(-\omega_1 \omega_2 \sin \theta) - (I_1 - I_2)(\omega_2 \sin \theta)$$

$$(\omega_1 + \omega_2 \cos \theta)] \, \bar{\jmath}$$

$$= \omega_2 \sin \theta [-I_1(\omega_1 + \omega_2 \cos \theta) + I_2 \omega_2 \cos \theta] \, \bar{\jmath}$$

Step 8 The moment equation in step 7 is satisfied when either $\omega_2 = 0$, or $\sin \theta = 0$, or

$$\omega_1 + \omega_2 \cos \theta = \frac{I_2}{I_1} \omega_2 \cos \theta$$

The first two possibilities mean that the only rotation is about the z axis. This is the axis of symmetry, so that $\bar{\omega}$ and $\bar{H}_G$ are then always in the z direction, making $\bar{H}_G$ a constant.

The more interesting possibility is the third one. Substituting the relationship between ω_1, ω_2, and θ for this case into the expression for $\bar{\omega}$ in step 7 gives

$$\bar{\omega} = \omega_2 \sin \theta \, \bar{\imath} + \frac{I_2}{I_1} \omega_2 \cos \theta \, \bar{k}$$

Then substituting these components of $\bar{\omega}$ and the moments of inertia into equation (5) for $\bar{H}_G$ gives

$$\bar{H}_G = I_2(\omega_2 \sin \theta)\bar{\imath} + I_1\left(\frac{I_2}{I_1} \omega_2 \cos \theta\right)\bar{k}$$

$$= I_2\omega_2(\sin \theta \, \bar{\imath} + \cos \theta \, \bar{k})$$

$$= I_2\omega_2\bar{K}$$

Thus $\bar{H}_G$ has the constant magnitude $I_2\omega_2$ in the direction of the Z axis, as required.

It is a simple matter to solve the foregoing relationship between ω_1 and ω_2 to find

$$\omega_2 = \frac{I_1\omega_1}{(I_2 - I_1)\cos \theta}$$

To determine the required angle β, we draw a sketch of the components of $\bar{\omega}$. This sketch shows that

$$\tan \beta = \frac{\omega_2 \sin \theta}{(I_2/I_1)\omega_2 \cos \theta} = \frac{I_1}{I_2} \tan \theta$$

From these results we see that the axis of symmetry, the fixed angular momentum $\bar{H}_G$, and the angular velocity $\bar{\omega}$ are coplanar, with constant angles between them. This type of motion is called *steady free precession;* ω_2 is the rate of precession. There are two kinds of free precession: regular ($\beta < \theta$ and $\omega_2/\omega_1 > 0$) and retrograde ($\beta > \theta$ and $\omega_2/\omega_1 < 0$, depending on whether I_1/I_2 is less than one or not. These topics are treated in depth in more advanced texts on dynamics.

HOMEWORK PROBLEMS

VII.54–VII.57 Each of the systems shown is rotating at $\omega = 200$ rev/min when a constant torque of 50 N-m is applied to slow the system. Determine the initial dynamic bearing reactions and the deceleration rate.

Prob. VII.54

Prob. VII.55

Prob. VII.56

Prob. VII.57

Prob. VII.58

Prob. VII.59

VII.58 A 20-lb force is applied normal to the plane of the 40-lb solid door to shut it. The design of the hinges is such that they only exert reaction forces to support the door. If the door is initially at rest when $\theta = 90°$, determine (a) the angular speed of the door as it closes ($\theta = 0$), (b) the hinge forces as a function of the angle θ as the door swings shut.

VII.59 Two identical 5-kg disks are welded to the rigid bent bar CD. This bar has a uniform cross section and a mass of 4 kg. The entire assembly is rolling without slipping at a constant speed v parallel to the horizontal surface. Determine the friction and normal forces exerted between the ground and each disk in terms of the value of v when bar CD is in the horizontal position shown.

Prob. VII.61

VII.60 Solve Problem VII.59 when the disks have rotated 90° from the position shown.

VII.61 The pad of a portable rotary sander rotates at 1800 rev/min. The pad may be approximated as a 1-lb thin disk of 8 in. diameter. When the plane of the pad is in the horizontal position shown, it is desired to rotate the sander (a) about the fixed X axis at 1 rad/s, (b) about the fixed Y axis at 1 rad/s. For each case of rotation, determine the couple that the person holding the sander must apply to produce the motion.

Prob. VII.62

VII.62 The cutting blade of a rotary lawn mower is rotating at 1200 rev/min, clockwise as viewed from above. The blade is 600 mm long, has a mass of 1 kg, and is dynamically balanced as it rotates about its center. The lawn mower is in the horizontal position when it is rotated backward about its rear wheels at a constant rate of 1 rad/s. Determine the gyroscopic moment that develops in terms of the angle θ.

VII.63 Two bodies in motion are interconnected but can move independently. The force system acting on each body is equivalent to the force-couple system $m_i \bar{a}_{G_i}$ and $\dot{H}_{G_i}$ acting at the center of mass G_i of each body. Derive expressions for the force-couple system acting at an arbitrary point P that is equivalent to the total system of external forces producing the motion of this system.

Prob. VII.63

VII.64 An automobile is executing a left turn on a road of radius of curvature $\rho = 60$ m while traveling at 72 km/hr. The 20-kg flywheel is being driven by the engine at 3000 rev/min clockwise as viewed from the front of the vehicle. The radius of gyration of the flywheel about its axis of symmetry is known to be 250 mm, but the other radius of gyration is unknown. The wheelbase of the vehicle is 3 m. Determine (a) the gyroscopic moment $\dot{H}_G$ for the flywheel, and (b) the amount that the normal force exerted by the ground on each of the front tires increases or decreases to provide this gyroscopic moment. (c) Explain why only one radius of gyration need be known to solve this problem.

Prob. VII.64

VII.65 The tires of the automobile in Problem VII.64 have a mass of 10 kg, a diameter of 700 mm, and a radius of gyration about their axle of 250 mm. (a) Determine $\dot{H}_G$ for each tire. (b) Determine the change in the normal reaction on each wheel necessary to produce the gyroscopic moment in part (a). The wheels on each axle are 2 m apart.

VII.66 A jet airplane at the bottom of a 4-mile-radius vertical circle is traveling at a constant speed of 1200 miles/hr. The 200-lb rotor of its single fanjet engine has a radius of gyration of 8 in. and rotates at 15,000 rev/min counterclockwise as viewed from the front of the airplane. The rotor is mounted by bearings in front and back. (a) Determine $\dot{H}_G$ for the rotor. (b) How are the aerodynamic control surfaces adjusted to provide this gyroscopic moment? (c) If the aerodynamic control surfaces were not adjusted, in which direction would the nose of the aircraft tend to shift?

$\rho = 4$ miles

$v = 1200$ miles/hr   15,000 rev/min—   —Engine

Prob. VII.66

VII.67 The blades of an oscillating electric fan rotate with the motor shaft at the constant angular speed ω_1, as the motor housing rotates about the vertical axis at angular speed ω_2. The angle θ between the motor shaft and the vertical is constant, but it may be adjusted. (a) Obtain an expression for $\dot{H}_G$ of the fan blades in terms of ω_1, ω_2, θ, the mass m of the blade assembly, and the principal radii of gyration. (b) By examining the $\bar{H}_G$ and $\bar{\omega}$ vectors, explain why only the radius of gyration of the fan blades about the motor shaft need be known in the case where $\theta = 90°$.

θ

ω_1

ω_2

Prob. VII.67

VII.68 A radar antenna rotates about the vertical axis at the constant rate of 1 rev/min as the angle of elevation at which the antenna is aimed varies according to $\theta = (\pi/6)(1 + \cos \pi t)$, where θ is in radians and t is in seconds. This antenna has a mass of 1000 kg, and its radii of gyration with respect to the principal axes shown are $k_x = 600$ mm, $k_y = k_z = 800$ mm. Determine the force-couple reaction at support A for this motion.

y 1 rev/min

θ G x

A

1 m

Prob. VII.68

VII.69 The boom AB of a crane may be rotated about the vertical axis at

ω_1 as the angle θ between the boom and the vertical axis is varied. The truck is stationary. The boom may be regarded as a slender rod weighing 5 tons that is supported only at point A. In the position shown, $\theta = 60°$, $\dot{\theta} = 0.5$ rad/s, and $\omega_1 = 0.3$ rad/s. If these rotation rates are constant values, determine the force-couple reaction at point A. Cable BC is slack because no load is being picked up by the hook.

Prob. VII.69

VII.70 Solve Problem VII.69 if the given value of ω_1 is increasing at 0.6 rad/s² and the given value of $\dot{\theta}$ is constant.

VII.71 The 30-kg bar AB is attached at both ends to collars that slide over the smooth fixed horizontal and vertical guides. The system is at rest in the position shown, and it is desired to give collar A an acceleration of 2 m/s² to the left. What force $\bar{F}$ should be applied to collar A parallel to the lower guide in order to obtain this acceleration? The mass of the collars is negligible. *Hint:* Use the body-fixed xyz reference frame shown. Refer to Section C.4 for the rotation transformation of inertia properties.

Prob. VII.71

VII.72 Solve Problem VII.71 if collar A is moving to the right at 4 m/s in the position shown.

VII.73 The homogeneous cone is rolling without slipping on a horizontal surface such that it makes 1 rev about the vertical axis every 2 seconds. The cone weighs 2 lb. Determine the force-couple system acting at the apex A that represents the reaction of the ground on the cone.

Prob. VII.73

VII.74 A coin of mass m and radius r rolls without slipping on a horizontal surface. The normal to the plane of the coin is inclined at a constant angle θ and the center of the coin follows a path of radius L. Determine the relationship between the constant speed v, the angle θ, and the radius r.

Prob. VII.74

Prob. VII.75

VII.75 A thin disk of radius r and mass m rotates freely at rate ω_1 about the lightweight rod AB. The rod is pinned to a vertical shaft that is rotating at the constant rate ω_2. The motion occurs in the region between two fixed horizontal surfaces, as shown in the cross-sectional view. The distance between these two surfaces is only slightly larger than the diameter of the disk, so rod AB may be considered always to be horizontal. Assuming that the disk is rolling without slipping over the lower surface, determine the value of the normal force exerted on the disk by the lower horizontal surface. Is there any value of ω_2 for which this force will be zero?

VII.76 The disk in Problem VII.75 is rotating without slipping over the upper surface, thereby reversing the direction of the rotation ω_1 of the disk relative to rod AB. Determine the normal force exerted on the disk by the upper horizontal surface. Is there a minimum value of ω_2 for this rotation?

Prob. VII.77

VII.77 The ore grinder consists of a 1000-lb disk mounted on arm AB, as shown. This arm is pinned to the vertical shaft, which is driven at a constant angular speed ω_1. The disk rotates freely on arm AB as it rolls without slipping over the interior surface of the vertical cylindrical container. Determine the minimum value for ω_1 required for contact to be maintained between the disk and the interior wall. Neglect the mass of arm AB.

VII.78 Solve Problem VII.77 if the disk is replaced by a 1000-lb sphere having a radius of 2 ft. The sphere is centered at point B, and the other dimensions given in the diagram are unchanged.

VII.79 A toy gyroscope consists of a spinning disk of radius r that spins about the lightweight arm AB at the constant angular speed ω_1. This arm is supported at point B and rotates about the vertical axis at angular speed ω_2. (a) If θ is constant, derive an equation for θ in terms of ω_1, ω_2, and the

Prob. VII.79

physical properties of the system. (This type of motion is a steady precession about the vertical axis.) (b) Determine the relationship between ω_1 and ω_2 for $\theta = 90°$.

VII.80 For the toy gyroscope in Problem VII.79, $l = 200$ mm, $r = 100$ mm, $\omega_2 = 2$ rad/s, and $\theta = 60°$. What is the value of ω_1?

VII.81 and VII.82 The body shown is pinned at point A to the vertical shaft which is rotating at the constant rate ω_1. (a) Considering the angle θ as a variable, derive a single differential equation of motion for this parameter. (b) Using the result of part (a), determine the relationship between ω_1 and the constant nonzero value that angle θ can be. (c) Letting θ be a very small angle, so that $\sin\theta \approx \theta$ and $\cos\theta \approx 1$, examine the result of part (a) to determine the maximum value of ω_1 for which θ will return to zero if slightly disturbed from the zero value. *Hint:* The equation in part (c) will resemble that for a pendulum.

Prob. VII.81

Prob. VII.82

VII.83 The assembly shown is rotating about the vertical axis at a constant rate ω_1. The slender bar CD of mass m is supported by a pin at point B. Derive an expression for the constant angle θ made by the bar.

Prob. VII.83

VII.84 In Problem VII.83, the slender bar CD is held at $\theta = 90°$ and then released as the assembly rotates about the vertical axis at the constant rate ω_1. Derive the differential equation of motion for the angle θ.

Prob. VII.85

VII.85 A 1200-kg light airplane is at the top of a vertical circle of 500-meter radius, flying at a constant speed of 180 km/hr. At this instant its wings are horizontal and it is performing an aerobatic stunt by rolling about the longitudinal x axis at the constant rate of 0.2 rad/s, as shown. The radii of gyration about the principal xyz coordinate system shown are $k_x = 2$ m, $k_y = 3$ m, $k_z = 3.5$ m. Determine the force-couple system acting at the center of mass G that is equivalent to the aerodynamic forces on the aircraft. The inertia of the propellers is negligible.

VII.86 In another maneuver, the aircraft in Problem VII.85 is rolling about its x axis at $\omega_x = 0.1$ rad/s, pitching about its y axis at $\omega_y = -0.2$ rad/s, and yawing about its z axis at $\omega_z = 0.05$ rad/s, where all rates are constant. Determine the aerodynamic moment about the center of mass for this maneuver. The inertia of the propeller is negligible.

VII.87 Due to the action of ocean waves, a ship is pitching and rolling as its center of mass moves with a constant velocity. At a certain instant these rotation rates about the axes of the body-fixed xyz coordinate system are maximum at $\omega_x = 0.02$ rad/s and $\omega_y = 0.10$ rad/s. This coordinate system forms a set of principal axes with origin at the center of mass G. The mass of the ship is 25,000 metric tons, and the radii of gyration are $k_x = 80$ m, $k_y = 10$ m, and $k_z = 75$ m. Determine the force-couple system at point G that is equivalent to the set of forces causing the motion.

Prob. VII.87

VII.88 In a different motion, the angles of rotation of the ship in Problem VII.87 about the pitch and roll axes are given by $\theta_x = 0.05 \sin(0.5t)$ and $\theta_y = 0.10 \cos(t)$, respectively, where the angles are measured in radians and t is in seconds. The center of mass has a constant velocity. Determine, as a function of time, the force-couple system at point G that is equivalent to the set of forces causing this motion.

ω = 150 rev/min

15°

10°

Prob. VII.89

VII.89 A football is thrown poorly. When released, the axis of symmetry is 10° above the horizontal and the angular velocity has a magnitude of 150 rev/min at an angle of 15° above the axis of symmetry, as shown. The radius of gyration of the football about its axis of symmetry is two-thirds that about an axis perpendicular to its axis of symmetry passing through the center of mass. Determine (a) the angle between the axis of precession and the horizontal direction, (b) the rate of precession, and (c) the rate of spin. Is this regular or retrograde precession? *Hint:* Refer to Illustrative Problem 4.

VII.90 A coin is thrown into the air. At a certain instant the coin is horizontal and the angular velocity is 450 rev/min in the direction shown. Determine (a) the angle between the precession axis and the centerline of the coin, (b) the rate of precession, and (c) the rate of spin. Is this regular or retrograde precession? *Hint:* Refer to Illustrative Problem 4.

60°

450 rev/min

Prob. VII.90

Note: Before attempting Problems VII.91–94, the kinematical approach presented in Illustrative Problem 5 of Module VI should be fully understood.

—100 mm

ω₂

θ

C

F

B

ω₁

150 mm

150 mm

A

Prob. VII.91

VII.91 The 2-kg disk rotates at a constant rate $\omega_1 = 2100$ rev/min about shaft AB, which is connected to the vertical shaft by a pin at point A. The vertical shaft may rotate freely about its bearings, and the mass of both shafts is negligible. Initially, the angle θ is held at 60°, and the rotation rate ω_2 of the vertical shaft is zero. A horizontal force $F = 50$ N is applied to the midpoint C of shaft AB, and the system is released. This force is always perpendicular to the shaft, as shown. Determine the values of $\ddot{\theta}$ and $\dot{\omega}_2$ at the instant after the system is released.

VII.92 Solve Problem VII.91 if initially the angle θ is 60°, but the vertical axis is rotating at $\omega_2 = 1500$ rev/min when the 50-N force F is applied and the system is released.

VII.93 Solve Problem VII.91 if initially the angle $\theta = 60°$, $\dot{\theta} = 10$ rad/s, and ω_2 is zero when the 50-N force is applied and the system is released.

VII.94 Considering θ and ω_2 as unknown functions of time in Problem VII.91, determine the two differential equations of motion for these variables in terms of the rotation rate ω_1 and the properties of the system.

F.   ENERGY AND MOMENTUM PRINCIPLES

1   Kinetic Energy of a Rigid Body in Three-Dimensional Motion

The work-energy principle for the motion of particles was derived in Module III, starting from Newton's second law. Then, in Module V we extended the applicability of this principle to the planar motion of rigid bodies by recognizing that the kinetic energy function for a particle, $T = \frac{1}{2}mv_P{}^2$, had to be rederived to account for the rotational characteristics of the body. Doing that, we obtained $T = \frac{1}{2}mv_G{}^2 + \frac{1}{2}I_{zz}\omega^2$. In this module we shall extend the field of application of the work-energy principle to the general three-dimensional motion of a rigid body by deriving the kinetic energy function for a three-axis rotation. Nevertheless, the basic form of the work-energy equation remains unchanged, being

$$\boxed{T_2 + V_2 = T_1 + V_1 + U^{(\mathrm{nc})}_{1\to2}} \tag{24}$$

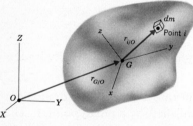

Figure 10

To derive the new expression for the kinetic energy, we have depicted a general body in Figure 10, where xyz is a body-fixed coordinate system, point i is the location of an infinitesimal element of mass, and point G is the center of mass of the rigid body. Summing the kinetic energy of each element of mass dm contained in the volume $\mathcal{V}$, we have

$$T = \tfrac{1}{2} \int_{\mathcal{V}} v_i{}^2 \, dm$$

We know that for a rigid body

$$\bar{v}_i = \bar{v}_G + \bar{\omega} \times \bar{r}_{i/G}$$

where $\bar{\omega}$ is the angular velocity of the rigid body. The kinetic energy expression then becomes

$$T = \tfrac{1}{2} \int_{\mathscr{D}} \bar{v}_i \cdot \bar{v}_i \, dm = \tfrac{1}{2} \int_{\mathscr{D}} (\bar{v}_G + \bar{\omega} \times \bar{r}_{i/G}) \cdot (\bar{v}_G + \bar{\omega} \times \bar{r}_{i/G}) \, dm$$

$$= \tfrac{1}{2} (\bar{v}_G \cdot \bar{v}_G) \int_{\mathscr{D}} dm + \bar{v}_G \cdot \left[\bar{\omega} \times \int_{\mathscr{D}} \bar{r}_{i/G} \, dm \right]$$

$$+ \tfrac{1}{2} \int_{\mathscr{D}} (\bar{\omega} \times \bar{r}_{i/G}) \cdot (\bar{\omega} \times \bar{r}_{i/G}) \, dm$$

In the foregoing we were able to remove $\bar{v}_G$ and $\bar{\omega}$ from the first and second integrals because these quantities are properties of the overall motion of the body, and thus depend only on time. The first integral is readily recognizable as the mass of the body, and the second integral is zero because it is the first moment of mass about the center of mass. Thus, the kinetic energy expression becomes

$$T = \tfrac{1}{2} m v_G{}^2 + \tfrac{1}{2} \int_{\mathscr{D}} (\bar{\omega} \times \bar{r}_{i/G}) \cdot (\bar{\omega} \times \bar{r}_{i/G}) \, dm$$

To proceed further, we express $\bar{\omega}$ and $\bar{r}_{i/G}$ in component form with respect to the body-fixed axes. Hence

$$\bar{\omega} = \omega_x \bar{\imath} + \omega_y \bar{\jmath} + \omega_z \bar{k}$$

$$\bar{r}_{i/G} = x \bar{\imath} + y \bar{\jmath} + z \bar{k}$$

$$\bar{\omega} \times \bar{r}_{i/G} = (\omega_y z - \omega_z y) \bar{\imath} + (\omega_z x - \omega_x z) \bar{\jmath} + (\omega_x y - \omega_y x) \bar{k}$$

$$(\bar{\omega} \times \bar{r}_{i/G}) \cdot (\bar{\omega} \times \bar{r}_{i/G}) = (y^2 + z^2) \omega_x{}^2 + (x^2 + z^2) \omega_y{}^2 + (x^2 + y^2) \omega_z{}^2$$

$$- 2yz\omega_y\omega_z - 2xz\omega_x\omega_z - 2xy\omega_x\omega_y$$

After substituting the last expression into the equation for kinetic energy, we utilize the fact that the components of $\bar{\omega}$ are functions only of time to write

$$T = \tfrac{1}{2} m v_G{}^2 + \tfrac{1}{2} \omega_x{}^2 \int_{\mathscr{D}} (y^2 + z^2) \, dm + \tfrac{1}{2} \omega_y{}^2 \int_{\mathscr{D}} (z^2 + x^2) \, dm$$

$$+ \tfrac{1}{2} \omega_z{}^2 \int_{\mathscr{D}} (x^2 + y^2) \, dm - \omega_x \omega_y \int_{\mathscr{D}} xy \, dm$$

$$- \omega_y \omega_z \int_{\mathscr{D}} yz \, dm - \omega_x \omega_z \int_{\mathscr{D}} xz \, dm$$

We now recognize the integrals as the inertia properties with respect to the *xyz* axes given in equations (6). Thus

$$T = \tfrac{1}{2}mv_G{}^2 + \tfrac{1}{2}I_{xx}\omega_x{}^2 + \tfrac{1}{2}I_{yy}\omega_y{}^2 + \tfrac{1}{2}I_{zz}\omega_z{}^2$$
$$- I_{xy}\omega_x\omega_y - I_{yz}\omega_y\omega_z - I_{xz}\omega_x\omega_z$$

(25a)

The (alternative) vector form of this equation is

$$T = \tfrac{1}{2}m\bar{v}_G \cdot \bar{v}_G + \tfrac{1}{2}\bar{\omega} \cdot \bar{H}_G$$

(25b)

The vector expression for the kinetic energy of a rigid body emphasizes that the reference system employed for the computation of the inertia properties has its origin at the mass center. Clearly, if the axes are principal axes, the last three terms in equation (25a) become zero.

From equation (25b) it is evident that, as was the case in planar motion, the total kinetic energy is the sum of the translational energy, $mv_G{}^2/2$, associated with the motion of the center of mass and the rotational energy, $\bar{\omega} \cdot \bar{H}_G/2$, associated with the rotation about the center of mass. In the case where the body is in planar motion parallel to the xy plane, the rotational kinetic energy reduces to $I_{zz}\omega^2/2$.

At this juncture let us retrace the steps in deriving equations (25). We began by selecting a body-fixed reference frame whose origin was the center of mass G. As a result, one of the terms in the initial expression for the kinetic energy was zero. Had we located the origin of the reference system at a general point, the derived expression for kinetic energy would be less convenient for computations. However, a special case arises when some point A, having a fixed position with respect to the body, has no velocity. In this case we may locate the origin of the body fixed reference frame at point A, with the result that

$$T = \tfrac{1}{2}I_{xx}\omega_x{}^2 + \tfrac{1}{2}I_{yy}\omega_y{}^2 + \tfrac{1}{2}I_{zz}\omega_z{}^2$$
$$- I_{xy}\omega_x\omega_y - I_{xz}\omega_x\omega_z - I_{yz}\omega_y\omega_z$$
$$= \tfrac{1}{2}\bar{\omega} \cdot \bar{H}_A$$

(26)

The vector form of equation (26) emphasizes that in this case we are considering the kinetic energy function to be the rotational kinetic energy about point A which is at rest, and equally important, that the origin of the xyz reference frame is located at point A.

EXAMPLE 9

A 100-kg model of a space capsule is being tested by mounting it in a ball-and-socket joint at point A. At the instant shown, its angular velocity is $\bar{\omega} = 2\bar{i} - 3\bar{j} + 4\bar{k}$ rad/s, where $\bar{i}, \bar{j}, \bar{k}$ are the unit vectors of the set of principal axes fixed to the capsule with origin at point A. It is known that, with respect to these axes, $I_{xx} = 30$ kg-m², $I_{yy} = 50$ kg-m², and $I_{zz} = 60$ kg-m². (a) Determine the kinetic energy of the capsule. (b) In a second experiment, point A is towed at a velocity $\bar{v}_A = -5\bar{i} + 2\bar{j} - \bar{k}$ m/s. Determine the kinetic energy of the capsule.

Solution

Part a

In this case the ball joint at point A is fixed, so we use equation (26).

$$T = \tfrac{1}{2}\bar{\omega} \cdot \bar{H}_A$$

The vector $\bar{\omega}$ was given as

$$\bar{\omega} = 2\bar{i} - 3\bar{j} + 4\bar{k} \text{ rad/s}$$

Equation (5) for the angular momentum gives

$$\begin{aligned}
\bar{H}_A &= I_{xx}\omega_x\bar{i} + I_{yy}\omega_y\bar{j} + I_{zz}\omega_z\bar{k} \\
&= 30(2)\bar{i} + 50(-3)\bar{j} + 60(4)\bar{k} \\
&= 60\bar{i} - 150\bar{j} + 240\bar{k} \text{ kg-m}^2\text{/s}
\end{aligned}$$

Therefore

$$T = \tfrac{1}{2}\bar{\omega} \cdot \bar{H}_A = \tfrac{1}{2}(120 + 450 + 960) = 765 \text{ J} \qquad \triangleleft$$

Part b

In this case the model is in general motion, so we use equation (25).

$$T = \tfrac{1}{2}m\bar{v}_G \cdot \bar{v}_G + \tfrac{1}{2}\bar{\omega} \cdot \bar{H}_G$$

To find $\bar{v}_G$ we write

$$\begin{aligned}
\bar{v}_G &= \bar{v}_A + \bar{v}_{G/A} \\
&= \bar{v}_A + \bar{\omega} \times \bar{r}_{G/A}
\end{aligned}$$

From the given diagram we see that

$$\bar{r}_{G/A} = 0.20\bar{i} - 0.10\bar{j} \text{ m}$$

so

$$\begin{aligned}
\bar{v}_G &= (-5\bar{i} + 2\bar{j} - \bar{k}) + (2\bar{i} - 3\bar{j} + 4\bar{k}) \times (0.20\bar{i} - 0.10\bar{j}) \\
&= -4.60\bar{i} + 2.80\bar{j} - 0.60\bar{k} \text{ m/s}
\end{aligned}$$

We denote as $\hat{x}\hat{y}\hat{z}$ the coordinate axes with origin at the center of mass whose axes are parallel to the given xyz reference frame. The coordinates of point A with respect to $\hat{x}\hat{y}\hat{z}$ are $(-0.20, 0.10, 0)$ m, so the parallel axis theorems yield

$$I_{\hat{x}\hat{x}} = I_{xx} - m(\hat{y}_A{}^2 + \hat{z}_A{}^2)$$

$$= 30 - 100[(0.10)^2 + 0] = 29.0 \text{ kg-m}^2$$

$$I_{\hat{y}\hat{y}} = I_{yy} - m(\hat{x}_A{}^2 + \hat{z}_A{}^2)$$

$$= 50 - 100[(-0.20)^2 + 0] = 46.0 \text{ kg-m}^2$$

$$I_{\hat{z}\hat{z}} = I_{zz} - m(\hat{x}_A{}^2 + \hat{y}_A{}^2)$$

$$= 60 - 100[(-0.20)^2 + (0.10)^2] = 55.0 \text{ kg-m}^2$$

$$I_{\hat{x}\hat{y}} = I_{xy} - m\hat{x}_A\hat{y}_A$$

$$= 0 - 100(-0.20)(0.10) = 2.0 \text{ kg-m}^2$$

$$I_{\hat{y}\hat{z}} = I_{\hat{x}\hat{z}} = 0$$

With the inertia properties determined, we can evaluate the angular momentum $\bar{H}_G$ using equation (5). This gives

$$\bar{H}_G = (I_{xx}\omega_x - I_{xy}\omega_y)\bar{i} + (I_{yy}\omega_y - I_{xy}\omega_x)\bar{j} + I_{zz}\omega_z\bar{k}$$

$$= [29(2) - 2(-3)]\bar{i} + [46(-3) - 2\,(2)]\,\bar{j} + 55(4)\bar{k}$$

$$= 64\bar{i} - 142\bar{j} + 220\bar{k}$$

Thus

$$T = \tfrac{1}{2}(100)[(-4.60)^2 + (2.80)^2 + (0.60)^2]$$

$$+ \tfrac{1}{2}[(2\bar{i} - 3\bar{j} + 4\bar{k}) \cdot (64\bar{i} - 142\bar{j} + 220\bar{k})]$$

$$= 2175 \text{ J}$$

The rotational portion of the kinetic energy is $\bar{\omega} \cdot \bar{H}_G/2 = 717$ J. Note that this is less than that found in part (a), even though the angular velocity is the same.

EXAMPLE 10

The 2-kg homogeneous cone is at rest on the 20° incline in the position shown. It is released and rolls without slipping down the incline. Determine the rate of rotation of the axis of the cone when it passes through the lowest position on the incline.

Solution

The solution will be formulated in terms of the work-energy principle because we wish to determine the motion of the cone in the final position for a given set of initial conditions. To evaluate the work term we draw a free body diagram of the cone in a general position.

The friction force $\bar{f}$, parallel to the incline, is the constraint force that prevents slipping. As was shown in planar motion, such a friction force does no work. Also, the contact force $\bar{N}$ is normal to the plane. Therefore, it also does no work, so $U_{1\to 2}^{(nc)} = 0$.

The remaining force is gravity, which is conservative. To compute the corresponding potential energy, let us draw sketches of the cone in its initial and final positions. Because the cone is not slipping, its apex remains fixed.

Choosing the elevation of the apex A as the datum, the initial and final elevations of the center of mass are

$$h_1 = 0.150 \sin 30° = 0.0750 \text{ m}$$

$$-h_2 = 0.150 \sin 10° = 0.02605 \text{ m}$$

Thus, the potential energy of gravity is

$$V_1 = mgh_1 = 2(9.806)(0.075) = 1.4709 \text{ J}$$

$$V_2 = mgh_2 = 2(9.806)(-0.02605) = -0.5108 \text{ J}$$

To evaluate the kinetic energy, we note that in the initial position the cone is at rest, so $T_1 = 0$. To find T_2 we locate the origin of the xyz reference frame at the fixed point A in order to employ equation (26). The x, y, and z axes are chosen to be parallel to the $\hat{x}$, $\hat{y}$, and $\hat{z}$ axes shown in the sketch on the next page because the inertia properties with respect to the latter system are known from Appendix B. The parallel axis theorems then give

$$I_{xx} = I_{\hat{x}\hat{x}} = \tfrac{3}{10}mr^2 = \tfrac{3}{10}(2)(0.2 \tan 10°)^2$$

$$= 7.462(10^{-4}) \text{ kg-m}^2$$

$$I_{yy} = I_{zz} = I_{\hat{y}\hat{y}} + md_y^2$$

$$= \tfrac{3}{80}(2)[4(0.2 \tan 10°)^2 + (0.2)^2]$$

$$+ (2)(0.15)^2 = 4.837(10^{-2}) \text{ kg-m}^2$$

$$I_{xy} = I_{xz} = I_{yz} = 0$$

To describe the rotation of the cone, we select a fixed reference system XYZ whose Y axis is normal to the incline and whose X axis is coincident with the line of contact, as shown in the foregoing sketch. Then the rotation of the cone consists of a superposition of rotations about the Y axis at angular speed ω_1 and about the x axis at angular speed ω_2. Note that ω_1 is the value we seek. From this we find that the general expression for the angular velocity is

$$\bar{\omega} = \omega_1 \bar{J} - \omega_2 \bar{\imath} \qquad \text{always}$$

Using the diagram to obtain the instantaneous expression, we have

$$\bar{\omega} = \omega_1(\sin 10° \; \bar{\imath} + \cos 10° \; \bar{\jmath}) - \omega_2 \bar{\imath}$$

$$= (0.1736\omega_1 - \omega_2)\bar{\imath} + 0.9848\omega_1 \, \bar{\jmath} \qquad \text{instantaneously}$$

To complete the kinematics study, we note that the X axis is the line of contact. Because of the restriction of no slipping, all points of the cone along this line are at rest. Hence, choosing point B at the base of the cone, we can write

$$\bar{v}_B = \bar{v}_A + \bar{\omega} \times \bar{r}_{B/A}$$

$$\bar{0} = \bar{0} + [(0.1736\omega_1 - \omega_2)\bar{\imath} + 0.9848\omega_1 \, \bar{\jmath})$$

$$\times (0.20\bar{\imath} - 0.20 \tan 10° \; \bar{\jmath})$$

$$\bar{0} = 0.20[-0.1763(0.172\omega_1 - \omega_2)$$

$$- 0.9848\omega_1]\bar{k}$$

We want to find ω_1, hence we solve this equation for ω_2, obtaining

$$\omega_2 = 0.1736\omega_1 + \frac{0.9848}{0.1763} \, \omega_1 = 5.7595\omega_1$$

It then follows that the angular velocity is

$$\bar{\omega} = (0.1736\omega_1 - 5.7595\omega_1)\bar{\imath} + 0.9848_1 \, \bar{\jmath}$$

$$= \omega_1(-5.586\bar{\imath} + 0.985\bar{\jmath})$$

We may now determine T_2 by substituting the inertia properties and the angular velocity components into equation (26). This gives

$$T_2 = \tfrac{1}{2}(I_{xx}\omega_x{}^2 + I_{yy}\omega_y{}^2 + I_{zz}\omega_z{}^2)$$

$$= \tfrac{1}{2}(7.462)(10^{-4})(-5.586\omega_1)^2 + \tfrac{1}{2}(4.837)(10^{-2})(0.985\omega_1)^2$$

$$= 3.511(10^{-2})\omega_1{}^2 \text{ J}$$

Finally, the energy principle for a conservative system gives

$$T_2 + V_2 = T_1 + V_1$$

$$3.511(10^{-2})\omega_1{}^2 - 0.5108 = 0 + 1.4709$$

$$\omega_1 = 7.51 \text{ rad/s}$$

HOMEWORK PROBLEMS

VII.95 The homogeneous ellipsoid shown has a mass of 100 kg. Its center of mass is fixed and its instantaneous angular velocity is $\bar{\omega} = 4\bar{i} + 3\bar{j} + 7\bar{k}$ rad/s. Determine the kinetic energy of the ellipsoid.

VII.96 The ellipsoid shown in Problem VII.95 has a mass of 100 kg. Point A at the end of the major axis is a fixed point. The instantaneous angular velocity of the ellipsoid is $\bar{\omega} = 4\bar{i} + 3\bar{j} + 7\bar{k}$ rad/s. Determine the kinetic energy of the ellipsoid (a) using equation (25), (b) using equation (26).

VII.97 The 20-kg gear A is welded to the 4-kg arm AB. Assuming the gear to be a uniform disk having a radius of 200 mm, determine the maximum angular velocity of arm AB if the system is released from rest at $\theta = 0°$.

Prob. VII.95

Prob. VII.97

Prob. VII.99

VII.98 The system in Problem VII.97 is started in motion at $\theta = 90°$. What is the minimum initial angular velocity of arm AB for which the system will reach the highest position?

VII.99 The 3-lb bar screws into the single threaded vertical shaft whose pitch is 0.25 in. The bar, which fits loosely onto the shaft, is given an initial angular speed of 20 rad/s to make it descend the shaft. After 32 rev the bar comes to rest. Determine the frictional couple opposing the motion of the bar.

VII.100 The 5-kg sphere rotates about arm AB at the constant rate $\omega_1 = 300$ rad/s. A 20 N-m torque is applied to the vertical shaft, causing it to rotate at the rate ω_2. The masses of the shafts are negligible, and the system is initially at rest. (a) Determine the value of ω_2 after the system has completed 5 rev about the vertical axis. (b) Compare the result of part (a) to that obtained when the angular speed ω_1 is zero.

Prob. VII.100

VII.101 The 15-kg bar AB is attached at both ends to lightweight collars that slide over the smooth horizontal and vertical guides. A constant 100-N force, parallel to the horizontal guide, is applied to collar A as shown. If the system is released from rest at $x = 0$, determine the speed of collar A when $x = 1.2$ m.

VII.102 The uniform steel block is moving at 16 ft/s along the smooth angle irons when it encounters the 10° bend in one of the angle irons. Determine the angular velocity of the block after it has moved through a horizontal distance of 4 in. beyond the bend. Assume that edge AB, which rides in the straight angle iron, remains horizontal. *Hint:* If edge AB remains horizontal, then the general motion of the block consists of a

Prob. VII.101

Prob. VII.102

rotation about this edge and a translation parallel to the edge. Also, the component of the velocity of corner C lying in the vertical plane must be parallel to the lower leg of the bent angle iron.

2 Impulse-Momentum Principles

In symbolic form the linear and angular impulse-momentum principles are the same as presented in Module III for a system of independently moving particles and in Module V for a rigid body in planar motion. Specifically, the linear impulse-momentum principle is

$$m(\bar{v}_G)_2 - m(\bar{v}_G)_1 = \int_{t_1}^{t_2} \Sigma \bar{F} \, dt \tag{27}$$

and the angular impulse-momentum principle is

$$(\bar{H}_A)_2 - (\bar{H}_A)_1 = \int_{t_1}^{t_2} \Sigma \bar{M}_A \, dt \tag{28}$$

In equation (28) we restrict point A to be one of the allowable points for summing moments (the center of mass or a point in the body that is fixed). Both of the foregoing relations express the fact that the increase in the respective momentum quantities equals the corresponding impulse. Clearly, when we substitute the general expression for $\bar{H}_A$ given by equation (5), the component form of equation (28) is quite lengthy.

Equations (27) and (28) are of particular importance in the case of impulsive forces, that is, very large forces that act over a short interval of time. As was seen when dealing with forces of this type in particle and planar motion problems, the major effect of any impulsive force is described by its impulse and not by its actual variation with time.

EXAMPLE 11

A disk of mass m is struck at point B by a bullet of mass m_b traveling perpendicular to the plane of the disk at a speed v. The disk is free to rotate about the ball joint connection at point A. Determine the instantaneous angular speed and the axis of rotation of the plate immediately after the impact.

Solution

Clearly, this is an impulse-momentum problem. We begin with free body diagrams of the disk and of the bullet, in which we neglect the nonimpulsive weight force in comparison to the impact force $\bar{F}_B$ exerted between the bullet and the disk.

In order to eliminate the impulsive reaction force at point A, we formulate the solution in terms of the angular momentum about point A. Furthermore, to eliminate the interaction force $\bar{F}_B$, we consider the system of particles formed by the disk and the bullet. Thus

$$(\Sigma \, \bar{H}_A)_2 - (\Sigma \, \bar{H}_A)_1 = \int_{t_1}^{t_2} \Sigma \, \bar{M}_A \, dt$$

where, for the principal xyz axes,

$$\Sigma \, \bar{H}_A = (\bar{H}_A)_{\text{disk}} + (\bar{H}_A)_{\text{bullet}}$$
$$= I_{xx}\omega_x \bar{\imath} + I_{yy}\omega_y \bar{\jmath} + I_{zz}\omega_z \bar{k} + \bar{r}_{B/A} \times m_b \bar{v}_{\text{bullet}}$$

Also, from the free body diagrams we note that the only impulsive external force acting on this system is $\bar{F}_A$, so $\Sigma \, \bar{M}_A = \bar{0}$. Hence, the angular momentum of the system about point A is conserved,

$$(\Sigma \, \bar{H}_A)_2 = (\Sigma \, \bar{H}_A)_1$$

To evaluate $\Sigma \, \bar{H}_A$ we need the mass moments of inertia. Letting $\hat{x}\hat{y}\hat{z}$ be the axes parallel to xyz with origin at the center of mass of the disk, the mass moments of inertia are found (with the aid of Appendix B and the parallel axis theorems) to be

$$I_{xx} = I_{\hat{x}\hat{x}} + m(\hat{y}_A{}^2 + \hat{z}_A{}^2) = \tfrac{1}{4}mr^2 + mr^2 = \tfrac{5}{4}mr^2$$
$$I_{yy} = I_{\hat{y}\hat{y}} + m(\hat{x}_A{}^2 + \hat{z}_A{}^2) = \tfrac{1}{2}mr^2 + mr^2 = \tfrac{3}{2}mr^2$$
$$I_{zz} \equiv I_{\hat{z}\hat{z}} = \tfrac{1}{4}mr^2$$

We next consider the kinematics of the motion. Initially, the disk is at rest ($\bar{\omega}_1 = \bar{0}$) and the velocity of the bullet is $v\bar{\jmath}$, so

$$(\Sigma \, \bar{H}_A)_1 = \bar{0} + [r \cos 30° \, \bar{\imath} - r(1 + \sin 30°)\bar{k}] \times m_b v\bar{\jmath}$$
$$= rm_b v[1.50\bar{\imath} + 0.8660\bar{k}]$$

The angular velocity of the disk following the impact is a general vector $\bar{\omega}_2$, but because the bullet is embedded in the disk, we know that

$$(\bar{v}_{\text{bullet}})_2 \equiv \bar{v}_B = \bar{v}_A + \bar{\omega}_2 \times \bar{r}_{B/A}$$
$$= 0 + (\omega_x \bar{\imath} + \omega_y \bar{\jmath} + \omega_z \bar{k}) \times (0.8660r\bar{\imath} - 1.50r\bar{k})$$
$$= r[-1.500\omega_y \bar{\imath} + (0.8660\omega_z + 1.500\omega_x)\bar{\jmath} - 0.8660\omega_y \bar{k}]$$

Substituting the inertia properties and the foregoing expression for

$(\bar{v}_{\text{bullet}})_2$ into the general equation for $\Sigma\,\bar{H}_A$, we find

$$(\Sigma\,\bar{H}_A)_2 = 1.25mr^2\omega_x\bar{i} + 1.50mr^2\omega_y\bar{j} + 0.25mr^2\omega_z\bar{k}$$

$$+\; r(0.8660\bar{i} - 1.50\bar{k}) \times m_b r\,[-1.50\omega_y\bar{i} + (0.8660\omega_z$$

$$+\; 1.50\omega_x)\bar{j} - 0.8660\omega_y\bar{k}]$$

$$=\; r^2(1.250m\omega_x + 1.2990m_b\omega_z + 2.250m_b\omega_x)\bar{i}$$

$$+\; r^2(1.50m\omega_y + 3.00m_b\omega_y)\bar{j}$$

$$+\; r^2(0.250m\omega_z + 0.750m_b\omega_z + 1.299m_b\omega_x)\bar{k}$$

Equating the corresponding components of $(\bar{H}_A)_2$ and $(\bar{H}_A)_1$ then yields

$\bar{i}$ component: $\quad r^2(1.250m\omega_x + 1.2990m_b\omega_z + 2.250m_b\omega_x)$

$$= 1.50rm_b v$$

$\bar{j}$ component: $\quad r^2(1.50m + 3.00m_b)\omega_y = 0$

$\bar{k}$ component: $\quad r^2(0.250m\omega_z + 0.750m_b\omega_z + 1.299m_b\omega_x)$

$$= 0.8660rm_b v$$

From the second equation, we see that $\omega_y = 0$. The simultaneous solution of the first and third equations is

$$\omega_x = \left(\frac{1.200m_b/m}{1 + 4.800m_b/m}\right)\frac{v}{r}$$

$$\omega_z = \left(\frac{3.464m_b/m}{1 + 4.800m_b/m}\right)\frac{v}{r}$$

Thus

$$\bar{\omega}_2 = \left(\frac{m_b/m}{1 + 4.80m_b/m}\right)\frac{v}{r}(1.20\bar{i} + 3.46\bar{k})$$

This vector defines the instantaneous axis of rotation, because point A is fixed. The sketch below gives the magnitude and orientation of the axis of rotation.

$$\omega_2 = |\bar{\omega}_2| = \left(\frac{3.67m_b/m}{1 + 4.80m_b/m}\right)\frac{v}{r}$$

$$\theta = \tan^{-1}\left(\frac{\omega_x}{\omega_z}\right) = 19.1°$$

Finally, we should note that an approximation is sometimes made for problems such as this by neglecting the momentum of the bullet following the impact. The result of such a procedure would be to make the denominator in the solution unity, rather than $1 + 4.80 m_b/m$. Certainly this is a reasonable approximation for this system if $m_b/m < 0.01$.

Prob. VII.103

HOMEWORK PROBLEMS

VII.103   The angular velocity of the 8-kg disk is initially $\bar{\omega}_1 = -4\bar{k}$ rad/s when it is hit by a 100-gram bullet at point B, as shown. The bullet becomes embedded in the disk, with the result that the disk has no angular velocity component about the z axis after the impact. Determine (a) the initial speed v of the bullet to cause this change, and (b) the angular velocity of the disk after the impact.

VII.104   The 12,000-kg artificial satellite is initially in pure translation when impacted by a 3-kg meteoroid traveling at a speed of 2 km/s relative to the space vehicle in the direction shown. The radii of gyration of the satellite are $k_x = k_z = 2$ m, $k_y = 1$ m. Determine the angular velocity of the satellite and the change in its velocity immediately after the impact.

VII.105   A square plate of mass m is suspended by a ball-and-socket joint at point A. It is initially at rest when subjected to a large impulsive force F normal to its plane applied at some point P. Determine the relationship between the x and y coordinates of point P if the reaction force at point A is zero.

Prob. VII.104

Prob. VII.105   Prob. VII.106

VII.106   A 2-lb football is in free flight, traveling horizontally at 60 ft/s as it spins about its x axis at 5 rad/s. A would-be receiver deflects the ball, applying a 10-lb force as shown for an interval of 0.10 s. Determine the linear and angular velocity of the football after it is deflected. The radii of gyration are $k_x = 2$ in., $k_y = k_z = 3$ in.

VII.107   Steering rockets are fired at points A, B, and C of the 12,000-kg artificial satellite, as shown. The radii of gyration of the satellite are $k_x = k_z = 2$ m, $k_y = 1$ m, and the thrust of each rocket is 1000 N. If the satellite is initially in pure translation, determine the change in the velocity of the center of mass G and the angular velocity after the rockets have been fired for 1 min.

Prob. VII.107

Prob. VII.108

VII.108 The 2-kg semicircular plate is translating downward at 5 m/s when corner A strikes the ledge. Assuming that the collision is perfectly plastic, determine the velocity of the center of mass G and the angular velocity of the plate immediately after the collision.

VII.109 Solve Problem VII.108 if the coefficient of restitution for the collision is 0.50.

VII.110 The 5-lb thin rectangular plate AB is pinned to the vertical shaft, which rotates freely with negligible resistance from friction. Initially, the shaft is rotating at $\omega_1 = 0.5$ rad/s and the angle θ is held constant at 0°. Plate AB is then released. Determine the values of ω_1 and $\dot{\theta}$ when $\theta = 45°$. The mass of the shaft is negligible. *Hint:* The system is conservative. Also, there is no moment exerted on the system about the axis of the shaft.

Prob. VII.110

Prob. VII.111

VII.111 The 10-kg disk spins about arm AB at a constant rate of 20 rad/s. Arm AB is pinned to the vertical shaft at point A, the shaft being free to rotate with negligible friction at the (variable) rate ω_1. Initially, $\omega_1 = 0$ and arm AB is released from rest at $\theta = 30°$. Determine the values of $\dot{\theta}$ and ω_1 in the position where $\theta = 90°$. All bodies except the disk have negligible mass. See the hint in Problem VII.110.

MODULE VIII
SPECIAL TOPICS

A. VIBRATIONS

The general concern in dynamics is with the motion of a system and the forces that cause that motion. This section is devoted to a specialized dynamics investigation entitled *vibrations,* which is the study of the oscillatory motion that may occur in systems having a stable position of static equilibrium. This branch of dynamics is so important that no industrial firm would think of marketing its products, be they household appliances, electronic equipment, airplanes, or automobiles, before thoroughly testing them for vibratory effects.

1 System Model

Consider any of the systems mentioned above, say an automobile. The objective in modeling a system is to create the simplest mathematical description that retains the essential features of the system. The automobile consists of a chassis suspended by springs, energy-dissipative shock absorbers, and pneumatic tires. A primary function of these components is to insulate the chassis from irregularities in the roadway.

As a first attempt to investigate the vertical motion of an automobile, let us model the automobile as a mass suspended by a spring and a viscous damping device, as shown in Figure 1. In this sketch $\bar{F}(t)$ is an excitational force that is causing the motion. The origin of this force could be an unbalanced rotation of the wheels or a rough highway. The damping device is idealized as a linear viscous *dashpot* which produces a force opposing the motion in proportion to its rate of extension. The subject of energy dissipation and material damping is beyond the scope of this book. Thus, suffice it to say that a linear dashpot has been shown by experiments to be a good model for the dissipation occurring in the vibrations of many systems.

A prominent feature of the model appearing in Figure 1 is that only one scalar quantity, the vertical displacement of the mass, need be known to locate the position of the body. Hence, the system of Figure 1 can only be used to treat the motion of the center of mass of the automobile in the vertical direction. A system where one geometric quantity is required to describe its movement is said to have *one degree of freedom.* In our studies here, we shall consider only one-degree-of-freedom systems.

Figure 1

2 General Equation of Motion

Consider the case of the system of Figure 1 resting in static equilibrium, as depicted in Figure 2 on the following page. In the equilibrium position the mass is not moving, so there is no damping force. From the free body diagram shown in Figure 2, we may write

$$mg = k\delta_{st} \tag{1}$$

Position where
spring is
unstretched

δ_{st}

Static
equilibrium
position

l_0

mg

$k\delta_{st}$

Figure 2

where k is the stiffness of the spring and δ_{st} is the static deflection of the mass.

Now let us draw a free body diagram for the situation where the system is in motion. To measure the displacement of the mass, we need a reference position. For this, let us choose the *static equilibrium position,* as shown in Figure 3. The length of the spring when the mass moves upward from this position is then $(l_0 - \delta_{st}) + x$, where l_0 is the unstretched length of the spring, shown in Figure 2. Thus, the extension of the spring is $[(l_o - \delta_{st}) + x] - l_o \equiv x - \delta_{st}$. Also, because δ_{st} is a constant, the rate of extension of the dashpot is $\dot{x}$.

$F(t)$

mg

$x \quad v = \dot{x} \quad a = \ddot{x}$

$k(x - \delta_{st})$ $c\dot{x}$

Static equilibrium
position

Figure 3

Note what was done to obtain the direction of each force vector in Figure 3. First, a positive x direction was *arbitrarily* assumed and the kinematical parameters of the motion of the block were expressed in terms of x. Noting that the spring force is opposite the sense of its elongation and the force in the dashpot is opposite the sense of its extensional velocity, Newton's second law gives

$$F(t) - k(x - \delta_{st}) - c\dot{x} - mg = m\ddot{x}$$

so that

$$m\ddot{x} + c\dot{x} + k(x - \delta_{st}) + mg = F(t)$$

Replacing δ_{st} from equation (1), we get

$$\boxed{m\ddot{x} + c\dot{x} + kx = F(t)} \qquad (2)$$

Equation (2) leads to an important observation, which turns out to be true for all systems where the springs are linear and gravity has only a static effect. We see that when we measure the displacement of a system from its static equilibrium position, we may conveniently ignore the weight of the system and the static spring force $k\delta_{st}$ in writing the equation

462

Static equilibrium position

Figure 4

of motion. Thus, given the system of Figure 1, all that is required is the free body diagram shown in Figure 4. Newton's second law for the free body diagram of Figure 4 also gives equation (2).

Equation (2) is the equation of motion. It is sometimes referred to as the equation of a forced, damped, linear oscillator. It is a second-order, linear ordinary differential equation with constant coefficients. As is true for the motion of any physical system, to determine the response fully we must know the initial conditions, that is, the displacement and velocity when the motion commences.

In its most general form, equation (2) shows the acceleration to be a function of position, velocity, and time. Clearly, we cannot solve this general differential equation by the method of separating variables, although we could use this method for the case where only c, k, or $F(t)$ is nonzero. In the following we shall employ another method, which is valid for all linear differential equations having constant coefficients.

The next three sections focus on three common problems that arise in connection with a system whose motion is defined by equation (2).

3   Free Undamped Motion

Here we shall investigate the motion of a system that is free of an external force, $F(t) = 0$, and where the damping is negligible, $c = 0$. Setting $F(t) = c = 0$ in equation (2) yields the equation of motion for a free undamped system,

$$m\ddot{x} + kx = 0 \qquad (3)$$

The theory of linear differential equations states that the solution to the homogeneous equation (3) must have the form

$$x = \mathcal{A}e^{\lambda t} \qquad (4)$$

Substituting equation (4) into equation (3) gives

$$(m\lambda^2 + k)\,\mathcal{A}e^{\lambda t} = 0$$

The solution must be valid for all instants t, so we drop the factor $e^{\lambda t}$. Also, we drop the factor $\mathcal{A}$, because $\mathcal{A} = 0$ is the trivial solution. Thus, because k and m are positive numbers, we have

$$\lambda^2 + \frac{k}{m} = 0 \qquad \lambda = \pm\sqrt{-\frac{k}{m}} = \pm i\sqrt{\frac{k}{m}} \qquad (5)$$

where $i = \sqrt{-1}$. Hence, there are two possible values of λ. Corresponding to each value of λ, there is a corresponding value of $\mathcal{A}$. As the most general solution of a linear differential equation is a sum of all possible solutions, we have

$$x = \mathcal{A}_1 \exp\left(i\sqrt{\frac{k}{m}}\ t\right) + \mathcal{A}_2 \exp\left(-i\sqrt{\frac{k}{m}}\ t\right) \tag{6}$$

We now recall the identity

$$e^{ip} \equiv \cos p + i \sin p$$

to write

$$x = \mathcal{A}_1\left(\cos\sqrt{\frac{k}{m}}t + i\sin\sqrt{\frac{k}{m}}\ t\right) + \mathcal{A}_2\left(\cos\sqrt{\frac{k}{m}}t - i\sin\sqrt{\frac{k}{m}}\ t\right) \tag{7}$$

Furthermore, x must be a real quantity. Noting that the coefficients of $\mathcal{A}_1$ and $\mathcal{A}_2$ are complex conjugates of each other, this requirement means that $\mathcal{A}_1$ and $\mathcal{A}_2$ must also be complex conjugates; that is

$$\mathcal{A}_1 = a + bi \qquad \mathcal{A}_2 = a - bi$$

where a and b are real numbers. Equation (7) now becomes

$$x = (a + bi)(\cos\sqrt{\frac{k}{m}}t + i\sin\sqrt{\frac{k}{m}}t)$$

$$+ (a - bi)\left(\cos\sqrt{\frac{k}{m}}\ t - i\sin\sqrt{\frac{k}{m}}\ t\right)$$

$$= 2a\cos\sqrt{\frac{k}{m}}\ t - 2b\sin\sqrt{\frac{k}{m}}\ t$$

Because a and b are arbitrary numbers, the solution may be rewritten as

$$x = A\cos\omega_n t + B\sin\omega_n t \tag{8}$$

where A and B are real numbers, and

$$\omega_n = \sqrt{\frac{k}{m}} \tag{9}$$

For reasons we shall soon discuss, ω_n is called the circular *natural frequency* of the system.

The values of the coefficients A and B are obtained by requiring that equation (8) satisfy the initial conditions for the problem. However, the following alternative form of equation (8), which also has two arbitrary coefficients, has the advantage of being easier to visualize and sketch,

Figure 5

$$x = C \sin(\omega_n t + \phi) \tag{10}$$

A typical x–t diagram is shown in Figure 5. As can be seen from Figure 5, the parameter C is the *amplitude* of the sinusoidal curve. The parameter ϕ is called the *phase angle;* it defines how far the response is shifted to the left of the origin compared to a simple sine curve.

Let us now consider the meaning of the parameter ω_n. From equation (9) we find that the units of ω_n are radians per unit time. As exhibited in Figure 5, the motion is cyclical, repeating itself when $\omega_n t$ is increased by 2π rad. The parameter $\tau = 2\pi/\omega_n$ is the *period,* which is the time required for the motion to pass through one cycle. Then, because it takes τ time units to complete one cycle, it follows that $1/\tau \equiv \omega_n/2\pi$ cycles will occur in one unit of time. The quantity $\omega_n/2\pi$ is the *cyclical natural frequency.* (The physical unit, cycles per second, has been given the name *hertz,* abbreviated as Hz.)

In contrast, as indicated after equation (9), the parameter ω_n is called the *circular natural frequency,* although the word circular is usually omitted. The reason for the word "natural" is that ω_n defines the frequency at which the system vibrates freely, that is, when it vibrates without any disturbance from external or dissipative forces.

It should be noted that in all computations a frequency ω, be it the natural frequency ω_n or more generally the angular velocity of a rotating body, should carry the units radians per unit time.

Continuing with the introduction of standard terminology, we have already used the term *vibration* and *oscillation,* which are synonomous. Any response $x(t)$, such as that in Figure 6, which represents a fluctuation about some mean value is a vibration (or oscillation).

Figure 6   **Figure 7**

When the vibration has the form of a repeated pattern with a constant time interval τ, the oscillation is termed *periodic*. A typical periodic vibration appears in Figure 7.

A periodic vibration that has the shape of a sinusoidal curve is said to be *harmonic*. Thus, Figure 5 represents a harmonic vibration about a zero mean value.

EXAMPLE 1

A pendulum consisting of an 80-g particle suspended by a 2-m cable is known to be returning to the vertical position with a speed of 100 mm/s when $\theta = 5°$. Determine the amplitude and the period of the motion.

Solution

To determine the equation of motion (which is invariably the first step in a vibrations analysis) we draw a free body diagram, using polar coordinates. Remember that $\bar{e}_\theta$ is oriented in the direction of increasing θ. Note that the influence of gravity on the system varies with the angle θ, so that gravity does not have a static effect that may be ignored in formulating the equation of motion.

We are not interested in the value of the cable tension T, so we only sum forces in the transverse direction. The radial distance R has the constant value of 2.0 m, so

$$\Sigma F_\theta = -0.080(9.806) \sin \theta = ma_\theta$$

$$= 0.080(R\ddot{\theta} + 2\dot{R}\dot{\theta}) = 0.080(2.0)\ddot{\theta}$$

Thus

$$\ddot{\theta} + 4.903 \sin \theta = 0$$

This differential equation of motion is nonlinear. We can find a linear approximation to it by limiting the allowable values of θ to be sufficiently small so that $\sin \theta \simeq \theta$. With this restriction, the equation of motion becomes

$$\ddot{\theta} + 4.903 \theta = 0$$

Comparing this equation to equation (3), we see that

$$\omega_n^2 = \frac{k}{m} = 4.903 \ (\text{rad/s})^2$$

$$\omega_n = 2.214 \ \text{rad/s}$$

The amplitude was expressly asked for, so it is most convenient to use the solution of the equation of motion in the form of equation (10) rather than that of equation (8). Thus

$$\theta = C \sin(\omega_n t + \phi)$$

$$= C \sin(2.214t + \phi)$$

To determine the unknown coefficients we return to the statement of the problem, which tells us that when $t = 0$, $\theta = 5°$, and $v = 100$ mm/s. Recalling that $v = R\dot{\theta}$ for motion in a circular path, we then have

$$\theta\big|_{t=0} = 5° = \frac{5}{180} \pi \ \text{rad}$$

$$\dot{\theta}\big|_{t=0} = \frac{v}{R} = 0.050 \ \text{rad/s}$$

Now, placing the initial conditions into the general solution gives

$$\theta\big|_{t=0} = C\sin(2.214t + \phi)\big|_{t=0}$$

$$= C\sin\phi = \frac{5}{180}\pi = 8.727(10^{-2})$$

$$\dot{\theta}\big|_{t=0} = 2.214C\cos(2.214t + \phi)\big|_{t=0}$$

$$= 2.214C\cos\phi = 0.050$$

Squaring and adding these two equations yields

$$C^2(\cos^2\phi + \sin^2\phi) \equiv C^2 = [(2.258)^2 + (8.727)^2]10^{-4}$$

and upon eliminating C from the two equations we get

$$\frac{\sin\phi}{\cos\phi} = \tan\phi = \frac{8.727}{2.258} = 3.865$$

Therefore the amplitude is

$$C = 9.014(10^{-2})\ \text{rad}$$

Although it was not requested, the phase angle is

$$\phi = \tan^{-1}(3.865) = 1.3176\ \text{rad} = 75.49°$$

We now can write the general solution to the pendulum problem as

$$\theta = 9.014(10^{-2})\sin(2.214t + 1.3176)$$

The period τ is the time interval required for the motion to repeat itself. This means that the argument of the trigometric function in the general solution must increase by 2π. Thus

$$\omega_n\tau = 2.214\tau = 2\pi$$

$$\tau = 2.838\ \text{s}$$

Note that, in this linear approximation, the period of the motion depends only on the physical properties of the system (k and m) and not on the amplitude of the motion.

Finally, to clarify a matter that can cause some confusion, note that although the natural frequency $\omega_n = 2.214$ has the units of radians per second, it is not the angular velocity of the pendulum. The angular velocity is $\dot{\theta}$, which is found to be

$$\dot{\theta} = (2.214)(9.014)(10^{-2})\cos(2.214t + 1.3176)\ \text{rad/s}$$

Prob. VIII.1

HOMEWORK PROBLEMS

VIII.1 Determine the ratio of k_2/k_1 required for the two systems shown to have the same natural frequency.

VIII.2 Three different ways of supporting a piston of mass m which rides in a smooth vertical cylinder are shown. Each system has two identical springs of stiffness k. Determine the natural frequency of each system.

Prob. VIII.2

Prob. VIII.3

VIII.3 The mass of block A and the stiffness of the spring are unknown. It is known that the system vibrates freely at 80 Hz. When a 4-kg block is fastened to block A the frequency drops to 50 Hz. Determine the mass of block A and the stiffness of the spring.

VIII.4 The 200-g collar slides on the smooth horizontal bar. The spring has a stiffness of 2 kN/m, and the system is in the static equilibrium position when $s = 100$ mm. (a) Determine the natural frequency in units of radians per second and hertz. (b) If the collar is released from rest at $s = 80$ mm, determine where it will again come to rest and the elapsed time between these two positions.

Prob. VIII.4

VIII.5 Two pretensioned springs are attached to the 0.5-kg block. The unstretched length of both springs is 60 mm, and their stiffnesses are $k_1 = k_2 = 1$ kN/m. Determine (a) the value of the distance s corresponding to static equilibrium, (b) the differential equation of motion in terms of the distance s, (c) the differential equation of motion in terms of the displacement x from the static equilibrium position, and (d) the natural frequency of the system. (e) Which, if any, of the foregoing answers were affected by the unstretched length of the springs?

Prob. VIII.5

VIII.6 Solve Problem VIII.5 for $k_1 = 1$ kN/m and $k_2 = 0.5$ kN/m.

VIII.7 A block of mass m is fastened to a lightweight (negligible mass) beam at its midpoint. Unloaded the beam is straight. With the block attached, the static downward displacement of the midpoint of the beam is δ_{st}, as shown. (a) Derive an expression for the period of free vibration of the beam-mass combination. (b) If the system is held in the position where the beam is straight and then released from rest, what will be the amplitude of the subsequent motion? *Hint:* The beam acts, in effect, like a spring.

Prob. VIII.7

Prob. VIII.8

VIII.8 Two 1-kg spheres are suspended from the roof of an elevator, as shown. When the elevator is at rest the period of extensional free vibration of the spring-supported sphere is the same as that for the rotation of the pendulum. (a) Determine the stiffness k of the spring. (b) Determine the period of each system when the elevator accelerates upward at $\dot{v} = 0.6g$.

Prob. VIII.9

VIII.9 The speed governor shown consists of a block of mass m which may slide within the radial groove in the disk as the disk rotates in the horizontal plane about a vertical axis passing through point O at angular speed Ω. The stiffness of the spring is k. Determine the equation of motion for the radial distance R and also determine the natural frequency of the system. For what value of the angular speed Ω does the block cease to be capable of simple harmonic motion? Explain your answer.

Prob. VIII.10

VIII.10 A 2-lb block slides on a smooth horizontal table under the restraint of the two cables. In the static equilibrium position the tensile force within the cables is 40 lb. Because of the (slight) elasticity of the cables, the block can execute small movements x in the direction perpendicular to line AB. (a) Determine the natural frequency of the system in terms of the length l of one of the cables. (b) Determine the value of l that minimizes the natural frequency and also determine this minimum frequency. *Hint:* Formulate the equation of motion in terms of the displacement x, considering the tension to be constant and the angles θ_1 and θ_2 to be very small.

VIII.11 The collar of mass m slides on the smooth horizontal bar. The spring, whose stiffness is k, has a pre-tension force T when the block is in the static equilibrium position. Derive the differential equation of motion that is valid for large values of x and then linearize this result to obtain the equation for small x. Does the value of k affect the latter equation? Explain your answer.

Prob. VIII.11

Prob. VIII.12

VIII.12 The 2-lb block slides on the smooth horizontal surface under the restraint of the spring whose stiffness is 40 lb/ft and whose unstretched length is 1 ft. The block is given an initial velocity of 5 ft/s to the right at s = 15 in. Determine (a) the response $x(t)$, where x is the displacement measured from the static equilibrium position, (b) the amplitude, period, and phase angle for the response in part (a). (c) Would the answers to parts (a) and (b) change if the system was inclined upward to the left at 30° above the horizontal?

VIII.13 A 20-gram sphere moves within the smooth curved tube, which is held stationary in the vertical plane. The block is released from rest at θ = −4.5°. (a) Without solving the equation of motion, determine how long it will take for the block to reach θ = 0°. (b) Determine the velocity, acceleration, and elapsed time from release corresponding to the positions where θ = −2.25° and where θ = 2.25°.

500 mm

Prob. VIII.13

VIII.14 For the system and initial conditions given in Problem VIII.13, locate the following positions: (a) where the acceleration component tangent to the path is maximum, (b) where the acceleration component normal to the path is maximum, (c) where the magnitude of the acceleration is maximum.

VIII.15 The 4-kg block rests on the spring, but is not attached to it. The block is held at rest 20 mm below the static equilibrium and then released, this distance being the largest initial displacement for which the block never loses contact with the spring. Determine (a) the stiffness of the spring, and (b) the displacement of the block from the static equilibrium position as a function of time.

Prob. VIII.15

VIII.16 A 6-oz sphere is to be catapulted into the air vertically by the piston, which rides within the smooth cylinder. The piston weighs 2 oz, and the stiffness of the spring is 80 lb/ft. The piston is held 3 in. below the static equilibrium position and then released. Determine the distance the

Prob. VIII.16

50 mm 150 mm
|← 400 mm →|← →|← →|

Prob. VIII.18

piston is above the static equilibrium position when the ball loses contact with the piston, and also the elapsed time from the instant of release.

VIII.17 For the system of Problem VIII.16, determine (a) the maximum height attained by the sphere after leaving the piston, (b) the time response of the piston after the sphere flies upward, measuring time t from the instant when the piston was released, and (c) the amplitude, frequency, and phase angle for the motion obtained in part (b).

VIII.18 Two identical 1-kg collars, which may slide on the smooth horizontal bar, are initially at rest in the positions shown. The spring that is attached to block A has a stiffness of 3.6 N/mm and an unstretched length of 500 mm. Collar A is released, causing this body to move to the right and hit collar B. This collision is perfectly elastic. Determine (a) the elapsed time from the instant when the cord is broken until the collars collide, (b) the velocity of collar B immediately after the collision, (c) the amplitude, frequency, and phase angle for the motion of block A after the collision, measuring t from the instant of release.

VIII.19 Solve Problem VIII.18 if the coefficient of restitution for the collision is 0.50.

4 Free Damped Vibrations

Here we wish to learn how damping affects the response of a system that is free of external forces when the system is given an initial disturbance. As the solution is developed, note that the natural frequency $\omega_n = \sqrt{k/m}$ of the undamped system is also an important parameter for the damped system.

The equation of motion for the free, damped system associated with Figure 3 is

$$m\ddot{x} + c\dot{x} + kx = 0 \tag{11}$$

The harmonic solution given by equation (10) does not satisfy this equation because of the presence of the first derivative resulting from damping. To solve equation (11) we use the basic form for the solution of a linear differential equation.

$$x = \mathcal{A}e^{\lambda t} \tag{12}$$

Substituting this expression into equation (11), dividing by m, and replacing k/m by $\omega_n{}^2$ yields

$$\left(\lambda^2 + \frac{c}{m}\lambda + \omega_n{}^2\right)\mathcal{A}e^{\lambda t} = 0$$

As we did in the undamped case, we equate the term multiplied by $\mathcal{A}e^{\lambda t}$ to zero. This yields a quadratic equation for λ, whose solution is

$$\lambda = -\frac{c}{2m} \pm \sqrt{\left(\frac{c}{2m}\right)^4 - \omega_n{}^2}$$

Clearly the sign of the quantity under the radical will affect the result. For convenience, we define a new parameter,

$$\beta = \frac{c}{2m\omega_n} = \frac{c}{2\sqrt{km}} \qquad (13)$$

so the two values for λ appearing above become

$$\lambda = -\omega_n\beta \pm \omega_n\sqrt{\beta^2 - 1} \qquad (14)$$

There are now three possibilities to consider, depending on whether β is less than, greater than, or equal to one.

Case a: $\beta < 1$ This is the situation where the coefficient of damping is small. For this case the square root in equation (14) is imaginary, and the two values of λ are complex conjugates of each other; that is,

$$\lambda = -\omega_n\beta \pm i\omega_n\sqrt{1 - \beta^2}$$

We leave it as a homework exercise to follow similar steps to those used in the study of undamped free vibrations to show that the corresponding solution for x can be written in either of the following equivalent forms.

$$x = e^{-\omega_n\beta t}(A \cos \omega_n\sqrt{1 - \beta^2}\, t$$
$$+ B \sin \omega_n\sqrt{1 - \beta^2}\, t)$$
$$= Ce^{-\omega_n\beta t} \sin(\omega_n\sqrt{1 - \beta^2}\, t + \phi) \qquad (15)$$

One cycle
$\tau = 2\pi/\omega_d$

Vibratory motion determined by $\sin(\omega_d\, t + \phi)$

Envelope determined by $Ce^{-\beta\omega_n t}$

C
$C\sin\phi$

Figure 8

Once again the unknown coefficients C and ϕ are found by satisfying the initial conditions. Analyzing equation (15), we see that the exponential multiplier means that the response will decay with increasing values of t, whereas the term in parentheses shows that the response is oscillatory. The quantity $\omega_d \equiv \omega_n(1 - \beta^2)^{1/2}$ is sometimes called the *damped natural frequency*. However, referring to the typical $x(t)$ plot in Figure 8 on the next page, note that the response is not periodic. The parameter ω_d is a measure of the number of times x changes sign (or alternatively, reaches an extreme value) in a unit interval of time. The corresponding period is $\tau = 2\pi/\omega_d$.

Case b: $\beta > 1$ The coefficient of damping in this situation is large. For this case the square root in equation (14) is real, and hence the two values of λ are real. Corresponding to these two values of λ are two solutions, the sum of which is the general solution

$$x = e^{-\omega_n \beta t}(Ae^{\omega_n \sqrt{\beta^2-1}\, t} + Be^{-\omega_n \sqrt{\beta^2-1}\, t}) \tag{16}$$

Noting that $(\beta^2 - 1)^{1/2}$ is less than β, the solution given by equation (16) will decay with increasing time. Also, because the solution is a combination of exponential terms, the response is not oscillatory. This is exhibited in Figure 9, where the three curves show the effect of different types of initial conditions on the response.

Figure 9

Case c: $\beta = 1$ This represents a transition between the lightly damped situation ($\beta < 1$) and the heavily damped case ($\beta > 1$). Equation (14) now yields only one value for λ, which is

$$\lambda = -\omega_n \beta \equiv -\omega_n$$

However, the theory of linear differential equations tells us that there must be two distinct solutions to a second-order equation. The second solution is obtained by multiplying the basic exponential solution by a factor t. Hence

$$x = e^{-\omega_n t}(A + Bt) \tag{17}$$

Recalling that

$$\lim_{t \to \infty} (te^{-\omega_n t}) = 0$$

Figure 10

it follows that the response in this case also decays to zero. In fact, Figure 10 shows that the response in this case strongly resembles in all respects the response of Figure 9 for a more heavily damped system, although the case when $\beta = 1$ gives the highest rate of decay for the response.

The terminology used to describe the foregoing three cases is to say that a system for which $\beta < 1$ is *underdamped,* whereas one for which $\beta > 1$ is *overdamped.* As $\beta = 1$ defines the transition between an oscillating response and an exponentially decaying one, we say that $\beta = 1$ is the state of *critical damping.* The corresponding critical damping coefficient is found from equation (13) by setting $\beta = 1$, which gives

$$c_{cr} = 2\sqrt{km}$$

(18a)

Substituting this expression back into equation (13) yields

$$\beta = \frac{c}{c_{cr}}$$

(18b)

In other words, the value of β for a system is the ratio of the actual coefficient of damping of the system to the critical coefficient; it is called the *ratio of critical damping.*

A common example is the case of the shock absorbers of an automobile. When new, they do not permit the vehicle to bounce continuously when it encounters an irregularity in the roadway. The automobile is then overdamped. As the shock absorbers wear from age and use, the automobile passes from the overdamped situation, through critical damping, and into the case of underdamping. Disturbances in the roadway now result in the automobile bouncing freely. It is then time to install new shock absorbers.

EXAMPLE 2

An elevator comes into contact with its supports at the bottom of the elevator shaft, as shown, with the result that the elevator vibrates. The period τ is timed electronically, and the displacements measured at two instants τ seconds apart are x_1 and x_2. Determine (a) the ratio of critical damping and the natural frequency in terms of τ, x_1, and x_2, and (b) the percent of critical damping by which the value of c must be increased to prevent the elevator from vibrating when it reaches the bumper.

Solution

We are measuring x from the static equilibrium position, so we ignore the static gravity force. The equation of motion is

$$m\ddot{x} + c\dot{x} + kx = 0$$

Thus we see that the system gives rise to an equation of motion of the form equation (11). We are told that the system is vibrating, so it must be underdamped. Thus the response is given by equation (15),

$$x = Ce^{-\omega_n \beta t} \sin(\omega_n \sqrt{1 - \beta^2}\, t + \phi)$$

where, from equations (9) and (13), we have

$$\omega_n = \sqrt{\frac{k}{m}} \quad \text{and} \quad \beta = \frac{c}{2\sqrt{km}}$$

The values of m, c, and k are unknown. However, we are told that the period is known. For an underdamped system,

$$\tau = \frac{2\pi}{\omega_d} = \frac{2\pi}{\omega_n \sqrt{1 - \beta^2}}$$

This expression provides one relationship between the unknown ratio of critical damping and the natural frequency. For an additional relationship we use the given information regarding the displacement at the two time instants.

We were not given initial conditions, but we can form a ratio of the two successive displacements x_1 and x_2 in the hope of eliminating the unknown constants C and ϕ. It is given that x_2 is the displacement τ seconds after x_1, hence

$$\frac{x_1}{x_2} = \frac{Ce^{-\omega_n \beta t} \sin(\omega_n \sqrt{1 - \beta^2}\, t + \phi)}{Ce^{-\omega_n \beta(t+\tau)} \sin[\omega_n \sqrt{1 - \beta^2}\, (t + \tau) + \phi]}$$

$$= e^{\omega_n \beta \tau} \frac{\sin(\omega_n \sqrt{1 - \beta^2}\, t + \phi)}{\sin[\omega_n \sqrt{1 - \beta^2}\, (t + \tau) + \phi]}$$

Substituting the expression for τ, we then have

$$\frac{x_1}{x_2} = e^{2\pi\beta/\sqrt{1-\beta^2}} \frac{\sin(\omega_n \sqrt{1 - \beta^2}\, t + \phi)}{\sin(\omega_n \sqrt{1 - \beta^2}\, t + 2\pi + \phi)}$$

$$= e^{2\pi\beta/\sqrt{1-\beta^2}}$$

From this result we see that the ratio x_1/x_2 is a constant (independent of time). Let us denote its natural logarithm by symbol δ, which we call the *logarithmic decrement* of the system. Thus

$$\delta \equiv \ln\left(\frac{x_1}{x_2}\right) = \ln(e^{2\pi\beta/\sqrt{1-\beta^2}})$$

$$= \frac{2\pi\beta}{\sqrt{1 - \beta^2}}$$

Solving for β in terms of the known parameter δ, we find that

$$\beta = \frac{\delta}{\sqrt{4\pi^2 + \delta^2}}$$

Knowing β, we now can use the expression for the known value τ to determine ω_n. This gives

$$\omega_n = \frac{2\pi}{\tau \sqrt{1 - \beta^2}}$$

Finally, because the system will not vibrate when critically damped (when $\beta = 1$), the necessary increase in β is

$$\Delta\beta = 1 - \beta = 1 - \frac{\delta}{\sqrt{4\pi^2 + \delta^2}}$$

We also know from equation (18b) that c is proportional to β. Thus, multiplying by 100 to obtain a percentage, we find

$$\%\left(\frac{\Delta c}{c_{\mathrm{cr}}}\right) = 100\left(1 - \frac{\delta}{\sqrt{4\pi^2 + \delta^2}}\right)\%$$

HOMEWORK PROBLEMS

VIII.20 An overdamped system for which $\omega_n = 50$ rad/s is released from rest with a positive initial displacement x_0. Determine the time required for the displacement to be $x_0/10$ if the critical damping ratio is (a) $\beta = 1.010$, (b) $\beta = 2.020$.

VIII.21 The result of electronic measurement of the displacement of a system is shown. Determine the damped natural frequency, the ratio of critical damping, and the displacement x_1 indicated.

VIII.22 A system is underdamped. (a) Show that the time interval between two adjacent maximum displacements is exactly equal to the period of damped free vibration. (b) Let x_1 and x_{n+1} denote the maximum displacements occurring at the beginning and end of n cycles of vibration and show that the logarithmic decrement may be obtained from $\delta = (1/n) \ln(x_1/x_{n+1})$.

VIII.23 The block slides on the horizontal surface, which is lubricated by oil, so the friction force is proportional to the velocity of the block. The system is underdamped. When the block is released from rest the spring is compressed a distance x_0. (a) What percentage of the mechanical energy initially given to the system is dissipated in the first cycle of vibration? (b) Does this percentage change for the second and subsequent cycles?

Prob. VIII.21

Prob. VIII.23

Prob. VIII.24

VIII.24   The 8-oz block is restrained by a spring whose stiffness is 80 lb/ft and a dashpot whose constant is 2 lb-s/ft. Initially, the spring is elongated by 0.2 ft and the block is given a velocity of 10 ft/s to the right. Determine (a) the response as a function of time, (b) the time at which the elongation of the spring first attains the zero value.

VIII.25   Following similar steps to those used to obtain the general undamped free vibration response, equation (10), derive equation (15) for damped free vibration.

VIII.26   A 125-gram block is suspended vertically by a spring whose stiffness is 5 N/mm, and also by a dashpot. The system is underdamped. The block is given an upward velocity of 2 m/s at the static equilibrium position. (a) Determine, in terms of the critical damping ratio β, the time when the upward displacement of the block is maximum and compare this time interval to the period of damped free vibration. (b) Compare the maximum displacement discussed in part (a) to the maximum that would be obtained if damping were not present. (c) Compute the results of parts (a) and (b) for critical damping ratios of 0.90 and 0.10.

VIII.27   In the shock absorbing mechanism for an automobile bumper shown, two springs act concentrically with two hydraulic dashpots. The stiffness of each spring is 3 kN/m, and the bumper has a mass of 40 kg. When subjected to the impulsive force $\bar{F}$, it is desired that the bumper compress in the shortest possible time, without subsequently oscillating when it rebounds. Determine the value of the damping constant μ required for each dashpot to accomplish this.

VIII.28   In Problem VIII.27, the damping constant for each dashpot is μ = 600 N-s/m. A collision force $\bar{F}$, which exerts an impulse of 1.20 kN-s over a very short time interval, is applied to the bumper. Determine (a) the displacement of the bumper as a function of time t, where t is measured from the instant when $\bar{F}$ is applied, (b) the maximum distance the springs are compressed and the corresponding time t, and (c) the maximum compression of the springs if damping were not present.

VIII.29   Solve Problem VIII.28 for μ = 300 N-s/m.

Prob. VIII.27

Bumper

Hydraulic
dashpots

F

5   Forced Vibration-Harmonic Excitation

Thus far we have concentrated on the free response of a system. We now wish to consider the effect of dynamic excitation of the system. The type of force we shall consider is a harmonic force, that is, one that varies sinusoidally with time. The reasons for studying such a forcing function are many. A prime reason is that if we know the response of the system to

harmonic excitation, it is possible to determine the response to any periodic excitation by using a Fourier series.

Letting $F(t) = F \sin \Omega t$, where Ω is the frequency of the excitation (having the units radians per unit time), the equation of motion for the system of Figure 1 is

$$m\ddot{x} + c\dot{x} + kx = F \sin \Omega t \qquad (19)$$

There are two types of solutions for equation (19). One, called the *particular solution* and denoted as x_p, corresponds to the nonhomogeneous term in the differential equation. To this solution we add the *complementary solution* x_c, which is the solution corresponding to $F(t) = 0$. For equation (19) the complementary solution has the form of the free response of an undamped, underdamped, overdamped, or critically damped system, depending on the values of m, c, and k. Thus, we shall focus attention here on the determination of the particular solution.

The most direct way to obtain a particular solution is to guess its form. Because x and its first and second derivatives must combine to give a sine term, we guess

$$\boxed{x_p = D \sin(\Omega t + \psi)} \qquad (20)$$

In contrast to the homogeneous solution (free vibration), the coefficients D and ψ are not arbitrary values. They are determined by requiring that x_p satisfy equation (19). Upon substituting equation (20) into equation (19), we find that

$$(-m\Omega^2 + k)D \sin(\Omega t + \psi) + c\Omega D \cos(\Omega t + \psi) = F \sin \Omega t$$

The easiest way to proceed at this point is to manipulate the term on the right side using the trigonometric identity for the sine of the difference of two angles, as follows.

$$(-m\Omega^2 + k)D \sin(\Omega t + \psi) + c\Omega D \cos(\Omega t + \psi)$$

$$= F \sin[(\Omega t + \psi) - \psi]$$

$$= F \sin(\Omega t + \psi) \cos \psi - F \cos(\Omega t + \psi) \sin \psi$$

Equating the coefficients of the sine and cosine terms gives

$$(-m\Omega^2 + k)D = F \cos \psi$$

$$c\Omega D = -F \sin \psi$$

Solving these equations for D and ψ in terms of F, Ω, m, c, and k yields

$$D = \frac{F}{\sqrt{(k - \Omega^2 m)^2 + c^2 \Omega^2}}$$

$$\tan \psi = \frac{c\Omega}{k - m\Omega^2}$$

This result is now simplified by replacing m, c, and k by the expressions for the natural frequency, $\omega_n = \sqrt{k/m}$, and the critical damping ratio, $\beta = c/2\sqrt{mk}$. Skipping the algebraic details, this yields

$$D = \frac{(F/k)}{\sqrt{(1 - \Omega^2/\omega_n^2)^2 + 4\beta^2\Omega^2/\omega_n^2}}$$

$$\psi = \tan^{-1}\left(\frac{2\beta\Omega/\omega_n}{1 - \Omega^2/\omega_n^2}\right) \tag{21}$$

There are a number of interesting results displayed in equation (21). For instance, notice that F/k is the static displacement that the mass would have if it were loaded by a static force of magnitude F. Thus, the fraction multiplying F/k in the equation for D is a *dynamic magnification factor* for the amplitude D. Furthermore, for a given set of system parameters, the amplitude and phase angle depend only on the ratio of the frequency of excitation to the natural frequency.

To study equations (21) further, let us sketch D and ψ as a function of Ω/ω_n, as is shown in Figure 11. Such plots are called amplitude-frequency (Figure 11a) and phase angle-frequency diagrams (Figure 11b). They are very useful for designing systems to meet dynamic criteria, such as limitations on the amplitude of vibration.

Focusing on Figure 11a, for an undamped system ($\beta = 0$), notice that the value of D becomes infinite as the frequency of excitation approaches the natural frequency ($\Omega/\omega_n \to 1$). This situation, which is called *resonance*, is obviously quite dangerous, although in reality the response can never be infinite, for all real systems exhibit some damping.

The phase angle for the undamped system is also of interest. When $\Omega/\omega_n < 1$, we see in Figure 11b that, for $\beta = 0$, $\psi = 0$ also. This means that the displacement x_p is positive whenever the force $F(t) = F \sin \Omega t$ is positive. We then say that the response and the excitation are *in phase*. As the resonance condition is passed, so that $\Omega/\omega_n > 1$, ψ suddenly changes to 180°, so the displacement x_p is negative whenever $F(t) = F \sin \Omega t$ is positive. The response and the excitation are then 180° *out of phase*. Another interesting feature of the phase angle-frequency diagram is that when $\Omega/\omega_n = 1$, the response x_p for a damped system is 90° out of phase with the force; that is, the magnitude of the response is a maximum when the force is zero and vice versa.

Let us now return to the problem of determining the total response for the system of equation (19). For this discussion we shall consider the

Figure 11a

Figure 11b

case of an underdamped system. Thus, the complementary solution is given by equation (15), which in combination with the particular solution of equation (20), gives

$$x = x_c + x_p$$

$$= e^{-\beta\omega_n t}C\,\sin(\omega_n\sqrt{1-\beta^2}\,t + \phi) + D\,\sin(\Omega t + \psi) \qquad (22)$$

A computational point to bear in mind is that the constants C and ϕ are found by using the full expression (22) to satisfy the initial conditions, not just the portion of equation (22) that is the complementary solution.

Examining equation (22), we see that as t becomes large the first term, the complementary solution, goes to zero. Because of this effect, dynamicists call x_c the *transient* solution and x_p the *steady state* solution. It is a steady state vibration in the sense that after the transient decays the system will maintain a steady harmonic oscillation at the amplitude D. The energy put in through the forcing function then balances the energy lost through damping. Note too that the steady state vibration takes place at the frequency of the forcing function. An exception occurs in the case of an undamped system, for then the complementary solution in equation (22) does not decay. In reality, this is of little concern, because all systems have some damping, which causes the transient solution eventually to decay to zero. Thus, we see that it is correct to approximate a system as having no damping only if we are observing it over a comparatively small time interval.

EXAMPLE 3
When $t = 0$, the illustrated system is at rest in its static equilibrium position, where the tension forces in each of the two springs both have the value T. At this instant a harmonic force $F\cos\Omega t$ is applied to the block. Neglecting friction, determine the subsequent motion of the block.

Solution
In the free body diagram let us arbitrarily choose x as the displacement of the block to the right from the static equilibrium position. Notice that a displacement x to the right reduces the tension forces in the right spring by kx, whereas it increases the tension force in the left springs by the same amount.

To obtain the equation of motion, we sum forces to obtain

$$\Sigma F_x = (T - kx) + F\cos\Omega t - (T + kx) = ma_x = m\ddot{x}$$

$$m\ddot{x} + 2kx = F\cos\Omega t$$

As can be seen, the tension force T does not appear in the equation of motion. This is another situation where a static force does not affect a dynamic displacement measured from the static equilibrium position.

The solution of the equation of motion is obtained by adding the complementary and particular solutions for an undamped oscillator. The natural frequency of this system is

$$\omega_n = \sqrt{\frac{2k}{m}}$$

and the complementary solution is found from equation (8) to be

$$x_c = A \cos \omega_n t + B \sin \omega_n t$$

The particular solution for harmonic excitation, given by equations (20) and (21), was derived for a sine force; the force in this problem is a cosine force. However, because this is simply a 90° difference in phase, we may directly employ the derived solution by also shifting the response by 90°, thus replacing the sine function by the cosine function. Hence

$$x_p = D \cos(\Omega t + \psi)$$

The amplitude D in equation (20) is positive, so for $\beta = 0$ we have

$$D = \frac{F/2k}{|1 - \Omega^2/\omega_n{}^2|}$$

$$\psi = 0 \qquad \text{if } \Omega < \omega_n$$

$$\psi = 180° \qquad \text{if } \Omega > \omega_n$$

It is useful in the case of no damping to remove the phase angle from the problem by setting $\psi \equiv 0$ and letting D be positive or negative. Thus, the equivalent of the foregoing is

$$x_p = D \cos \Omega t$$

$$D = \frac{F/2k}{1 - \Omega^2/\omega_n{}^2}$$

Finally, we satisfy the initial conditions. From the given information we know that $x = 0$ and $\dot{x} = 0$ when $t = 0$. This gives

$$x(0) = 0 = (x_c + x_p)\big|_{t=0}$$

$$= (A \cos \omega_n t + B \sin \omega_n t + D \cos \Omega t)\big|_{t=0}$$

$$= A + D$$

$$\dot{x}(0) = 0 = (\dot{x}_c + \dot{x}_p)\big|_{t=0}$$

$$= [\omega_n(-A \sin \omega_n t + B \cos \omega_n t)$$

$$- \Omega D \sin \Omega t]_{t=0} = \omega_n B$$

Thus, $A = -D$ and $B = 0$, which gives

$$x = D(\cos \Omega t - \cos \omega_n t)$$

$$= \left(\frac{F/2k}{1 - \Omega^2/\omega^2}\right)(\cos \Omega t - \cos \omega_n t)$$

Notice that the transient solution is not zero even though the initial displacement and velocity were zero, thus emphasizing the fact that the initial conditions must be satisfied after first combining the complementary and particular solutions.

EXAMPLE 4

The center of mass G of the 0.5-kg flywheel of an electric motor is at a distance of 20 mm from the motor shaft. The body of the motor is attached to a rigid framework which is supported by four springs, as shown. The framework is capable only of translatory motion in the vertical direction, and the combined mass of the body of the motor and its framework is 4.5 kg. It is desired that the steady state displacement not exceed 2.0 mm when the flywheel is rotating at the natural frequency of the undamped system. Determine the minimum value of the damping coefficient c required to satisfy this criterion.

Solution

This problem illustrates a common source of harmonic excitation for a system: an unbalanced rotating mass. The determination of the equation of motion, which is our first task, is not as trivial as in the prevous developments.

The system is composed of two moving bodies, one being the body of the motor and the framework, the other being the flywheel. Therefore, let us draw free body diagrams for both systems. The interaction force exerted between them is the reaction $\bar{F}$ at the shaft.

Notice that we are measuring the vertical displacement y from the static equilibrium position, so we ignore the static effect of gravity. The equation for vertical motion of the motor and its framework is

$$\Sigma\, F_y = -4ky - c\dot{y} + F_y = Ma_y = M\ddot{y}$$

where, from the given information, $M = 4.5$ kg and $k = 125$ N/m.

To determine the reaction component F_y, we sum forces in the vertical direction for the flywheel. This gives

$$\Sigma\, F_y = -F_y = m(a_G)_y = m\ddot{y}_G$$

where $m = 0.50$ kg and the negative sign arises because of Newton's third law. From the geometry of the free body diagram we can write

$$y_G = y + 0.15 + \left|\bar{r}_{G/O}\right| \sin\theta \text{ meters}$$

$$= y + 0.15 + 0.020 \sin\omega t$$

Hence

$$\ddot{y}_G = \ddot{y} - 0.020\omega^2 \sin\omega t$$

and we have

$$F_y = -m\ddot{y} + 0.020m\omega^2 \sin\omega t$$

Now, replacing F_y in the first equation of motion, the basic equation of motion for the system becomes,

$$M\ddot{y} = -4ky - c\dot{y} + (-m\ddot{y} + 0.020m\omega^2 \sin\omega t)$$

$$(M + m)\ddot{y} + c\dot{y} + 4ky = 0.020m\omega^2 \sin\omega t$$

We can determine the response of the system by solving this differential equation, which we recognize to be of the form of equation (19). Because initial conditions are not given, we consider only the steady state response. Let us denote the symbols appearing in the basic differential equation (19) by primes. Then, in comparison with the equation of motion in this problem, we have

$$m' = M + m = 4.5 + 0.5 = 5.0 \text{ kg}$$

$$c' = c \text{ N-s/m}$$

$$k' = 4k = 4(125) = 500 \text{ N/m}$$

$$F' = 0.020m\omega^2 = 0.010\omega^2 \text{ N}$$

$$\Omega = \omega \text{ rad/s}$$

Hence, the natural frequency and ratio of critical damping are

$$\omega_n = \sqrt{\frac{k'}{m'}} = \sqrt{\frac{500}{5.0}} = 10.0 \text{ rad/s}$$

$$\beta = \frac{c'}{2\sqrt{k'm'}} = \frac{c}{2\sqrt{500(5)}} = \frac{c}{100}$$

We may now use equations (20) and (21) to determine the steady state solution. Replacing the coefficients in these equations by their equivalents here, we find that

$$y_p = D \sin (\omega t + \psi)$$

where

$$D = \frac{(0.010\omega^2/500)}{\sqrt{[1 - \omega^2/(10.0)^2]^2 + 4\beta^2\omega^2/(10.0)^2}}$$

$$\psi = \tan^{-1}\left[\frac{2\beta\omega/10.0}{1 - \omega^2/(10.0)^2}\right]$$

To find the value of β, we satisfy the given limitation on the magnitude of the displacement by setting the amplitude D equal to 0.002 meters and $\omega = \omega_n = 10.0$ rad/s. This yields

$$0.002 = \frac{(0.010)(10.0)^2/500}{\sqrt{(1.0 - 1.0)^2 + 4\beta^2(1.0)}} = \frac{0.002}{2\beta}$$

$$\beta = 0.5$$

In other words, the damping in the system must be at least half the critical value. The corresponding minimum value of c is now found from the earlier expression, $\beta = c/100$, to be

$$c = 100\ \beta = 50.0 \text{ N-s/m}$$

HOMEWORK PROBLEMS

Prob. VIII.30

VIII.30 The 4-kg block is restrained by a spring whose stiffness is 4 kN/m. The external force varies harmonically according to $F = 12 \sin 16t$, where F is in units of newtons and t is in seconds. At $t = 0$ the block is at the static equilibrium position and has a velocity of 4 m/s to the left. Determine the response of the block and identify the transient and steady state portions.

VIII.31 Solve Problem VIII.30 if the external force is a combination of two harmonics given by $F = 12(\sin 8t + \sin 16t)$.

VIII.32 The harmonic force $F \cos \Omega t$ is applied to the block at $t = 0$. Initially, the block was at rest in the position where the compressive force

$F \cos \Omega t$

Prob. VIII.32

in the series-connected spring is F. Determine (a) the response of the block, (b) the amplitude and phase angle of the steady state response for $\Omega = \sqrt{k/m}$ and $\Omega = 2\sqrt{k/m}$.

VIII.33 and VIII.34 A 4-lb collar on a smooth horizontal rod is restrained by a spring whose stiffness is 2 lb/in. The collar is initially at rest in the static equilibrium position when it is subjected to the tangential force F shown in the graph. Determine and plot the displacement $x(t)$ (a) for $0 \le t \le 1$, (b) for $1 \le 1 \le 2$.

Prob. VIII.33   **Prob. VIII.34**

VIII.35 A 9-kg electric motor is mounted on a lightweight beam. The flywheel of the motor has a mass of 1kg, and its center of mass is 6 mm from the axis of rotation. Knowing that the static deflection of the beam when the motor and flywheel are at rest is 5 mm, determine (a) the angular velocity of the flywheel that will cause the system to resonate, (b) the amplitude of the steady state displacement when the angular speed of the flywheel is eight-tenths of the resonant speed found in part (a).

Prob. VIII.35

VIII.36 It is desired to operate the electric motor in Problem VIII.35 at the angular velocity that would cause the system to resonate. To do this, a dashpot is attached to the beam from the ground, directly under the motor. If the steady state displacement of the motor in this condition is to be 10 mm, determine the required constant for the dashpot.

VIII.37 A simple pendulum formed from a particle of mass m at the end of a cable of length l is attached to collar A. The collar is given a harmonic displacement $\delta \cos \Omega t$. For what range of values of Ω will the steady state amplitude of the horizontal displacement x of the mass be (a) at least three times larger than δ? (b) less than δ?

Prob. VIII.37

VIII.38 The runway of an airport can be approximated as a sinusoidal surface whose vertical elevation h is defined by $h = h_0 \sin(\pi x/l)$. An airplane of mass m lands smoothly on the runway with a speed v. Use the model shown to determine the critical landing speed at which the airplane would resonate. *Hint:* Formulate the equation of motion in terms of the absolute vertical displacement y and let $x = vt$ in the expression for the elevation h.

Prob. VIII.38

VIII.39 The automobile gives the 500-kg trailer a constant horizontal velocity component v as the wheels of the trailer follow a wavy road. The elevation of the road surface is defined by $h = 40 \sin \pi x/2$, where h is in millimeters and x is the horizontal distance traveled by the trailer in meters. The trailer does not have shock absorbers, and each of the two springs supporting the trailer has a stiffness of 40 kN/m. Neglecting the mass and the radius of the wheels determine (a) the differential equation of motion for the vertical displacement of the trailer, (b) the speed v at which the trailer would resonate, (c) the amplitude of the vibration of the trailer when $v = 10$ m/s. See the hint given in Problem VIII.38.

Prob. VIII.39

VIII.40 Solve Problem VIII.39 if, in addition to the two springs, two shock absorbers whose constants are 5 kN-s/m each are also connected between the axle and the body of the trailer.

VIII.41 A system whose natural frequency is ω_n and whose critical damping ratio is β is subjected to a harmonic excitation at frequency Ω. (a) Determine the value of Ω at which the amplitude of the steady state displacement is a maximum. (b) What are the amplitude and phase angle at the frequency found in part (a)? (c) For what range of values of β will an increase in the value of Ω always result in a decrease in the steady state amplitude?

VIII.42 In order to isolate an electronic instrument from the vibratory motion of the floor, it is mounted on a platform that is supported by four springs (only two are shown). The mass of the instrument and the platform is 25 kg and the stiffness of each spring is 12.5 kN/m. The upward displacement of the floor is $h = 5 \sin \Omega t$, where h is in millimeters. (a)

Letting x denote the upward displacement of the platform from the static equilibrium position when the floor is at rest, determine the steady state expression for x. (b) Determine the maximum acceleration of the instrument when the floor is vibrating at 6.5 Hz and at 7.5 Hz.

VIII.43   Solve Problem VIII.42 if each of the springs is concentric with a dashpot whose constant is 250 N-s/m.

VIII.44   Instrumentation within the accelerometer shown measures the displacement x of the block relative to the surrounding case as an indication of the vertical displacement δ of the floor. In a certain situation the block is initially at rest in the static equilibrium position when the base is given a constant upward acceleration $\ddot{\delta}$. Determine the expression for $x(t)$ in this case and plot the nondimensional displacement parameter $x\omega_n{}^2/\ddot{\delta}$ in terms of the nondimensional time parameter $\omega_n t$ (a) for no damping, (b) for 50% of critical damping.

VIII.45   In another case of excitation, the base of the accelerometer in Problem VIII.44 is given a harmonic motion $\delta = \delta_0 \cos \Omega t$. Determine the steady state displacement x and sketch diagrams of the amplitude magnification factor x_0/δ_0 and the phase angle ψ in terms of the frequency parameter Ω/ω_n. The system is lightly damped ($\beta \ll 1$).

VIII.46   A system whose natural frequency is ω_n and whose ratio of critical damping is β is subjected to a harmonic force $F_0 \cos \Omega t$. Determine (a) the instantaneous power developed by the force as a function of time in the steady state motion of the system, (b) the total work done by the force in one cycle, (c) the average power developed by the force over a complete cycle. Compare the results of parts (a) and (c).

6   Vibrations of Rigid Bodies

The vibrating systems we have treated so far were successfully modeled as particles. However, as we saw in the basic studies of kinetics, in order to retain the essential features of a physical system it is frequently necessary to model the system as a rigid body. The techniques for investigating the vibration of rigid bodies about their static equilibrium position are essentially the same as for particles, consisting of the formulation and solution of the equation of motion.

There is a small complication that arises in account for the force of a spring that is attached to a specific point in a rotating body. In some respects this complication resembles that encountered in the study of the pendulum in Example 1. We say there that, for large angles of rotation of the pendulum, the equation of motion is nonlinear, and thus difficult to solve. By restricting the angle of rotation to be a small value, we were able to linearize the equation. Similarly, when dealing with rigid bodies that rotate, it will be necessary for us to restrict our attention to the case

Figure 12a   Figure 12b

Figure 8.13

of small rotations and displacements in order to have a linear equation of motion.

To see how such an approximation is made, consider the system in Figure 12a, where a slender bar may rotate in the vertical plane about the pin A. For small angles of rotation θ of the bar from the vertical position, shown in Figure 12a, we may consider point C to move only horizontally, so, in essence, the springs maintain their horizontal orientations. Furthermore, for small θ, $\tan \theta \simeq \sin \theta \simeq \theta$. Hence, according to Figure 12b, we may approximate the deformation Δ of the springs as

$$\Delta \simeq \frac{3L}{4}\theta$$

For this approximate analysis, the free body diagram of the bar is as shown in Figure 13. Notice that the free body diagram was drawn for a general position of the body, but all linearizing approximations regarding the smallness of the motion are employed.

Continuing, we obtain the equation of motion by summing moments about the z axis passing through the fixed point A, thereby eliminating the unknown reaction forces $\bar{A}_x$ and $\bar{A}_y$. Using the approximation for the length of the lever arm of the gravity force appearing in the free body diagram, this yields

$$\Sigma M_{Az} = -mg\left(\frac{L}{2}\theta\right) - 2k\left(\frac{3L}{4}\theta\right)\left(\frac{3L}{4}\right) = I_{zz}\alpha = \tfrac{1}{3}mL^2(\ddot{\theta})$$

$$\tfrac{1}{3}mL^2\ddot{\theta} + (\tfrac{9}{8}kL^2 + \tfrac{1}{2}mgL)\,\theta = 0$$

Thus, in this approximation, the system is equivalent to the model of an undamped freely vibrating oscillator, equation (3).

EXAMPLE 5

The gear shown in the diagram has a mass M and a radius of gyration r. It rolls over a fixed rack and is restrained by a spring. The gear is in static equilibrium in the position shown. Determine the natural frequency of the system.

Static equilibrium
position of point G

Solution

For small motions of the gear, we can consider the spring to remain horizontal. Let us denote the displacement of the center of mass G by the variable s, with positive s representing a movement to the right. Such a movement will result in a compressive deformation of the spring by the quantity Δ. Hence in the free body diagram shown, the spring force is $k\Delta$ to the left.

The gear is in general motion, so we sum moments about the center of mass. The basic equations of motion are therefore

$$\Sigma F_x = -k\Delta - f = M(a_G)_x = M\ddot{s}$$

$$\Sigma F_y = N - Mg = M(a_G)_y = 0$$

$$\Sigma M_{Gz} = (k\Delta)b - fa = I_{zz}\alpha = \tfrac{1}{2}Mr^2\alpha$$

In these three equations there are five unknowns: Δ, f, s, N, and α. To find additional relations we use the kinematical constraint that the gear rolls without slipping, which enables us to relate the rotation and the deformation of the springs to the displacement s. We do this by sketching the displaced position of the gear.

Note that in this sketch we employed the restriction of small displacement, allowing us to approximate the horizontal movement of point A with respect to point G as $b\theta$. Then, for rolling without slipping,

$$\theta = \frac{s}{a} \qquad \Delta = s + b\theta = s\left(1 + \frac{b}{a}\right)$$

Furthermore, because a positive θ is clockwise and a positive α is counterclockwise, in accord with the right-hand rule with respect to the z axis we have

$$\alpha = -\ddot{\theta} = -\frac{\ddot{s}}{a}$$

Substituting these expressions into the basic equations of motion yields

$$-ks\left(1 + \frac{b}{a}\right) - f = M\ddot{s}$$

$$N = Mg$$

$$ks\left(1 + \frac{b}{a}\right)b - fa = -\frac{1}{2}Mr^2\frac{\ddot{s}}{a}$$

Solving for f in the first equation and placing it into the third equation, we find that

$$f = -\left[ks\left(1 + \frac{b}{a}\right) + M\ddot{s}\right]$$

$$ks\left(1 + \frac{b}{a}\right)b + ks\left(1 + \frac{b}{a}\right)a = -\frac{1}{2}Mr^2\frac{\ddot{s}}{a}$$

We now simplify this result, and thus obtain the differential equation of motion.

$$\tfrac{1}{2}M\left(\frac{r}{a}\right)^2\ddot{s} + k\left(1 + \frac{b}{a}\right)^2 s = 0$$

This equation has the same form as that for the undamped free vibration of the system model in Figure 1. Hence, the natural frequency is

$$\omega_n = \sqrt{\frac{k(1 + b/a)^2}{M(r/a)^2}} = \left(\frac{1 + b/a}{r/a}\right)\sqrt{\frac{k}{M}}$$

$$= \frac{a + b}{r}\sqrt{\frac{k}{M}}$$

HOMEWORK PROBLEMS

VIII.47–VIII.54   Derive the linear differential equation of motion for the (small) angle θ measuring the rotation of the system shown from its static equilibrium position. Wherever applicable, assume that there is no slipping.

Homogeneous
semicircular
disk of mass m

Prob. VIII.47

Homogeneous
semicircular
disk of mass m

Prob. VIII.48

1.5 m

20 kg uniform bar

50 kN/m

Prob. VIII.49

500 mm   500 mm   500 mm

20 kg
uniform bar

50 kN/m

Prob. VIII.50

a

k Massless
rigid bar

c

l

Particle of
mass m

Prob. VIII.51

k

a

Mass/unit length of
slender bars is m

b

Prob. VIII.52

Prob. VIII.53

Prob. VIII.54

VIII.55–VIII.58   The system is shown in its static equilibrium position. Determine its natural frequency.

Prob. VIII.55

Prob. VIII.56

Prob. VIII.57

Prob. VIII.58

Prob. VIII.59

VIII.59   A flywheel is suspended from its center by a metal cable. When the flywheel is rotated by an angle θ about the axis of the cable, the cable exerts a restoring torque $k\theta$ opposite the sense of the rotation, due to its elasticity. The free vibration of the system is observed to have a period τ. Derive an expression for the mass moment of inertia of the flywheel about the axis of the cable.

VIII.60 and VIII.61   Determine the natural frequency of each of the systems shown. Are there any values of the spring stiffness k for which either system will not oscillate freely about the vertical position? Discuss the meaning of these values of k with regard to the static stability of the vertical position.

Prob. VIII.60

Prob. VIII.61

B. ORBITAL MOTION

Newton's gravitational law postulates that the force of attraction between two bodies is a function of the distance between them. In our earlier studies we investigated the ballistic motion of a particle for the case where the particle was close to the earth's surface. Thus, the distance from the earth's surface to the particle was small compared to the radius of the earth. This allowed us to approximate gravity as a constant force, thereby simplifying the formulation and solution of the equations of motion.

Our aim here is to account for the varying magnitude and direction of the gravitational force acting on a particle that is located sufficiently far from the attracting body's mass center to make the force a variable quantity.

To place the problem in perspective, consider the gravitational force $\bar{F}_G$ exerted between the two bodies shown in Figure 14.

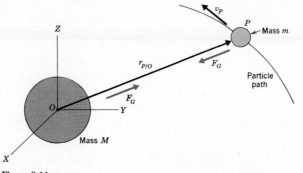

Figure 8.14

According to Newton's law of gravitation,

$$|\bar{F}_G| = \frac{GMm}{|\bar{r}_{P/O}|^2}$$

where G is the universal gravitational constant. Also, the attractive force $\bar{F}_G$ acting on the particle at point P is in the opposite direction of the position vector $\bar{r}_{P/O}$. Putting these two facts together, we have

$$\bar{F}_G = \frac{GMm}{|\bar{r}_{P/O}|^2}\left(-\frac{\bar{r}_{P/O}}{|\bar{r}_{P/O}|}\right) = -\frac{GMm}{|\bar{r}_{P/O}|^3}\bar{r}_{P/O} \tag{23}$$

Let us restrict the discussion to the case where the mass M is sufficiently large compared to the smaller mass m that the gravitational force $-\bar{F}_G$ exerted on M by m has a negligible effect on the body M. This allows us to consider the center point O of body M to be fixed. As a result, the line of action of the gravitational force acting on mass m always passes through the fixed point O. The model we now have, where the force of attraction between two bodies is along the line of their centers, with one center fixed, is called *central force motion*.

EXAMPLE 6
Determine the value of GM in Newton's gravitational law when the attracting body is the earth.

Solution
Newton's gravitational law is

$$|\bar{F}_G| = \frac{GMm}{|r_{P/O}|^2}$$

If the attracting body is the earth, then, at the surface of the earth, $|\bar{F}_G|$ is also mg, and $|\bar{r}_{P/O}| = R_e$, where R_e is the mean radius of the earth, $R_e = 6370$ km $= 3960$ miles. Thus

$$mg = \frac{GMm}{R_e^2}$$

so

$$GM = gR_e^2 = (9.806)(6370)^2(1000)^2 = 3.979(10^{14})\,\frac{m^3}{s^2} \qquad \triangleleft$$

$$= (32.17)(3960)^2(5280)^2 = 1.4064(10^{16})\,\frac{ft^3}{s^2} \qquad \triangleleft$$

The value of GM is called the *intensity of the force*.

Figure 15

1 Central Force Motion

In this section we shall consider the motion of a particle under the influence of a general central force $\bar{F}$, whose magnitude may be a variable quantity. In Figure 15, point O is the fixed point through which this force always acts, and point P is the location of the particle.

Let the initial position and velocity of the particle be arbitrary and known. Given these initial conditions, we wish to determine the path the particle follows and the relationships that exist among its kinematical variables.

We begin by noting that in central force motion the moment of $\bar{F}$ about the origin O is zero. Hence, the angular impulse-momentum principle for a single particle, formulated about the fixed point O, yields

$$\Sigma \bar{M}_O \equiv \bar{0} \equiv \dot{\bar{H}}_O = \frac{d}{dt}(\bar{r}_{P/O} \times m\bar{v}_P) \tag{24}$$

It immediately follows from equation (24) that the angular momentum $\bar{H}_O$ is constant throughout the motion. For notational convenience let us divide $\bar{H}_O$ by the constant mass m to obtain

$$\bar{h}_O \equiv \frac{\bar{H}_O}{m} = \bar{r}_{P/O} \times \bar{v}_P \equiv \text{constant} \tag{25}$$

where $\bar{h}_O$ is called the angular momentum per unit mass.

In order for the vector $\bar{h}_O$ in equation (25) to be constant, both its magnitude and direction must be constant. Recalling the definition of the direction of the cross product, from equation (25) we see that this constant vector is normal to the plane formed by $\bar{r}_{P/O}$ and $\bar{v}_P$. It follows then that $\bar{h}_O$ can have a constant orientation only if the plane of $\bar{r}_{P/O}$ and $\bar{v}_P$ is constant. Hence, we deduce that

> A particle moving under the influence of a central force is in planar motion.

Considering that the central force in this planar motion is always parallel to the radial line from point O to point P, it is logical that, for further study, we employ cylindrical coordinates with $\bar{e}_R$ and $\bar{e}_\phi$ lying in the plane, as illustrated in Figure 16.

Setting $Z \equiv 0$ for planar motion, the kinematical formulas for the position and velocity of particle P in terms of cylindrical coordinates are

Figure 16

$$\bar{r}_{P/O} = R\bar{e}_R \qquad \bar{v}_P = \dot{R}\bar{e}_R + R\dot{\phi}\bar{e}_\phi$$

Substituting these expressions into equation (25) yields

$$\bar{h}_O = R\bar{e}_R \times (\dot{R}\bar{e}_R + R\dot{\phi}\bar{e}_\phi) = R^2\dot{\phi}\bar{K}$$

from which we note that h_O has a single component in the (fixed) $\bar{K}$ direction, which is in agreement with the earlier discussion. Equally important, from the initial conditions we can evaluate the magnitude h_O of the vector $\bar{h}_O$, which is a constant for the entire motion.

$$h_O = |\bar{h}_O| = R_1^2 \dot{\phi}_1 = R^2 \dot{\phi} \tag{26}$$

This result has a geometric interpretation. Consider the particle in two positions that are very close to each other, as illustrated in Figure 17. Let us compute the shaded area $d\mathcal{A}$ in Figure 17, which is the area swept out by the position vector in the time interval from t to $t + dt$. As shown in the figure, the base of the triangle is R and the height is $R\,d\phi$, so that

$$d\mathcal{A} = \tfrac{1}{2}R(R\,d\phi)$$

Dividing by dt and comparing the result to equation (26) gives

$$\frac{d\mathcal{A}}{dt} = \frac{1}{2}R^2\frac{d\phi}{dt} = \frac{1}{2}R^2\dot{\phi} = \frac{1}{2}h_O \tag{27}$$

Figure 17

The quantity $d\mathcal{A}/dt$ is called the *areal velocity*. It is the rate at which the position vector sweeps out area. Hence, because h_O is a constant, we see that the areal velocity is constant in central force motion.

We now wish to determine the precise planar path of a particle undergoing central force motion. The central force $\bar{F}$ has only one component (in the radial direction), so from Newton's second law we find the scalar equations of motion to be

$$\Sigma F_R = F = ma_R = m(\ddot{R} - R\dot{\phi}^2)$$
$$\Sigma F_\phi = 0 = ma_\phi = m(R\ddot{\phi} + 2\dot{R}\dot{\phi})$$
$$\Sigma F_Z \equiv 0 \tag{28}$$

Equation (26) may now be used to eliminate $\dot{\phi}$ from the two nontrivial equations of motion. Solving for $\dot{\phi}$ we have

$$\dot{\phi} = \frac{h_O}{R^2}$$

Upon differentiating, this yields

$$\ddot{\phi} = -\frac{2h_O}{R^3}\dot{R}$$

Thus the equations of motion become

$$F = m\left[\ddot{R} - R\left(\frac{h_0}{R^2}\right)^2\right]$$

$$0 = m\left[R\left(-\frac{2h_0}{R^3}\dot{R}\right) + 2\dot{R}\left(\frac{h_0}{R^2}\right)\right] \equiv 0$$

Note that the second equation (the transverse force component) is also trivial. This is a consequence of the fact that the angular impulse-momentum principle for a single particle is equivalent to the transverse component of Newton's second law (see Module III). We are now left with a single scalar equation of motion, the solution to which defines the particle's path.

$$\ddot{R} - \frac{h_0^2}{R^3} = \frac{F}{m} \tag{29}$$

Equation (29) is a nonlinear differential equation that is not readily solvable. To facilitate its solution let us transform the equation from its present form, relating R and t, to a simpler one relating R and ϕ.

Considering R to be a function of ϕ, from equation (26) and the chain rule we can form

$$\dot{R} \equiv \frac{dR}{dt} = \frac{d\phi}{dt}\frac{dR}{d\phi} = \frac{h_0}{R^2}\frac{dR}{d\phi}$$

and

$$\ddot{R} \equiv \frac{d}{dt}(\dot{R}) = \frac{d\phi}{dt}\frac{d\dot{R}}{d\phi} = \frac{h_0}{R^2}\frac{d}{d\phi}\left(\frac{h_0}{R^2}\frac{dR}{d\phi}\right)$$

For convenience, we introduce a new variable $q = 1/R$ so that

$$\frac{dR}{d\phi} = \frac{d}{d\phi}\left(\frac{1}{q}\right) = -\frac{1}{q^2}\frac{dq}{d\phi}$$

The expression for $\ddot{R}$ now becomes

$$\ddot{R} = h_0 q^2 \frac{d}{d\phi}\left[h_0 q^2\left(-\frac{1}{q^2}\frac{dq}{d\phi}\right)\right] = -h_0^2 q^2 \frac{d}{d\phi}\left(\frac{dq}{d\phi}\right)$$

Now substituting these expressions for R and $\ddot{R}$ into equation (29) yields

$$-h_0^2 q^2 \frac{d^2q}{d\phi^2} - h_0^2 q^3 = \frac{F}{m}$$

$$\frac{d^2q}{d\phi^2} + q = -\frac{F}{mh_0^2 q^2} \tag{30}$$

Equation (30) is the general form of the equation for the path of a particle undergoing central force motion. To proceed further we must define the functional dependence of the central force. Our concern here is with the motion of a particle under the influence of gravity.

2   Gravitational Central Force Motion — Conic Sections

Restricting our attention to the case where the central force is the gravitational force defined by equation (23), in cylindrical coordinates we have

$$\bar{F} = -\frac{GMm}{R^2}\,\bar{e}_R$$

Upon replacing $1/R$ by q and substituting this force expression into the basic equation (30), we obtain

$$\frac{d^2q}{d\phi^2} + q = -\frac{1}{mh_o{}^2q^2}(-GMmq^2) = \frac{GM}{h_o{}^2} \tag{31}$$

This is a linear, nonhomogeneous, second-order differential equation with constant coefficients. Its solution is a principal topic in a first course on differential equations. It also is essentially the same as the equation of motion of the vibrating systems considered in Section A of this module, so we shall not repeat the steps in the derivation of the solution. Rather, we call attention to the fact that it is readily verified by direct substitution that the general solution to equation (31) is

$$q = \frac{1}{R} = C\cos(\phi - \beta) + \frac{GM}{h_o{}^2} \tag{32}$$

This is the equation for the *trajectory* of a particle undergoing gravitational central force motion. The constants C and β are constants of integration that, as we shall see, are determined from the initial conditions.

Although it is not obvious, equation (32) is the equation in polar coordinates of the curve formed from a conic section. Hence, it will prove beneficial to review briefly some properties of a conic section. To construct the conic section of Figure 18 on the following page, we first choose a fixed point O, called the *focus*, and a fixed line, called the *directrix*. The line OQ', perpendicular to the directrix and passing through the focus, is the axis of symmetry of the conic section.

A conic section has the property that the radial distance R from the focus to any point P on the curve is in a constant ratio to the perpendicular distance d from the directrix to point P; that is,

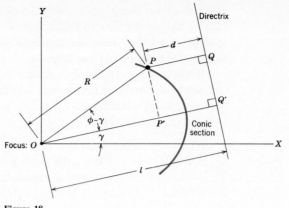

Figure 18

$$e = \frac{R}{d} \tag{33}$$

where the constant e, called the *eccentricity*, is always greater than or equal to zero.

For a specific choice of the focus and directrix, we can determine the constant angle γ and the constant distance l shown in Figure 18 by constructing the line OQ'. Hence, let us express the distance d in terms of these parameters. From the figure we see that

$$l = \left|\bar{r}_{P'/O}\right| + \left|\bar{r}_{Q'/P'}\right| = R\cos(\phi - \gamma) + d$$

Solving this expression for d and substituting the result into equation (33) yields

$$e = \frac{R}{l - R\cos(\phi - \gamma)}$$

or

$$\frac{1}{R} = \frac{1}{el}\left[1 + e\cos(\phi - \gamma)\right] \tag{34}$$

Comparing this expression for R to equation (32) for the gravitational motion gives

$$\boxed{\gamma = \beta \quad , \quad \frac{1}{l} = C \quad , \quad \frac{1}{el} = \frac{GM}{h_0^2}} \tag{35}$$

Thus, once the constants, h_0, β, and C are evaluated from the initial conditions on the motion, we can determine the eccentricity e, the distance l from the focus to the directrix, and the angle γ between the axis of symmetry of the conic section and the X axis. These computations fully define the path of the particle.

Before we perform this determination, let us recall some results of the study of analytic geometry, where it is shown that the type of conic section can be determined by the value of the eccentricity e. We have the following possibilities:

1   When $e = 0$ the path is a circle.
2   When $0 \le e < 1$ the path is an ellipse.
3   When $e = 1$ the path is a parabola.
4   When $e > 1$ the path is a hyperbola.

Figure 19

The ellipse is the only closed conic section (considering the circle as a special ellipse). It is of particular interest to us because many space vehicles, as well as the planets, follow elliptical *orbits*. Basic properties of an elliptical path are the lengths of the semimajor and semiminor axes, which are the values a and b shown in Figure 19. From this figure we see that the minimum and maximum values of R occur when the particle is at point A ($\phi = \gamma$) and when it is at point B ($\phi = 180° + \gamma$), and also that $2a = R_A + R_B$. Solving equation (34) for the values of R_A and R_B yields

$$\frac{1}{R_A} = \frac{1}{el}(1 + e) \qquad \frac{1}{R_B} = \frac{1}{el}(1 - e)$$

Now, using equation (35) to eliminate the term el gives

$$R_A = \frac{h_0{}^2}{GM}\left(\frac{1}{1 + e}\right) \qquad R_B = \frac{h_0{}^2}{GM}\left(\frac{1}{1 - e}\right)$$

Hence

$$\boxed{\begin{array}{l} a = \dfrac{1}{2}(R_A + R_B) = \dfrac{h_0{}^2}{GM}\left(\dfrac{1}{1 - e^2}\right) \\[2mm] R_A = a(1 - e) \qquad R_B = a(1 + e) \end{array}}$$

(36a)

The value of b can then be shown to be

$$\boxed{b = a\sqrt{1 - e^2}}$$

(36b)

In the case of orbital motion about the earth, point A, where R is a minimum, is called the *perigee*, and point B, where R is a maximum, is called the *apogee*.

3 Initial Conditions

To aid in our understanding of the role of the initial conditions, consider the problem of placing a satellite into orbit, as shown in Figure 20. At the completion of the powered portion of the flight, the space vehicle is at a radial distance R_1 and has a velocity $\bar{v}_1$ directed at an angle θ_1 from the radial line.

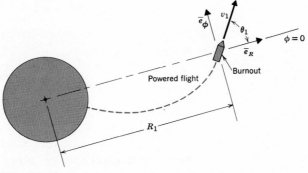

Figure 20

As shown in the figure, ϕ is measured from the initial position, so the initial conditions are stated as

$$R = R_1 \quad \text{and} \quad \bar{v} = v_1(\cos \theta_1 \, \bar{e}_R + \sin \theta_1 \, \bar{e}_\phi) \qquad \text{when } \phi = 0$$

However, when we match the above velocity expression to the cylindrical coordinate expression for velocity, we may rewrite the initial conditions in scalar form as

$$R = R_1 \quad , \quad \dot{R}_1 = v_1 \cos \theta_1 \quad , \quad \dot{\phi}_1 = \frac{v_1 \sin \theta_1}{R_1} \qquad \text{when } \phi = 0 \tag{37}$$

The task now is to determine the constants in the general solution for the trajectory, equation (32), which match these initial conditions. We do this by first using equation (26) to determine the angular momentum per unit mass. This gives

$$\boxed{h_O = R_1^2 \, \dot{\phi}_1 = R_1 v_1 \sin \theta_1} \tag{38}$$

Next, differentiating equation (32) yields an expression for $\dot{R}$.

$$\frac{d}{dt}\left(\frac{1}{R}\right) \equiv -\frac{1}{R^2}\dot{R} = \frac{d}{dt}\left[C \cos(\phi - \beta) + \frac{GM}{h_O{}^2}\right]$$

$$= -C\dot{\phi}\sin(\phi - \beta) = -C\left(\frac{h_O}{R^2}\right)\sin(\phi - \beta)$$

When we set $\phi = 0$ and substitute equations (37) for R and $\dot{R}$ at the initial position, we obtain

$$\frac{1}{R_1} = C \cos \beta + \frac{GM}{h_0^2}$$

$$-\frac{1}{R_1^2} v_1 \cos \theta_1 = C\left(\frac{h_0}{R_1^2}\right) \sin \beta$$

Employing equation (38) and rearranging terms, the foregoing expressions become

$$C \cos \beta = \frac{1}{R_1} - \frac{GM}{h_0^2}$$

$$C \sin \beta = -\frac{v_1}{h_0}\cos \theta_1 = -\frac{1}{R_1}\cos \theta_1$$ (39)

These expressions may be solved for the values of C and β. In doing so, special care must be taken to place the value of β in the correct quadrant.

Once the value of C is computed, we may determine the properties of the resulting trajectory. In particular, we are interested in the eccentricity. From equations (35) we have

$$e = \frac{h_0^2}{GM} \frac{1}{l} = \frac{h_0^2}{GM} C$$

A formula for e in terms of the initial conditions may now be found by squaring each of the equations (39) and adding them to solve for C^2. This gives

$$C^2(\cos^2 \beta + \sin^2 \beta) = C^2 = \left(\frac{1}{R_1} - \frac{GM}{h_0^2}\right)^2 + \frac{1}{R_1^2} \cot^2 \theta_1$$

$$= \frac{1}{R_1^2} \frac{1}{\sin^2 \theta_1} - \frac{2GM}{h_0^2} \frac{1}{R_1} + \left(\frac{GM}{h_0^2}\right)^2$$

Using equation (38) we then have

$$C = \sqrt{\frac{v_1^2}{h_0^2} - 2\frac{GM}{h_0^2} \frac{1}{R_1} + \left(\frac{GM}{h_0^2}\right)^2}$$

The earlier expressions for the eccentricity then becomes

$$e = \frac{h_0^2}{GM} C = \frac{h_0}{GM}\sqrt{v_1^2 - 2\frac{GM}{R_1} + \left(\frac{GM}{h_0}\right)^2}$$ (40)

v = 5 km/s

30°

300 km

Earth

EXAMPLE 7

A satellite is placed into its orbit at an altitude of 300 km above the earth's surface at a speed of 5 km/s and an angle $\theta = 30°$, as illustrated. Determine the equation for its trajectory and show that its orbit is elliptical. Also determine the perigee and apogee distances.

Solution

Remember that the radial distance R is the distance from the focus of the central force. Taking the earth's radius to be 6370 km, the given initial conditions are that, when $\phi = 0$,

$$R_1 = 6370 + 300 = 6670 \text{ km}$$

$$v_1 = 5 \text{ km/s} \qquad \theta_1 = 30°$$

The two key parameters for determining the orbit are the angular momentum h_O and the eccentricity e. From equation (38) we have

$$h_O = R_1 v_1 \sin \theta_1 = 6670(5) \sin 30° = 16675 \text{ km}^2/\text{s}$$

Equation (40) then gives the eccentricity and the value of the coefficient C.

$$e = \frac{h_O}{GM} \sqrt{v_1{}^2 - 2\frac{GM}{R_1} + \left(\frac{GM}{h_O}\right)^2} = \frac{h_O{}^2}{GM} C$$

Using the value of GM for the earth determined in Example 6, the eccentricity is

$$e = \frac{16675}{3.979(10^5)} \sqrt{(5)^2 - \frac{2(3.979)(10^5)}{6670} + \frac{(3.979)^2(10^{10})}{(16675)^2}} = 0.9134$$

Hence, because $0 < e < 1$, the trajectory is an ellipse.

To describe the equation of the ellipse we need the values of C and β. From the foregoing,

$$C = \frac{GM}{h_O{}^2} e = \frac{3.979(10^5)}{(16675)^2}(0.9134) = 1.3071(10^{-3}) \text{ 1/km}$$

and from equations (39),

$$\cos \beta = \frac{1}{C}\left(\frac{1}{R_1} - \frac{GM}{h_O{}^2}\right) = -0.9801$$

$$\sin \beta = -\frac{v_1 \cos \theta_1}{h_O C} = -0.1987$$

Then, because the sine and cosine are both negative, β is in the third quadrant, so

$$\beta = -168.54°$$

Equation (32) for the orbit then gives

$$\frac{1}{R} = C\cos(\phi - \beta) + \frac{GM}{h_0{}^2}$$

$$= 1.3071(10^{-3})\cos(\phi + 168.54°) + 1.4310(10^{-3})\ 1/\text{km} \qquad \triangle$$

Finally, we use equations (36a) to find the perigee and apogee distances. Thus,

$$a = \frac{h_0{}^2}{GM}\frac{1}{(1 - e^2)} = \frac{1}{1.4310(10^{-3})}\frac{1}{[1-(0.9134)^2]} = 4217\ \text{km}$$

$$R_{\text{perigee}} = a(1 - e) = 4217(1 - 0.9134) = 365\ \text{km} \qquad \triangle$$

$$R_{\text{apogee}} = a(1 + e) = 4217(1 + 0.9134) = 8069\ \text{km} \qquad \triangle$$

The fact that the perigee distance is smaller than the radius of the earth means that the satellite will crash into the earth before it reaches its perigee; the launch was unsuccessful.

HOMEWORK PROBLEMS

VIII.62   A satellite is launched at an altitude of 500 km above the earth's surface with a velocity of 9 km/s parallel to the earth's surface. Determine the eccentricity of the orbit and the maximum and minimum altitudes for the satellite's trajectory.

Prob. VIII.62

VIII.63   The altitude of a satellite at apogee is 4040 miles and at perigee it is 540 miles. Determine the equation of the trajectory and the velocity of the satellite at apogee and at perigee.

VIII.64   A satellite in circular orbit at an altitude of 1130 km is to be given a new orbit. To do this the engines are fired tangent to the orbit, imparting an additional velocity of 4 km/s to the satellite. Determine the equation for the new orbit and the eccentricity of the orbit. Is the orbit open or closed?

Prob. VIII.64

VIII.65   Solve Problem VIII.64 if the engines are aligned radially, giving the satellite a radial velocity component of 4 km/s outward.

VIII.66   A satellite in circular orbit at an altitude of 240 miles is to be moved to a higher coplanar circular orbit at an altitude of 6040 miles. To accomplish this the engines are fired tangentially at point P_1, increasing the speed by Δv_1, thus placing the satellite into an elliptical orbit. (This orbit is called the Hohmann transfer ellipse.) At point P_2 the rockets are again fired tangentially, increasing the speed by Δv_2, and thus placing the satellite in the desired circular orbit. Determine Δv_1 and Δv_2.

Prob. VIII.67

Prob. VIII.66

Prob. VIII.68

VIII.67 A satellite is in a circular orbit at an altitude of 630 km when retrorockets are fired tangent to the orbit, slowing the rocket by 5000 km/hr. Determine the angle ϕ locating the point S where the satellite will splash down. Neglect air resistance.

VIII.68 At an altitude of 2040 miles above the earth's surface, a space probe is launched into a hyperbolic trajectory, giving it a velocity of $3(10^4)$ miles/hr tangent to the earth's surface. Determine the largest value of the angle ϕ that is possible for this trajectory.

Prob. VIII.69

VIII.69 At an altitude of 30,000 km above the surface of the earth, a meteorite has a velocity of 15 km/s at $\theta = 120°$. Determine (a) the eccentricity of the trajectory of the meteorite, (b) the minimum altitude for the trajectory, (c) the angle ϕ at the position of closest approach. Will the meteorite enter the earth's atmosphere?

Prob. VIII.70

VIII.70 The engine of a missile is shut off at an altitude of 150 km, at which time the velocity is 6 km/s at an angle $\theta = 60°$. Determine the equation for the trajectory, the maximum altitude H attained by the missile, and the polar angle ϕ at this location.

VIII.71 Determine the angle ϕ defining the location where the missile in Problem VIII.70 returns to the surface of the earth. Air resistance is negligible.

Prob. VIII.71

VIII.72 A bullet is fired from the surface of the earth at 600 m/s at an angle of elevation of 45°. (a) Using the orbit equations, determine the distance d locating where the bullet returns to the earth. (b) Compare the result of part (a) to that obtained when the earth is considered flat and the gravity force is considered constant.

4 Energy

In the preceding section we saw how the trajectory of a particle is determined. The evaluation of the velocity of the particle when it is at a specific position along its trajectory requires further development. There are many ways in which we may proceed. Here we choose the work-energy principles, for aside from being a direct approach, it is a good tool for gaining further insight into the orbital motion problem.

When the gravitational force is defined by the inverse square law, equation (23), the potential energy of this conservative force was shown in Module III to be

$$V_{gr} = - \frac{GMm}{R}$$

The only force acting on the particle is gravity, and hence mechanical energy is conserved; that is,

$$T + V_{gr} = \text{constant}$$

Denoting the constant energy per unit mass as E, we then have

$$E \equiv \frac{1}{m}(T + V_{gr}) = \frac{1}{2}v_P{}^2 - \frac{GM}{R} \qquad (41)$$

The value of E may be obtained from the initial conditions. Then the energy relation, equation (41), will yield the particle speed at any other value of R. The radial component of velocity v_R may then be determined by noting that the speed is the magnitude of the velocity vector. Thus

$$v_P{}^2 = v_R{}^2 + v_\phi{}^2$$

Because $v_\phi = R\dot\phi$, we find from equation (26) that

$$v_\phi = R\dot\phi = R\left(\frac{h_o}{R^2}\right) = \frac{h_o}{R} \qquad (42)$$

$$v_R = \pm \sqrt{v_P{}^2 - \left(\frac{h_o}{R}\right)^2}$$

Hence, equations (41) and (42) define the velocity as a function of R.

The parameter E assumes greater meaning when we substitute E from equation (41) into equation (40) for the eccentricity of the orbit, for we then find that

$$e = \frac{h_o}{GM} \sqrt{2E + \left(\frac{GM}{h_o}\right)^2}$$

or

$$e = \sqrt{1 + 2E\left(\frac{h_0}{GM}\right)^2}$$

(43)

Recalling the types of trajectories associated with the different ranges of values of the eccentricity, we see that:

1 Circular trajectory: $e = 0$, so $E = -0.5(GM/h_0)^2$
2 Elliptical trajectory: $0 < e \le 1$, so $-0.5(GM/h_0)^2 \le E < 0$
3 Parabolic trajectory: $e = 1$, so $E = 0$
4 Hyperbolic trajectory: $e > 1$, so $E > 0$

Thus, the energy level E is a direct indicator of the type of trajectory.

An important consequence of the foregoing is the concept of the *escape speed*, which is the minimum speed that a body must have in order to be in an open trajectory that extends to an infinite value of R. Because the parabola is the open trajectory having the lowest energy level, we set $E = 0$ to find v_{escape}. This yields

$$v_{\text{escape}} = \sqrt{\frac{2GM}{R}}$$

(44)

EXAMPLE 8

The system and initial conditions presented in Example 7 are repeated in the accompanying diagram. Once again, determine the perigee and apogee distances, but now use the energy and angular momentum equations derived in this section.

Solution

The given initial conditions allow us to compute the energy parameter E and angular momentum parameter h_0, as follows.

$$E = \frac{1}{2}v_1^2 - \frac{GM}{R_1} = \frac{1}{2}(5)^2 - \frac{3.979(10^5)}{(6370 + 300)}$$

$$= -47.16 \text{ km}^2/\text{s}^2$$

$$h_0 = R_1 v_1 \sin \theta_1 = (6370 + 300)5 \sin 30°$$

$$= 16675 \text{ km}^2/\text{s}$$

Equation (41) then gives

$$-47.16 = \frac{1}{2}v_P^2 - \frac{3.979(10^5)}{R}$$

whereas equation (42) gives

$$v_\phi = \frac{16675}{R}$$

Now, imposing the condition that the value of R corresponds to the perigee or apogee (referring to Figure 19), we see that the tangent to the path at these two positions is perpendicular to the radial line. Hence, at these locations $v_R = 0$, so $v_P = v_\phi$. Thus, we have

$$v_P = \frac{16675}{R}$$

Substituting this into the energy equation gives

$$-47.16 = \frac{1}{2}\left(\frac{16675}{R}\right)^2 - \frac{3.979(10^5)}{R}$$

This expression simplifies to a quadratic equation, called the *absidial quadratic,* whose roots are the perigee and apogee distances.

$$R^2 - \frac{3.979(10^5)}{47.16}R + \frac{1}{2}\frac{(16675)^2}{47.16} = 0$$

$$R = 365 \text{ or } 8073 \text{ km}$$

The smaller value corresponds to perigee. To the accuracy of our calculations these values are identical to those obtained in Example 7.

As an aside, it should be mentioned that the multitude of relations that have been derived often means that there is a choice of methods for solving a problem. No single approach is, in general, optimal.

EXAMPLE 9
A space probe is injected into orbit at position A (altitude of 100 miles) with a velocity that is perpendicular to the radial line and is 10% greater than the local escape velocity. Determine the altitude and velocity of the probe at position B, where the radial line from the center of the earth has swept out a 90° angle.

Solution
First, to avoid any error in units, let us convert the value of GM found in Example 6 to units of miles and hours. Thus,

$$GM = 1.4064(10^{16})\frac{\text{ft}^3}{\text{s}^2}\left(\frac{3600\text{s}}{1\text{ hr}}\right)^2\left(\frac{1\text{ mile}}{5280\text{ ft}}\right)^3$$

$$= 1.2383(10^{12})\text{ miles}^3/\text{hr}^2$$

We now list the given initial conditions, using equation (44) for v_escape.

Hence when $\phi = 0$,

$$R_1 = (3960 + 100) = 4060 \text{ miles}$$

$$v_1 = (v_\phi)_1 = 1.10 v_{\text{escape}}$$

$$= 1.10 \sqrt{\frac{2GM}{R_1}} = 2.717(10^4) \text{ miles/hr}$$

With these values we can determine the energy and angular momentum parameters, as follows.

$$E = \frac{1}{2} v_1^2 - \frac{GM}{R_1} = 6.410(10^7) \text{ miles}^2/\text{hr}^2$$

$$h_O = R_1(v_\phi)_1 = 4060(27170) = 1.1031(10^8) \text{ miles}^2/\text{hr}$$

To determine the distance for a specific value of ϕ we must first determine the coefficients C and β appearing in equation (32) for the trajectory. Because the probe is launched perpendicularly to the radial line ($\theta_1 = 90°$), the launch point A is on the axis of symmetry and $\beta = 0$. This is readily verified from the second of equation (39). Solving equation (40) for e and then for C yields

$$e = \frac{1.1031(10^8)}{1.2383(10^{12})} \left[(2.717)^2(10^8) - 2\frac{1.2383(10^{12})}{4060} + \frac{(1.2383)^2(10^{24})}{(1.1031)^2(10^{16})} \right]^{1/2} = 1.4204$$

$$C = \frac{1.2383(10^{12})}{(1.1031)^2(10^{16})} (1.4204) = 1.4455(10^{-4}) \text{ 1/miles}$$

Equation (32) for the trajectory then becomes, with $\beta = 0$,

$$\frac{1}{R} = C \cos \phi + \frac{GM}{h_O^2}$$

$$= 1.4455(10^{-4}) \cos \phi + 1.0176(10^{-4})$$

To find the radial distance of point B we simply substitute $\phi = 90°$. This gives

$$\frac{1}{R_B} = 1.0176(10^{-4}) \qquad R_b = 9827 \text{ miles}$$

so

$$\text{altitude} = 9827 - 3960 = 5867 \text{ miles} \qquad \triangleleft$$

We can now determine the velocity at this location by employing equations (41) and (42). The former equation yields

$$E = 6.410(10^7) \frac{\text{miles}^2}{\text{hr}^2} = \frac{1}{2} v_B{}^2 - \frac{GM}{R_B}$$

$$= \frac{1}{2} v_B{}^2 - \frac{1.2383(10^{12})}{9827}$$

Solving this equation, we obtain

$$v_B = 19500 \text{ miles/hr}$$

The first of equations (42) then yields

$$(v_\phi)_B = \frac{h_O}{R_B} = \frac{1.1031(10^8)}{9827} = 11,225 \text{ miles/hr}$$

Now, substituting this result into the second of equations (42), we find that

$$(v_R)_B = \sqrt{v_B{}^2 - (v_\phi)_B{}^2} = 15,945 \text{ miles/hr}$$

The resulting velocity is sketched below. The angle between $\bar{v}_B$ and the radial line is $\theta_B = \tan^{-1}[(v_\phi)_B/(v_R)_B] = 35.14°$

HOMEWORK PROBLEMS

VIII.73   An earth satellite has a speed of 5.5 miles/s at its minimum altitude of 400 miles. Determine (a) its angular momentum per unit mass, (b) its energy per unit mass, (c) that its trajectory is an ellipse, (d) its speed at apogee, and (e) its altitude at apogee.

VIII.74   A weather satellite is placed into orbit at an altitude of 330 km above the earth's surface, moving parallel to the earth's surface. Determine (a) the maximum initial velocity that will keep the satellite in the earth's gravitational field. (b) The initial speed that will cause the orbit to be a circle. (c) The minimum initial speed for a successful orbit. (The perigee for this orbit is the radius of the earth, so it is called the *grazing orbit*). Determine the eccentricity and energy per unit mass for such an orbit. (d) The initial speed that will result in a second (different) orbit of the same eccentricity as that of part (c), with the launch position now being the perigee. Find its energy per unit mass and altitude at apogee.

VIII.75   A meteorite at a distance of 60,000 km from the center of the earth has a velocity of 50,000 km/hr at $\theta = 165°$. Determine the velocity of the meteorite at a radial distance of 25,000 km.

Prob. VIII.75

Prob. VIII.77

Prob. VIII.78

Prob. VIII.79

Prob. VIII.80

VIII.76   Determine the velocity of the meteorite in Problem VIII.75 at its closest approach to the earth.

VIII.77   A satellite is in a polar orbit of eccentricity 0.40 for which perigee occurs over the north pole at an altitude of 630 km. Determine the velocity of the satellite as it passes over (a) the south pole, (b) the equator.

VIII.78   A space probe is launched into a parabolic trajectory by shutting down the engines at an altitude of 90 miles, at which point the velocity is tangent to the earth's surface. Determine the velocity of the probe when it reaches an altitude of 10,000 miles.

VIII.79   Determine the rate of change of the speed of the satellite in Problem VIII.78 and the radius of curvature of the trajectory at an altitude of 10,000 miles.

VIII.80   A space probe is following a hyperbolic trajectory of eccentricity 1.50. The engines were shut down at an altitude of 230 km, at which point the velocity was 80° from the outward radial direction. Determine the velocity of the space probe when it is (a) 1000 km above the earth's surface, (b) 10,000 km above the earth's surface.

VIII.81   For the space probe of Problem VIII.80, determine the rate of change of the speed and the radius of curvature of the trajectory at each altitude.

VIII.82   At its closest approach to the earth a satellite is at an altitude of 530 km and has a speed of 36,000 km/hr. Determine the velocity of the satellite when $\phi = 60°$.

VIII.83   Solve Problem VIII.82 for $\phi = 120°$.

5   The Period of an Orbit

Our study of the dynamics of gravitational central force motion is almost completed. We have seen how the position parameters R and ϕ are related through the equations of the trajectory. We have also seen how the velocity at any position along the trajectory may be evaluated. An important parameter we have not yet discussed is time.

In general, the easiest way to bring time into the problem is to use the fact that the angular momentum is constant. From equation (26) we have

$$\dot{\phi} \equiv \frac{d\phi}{dt} = \frac{h_o}{R^2}$$

Eliminating R between this result and equation (32), we obtain a first-order differential equation for $\phi(t)$.

$$\frac{d\phi}{dt} = h_0 \left[C \cos(\phi - \beta) + \frac{GM}{h_0{}^2} \right]^2$$

where the values of C and β are known from the initial conditions. This equation is solved by separating variables and integrating to obtain

$$t = \frac{1}{h_0} \int_0^\phi \frac{d\phi}{[C \cos(\phi - \beta) + GM/h_0{}^2]^2} \tag{45}$$

The integral may be evaluated with the aid of a table of integrals, but the result is too lengthy to present here.

A very important time parameter is the *period* of an elliptical orbit, that is, the length of time required for a particle to return to its initial position, thus completing one orbit. For this determination we recall equation (27), which shows that h_0 is twice the areal velocity $d\mathcal{A}/dt$. Thus $d\mathcal{A}/dt = \frac{1}{2}h_0$, and upon integrating.

$$\boxed{\mathcal{A} = \tfrac{1}{2}h_0 t} \tag{46}$$

where $\mathcal{A}$ is the sectorial area swept out in the time interval t. For an elliptical trajectory, the area swept out in a full period τ is the area of an ellipse, which is

$$\mathcal{A} = \pi a b$$

where a and b are the lengths of the semimajor and semiminor axes, respectively. Their values may be found from equations (36). Using the foregoing area and setting $t = \tau$ in equation (46) yields

$$\tau = \frac{2\pi a b}{h_0}$$

At this juncture it is interesting to note the work of the early astronomer Johannes Kepler (1571–1630), who painstakingly studied our solar system. He postulated the following three laws governing planetary motion:

1   The orbit of the planets are ellipses for which the sun is one focus.
2   The areal velocity is constant.
3   The square of the period of the elliptical orbit is in a constant ratio to the cube of its semimajor axis.

We have already proven the first two laws, and the third law may be obtained by using equations (36) to express τ in terms of a. This gives

$$\tau = \frac{2\pi ab}{h_O} = \frac{2\pi a}{h_O}a\sqrt{1-e^2} = \frac{2\pi a^2}{h_O}\left(\frac{h_O}{\sqrt{GMa}}\right)$$

$$\boxed{\tau = \frac{2\pi}{\sqrt{GM}}\, a^{3/2}} \qquad (47)$$

The remarkable aspect of Kepler's laws is that, although we obtained them as mathematical consequences of Newton's laws of gravity and motion, Kepler's work preceded Newton and was the result of careful observation and empiricism.

EXAMPLE 10

The perigee and apogee altitudes for an artificial satellite are 430 and 14,030 km, respectively. Determine the time required for the satellite to travel from point C on the semiminor axis of its elliptical orbit, past the perigee A, and then on to point D, which is the other point on the semiminor axis. Compare this result to the period of the orbit.

Solution

This problem could be solved by integrating equation (45), but a shortcut, based on the concept of the areal velocity, is available. This is so because the sectorial area bounded by the arc CAD and the radial lines OC and OD can be evaluated. In this approach we first determine the eccentricity e and the lengths a and b of the semiaxes of the ellipse.

It is given that

$$R_A = 6,370 + 430 = 6,800$$

$$R_B = 6,370 + 14,030 = 20,400 \text{ km}$$

From equations (36) we then find

$$a = \frac{1}{2}(R_A + R_B) = 13,600 \text{ km}$$

$$e = \frac{R_B}{a} - 1 = 0.5$$

$$b = a\sqrt{1-e^2} = 11,778 \text{ km}$$

The first of equations (36a) provides the value of h_O. Thus

$$h_O = \sqrt{GMa(1-e^2)} = \sqrt{3.974(10^5)13600(1-0.25)}$$

$$= 6.371(10^4) \text{ km}^2/\text{s}$$

The last step before using equation (46) to compute the time is the evaluation of the sectorial area $\mathcal{A}$, which is shaded in the sketch.

430 km   14,030 km

We find $\mathcal{A}$ by deducting the triangular area OCD from the area of the semiellipse. Thus

$$\mathcal{A} = \tfrac{1}{2}\pi ab - \tfrac{1}{2}(2b)(a - R_A) = 1.7152(10^8) \text{ km}^2$$

Denoting the desired time as t^*, equation (46) yields

$$t^* = \frac{2\mathcal{A}}{h_O} = 5384 \text{ s} = 1.496 \text{ hr}$$

In contrast, the orbital period is

$$\tau = \frac{2(\pi ab)}{h_O} = 15{,}797 \text{ s} = 4.38 \text{ hr}$$

Thus the time t^* is 32% less than $\tau/2$. Why isn't $t^* = \tau/2$?

EXAMPLE 11

The engines of the launch vehicle burn out at the correct altitude to inject a weather satellite into a stationary equatorial orbit about the earth. The satellite also has the proper speed at burnout for a stationary orbit, but due to a malfunction, the velocity at burnout is only at an angle of 80° with the radial line. Determine the eccentricity and period of the orbit.

Solution

We must first determine the initial conditions by studying the desired stationary circular orbit, which is an orbit whose period is 1 day. For a circular orbit at radius R_{st} and speed v_{st}, Newton's second law for the radial direction gives

$$\Sigma F_R = -\frac{GMm}{R_{st}^2} = ma_R = -m\frac{v_{st}^2}{R_{st}}$$

so

$$v_{st} = \sqrt{\frac{GM}{R_{st}}}$$

Furthermore, for a circular orbit $a \equiv R_{st}$ and for a stationary orbit $\tau = 23.93$ hr. Hence, setting $GM = 3.979(10^5) \text{ km}^3/\text{s}^2$, equation (47) yields

$$(R_{st})^{3/2} = \frac{\tau\sqrt{GM}}{2\pi} = \frac{23.93(3600)\sqrt{3.979(10^5)}}{2\pi}$$

$$R_{st} = 4.213(10^4) \text{ km}$$

$$v_{st} = \sqrt{\frac{GM}{R_{st}}} = 3.073 \text{ km/s}$$

These values are also the initial distance and speed for the actual orbit. Due to the 80° launch angle, the initial transverse speed is

$$(v_\phi)_1 = v_{st} \sin 80° = 3.026 \text{ km/s}$$

Thus the value of h_0 for the orbit is

$$h_0 = R_1(v_\phi)_1 = 4.213(10^4)(3.026)$$

$$= 1.2749(10^5) \text{ km}^2/\text{s}$$

and the corresponding value of E is

$$E = \frac{1}{2}(v_1)^2 - \frac{GM}{R_1} = \frac{1}{2}(3.073)^2 - \frac{3.979(10^5)}{4.213(10^4)}$$

$$= -4.723 \text{ km}^2/\text{s}^2$$

We may now compute the eccentricity from equation (43). This yields

$$e = \sqrt{1 + 2(-4.723)\left(\frac{1.2749}{3.979}\right)^2} = 0.1740 \qquad \triangle$$

To compute the period, we first evaluate the semimajor axis using equation (36a).

$$a = \frac{h_0^2}{GM(1 - e)^2} = \frac{(1.2749)^2(10^{10})}{3.979(10^5)[1 - (0.1740)^2]}$$

$$= 4.213(10^4) \text{ km}$$

The period is then found from equation (47).

$$\tau = \frac{2\pi}{\sqrt{3.979(10^5)}}[4,213(10^4)]^{3/2}$$

$$= 8.615(10^4) \text{ s} = 23.93 \text{ hr} \qquad \triangle$$

The fact that a is the same as R_{st}, thus giving the same period for the stationary and elliptical orbits, is not coincidental. By using equation (43) to eliminate the eccentricity from equation (36a), we can show that the value of a depends only on the energy level E. Hence, we can conclude that the period is a function only of the magnitude of the initial velocity, not its direction. Nevertheless, the elliptical orbit in this problem is not stationary. Why?

HOMEWORK PROBLEMS
VIII.84 The altitude of an artificial satellite at apogee is 1230 km and at perigee it is 330 km. Determine the period of the orbit.

VIII.85 A satellite is in an elliptical orbit around the earth whose eccentricity is 0.90. The perigee altitude is 540 miles. Determine the period of the orbit.

VIII.86 The eccentricity of the earth's orbit is 0.0167 and its mean distance from the sun (the average of the perigee and apogee distance) is $1.495(10^8)$ km. Taking the earth's period as 365.3 days, determine (a) the angular momentum per unit mass for the earth, and (b) the maximum and minimum distances from the earth to the sun.

VIII.87 Determine the approximate time required for the earth to traverse from one end of the latus rectum of its orbit (point 2) through the perihelion (point A) to the other end of the latus rectum (point 3) sweeping out the shaded area shown. The eccentricity of the earth's orbit is 0.0167. *Hint:* Approximate the shape of area 1-2-3-4 as a trapezoid. The mass of the sun is $3.328(10^5)$ times that of the earth.

Prob. VIII.87   **Prob. VIII.88**

VIII.88 When the engines of the launch vehicle of a satellite are shut off, the altitude is 530 km and the velocity is 9 km/s at $\theta = 90°$ from the outward radial line. Determine the period of the resulting orbit.

VIII.89 Solve Problem VIII.88 if $\theta = 85°$ for the launch and all other parameters are unchanged.

VIII.90 A satellite is in a circular orbit at an altitude of 1000 miles. The engines are fired tangentially at point A, placing the satellite into an elliptical orbit whose apogee is at point B, in order to place it in a new circular orbit at an altitude of 10,000 miles. Determine the length of time after the firing required for the satellite to reach (a) point B, (b) point C on the semiminor axis.

VIII.91 A satellite is destined for a rendezvous with a space station in a stationary orbit. The engines of the launch vehicle are shut down at point A, where the altitude is 130 km and the velocity is perpendicular to the radial line from the center of the earth. At what angle ϕ should the space station be when the engines of the satellite are turned off if rendezvous is to be accomplished at point B?

Prob. VIII.90

Prob. VIII.91

C. SYSTEMS WITH PARTICLE FLOW

Many engineering problems involve the flow of a system of particles, such as the flow of a fluid through or over a body. Depending on the nature of the information required, and hence the approximations made, a fluid dynamics analysis can be quite complicated. In part, this section is an introduction to the subject of fluid dynamics.

1 Basic Equations

Figure 21 depicts a situation where particles are continually entering and leaving a region of fixed shape, such as the interior of a hollow body. This region is called the control volume, $\mathscr{C}\mathscr{V}$.

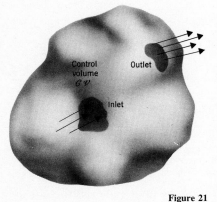

Figure 21

We wish to derive the equations of motion for the $\mathscr{C}\mathscr{V}$ system. To accomplish this we shall use the linear and angular impulse-momentum principles for a system of particles. As in the study of all dynamic systems, we must first decide on a model. Here we begin by considering the question of what details of the motion of the particles we must know to write the equations of motion.

To deal with this question, we make some simplifying assumptions in modeling a time interval of flow from t to $t + \Delta t$. In this interval a small group of particles having mass Δm_1 enters the $\mathscr{C}\mathscr{V}$ as another group of particles Δm_2 leaves. Letting m be the mass of the $\mathscr{C}\mathscr{V}$ at time t, it follows that the mass of the $\mathscr{C}\mathscr{V}$ at time $t + \Delta t$ must be $m + \Delta m_1 - \Delta m_2$.

Let us make the following assumptions about the kinematics of the particles as they enter and leave the $\mathscr{C}\mathscr{V}$.

1 Assume that all particles forming Δm_1 have the same velocity $\bar{v}_1$.
2 Assume that all particles forming Δm_2 have the same velocity $\bar{v}_2$.
3 Assume that all particles contained within the $\mathscr{C}\mathscr{V}$ have no motion relative to the $\mathscr{C}\mathscr{V}$. In other words, the $\mathscr{C}\mathscr{V}$ and the particles it contains form a rigid body.

Figure 22a

Figure 22b

The first two assumptions follow reasonably from the concept of the motion of a mass center. The third is more difficult to visualize. Experimental data have shown the assumptions to be reasonably accurate in a significant number of engineering problems.

Using these assumptions, in Figure 22 we have the linear momentum for the particles of the $\mathscr{C}\mathscr{V}$ entering at time t and upon leaving the $\mathscr{C}\mathscr{V}$ at time $t + \Delta t$. Note that the velocities depicted in Figure 22b are written as $\bar{v}_G + \Delta\bar{v}_G$ for the center of mass and $\bar{v}_2 + \Delta\bar{v}_2$ for the exiting mass because the velocities need not be constant.

Writing the linear momentum vectors for each of the systems of Figures 22a and 22b, we have

$$(\Sigma \ m_i \bar{v}_i)_t = m\bar{v}_G + \Delta m_1 \bar{v}_1$$

$$(\Sigma \ m_i \bar{v}_i)_{t+\Delta t} = (m + \Delta m_1 - \Delta m_2)(\bar{v}_G + \Delta \bar{v}_G) + (\Delta m_2)(\bar{v}_2 + \Delta \bar{v}_2)$$

Letting $\Sigma \ \bar{F}$ denote the total external force acting on the system, the linear impulse-momentum equation then gives

$$(\Sigma \ m_i v_i)_{t+\Delta t} - (\Sigma \ m_i v_i)_t = \int_t^{t+\Delta t} \Sigma \ \bar{F} \ dt$$

$$m(\Delta \bar{v}_G) + (\Delta m_1)\bar{v}_G + (\Delta m_1)(\Delta \bar{v}_G) - (\Delta m_2)\bar{v}_G - (\Delta m_2)(\Delta \bar{v}_G)$$

$$+ (\Delta m_2)(\bar{v}_2) + (\Delta m_2)(\Delta \bar{v}_2) - (\Delta m_1)\bar{v}_1$$

$$= \int_t^{t+\Delta t} \Sigma \ \bar{F} \ dt$$

Collecting terms and retaining only first-order terms, this reduces to

$$m \ \Delta \bar{v}_G - \Delta m_1(\bar{v}_1 - \bar{v}_G) + \Delta m_2(\bar{v}_2 - \bar{v}_G) = \int_t^{t+\Delta t} \Sigma \ \bar{F} \ dt$$

Dividing this expression by Δt and taking the limit as Δt approaches zero yields

$$\boxed{m\bar{a}_G - \dot{m}_1(\bar{v}_1 - \bar{v}_G) + \dot{m}_2(\bar{v}_2 - \bar{v}_G) = \Sigma \ \bar{F}} \qquad (48)$$

where $\dot{m}_1$ is the rate at which mass is entering the inlet and $\dot{m}_2$ is the rate at which it exists in the $\mathscr{C} \mathscr{V}$.

To find the equation for the resultant moment of the force system, we formulate the angular impulse-momentum of the system relative to a point A. For convenience, we restrict point A to be one of the allowable points for summing moments for the rigid body. Also, for simplicity, we shall consider only the case where the $\mathscr{C} \mathscr{V}$ is in planar motion. Thus, the angular momentum for the $\mathscr{C} \mathscr{V}$ is simply $I_{zz}\omega$, where the z axis is perpendicular to the plane of motion passing through point A.

Relative to point A, the velocities of the incoming and outgoing elements of mass are $\bar{v}_1 - \bar{v}_A$ and $\bar{v}_2 - \bar{v}_A$, respectively. Using this notation, the relative and angular momentum vectors for the instant t and $t + \Delta t$ are as shown in Figure 23. The points G_1 and G_2 are the centers of mass of Δm_1 and Δm_2. The thickness of these elements approach zero as $\Delta t \to 0$, so that in the limit, points G_1 and G_2 coincide with the centroids of the inlet and outlet areas, which are points P_1 and P_2. Recalling that, in general,

$$\bar{H}_A = \Sigma \ \bar{r}_{i/A} \times m_i \bar{v}_{i/A}$$

it follows that the angular momentum vectors are

P_1: centroid of inlet area

$I_{zz}\bar{\omega}$

Δm_1

G_1

$\mathscr{C} \mathscr{V}$

$\Delta m_1(\bar{v}_1 - \bar{v}_A)$

Figure 23a

P_2: centroid of outlet area

Δm_2

G_2

$\bar{v}_2$

$I_{zz}(\bar{\omega} + \Delta \bar{\omega})$

$\mathscr{C} \mathscr{V}$

$\Delta m_2(\bar{v}_2 + \Delta \bar{v}_2 - \bar{v}_A - \Delta \bar{v}_A)$

Figure 23b

$$(\bar{H}_A)_t = I_{zz}\bar{\omega} + \bar{r}_{P_1/A} \times \Delta m_1(\bar{v}_1 - \bar{v}_A)$$

$$(\bar{H}_A)_{t+\Delta t} = I_{zz}(\bar{\omega} + \Delta\bar{\omega}) + \bar{r}_{P_2/A} \times \Delta m_2(\bar{v}_2 + \Delta\bar{v}_2 - \bar{v}_A - \Delta\bar{v}_A)$$

Thus, the angular impulse-momentum principle gives

$$(\bar{H}_A)_{t+\Delta t} - (\bar{H}_A)_t = \int_t^{t+\Delta t} \Sigma \bar{M}_A \, dt$$

$$I_{zz}\Delta\bar{\omega} + \bar{r}_{P_2/A} \times \Delta m_2(\bar{v}_2 + \Delta\bar{v}_2 - \bar{v}_A - \Delta\bar{v}_A)$$

$$- \bar{r}_{P_1/A} \times \Delta m_1(\bar{v}_1 - \bar{v}_A)$$

$$= \int_t^{t+\Delta t} \Sigma \bar{M}_A \, dt$$

Again neglecting higher-order terms, dividing by Δt and evaluating the limit as $\Delta t \rightarrow 0$ yields

$$\boxed{\begin{aligned} I_{zz}\bar{\alpha} - \bar{r}_{P_1/A} &\times \dot{m}_1(\bar{v}_1 - \bar{v}_A) \\ &+ \bar{r}_{P_2/A} \times \dot{m}_2(\bar{v}_2 - \bar{v}_A) = \Sigma \bar{M}_A \end{aligned}} \tag{49}$$

where $\bar{\alpha} \equiv \dot{\bar{\omega}}$ is the angular acceleration of the control volume.

Equations (48) and (49) are the basic equations of motion for the control volume. To understand their meaning further, we can transpose the terms containing $\dot{m}_1$ and $\dot{m}_2$ to the right-hand side of the equations and think of them as forces and moments. Then in equation (48) we see that the terms $\dot{m}_1(\bar{v}_1 - \bar{v}_G)$ and $-\dot{m}_2(\bar{v}_2 - \bar{v}_G)$ add to the sum of the external forces, whereas in equation (49) the terms $\bar{r}_{P_1/A} \times \dot{m}_1(\bar{v}_1 - \bar{v}_A)$ and $-\bar{r}_{P_2/A} \times \dot{m}_2(\bar{v}_2 - \bar{v}_A)$ add to the moment of the external forces about point A. Hence, in the situation where point A is chosen as the center of mass G, the equivalent force system representing the combined effect of the external forces and the particle flow is as depicted in Figure 24.

The term $\dot{m}_1(\bar{v}_1 - \bar{v}_G)$ represents the force the incoming particles exert on the $\mathscr{C}\mathscr{V}$. It is the reaction to the force exerted by the $\mathscr{C}\mathscr{V}$ on the particles in order to capture them as they enter the inlet. Similarly, the term $-\dot{m}_2(\bar{v}_2 - \bar{v}_G)$ is the force the particles leaving the outlet exert upon the $\mathscr{C}\mathscr{V}$ in reaction to the increase of momentum imparted to them by the $\mathscr{C}\mathscr{V}$. Notice that these equivalent forces act at the centroids of the inlet and outlet areas.

In summary, equations (48) and (49) describe the general effect of mass flow through a control volume. After studying an example, we shall consider two important cases: the flow of fluids and propulsion by rocket and jet engines.

Figure 24

EXAMPLE 12

When the bottom of a hopper car is opened, coal falls out at the rate of 500 kg/s with an absolute velocity of 4.0 m/s downward, both values being constant. If the hopper car has a velocity of 2.0 m/s to the right on a level track when the bottom is first opened, what is the velocity of the car when it is completely emptied? The car has a mass of 10,000 kg and was loaded with 50,000 kgs of coal.

Solution

Clearly, it is best to describe the vectors in equation (48) with respect to an XYZ coordinate system having horizontal and vertical axes, as shown in the free body diagram.

Treating the hopper car as the control volume, the given information is that mass is flowing out at $\dot{m}_2 = 500$ kg/s with a velocity $\bar{v}_2 = -4\bar{J}$ m/s. When the car is fully loaded, the combined mass of the car and the coal is 60,000 kg. Thus, accounting for the loss of mass, the mass within the control volume at an instant t after opening is

$$m = 60{,}000 - 500t \text{ kg}$$

As the empty mass of the car is 10,000 kg, the time t_f required to empty the car is

$$10{,}000 = 60{,}000 - 500t_f \quad \text{or} \quad t_f = 100 \text{ seconds}$$

Also, the control volume is translating to the right, so $\bar{v}_G = v\bar{I}$.

Using the free body diagram to evaluate the force sum, and substituting the foregoing terms, equation (48) becomes

$$m\bar{a}_G + \dot{m}_2(\bar{v}_2 - \bar{v}_G) = \Sigma\,\bar{F}$$

$$(60,000 - 500t)\dot{v}\bar{I} + 500(-4\bar{J} - v\bar{I}) = (N_{\text{front}} + N_{\text{rear}} - mg)\bar{J}$$

The Y component of this equation involves the values of the normal forces, which we do not need. The X component gives

$$(60,000 - 500t)\dot{v} - 500v = 0$$

To find the value of v at t_f, we separate variables in this differential equation, as follows.

$$(60,000 - 500t)\frac{dv}{dt} = 500v$$

$$\frac{dv}{v} = \frac{500\,dt}{(60,000 - 500t)} = \frac{dt}{(120 - t)}$$

The initial speed is 2 m/s, so

$$\int_2^v \frac{dv}{v} = \int_0^t \frac{dt}{120 - t}$$

$$\ln v - \ln 2 = -\ln(120 - t) + \ln(120)$$

$$v = 2\left(\frac{120}{120 - t}\right)$$

Hence, for $t_f = 100$ seconds,

$$v_f = 12.0 \text{ m/s}$$

HOMEWORK PROBLEMS

VIII.92   Solve Example 12 for the case where the given velocity of the coal as it falls out of the hopper car is 4 m/s downward relative to the car.

VIII.93   Sand is dropped from a bin into a 10,000-kg hopper car at a rate of 250 kg/s, attaining a downward velocity of 10 m/s before landing in the car. A horizontal force $\bar{F}$ is applied to the car to give it a constant velocity of 4 m/s to the right. Determine (a) the magnitude of $\bar{F}$, (b) the increase in the total reaction force between the wheels and the tracks due to the action of the sand falling into the car.

Prob. VIII.93

VIII.94   In Problem VIII.93, the force $\bar{F} = \bar{0}$ and the initial velocity of the hopper car is 4 m/s to the right. Determine the velocity of the hopper car after 20,000 kg of sand have been loaded.

VIII.95   Water from a hose is being collected in a 20-lb cart that rolls freely. The stream of water is discharged from the stationary nozzle at 4 lbs/s with a velocity of 30 ft/s in the direction shown. If the cart is initially at rest, determine (a) the differential equation for the speed v of the cart, (b) the speed of the cart after 0.5 s and 1.0 s.

Prob. VIII.95

VIII.96   When $\theta = 20°$ it takes 3 min to dump 6000 kg of sand from the dump truck. The average velocity relative to the truck of the sand as it leaves is 4 m/s tangent to the bed. Determine the traction force that the wheels must supply to move the truck at 10 km/hr to the right.

Prob. VIII.96

Prob. VIII.97

VIII.97   Portland cement falls from the bin onto the conveyor belt with a velocity of 5 ft/s at a rate of 80 lb/s. Determine the total force F that must be applied to the belt to give it a constant speed of 18 ft/s. Assume that the cement falling off the upper end of the belt is moving parallel to the inclined portion of the belt.

F

Prob. VIII.98

VIII.98   A chain whose mass per unit length is σ is piled on the ground when a force $\bar{F}$ is applied to its end, with the result that the end is pulled upward with a constant speed v. Determine the magnitude of $\bar{F}$ in terms of the length s of the suspended portion of the chain. *Hint:* For a chain whose links are loose, the links piled on the ground can be considered to change velocity instantaneously when they are pulled. For this reason, these links exert a negligible force on the suspended links.

VIII.99   A chain whose mass per unit length is σ and whose length is l is initially suspended vertically by the force $\bar{F}$. The magnitude of $\bar{F}$ is then adjusted with the result that the end of the chain is lowered with a constant speed v. In terms of the length s of the suspended portion of the chain, determine (a) the magnitude of the normal force exerted on the portion of the chain which is piled on the ground, (b) the magnitude of $\bar{F}$. See the hint for Problem VIII.98.

VIII.100   A chain whose mass per unit length is σ is piled on a table at point A with a section overhanging the edge. The chain is released and slides off the table. In terms of the dimensions d and s and the speed v of end B, derive an expression for the acceleration $\dot{v}$ (a) if the table is smooth, (b) if the coefficient of kinetic friction is μ. See the hint for Problem VIII.98.

Prob. VIII.100

2   Fluid Flow

The application of the basic force and moment equations (48) and (49) to describe the motion of fluids frequently requires that we first determine the values of $\dot{m}_1$ and $\dot{m}_2$ in terms of the properties of the system. To see how this is done, let us consider the element of fluid Δm_1 which enters the inlet during a small time interval Δt. Clearly, for the fluid to enter the inlet, the relative velocity, $\bar{u}_1 \equiv \bar{v}_1 - \bar{v}_{P_1}$ of the fluid with respect to the centroid of the inlet P_1 must be inward. We then have the

Figure 25

situation depicted in Figure 25, where the x axis was chosen, for convenience, to be parallel to $\bar{u}_1$.

Letting $\mathcal{A}_1$ be the area of the inlet projected onto the yz plane, that is, the inlet area perpendicular to the incoming flow, the volume of the mass element depicted in Figure 25 is $\mathcal{A}_1 \Delta x$, which means that

$$\Delta m_1 = \rho_1 \mathcal{A}_1 \Delta x$$

where ρ_1 is the density of the fluid at the inlet. As all of the particles forming Δm_1 must cross the yz plane in the time interval Δt, for small Δt we have

$$\Delta x \simeq u_1 \Delta t$$

Thus

$$\Delta m_1 \simeq \rho_1 \mathcal{A}_1 u_1 \Delta t$$

Dividing by Δt and taking the limit as $\Delta t \to 0$, we find that

$$\boxed{\dot{m}_1 = \rho_1 \mathcal{A}_1 u_1} \tag{50a}$$

By similar reasoning, we can show that the outlet mass flow is given by

$$\boxed{\dot{m}_2 = \rho_2 \mathcal{A}_2 u_2} \tag{50b}$$

where $\bar{u}_2$ is the relative velocity of the fluid leaving the control volume with respect to the outlet, $\bar{u}_2 \equiv \bar{v}_2 - \bar{v}_{P_2}$, and ρ_2 is the density of the fluid at the outlet.

In many cases the fluid merely passes through the control volume, so that the mass of the control volume remains constant and

$$\boxed{\dot{m}_1 = \dot{m}_2} \tag{51}$$

In addition to the simplifications in equations (48) and (49) that result from equation (51), the equation provides a useful formula relating the incoming and outgoing flow velocities. Specifically, placing equations (50) into equation (51) gives

$$\boxed{\rho_1 \mathcal{A}_1 u_1 = \rho_2 \mathcal{A}_2 u_2} \tag{52}$$

For a large class of engineering problems we may consider all liquids, as well as unheated gases flowing at a low speed, to be incompressible, which means that ρ is constant. Hence for this class of problems, equation (52) quantifies the simple fact that, for a fixed mass flow, a fluid must move faster to pass through a smaller opening.

EXAMPLE 13

The blade shown in the diagram is adjusted to scoop water from the surface to a depth of 3 in. by moving to the left at the constant speed of 10 ft/s. The width of the blade (into the diagram) is 18 in. Determine the resultant force exerted by the water on the blade and the power required for this motion. Assume that the layer of water over the blade has constant thickness.

Solution

The desired force is the reaction corresponding to the force the blade applies to the water passing over it. We therefore draw a free body diagram of the layer of water. The sketch does not include the hydrostatic pressure at the inlet A and the weight of the control volume because they are negligible compared to the force required to change the momentum of the fluid. Recall that similar reasoning prevailed when we studied impulsive forces in Module III.

We must first determine the rate at which mass is flowing through the control volume. At the inlet, we know that the fluid is at rest, $\bar{v}_1 = \bar{0}$, so the relative velocity of the fluid with respect to the blade is 10 ft/s to the right. For an 18-in.-wide blade the inlet area is

$$\mathscr{A}_1 = (\tfrac{3}{12})(\tfrac{18}{12}) = 0.375 \text{ ft}^2$$

Hence

$$\dot{m}_1 = \rho_{\text{water}} \mathscr{A}_1 u_1 = \left(1.9397 \frac{\text{lb-s}^2}{\text{ft}^4}\right)\left(0.375\text{ft}^2\right)\left(10 \ \frac{\text{ft}}{\text{s}}\right)$$

$$= 7.274 \text{ lb-s/ft}$$

The control volume has constant mass, so $\dot{m}_2 = \dot{m}_1$. The thickness of the layer of water is constant, so the outlet area is the same as the inlet area, which means that $u_2 = u_1$. Thus, the relative velocity of the water with respect to the blade as it leaves the control volume is 10 ft/s tangent to the blade, that is, at a 45° angle above the horizontal.

$$\bar{u}_2 = 10(\cos 45° \ \bar{I} + \sin 45° \ \bar{J}) = 7.071\bar{I} + 7.071\bar{J}$$

The control volume is translating, which means that the outlet is moving with velocity $\bar{v}$. Hence, we may find $\bar{v}_2$ from the relative velocity formula, as follows.

$$\bar{v}_2 - \bar{v} = \bar{u}_2$$

$$\bar{v}_2 = \bar{v} + \bar{u}_2 = -10\bar{I} + 7.071\bar{I} + 7.071\bar{J}$$

$$\bar{v}_2 = -2.929\bar{I} + 7.071\bar{J} \text{ ft/s}$$

The free body diagram shows that $\bar{R}$ is the only external force. Thus, setting $\bar{v}_G = \bar{v}$ and $\bar{a}_G = \bar{0}$, equation (48) yields

$$-\dot{m}_1(\bar{v}_1 - \bar{v}) + \dot{m}_1(\bar{v}_2 - \bar{v}) \equiv \dot{m}_1(\bar{v}_2 - \bar{v}_1) = \bar{R}$$

$$7.274(-2.929\bar{I} + 7.071\bar{J} - \bar{0}) = \bar{R}$$

$$\bar{R} = -21.31\bar{I} + 51.43\bar{J} \text{ lb}$$

The force applied to the blade is $-\bar{R}$, as shown in the diagram.

The component of fluid force opposite the velocity of a body is called the *drag force*. Thus, a horizontal force of 21.31 lb must be applied in opposition to the drag force to move the blade with a constant velocity of 10 ft/s. The power required to produce this motion is

$$\text{Power} = (21.31)(10) = 213.1 \text{ ft-lb/s} = 0.387 \text{ hp}$$

EXAMPLE 14

The reducing elbow connects pipe 1 whose inside diameter is 100 mm to pipe 2 whose diameter is 50 mm. The centroidal axis of the system coincides with the horizontal plane. The velocity of the inlet water in pipe 1 is 16 m/s to the right. Assuming that the water pressure is constant at 200 kilopascals, determine the force-couple system at point B representing the influence of the water on the piping system.

Solution

Because we want to study the interaction of the water and the pipe, let the water within the elbow be the control volume. The force-couple system we seek is shown in the free body diagram as a force $\bar{R}$ at point B and a couple $\bar{M}_B$. The pressure forces at both ends of the elbow are also accounted for in the free body diagram. Recall that, by definition, a pascal is 1 N/m².

Using the given information that the water in pipe 1 has a speed of 15 m/s, we find that

$$\dot{m}_1 = \rho_{\text{water}} \, \mathcal{A}_1 v_1$$

$$= \left(1000 \, \frac{\text{kg}}{\text{m}^3}\right)\left(\frac{\pi}{4}\right) (0.10 \text{ m})^2 \left(16 \, \frac{\text{m}}{\text{s}}\right)$$

$$= 125.66 \text{ kg/s}$$

The control volume is always filled with water, so we have

$$\dot{m}_2 = \dot{m}_1$$

Equation (52) then gives

$$\rho_{\text{water}} \, \mathcal{A}_2 v_2 = \rho_{\text{water}} \, \mathcal{A}_1 v_1$$

$$v_2 = \frac{\mathcal{A}_1}{\mathcal{A}_2} \, v_1 = 4(16) = 64 \text{ m/s}$$

Thus, the velocity vectors are

$$\bar{v}_1 = 16\bar{I} \qquad \bar{v}_2 = 64\bar{J} \text{ m/s}$$

Noting that the control volume is stationary, we have $\bar{v}_G \equiv \bar{0}$. Thus, using equation (48) to determine the force $\bar{R}$ yields

$$-\dot{m}_1 \bar{v}_1 + \dot{m}_1 \bar{v}_2 = 125.66(-16\bar{I} + 64\bar{J}) = 1570.8\bar{I} - 392.7\bar{J} + \bar{R}$$

$$\bar{R} = -3581\bar{I} + 8435\bar{J} \text{ N} \qquad \triangle$$

To evaluate the couple $\bar{M}_B$ we sum moments. Because the control volume is stationary, $\bar{\alpha} = \bar{0}$ and we may sum moments about any point. In order to eliminate the moment of $\bar{R}$, we choose point B. Equation (49) then gives

$$\dot{m}_1 \bar{r}_{A/B} \times (-\bar{v}_1) + \dot{m}_1 \bar{r}_{B/B} \times \bar{v}_2 = \Sigma \, \bar{M}_B$$

$$125.66(-0.10\bar{I} - 0.10\bar{J}) \times (-16\bar{I}) + \bar{0} = (-0.10\bar{I} - 0.10\bar{J}) \times 1570.8\bar{I} + M_{BZ}\bar{K}$$

$$-201.1\bar{K} = (157.1 + M_{BZ})\bar{K}$$

$$M_{BZ} = -358.2$$

$$\bar{M}_B = -358.2\bar{K} \text{ N-m} \qquad \triangle$$

$-\bar{M}_B = 358.1 \, \bar{K}$ N-m

$-\bar{R} = 3581 \, \bar{I} - 8435 \, \bar{J}$ N

The force and couple exerted by the water on the pipe are the reactions, $-\bar{R}$ and $-\bar{M}_B$, respectively. These are illustrated in the diagram.

150 mm

45°

ω

EXAMPLE 15

A rotary lawn sprinkler has three equally spaced curved arms of 5 mm inside diameter. Water flows into the sprinkler at the center from a garden hose, and is divided equally among the arms. Neglecting any resistance from friction, determine the constant rate of rotation ω for a flow rate of 1.20 l/s.

Solution

We consider the entire sprinkler as the control volume in order to study its motion. In the absence of friction, no forces are exerted in the horizontal plane (assuming the sprinkler head is balanced in its rotation). Hence, we do not need a free body diagram; the given sketch will provide the required geometry.

The given flow rate is in terms of volume (1000 liters = 1 m³). The mass rate is then obtained by multiplying by the density of the water.

$$\dot{m}_1 = \left(1000 \,\frac{\text{kg}}{\text{m}^3}\right)\left(1.20 \,\frac{\text{l}}{\text{s}}\right)\left(\frac{1}{1000} \,\frac{\text{m}^3}{\text{l}}\right)$$

$$= 1.20 \text{ kg/s}$$

This flow is divided equally among the three arms. The outlet has a diameter of 5 mm, so

$$\mathscr{A}_2 = \left(\frac{\pi}{4}\right)(0.005)^2 = 1.9635(10^{-5}) \text{ m}^2$$

Thus, the mass leaving each nozzle is

$$\rho_{\text{water}} \mathscr{A}_2 u_2 = \tfrac{1}{3}\dot{m}_1 = 0.40 \text{ kg/s}$$

from which we obtain

$$u_2 = \frac{0.40}{(1000)(1.9635)(10^{-5})} = 20.37 \text{ m/s}$$

The rotation of the system is defined by the moment equation (49), but this equation was derived for the case where fluid is leaving the control volume at only one outlet; the system considered here involves three identical outlets. To handle this we shall consider the mass flow through one arm and multiply the result by three. The control volume we have chosen is the sprinkler. Because it is in fixed-axis rotation about its center O, we shall sum moments about this point. The angular velocity $\bar{\omega}$ is constant, therefore $\bar{\alpha} = \bar{0}$, and we also know that $\bar{v}_G \equiv \bar{0}$. Furthermore, noting that the water enters the sprinkler at its center, the momentum of the incoming fluid has a zero moment arm, and hence drops out of equation (49). Denoting the tip of an arm as point P, equation (49) then reduces to

$$3\bar{r}_{P/O} \times (\tfrac{1}{3}\dot{m}_1\bar{v}_2) = \Sigma \, \bar{M}_O = \bar{0}$$

This equation requires that the rotation of the system be such that $\bar{v}_2$ is parallel to $\bar{r}_{P/O}$. In other words, the reaction force from each stream of water must pass through the center point O. To describe the vectors we employ cylindrical coordinates, as sketched here.

From the sketch we see that

$$\bar{r}_{P/O} = 0.150\bar{e}_R \text{ meters}$$

Earlier, we found that $u_2 = 20.37$ m/s. The vector $\bar{u}_2$ is tangent to the arm, as shown in the sketch. Therefore

$$\bar{u}_2 = 20.37(\cos 45°\,\bar{e}_R - \sin 45°\,\bar{e}_\phi) = 14.404(\bar{e}_R - \bar{e}_\phi) \text{ m/s}$$

We also know that $\bar{u}_2$ is the velocity of the water relative to the exit area, which means that $\bar{u}_2 = \bar{v}_2 - \bar{v}_P$. Point P is following a circular path of 150 mm radius. Hence we may solve for $\bar{v}_2$.

$$\bar{v}_2 = \bar{v}_P + \bar{u}_2 = 0.15\omega\bar{e}_\phi + 14.404(\bar{e}_R - \bar{e}_\phi)$$

$$= 14.40\bar{e}_R + (0.15\omega - 14.404)\bar{e}_\phi$$

Earlier we showed that to satisfy the moment sum the transverse component of $\bar{v}_2$ must be zero (in order for $\bar{v}_2$ to be parallel to $\bar{r}_{P/O}$). Hence

$$0.15\omega - 14.404 = 0$$

$$\omega = 96.03 \text{ rad/s} = 917 \text{ rev/min}$$

In the following problems a fluid exposed to the atmosphere may be considered to be at atmospheric pressure.

HOMEWORK PROBLEMS

VIII.101 and VIII.102   The nozzle of a 50-mm-diameter hose has an outlet whose opening is 10 mm in diameter. If the water within the hose is flowing at 10 m/s and the internal pressure is 100 kilopascals above atmospheric pressure, determine the force-couple system at point A representing the reaction between the hose and the nozzle.

Prob. VIII.101

Prob. VIII.102

Prob. VIII.103

Prob. VIII.104

Prob. VIII.106

40 m/s

500 mm

30°

Prob. VIII.107

VIII.103   A 20-mm garden hose is connected to a right-angle nozzle whose opening is 5 mm in diameter. The water within the hose is flowing at 1 l/s. The mass of the nozzle is 100 grams and that of the hose is 0.5 kg/m. Neglecting the rigidity of the hose, what length l of the hose can be supported by a person holding the hose horizontally at point A?

VIII.104   Water flows through a $\frac{5}{8}$-in. inside diameter hose at 20 ft/s, then emerges from the nozzle whose opening is $\frac{1}{4}$ in. in diameter. Determine the magnitude of the force $\bar{F}$ required to hold plate AB stationary.

VIII.105   In Problem VIII.104, determine the magnitude of the force $\bar{F}$ required to translate plate AB at 15 ft/s (a) to the right, (b) to the left.

VIII.106   Water flows between two parallel plates A and B at a speed v. The spacing between these plates is h. The stream is broken into two sheets by a divider at point C on the smooth horizontal plate. Assuming that both sheets also flow at the speed v, determine (a) the force the water exerts on the horizontal plate, (b) the depths h_1 and h_2 of both sheets of water.

VIII.107   Water flowing through a hose at 1000 l/min emerges from the nozzle at 40 m/s. The stream is deflected downward by a vane in the shape of a circular arc of 500 mm radius. Determine the magnitude, direction, and point of application of the force the water exerts on the vane.

VIII.108   Water emerges from a nozzle at the rate of 15 ft³/s with a speed of 100 ft/s and flows at a constant speed over the curved vane. Determine the reactions at supports A and B.

Prob. VIII.108

VIII.109   Grain is carried on the horizontal conveyor belt at 150 kg/s at a speed of 10 m/s. It is then transfered to a curved conveyor belt CD, which is moving at 20 m/s. Determine the components of the reactions at supports A and B. The total mass of the grain on conveyor belt CD and the structure is 1000 kg, and the center of mass is point G.

VIII.110 and VIII.111   Iron ore falls from a hopper at the rate of 300 kg/s, hitting the conveyor chute at point B at a speed of 10 m/s. The ore leaves

Prob. VIII.109

Prob. VIII.110   **Prob. VIII.111**

the chute at point C with a speed of 12 m/s. The total mass of the chute and the ore being transported is 1000 kg, and the center of mass is point G. Determine the reactions at supports A and C.

VIII.112 The slipstream of an electric fan has a diameter of 2 ft and a speed of 40 ft/s. The air weighs 0.07651 lb/ft³. Determine the thrust force developed by the fan blades. *Hint:* The air approaching the blades, being part of the atmosphere, may be considered to be at rest.

Prob. VIII.112

VIII.113 When the helicoper hovers, the slipstream has a diameter of 10 m and the air within it is flowing at 20 m/s downward. Knowing that the density of the air is 1.0 kg/m³, determine the thrust being developed by the blades. See the hint for Problem VIII.112.

VIII.114 The slipstream of each engine of a twin-engine propeller airplane has a diameter of 1 m. The airplane has a speed of 80 m/s, and the engines are developing 400 kW of power. If the air has a density of 0.9820 kg/m³, what is the velocity of the air in the slipstream? See the hint for Problem VIII.112.

Prob. VIII.113

Prob. VIII.114

VIII.115 In Example 13, the blade is moving to the left at the speed v. Determine the required power. Is there a value of v for which the power is maximum?

VIII.116 The Pelton wheel turbine is rotating at angular speed ω due to the impulse of the stream that is flowing at an absolute speed v and volume flow rate q. The tangent to the blade is at angle θ from the incoming stream, as shown. All of the water that strikes a blade is deflected outward tangent to the moving blade. Derive an expression for the couple M resisting the rotation of the turbine in terms of the other parameters.

Prob. VIII.116

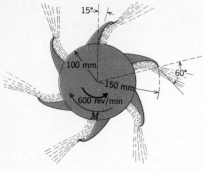

Prob. VIII.118

VIII.117   Solve Example 15 for the case where there is a frictional torque of 0.5 N-m opposing the rotation of the arms of the sprinkler.

VIII.118   The centrifugal-flow turbine is turning counterclockwise at 600 rev/min. Water flows into the center at 20 m³/s and flows outward over each of the six vanes, moving with a speed of 50 m/s relative to the vanes. Determine the torque M opposing the rotation.

3   Jet and Rocket Propulsion

The propulsion of a system by a jet of fluid is accomplished by accelerating the fluid as it passes through the system's engine. In rocket and jet engines this acceleration is achieved by means of the combustion of fuel. One primary difference between a rocket and a jet engine is that the oxidant for the combustion process is carried inside the vehicle for rocket propulsion, whereas it is gathered in from the surrounding atmosphere in the case of a jet engine.

Let us compare the rocket-propelled vehicle in Figure 26a with the jet aircraft in Figure 26b. In normal operation the angular velocities are small, so we shall consider both vehicles to be translating bodies.

Figure 26a   Figure 26b

For both vehicles, $\Sigma \bar{F}$ denotes the combination of the aerodynamic and weight forces. Also, for both vehicles, the gases are exhausted with velocity $\bar{u}_2$ relative to the vehicle, which means that $\bar{v}_2 = \bar{v} + \bar{u}_2$.

Let $\dot{m}_P$ be the rate at which the propellant carried within the vehicle is consumed. Then, for steady operation of the engine, the instantaneous mass of the vehicle is

$$m = m_0 - \dot{m}_P t$$

where m_0 is the initial mass.

In the case of the rocket, the exhaust gases are obtained only from the propellant, so that $\dot{m}_1 = 0$ and $\dot{m}_2 = \dot{m}_P$. Specializing equation (48) for the case of the rocket vehicle then yields

$$(m_O - \dot{m}_P t)\bar{a} + \dot{m}_P \bar{u}_2 = \Sigma \bar{F} \tag{53}$$

On the other hand, for an air-breathing engine, air enters the inlet at a mass rate $\dot{m}_1 = \dot{m}_{\text{air}}$. The velocity of the air at the inlet is usually negligible, $\bar{v}_1 \simeq \bar{0}$, because it is at atmospheric conditions. The propellant consumed by the engine is the fuel, which is combined in the combustion chamber with the air. Thus, $\dot{m}_2 = \dot{m}_{\text{air}} + \dot{m}_P$, so for an air-breathing engine equation (48) gives

$$(m_O - \dot{m}_P t)\bar{a} + \dot{m}_{\text{air}}\bar{v} + (\dot{m}_{\text{air}} + \dot{m}_P)\bar{u}_2 = \Sigma \bar{F} \tag{54}$$

In normal aircraft operation the fuel is expended at a slow rate compared to $\dot{m}_{\text{air}}$. For this reason, equation (54) can, at times, be simplified by setting $\dot{m}_P = 0$.

EXAMPLE 16
A single-stage rocket, weighing 1000 lb when unfueled, follows a vertical path. The engine burns propellant at the rate of 300 lb/s, and the gases are exhausted at 5000 ft/s. Determine the maximum speed attained by the rocket if it is launched from rest with 15,000 lb of propellant. Neglect air resistance and the variation in the gravitational attraction resulting from the change of altitude.

Solution
We begin with a free body diagram of the system. It is given that air resistance is negligible, and we shall assume that the pressure at the nozzle is also negligible. Thus, the only force acting on the system is gravity.

From the given information we know that $\dot{m}_P = 300/32.17 = 9.325$ lb-s/ft (be careful to express $\dot{m}_P$ in mass units) and that $u_2 = 5000$ ft/s. In terms of components with respect to the coordinate axes shown, we have

$$\bar{u}_2 = \bar{v}_2 - \bar{v}_G = -5000\bar{J} \text{ ft/s}$$

Also from the given information, we know that the take-off weight is 16,000 lb, which means that $m_O = 16,000/32.17 = 497.4$ lb-s^2/ft.

Substituting these terms into equation (53) and setting $\bar{a} = \dot{v}\bar{J}$ for vertical motion gives

$$(m_0 - \dot{m}_P t)\bar{a} + \dot{m}_P(\bar{v}_2 - \bar{v}_G) = -(m_0 - \dot{m}_P t)g\bar{J}$$

$$(497.4 - 9.325t)\dot{v}\bar{J} + 9.325(-5000)\bar{J} = -(497.4 - 9.325t)(32.17)\bar{J}$$

Rearranging terms, this expression yields a differential equation in v and t. The value of v is maximum when all of the propellant has been consumed. Hence the corresponding value of t is given by

$$(\dot{m}_P g)t_{max} = 15,000 \text{ lb} \qquad t_{max} = 50 \text{ seconds}$$

We can solve the differential equation of motion by separating variables.

$$(497.4 - 9.325t)\dot{v} = 46,625 - (497.4 - 9.325t)(32.17)$$

$$dv = \left(\frac{46,625}{497.4 - 9.325t} - 32.17 \right) dt$$

$$\int_0^{v_{max}} dv \equiv v_{max} = \int_0^{50} \left[\frac{5000}{(53.34 - t)} - 32.17 \right] dt$$

$$= 5000[-\ln(53.34 - 50) + \ln(53.34)] - 32.17(50)]$$

$$v_{max} = 12,245 \text{ ft/s}$$

EXAMPLE 17

The two-engine jet aircraft shown in the diagram, whose mass is $4(10^4)$ kg, is in level flight at a constant speed of 600 m/s. Its nose is pitched upward such that the engine exhaust is at an angle of 10° below the horizontal. The inlet area for each engine is 0.40 m², and the density of the atmosphere is 1.10 kg/m³. The exhaust has a speed of 1200 m/s realtive to the aircraft, corresponding to a rate of fuel consumption of 4.0 kg/s. Determine (a) the thrust of the engine, (b) the lift force (the aerodynamic force perpendicular to the velocity), (c) the drag force (the aerodynamic force parallel to the velocity).

Solution

The external forces acting on the aircraft are the weight and the unknown aerodynamic force, as shown in the free body diagram.

Choosing the XYZ coordinate system shown in the sketch, we may write

$$\bar{v} = 600\bar{I} \text{ m/s}$$

$$\bar{u}_2 = 1200(-\cos 10° \ \bar{I} - \sin 10° \ \bar{J})$$

$$= -1181.8\bar{I} - 208.4\bar{J} \text{ m/s}$$

To account for both engines, we double the given fuel consumption rate to get

$$\dot{m}_P = 8 \text{ kg/s}$$

The rate at which air is taken in is not given, but it can be determined from the formula

$$\dot{m}_{\text{air}} \equiv \dot{m}_1 = \rho_1 \mathcal{A}_1 u_1$$

Because $\bar{u}_1 = \bar{v}_1 - \bar{v}$ and the air at the inlet is at rest in the atmosphere, $u_1 = v = 600$ m/s. Doubling the given inlet area (two engines) and using the given value for density, we compute

$$m_{\text{air}} = \left(1.10 \ \frac{\text{kg}}{\text{m}^3}\right)(2)(0.40 \text{ m}^2)\left(600 \ \frac{\text{m}}{\text{s}}\right) = 528 \text{ kg/s}$$

Hence, equation (54), for the condition $\bar{a}_G = \bar{0}$ gives

$$\dot{m}_{\text{air}}\bar{v} + (\dot{m}_{\text{air}} + \dot{m}_P)\bar{u}_2 = \Sigma \ \bar{F}$$

$$528(600\bar{I}) + (528+8)(-1181.8\bar{I} - 208.4\bar{J}) = \bar{F}_{\text{aero}} - 3.922(10^5)\bar{J}$$

$$-3.166(10^5)\bar{I} - 1.117(10^5)\bar{J} = \bar{F}_{\text{aero}} - 3.922(10^5)\bar{J} \text{ N}$$

In our earlier discussion we saw that the mass flow terms on the left side of equation (48) were the negative of the reaction forces corresponding to the flow. Therefore

$$\bar{F}_{\text{engine}} = 3.166(10^5)\bar{I} + 1.117(10^5)\bar{J}$$

$$|\bar{F}_{\text{engine}}| = 3.357(10^5) \text{ N}$$

Solving the foregoing force equation for $\bar{F}_{\text{aero}}$ yields

$$\bar{F}_{\text{aero}} = -3.166(10^5)\bar{I} + (3.922 - 1.117)(10^5)\bar{J}$$

$$= -3.166(10^5)\bar{I} + 2.805(10^5)\bar{J} \text{ N}$$

The velocity is in the X direction, so the drag and lift forces are the X and Y components of $\bar{F}_{\text{aero}}$. Hence

$$F_{\text{lift}} = 2.805(10^5) \text{ N}$$

$$F_{\text{drag}} = 3.166(10^5) \text{ N}$$

The aerodynamic and thrust forces are shown in the sketch. Notice that because of the pitched-up condition of the nose, the lift force is substantially less than the weight. It is also noteworthy that in this problem the engine thrust is not parallel to the direction of the exhaust velocity.

HOMEWORK PROBLEMS

VIII.119 At take-off a rocket has a weight of 2000 lb, including 1500 lb of fuel. The exhaust speed is a constant 10,000 ft/s relative to the rocket. Compare the thrust of the rocket at launching and at burnout if the fuel is consumed (a) at 30 lb/s, (b) at 60 lb/s.

VIII.120 The rocket in Problem VIII.119 is launched from an airplane moving in level flight at 800 ft/s and follows a horizontal path. Neglecting air resistance, determine the maximum speed attained by the rocket for both rates of fuel consumption.

VIII.121 An atmospheric sounding rocket is fired vertically upward. When launched, the rocket has mass m_0. The exhaust gases from the engine have a constant speed u relative to the vehicle. If the rocket is to have a constant acceleration $\dot{v}$, derive an expression for the required rate of consumption of fuel $\dot{m}_P$ in terms of time. Neglect air resistance.

VIII.122 Two missiles are to be launched at the same time from the ground. The first is a single-stage rocket having a mass of 20,000 kg at take-off, 16,000 of which is fuel. The second is a two-stage rocket, each stage having a mass of 10,000 kg at take-off, including 8000 kg of fuel. The engines of both missiles consume fuel at the rate of 200 kg/s, exhausting them at 3000 m/s relative to the rocket. Compare the maximum speed attained by the two missiles in vertical flight.

VIII.123 Compare the height at burnout for the two missiles in Problem VIII.122.

VIII.124 A 10,000 kg, two-engine jet aircraft whose speed is 250 m/s takes in air at the total rate of 150 kg/s and has a total fuel consumption rate of 1 kg/s. The combustion gases are exhausted at 800 m/s relative to the aircraft in a direction opposite the velocity of the aircraft. (a) If the maximum angle of climb at which the aircraft can maintain its speed is 20°, determine the air resistance. (b) For the air resistance obtained in part (a), determine the rate at which the aircraft gains speed in level flight at an instantaneous speed of 250 m/s.

Prob. VIII.124

Prob. VIII.125

Prob. VIII.126

VIII.125 A jet aircraft may be decelerated by deflecting the exhaust, as shown. Each engine has an intake area of 5 ft², consumes fuel at the rate of 1.5 lb/s, and exhausts the combustion gases at a speed $u = 1500$ ft/s. Determine the thrust of each engine for a ground speed of 220 ft/s when the exhaust is impeded by the deflectors. The air weighs 0.07631 lb/ft³.

VIII.126 A 15,000-kg VTOL aircraft is powered by two jet engines, each of which scoops air in horizontally at 110 kg/s. The inlet area of each engine is 0.5 m². The gases are exhausted at 800 m/s, and each engine consumes fuel at 1 kg/s. At take-off the gases are deflected downward, relative to the aircraft, as shown. Determine the initial acceleration of the aircraft at take-off if the density of the air is 1.10 kg/m. Note: For this type of engine, it is not allowable to consider the air at the inlet to be at rest.

VIII.127 The aircraft in Example 17 is traveling horizontally at 400 m/s. The aerodynamic drag is 150 kN. If all other parameters are as given in the example, determine the acceleration of the aircraft.

APPENDIX A
SI UNITS

Table 1   Conversion factors from British units

PHYSICAL QUANTITY	U.S.-BRITISH UNIT	= SI EQUIVALENT
		BASIC UNITS
Length	1 foot (ft)	$= 3.048(10^{-1})$ meter (m)*
	1 inch (in.)	$= 2.54(10^{-2})$ meter (m)*
	1 mile (U.S. statute)	$= 1.6093(10^3)$ meter (m)
Mass	1 slug (lb-s²/ft)	$= 1.4594(10)$ kilogram (kg)
	1 pound mass (lbm)	$= 4.5359(10^{-1})$ kilogram (kg)
		DERIVED UNITS
Acceleration	1 foot/second² (ft/s²)	$= 3.048(10^{-1})$ meter/second² (m/s²)
	1 inch/second² (in./s²)	$= 2.54(10^{-2})$ meter/second² (m/s²)
Area	1 foot² (ft²)	$= 9.2903(10^{-2})$ meter² (m²)
	1 inch² (in.²)	$= 6.4516(10^{-2})$ meter² (m²)*
Density	1 slug/foot³ (lb-s²/ft⁴)	$= 5.1537(10^2)$ kilogram/meter³ (kg/m³)
	1 pound mass/foot³ (lbm/ft³)	$= 1.6018(10)$ kilogram/meter³ (kg/m³)
Energy and Work	1 foot-pound (ft-lb)	$= 1.3558$ joules (J)
	1 kilowatt-hour (kW-hr)	$= 3.60(10^6)$ joules (J)*
	1 British thermal unit (Btu)	$= 1.0551(10^3)$ joules (J)
	(1 joule $\equiv$ 1 meter-newton)	
Force	1 pound (lb)	$= 4.4482$ newtons (N)
	1 kip (1000 lb)	$= 4.4482(10^3)$ newtons (N)
	(1 newton $\equiv$ 1 kilogram-meter/second²)	
Power	1 foot-pound/second (ft-lb-s)	$= 1.3558$ watt (W)
	1 horsepower (hp)	$= 7.4570(10^2)$ watt (W)
	(1 watt $\equiv$ 1 joule/second)	
Pressure and Stress	1 pound/foot² (lb/ft²)	$= 4.7880(10)$ pascal (Pa)
	1 pound/inch² (lb/in.²)	$= 6.8948(10^\cdot)$ pascal (Pa)
	1 atmosphere (standard, 14.7 lb/in.²)	$= 1.0133(10^5)$ pascal (Pa)
	(1 pascal $\equiv$ 1 newton/m²)	
Speed	1 foot/second (ft/s)	$= 3.048(10^{-1})$ meter/second (m/s)*
	1 mile/hr	$= 4.4704(10^{-1})$ meter/second (m/s)
	1 mile/hr	$= 1.6093$ kilometer/hr (km/hr)
Volume	1 foot³ (ft³)	$= 2.8317(10^{-2})$ meter³ (m³)
	1 inch³ (in.³)	$= 1.6387(10^{-5})$ meter³ (m³)
	1 gallon (U.S. liquid)	$= 3.7854(10^{-3})$ meter³ (m³)

*Denotes an exact factor.

Table 2 Conversion factors from "old" metric units

PHYSICAL QUANTITY	"OLD" METRIC UNIT	= SI EQUIVALENT
Energy	1 erg	$= 1.00(10^{-7})$ joule (J)*
Force	1 kyne	$= 1.00(10^{-5})$ newton (N)*
Length	1 angstrom	$= 1.00(10^{-10})$ meter (m)*
	1 micron	$= 1.00(10^{-6})$ meter (m)*
Pressure	1 bar	$= 1.00(10^{5})$ pascal (Pa)*
Volume	1 liter	$= 1.00(10^{-3})$ meter^3 (m^3)*

*Denotes an exact factor

Table 3 SI prefixes

PREFIX	SYMBOL	FACTOR BY WHICH UNIT IS MULTIPLED
tera*	T	10^{12}
giga*	G	10^{9}
mega*	M	10^{6}
kilo*	k	10^{3}
hecto	h	10^{2}
deka	da	10
deci	d	10^{-1}
centi	c	10^{-2}
milli*	m	10^{-3}
micro*	μ	10^{-6}
nano*	n	10^{-9}
pico*	p	10^{-12}
femto*	f	10^{-15}
atto*	a	10^{-18}

*Denotes preferred prefixes.

APPENDIX B
INERTIA PROPERTIES

SHAPE	VOLUME	MOMENTS OF INERTIA

Uniform slender bar

$V = LA$

$I_{xx} = I_{zz} = \frac{1}{12}mL^2$

$I_{yy} = 0$

Thin circular ring

Cross–sectional area $\mathcal{A}$

$V = 2\pi rA$

$I_{xx} = I_{yy} = \frac{1}{2}mr^2$

$I_{zz} = mr^2$

Circular plate

$V = \pi r^2 h$

$I_{xx} = I_{yy} = \frac{1}{4}mr^2$

$I_{zz} = \frac{1}{2}mr^2$

Quarter circular plate

$V = \frac{\pi}{4}\, r^2 h$

$I_{xx} = I_{yy} = \left(\frac{9\pi^2 - 64}{36\pi^2}\right)mr^2$

$I_{zz} = \left(\frac{9\pi^2 - 64}{18\pi^2}\right)mr^2$

$I_{xy} = \left(\frac{9\pi - 32}{18\pi}\right)mr^2$

Elliptical plate

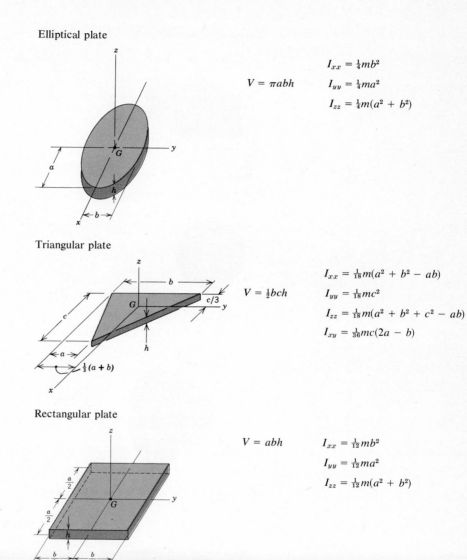

$$V = \pi abh$$

$$I_{xx} = \tfrac{1}{4}mb^2$$
$$I_{yy} = \tfrac{1}{4}ma^2$$
$$I_{zz} = \tfrac{1}{4}m(a^2 + b^2)$$

Triangular plate

$$V = \tfrac{1}{2}bch$$

$$I_{xx} = \tfrac{1}{18}m(a^2 + b^2 - ab)$$
$$I_{yy} = \tfrac{1}{18}mc^2$$
$$I_{zz} = \tfrac{1}{18}m(a^2 + b^2 + c^2 - ab)$$
$$I_{xy} = \tfrac{1}{36}mc(2a - b)$$

Rectangular plate

$$V = abh$$

$$I_{xx} = \tfrac{1}{12}mb^2$$
$$I_{yy} = \tfrac{1}{12}ma^2$$
$$I_{zz} = \tfrac{1}{12}m(a^2 + b^2)$$

Rectangular parallelipiped

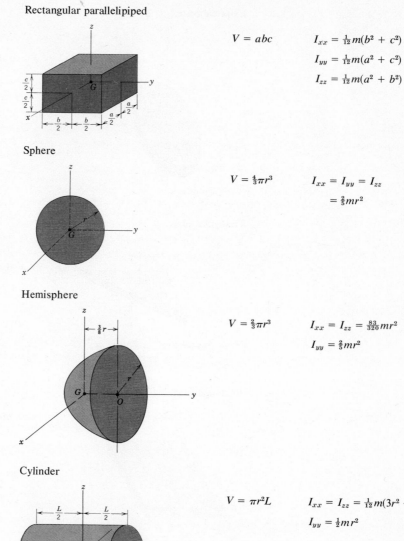

$$V = abc$$

$$I_{xx} = \tfrac{1}{12}m(b^2 + c^2)$$

$$I_{yy} = \tfrac{1}{12}m(a^2 + c^2)$$

$$I_{zz} = \tfrac{1}{12}m(a^2 + b^2)$$

Sphere

$$V = \tfrac{4}{3}\pi r^3$$

$$I_{xx} = I_{yy} = I_{zz}$$

$$= \tfrac{2}{5}mr^2$$

Hemisphere

$$V = \tfrac{2}{3}\pi r^3$$

$$I_{xx} = I_{zz} = \tfrac{83}{320}mr^2$$

$$I_{yy} = \tfrac{2}{5}mr^2$$

Cylinder

$$V = \pi r^2 L$$

$$I_{xx} = I_{zz} = \tfrac{1}{12}m(3r^2 + L^2)$$

$$I_{yy} = \tfrac{1}{2}mr^2$$

Semicylinder

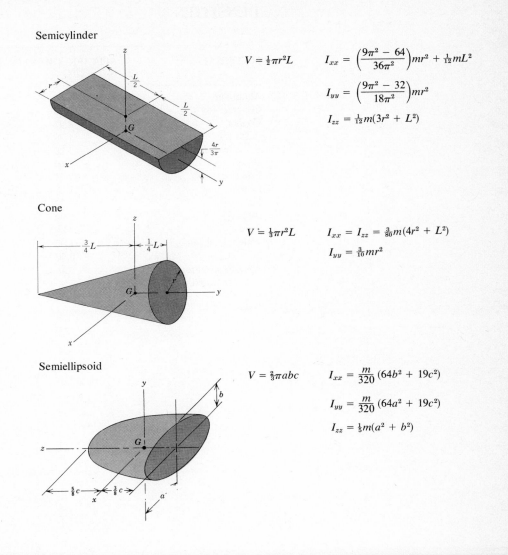

$$V = \tfrac{1}{2}\pi r^2 L \qquad I_{xx} = \left(\frac{9\pi^2 - 64}{36\pi^2}\right)mr^2 + \tfrac{1}{12}mL^2$$

$$I_{yy} = \left(\frac{9\pi^2 - 32}{18\pi^2}\right)mr^2$$

$$I_{zz} = \tfrac{1}{12}m(3r^2 + L^2)$$

Cone

$$V = \tfrac{1}{3}\pi r^2 L \qquad I_{xx} = I_{zz} = \tfrac{3}{80}m(4r^2 + L^2)$$

$$I_{yy} = \tfrac{3}{10}mr^2$$

Semiellipsoid

$$V = \tfrac{2}{3}\pi abc \qquad I_{xx} = \frac{m}{320}(64b^2 + 19c^2)$$

$$I_{yy} = \frac{m}{320}(64a^2 + 19c^2)$$

$$I_{zz} = \tfrac{1}{5}m(a^2 + b^2)$$

APPENDIX C
DENSITIES

MATERIAL	SPECIFIC GRAVITY $(\rho/\rho_{water})^*$
Aluminum	2.69
Concrete (av.)	2.40
Copper	8.91
Earth (av. wet)	1.76
(av. dry)	1.28
Glass	2.60
Iron (cast)	7.21
Lead	11.38
Mercury	13.57
Oil (av.)	0.90
Steel	7.84
Water (fresh, liquid)	1.00
(ice)	0.90
Wood (soft pine)	0.48
(hard oak)	0.80

$^*\rho_{water} = 1.00(10^3)$ kg/m^3.

$\gamma_{water} = \rho_{water}g = 62.4$ lb/ft^3.

ANSWERS TO ODD-NUMBERED PROBLEMS

I.1 (a) $1 \text{ kg} - \text{m}^2/\text{s}^2$
  (b) $1 \text{ kg} - \text{m}^2/\text{s}^3$
  (c) $1 \text{ kg} - \text{m}^2/\text{s}^3$

I.3 (a) $6.45(10^{-4}) \text{ m}^2$
  (b) $2.83(10^{-2}) \text{ m}^3$
  (c) 0.278 N
  (d) $6.90(10^3) \text{ N/m}^2$
  (e) 47.9 N/m^2
  (f) 0.305 m/s
  (g) $7.06(10^{-6}) \text{ m/s}$
  (h) $1.242(10^{-4}) \text{ m/s}^2$

I.5 (a) 91.44 m
  (b) 9 seconds
  (c) 0.450 N

I.7 L^2/T

I.9 $-53.3\bar{I} + 53.3\bar{J} \text{ N-m}$

I.11 $\bar{R} = -5576\bar{I} - 4082\bar{J} + 673\bar{K} \text{ N}$
  $M_D = -217\bar{I} - 313\bar{J} - 1115\bar{K} \text{ N}$
  $(\bar{I} = \bar{e}_{B/A}, \bar{J} = \bar{e}_{D/A})$

I.13 $\theta_A = \theta_B = 55.55°, \theta_C = 68.90°$
  $\bar{e}_n = 0.5145\bar{I} + 0.5145\bar{J} + 0.6860\bar{K}$

I.15 $\bar{R} = 250\bar{I} + 86.6\bar{J} \text{ N}$
  (a) $M_{OZ} = 45 \text{ N-m}$
  (b) $M_{CZ} = 91.3 \text{ N-m}$
  $(\bar{I} \text{ rt}, \bar{K} \text{ out})$

I.17 $R = 4348 \text{ m}, \theta = 50.04°$

I.19 $\bar{v}_P = -7.9\bar{I} - 19.5\bar{J} \text{ m/s}$
  $(\bar{I} \text{ rt}, \bar{J} \text{ up})$

I.21 $\bar{a}_P = -0.7\bar{I} - 0.9\bar{J} \text{ m/s}^2$
  $(\bar{I} \text{ rt}, \bar{J} \text{ dn})$

II.1 139.4 ft

II.3 (a) -14.71 m/s^2
  (b) $\bar{e}_n$ is 30° lt of vert, up
  (c) 1938 m

II.5 (a) $s = 125 \text{ ft}$
  $v_P = 45 \text{ ft/s}$
  $\dot{v}_P = 6 \text{ ft/s}^2$
  (b) $s = 675 \text{ ft}$
  $v_P = 45 \text{ ft/s}$
  $\dot{v}_P = -6 \text{ ft/s}^2$

II.7 $2.83 \, (gl)^{1/2}$

II.9 (a) $2\bar{e}_t + \bar{e}_n \text{ ft/s}^2$
  (b) $2\bar{e}_t + 6.25\bar{e}_n \text{ ft/s}^2$

II.11 (a) $s = 1250 \text{ m}$
  (b) $s = 1229 \text{ m}$

II.13 $-4.903\bar{e}_t + 0.870\bar{e}_n \text{ m/s}^2$

II.15 $\bar{e}_t = -0.930\bar{I} + 0.163\bar{J} + 0.332\bar{K}$
  $\bar{e}_n = -0.208\bar{I} - 0.972\bar{J} - 0.105\bar{K}$
  $\bar{e}_b = 0.306\bar{I} - 0.167\bar{J} + 0.938\bar{K}$

II.17 (a) $\bar{v}_P = 4\bar{I} + 3.46\bar{J} \text{ m/s}$
  $\bar{a}_P = -3.46\bar{J} \text{ m/s}^2$
  (b) $\bar{v}_P = 4\bar{I} + 1.55\bar{J} \text{ m/s}$
  $\bar{a}_P = -0.31\bar{J} \text{ m/s}^2$

II.19 $\bar{v}_P = 6.0\bar{I} + 4.2\bar{J} + .6.36\bar{K} \text{ m/s}$
  $\bar{a}_P = 180\bar{I} + 126\bar{J} + 159\bar{K} \text{ m/s}^2$

II.21 $\bar{e}_t = -\frac{2}{3}\bar{I} - \frac{2}{3}\bar{J} - \frac{1}{3}\bar{K}$
  $\bar{e}_n = -0.447\bar{J} + 0.894\bar{K}$
  $\dot{v}_P = 0$

II.23 $\bar{v}_P = 0.544\bar{I} - 0.238\bar{J} \text{ m/s}$
  $\bar{a}_P = -4.93\bar{I} - 5.54\bar{J} \text{ m/s}^2$

II.25 $\dot{R} = -2.07 \text{ m/s}$
  $\dot{\phi} = -3.86 \text{ rad/s}$
  $\ddot{R} = 12.48 \text{ m/s}^2$
  $\ddot{\phi} = 12.99 \text{ rad/s}^2$

II.27 $\dot{\theta} = 0.5 \text{ rad/s}$
  $\ddot{\theta} = -0.573 \text{ rad/s}^2$

II.29 $\bar{v}_P = 80.8\bar{I} + 9.7\bar{J} \text{ in./s}$
  $\bar{a}_P = 880\bar{I} - 2017\bar{J} \text{ in./s}^2$
  $(\bar{I} \text{ lt}, \bar{J} \text{ up})$

II.31 $-5\bar{e}_R + 1.60\bar{e}_\theta + 1.27\bar{e}_h \text{ m/s}^2$

II.33 $\bar{v}_P = 4.3\bar{e}_R - 130.9\bar{e}_\theta - 2.5\bar{e}_h \text{ mm/s}$
  $\bar{a}_P = -1370\bar{e}_R - 91\bar{e}_\theta - \bar{e}_h \text{ mm/s}^2$

II.35 $\bar{v}_A = -8.78\bar{e}_t \text{ m/s}$
  $\bar{a}_A = -24.6\bar{e}_t \text{ m/s}^2$

II.37 (a) $\bar{v}_B = 3\bar{e}_R - 3.46\bar{e}_\theta \text{ in./s}$
  $\bar{a}_B = -0.50\bar{e}_R - 2.98\bar{e}_\theta \text{ in./s}^2$
  (b) $\dot{\phi} = -0.1455 \text{ rad/s}$
  $\ddot{\phi} = -0.0886 \text{ rad/s}^2$

II.39 $\bar{v}_A = 5000\bar{e}_t \text{ in./s}$
  $\bar{a}_A = 10^5[-7.5\bar{e}_t + 10\bar{e}_n] \text{ in./s}^2$

II.41 $\dot{\phi} = 0.1483 \text{ rad/s}$
  $\bar{v}_A = 6.69 \text{ m/s}$

II.43 $\bar{a}_P = 10\bar{e}_t \text{ in./s}^2: t < 0.465 \text{ seconds}$
  $= 10\bar{e}_t + 6.67(3 + t)^2\bar{e}_n: 0.465 < t < 0.789$
  $= 10\bar{e}_t + 3.33(3 + t)^2\bar{e}_n: t > 0.789$

II.45 $\bar{v}_P = 20\bar{e}_R + 1.57\bar{e}_\phi \text{ m/s}^2$
  $\bar{a}_P = -4.93\bar{e}_R + 125.7\bar{e}_\phi \text{ m/s}^2$
  $\rho = 3.20 \text{ m}$

II.47 $-(K_1 x + K_2 x^3)\bar{e}_t$

II.49 $v_P = 20.7$ ft/s
$|\bar{a}_P| = 138.5$ ft/s^2

II.51 $-8.65\bar{e}_R + 4.35\bar{e}_\phi - 0.22\bar{e}_h$ m/s^2

II.53 $v_A = 68.6$ m/s
$v_B = 52.3$ m/s
$v_C = 24.3$ m/s

II.55 $\ddot{d} = 12.06$ m/s^2
$\ddot{\phi} = 0.0150$ rad/s^2

II.57 $\bar{v}_P = 4t[(1 - \cos 4t^2)\bar{I} + \sin 4t^2\bar{J}]$ ft/s
$\bar{a}_P = 4[(1 - \cos 4t^2 + 8t^2 \sin 4t^2)\bar{I}$
$+ (\sin 4t^2 + 8t^2 \cos 4t^2)\bar{J}]$ ft/s^2

II.59 $\bar{a}_P = -\dfrac{\pi A c}{L}\left\{1 - \dfrac{A}{c}\ \sin\left[\dfrac{\pi}{L}(X - ct)\right]\right\}\cos\left[\dfrac{\pi}{L}(X - ct)\right]\bar{I}$

II.61 $v_P = 20\pi\ (1 + \sin^2 \pi t/20)^{1/2}$ mm/s
$\dot{v}_P = (\pi^2 \sin \pi t/20)(\cos \pi t/20)(1 + \sin^2 \pi t/20)^{-1/2}$
$\rho = 400(1 + \sin^2 \pi t/20)^{3/2} (5 + 3 \sin^2 \pi t/20)^{-1/2}$

II.63 $\bar{a}_P = -\dfrac{K^2}{a^2 b^2} R\ \bar{e}_R$

II.65 $-\dfrac{A^2}{R^3}\ \bar{e}_R$; circle

II.67 $|\bar{v}_{P/air}| = 273$ km/hr
$|\bar{v}_P| = 193$ km/hr

II.69 $\dot{R} = -7.24$ m/s
$\ddot{R} = 3.12$ m/s^2

II.71 $\bar{v}_{A/B} = -150.5\bar{i} - 40.3\bar{j}$ ft/s
$\bar{a}_{A/B} = -6.51\bar{i} + 0.43\bar{j}$ ft/s^2
$[\bar{i} = (\bar{e}_t)_B, j = (\bar{e}_n)_B]$

II.73 $\bar{r}_{B/A} = -107.21\bar{i}$ m
$\bar{v}_{B/A} = -6\bar{i} + 10.39\bar{j}$ m/s
$\bar{a}_{B/A} = -0.373\bar{i} + 0.246\bar{j}$ m/s^2
$(\bar{i}$ rt, $\bar{j}$ up)

II.75 (a) $\alpha = \cos^{-1}\left[-\dfrac{v_w}{v} \sin^2 \theta\right.$
$\left. \pm \cos \theta \left(1 - \dfrac{v_w^2}{v^2} \sin^2\theta\right)^{1/2}\right]$ from $\bar{I}$
(b) $v_0 = v_w \cos \theta \pm (v^2 - v_w^2 \sin^2 \theta)^{1/2}$
(c) $t = R/v_0$
(d) $v_{max} = v/\sin \theta$ for $\theta < 90°$, $v_{max} = v$ for $\theta > 90°$

II.77 $\bar{a}_{B/A} = 21.12\bar{j} - 6.42\bar{k}$ ft/s^2

II.79 $s_{car} = 20$ m
$v_{car} = 3.5$ m/s (up)

II.81 0.25 m/s (up)

II.83 (a) 3.2 ft/s (up)
(b) 2.304 ft/s^2 (up)

III.3 $\dot{v}_{max} = 128.7$ ft/s
$s_{max} = 1222$ ft

III.5 (a) $v_{n+1} = v_n - 0.1\ g$

III.7 (a) 2.60 m/s (up)
(b) 1.799 m/s (up)

III.9 $v_P = 801.9$ ft/s for variable gravity
$v_P = 802.1$ ft/s for constant gravity

III.11 12.65 m/s (rt)

III.13 (a) 0.862 seconds
(b) 4.80 ft

III.15 $v = v_0 \exp(-kt/m)$
$v = v_0 - \dfrac{k}{m} s$
$s = \dfrac{mv_0}{k} [1 - \exp(-kt/m)]$

III.17 (a) 30.9 m
(b) 24.2 m/s
(c) 4.92 seconds

III.19 (a) 98.5 m
(b) 44.2 m/s

III.21 $t = 1.355$ seconds
$s_{el} = 4.06$ m

III.23 (a) $\dot{v}_P = 5.51$ ft/s^2
$v_P = 141.3$ ft
(b) $\dot{v}_P = -1.126$ ft/s^2
$v_P = 8.0$ ft/s

III.25 (a) 133.0 ft/s^2
(b) $\rho = 9.95(10^5)$ ft

III.27 $T = \dfrac{mg}{\cos \theta}$, $v_P = \left(\dfrac{gl \sin^2\theta}{\cos \theta}\right)^{1/2}$

III.29 $F_{hor} = 1.207m$ newtons
$\theta = 110.19°$ from $\bar{e}_t$, 20.19° from $\bar{e}_n$

III.31 $h = 3.67$ m
$d = 19.26$ m

III.33 $Y = (\tan \alpha)X - \dfrac{g}{2v_0^2 \cos^2\alpha} X^2$
$v_P = v_0 \cos \alpha\ \bar{I} + \left(v_0 \sin \alpha - \dfrac{gX}{v_0 \cos \alpha}\right)\bar{J}$

III.35 $N_{cir} = 1956$ newtons
$N_{arm} = 1173$ newtons

III.37 $T = mR\omega^2 \cos \alpha$
$N = mR\omega^2 \sin \alpha$

III.39 0.982 rad/s

III.41 158.1 rad/s

III.43 0.0922 m

III.45 $t = 5.02$ s
$f = 6000$ lb

III.47 $s = 34.1$ m
$t = 6.82$ s

III.49 $F_{tr} = 1.633(10^4)$ N
$C_1 = 8.67(10^3)$ N
$C_2 = 4.33(10^3)$ N

III.51 (a) 2.77 m/s² (up)
(b) 660 N

III.53 $s = 133.8$ ft

III.55 $F = \frac{1}{2} mg\left(1 + \frac{v_0^2 L^2}{16gh^3}\right)\left(1 + \frac{L^2}{4h^2}\right)$

III.57 15.26°

III.59 $v_P = [g(r^2 - h^2)/h]^{1/2}$

III.61 $2.21 < v_P < 2.91$ m/s

III.63 $245\bar{e}_t + 196.1\bar{e}_n + 2\bar{e}_b$ kN

III.65 (a) 126.8 ft/s
(b) -11.68 ft/s²

III.67 $603\bar{e}_t + 627\bar{e}_n$ lb

III.69 (a) & (b) 8.86 m/s

III.71 (a) 7.84 N
(b) 5.79 N

III.73 $-3.01m \, \bar{e}_t + 4.33m \, \bar{e}_n$ N

III.75 $-0.5/X^2$ N

III.77 (a) $\ddot{R} = 100R$
(b) $\dot{R} = 10.01$ m/s
$\bar{v}_P = 10.01\bar{e}_R + 10\bar{e}_\phi$ m/s

III.79 (a) 13.86 m
(b) 3.33 seconds

III.81 (a) $58.7\bar{I} + 44\bar{J}$ ft/s
(b) $\Delta s = 15$ ft
(c) $\Delta h = 1.869$ ft
$(\bar{I}$ rt, $\bar{J}$ up)

III.83 $X = 29.5$ m
$X = 40.8$ m for $k = 0$

III.85 (a) 0.639 seconds
(b) 3.19 m

III.87 $\dfrac{eV_0 l}{mbu^2} \left(\dfrac{l}{2} + h\right)$

III.89 10.52 J

III.91 $-1599 \, \dfrac{W}{g}$ ft-lb

III.93 10.36 J

III.95 54.7 J

III.97 (a) 49.02 kJ
(b) 12.26 kJ

III.99 $-mgl \cos \theta$

III.101 0.0922 m

III.103 6.44 m/s

III.105 Cannot reach point B

III.107 (a) 2.52 m/s
(b) 94.44°

III.109 $d = 0.6l$

III.111 $m(999e^{0.2\theta} - 79 \cos \theta + 396 \sin \theta - 1343)$ J

III.113 2.60 m/s

III.115 $\Delta = 1.773$ ft
$F = 42.6$ lb

III.117 2474 N

III.119 (a) 1.609 in.
(b) 5.44 in.

III.121 14.1 m/s

III.123 (a) 40.5 ft/s
(b) does not reach point B

III.125 3.63 m/s

III.127 179.9 N

III.129 4.22 rad/s (cw)

III.131 (a) & (b) 11.34 ft/s

III.133 126 in./s (rt)

III.135 0.976 m/s²

III.137 (a) 122.2 ft/s
(b) 225 seconds

III.139 78.5 J

III.141 $\frac{1}{2} mgv \left(1 + \dfrac{l^2 v^2}{4gd^3}\right)\left(1 + \dfrac{l^2}{d^2}\right)^{1/2}$

III.143 (a) & (b) 2.84(10⁴) lb

III.145 1500 N

III.147 (a) $\bar{v}_{bl} = 13.86\bar{I} - 12.47\bar{J}$ ft/s
$\bar{v}_C = -2.77\bar{I}$ ft/s
(b) $-1.333\bar{I}$ ft
$(\bar{I}$ rt, $\bar{J}$ up)

III.149 (a) 2.86 m/s (uphill)
(b) 38.2 m/s (uphill)

III.151 (a) $-46.7\bar{I} + 50\bar{J} + 30\bar{K}$ mN
(b) 113.0 mN

III.153 (a) 71.4 lb
(b) 14.14 ft/s
(c) 3.11 ft

III.155 1.112 m

III.157 (a) 559 N
(b) 791 N

III.159 $v_{final} = 0.667$ m/s
$F_{AB} = 30$ kN, then 10 kN
$F_{BC} = 0$, then 20 kN

III.161 3 collisions
$\bar{v}_A = \frac{1}{8}v(3 - 3\epsilon + \epsilon^2 - \epsilon^3)\bar{I}$
$\bar{v}_B = \frac{1}{8}v(3 - \epsilon - 3\epsilon^2 + \epsilon^3)\bar{I}$
$\bar{v}_C = \frac{1}{4}v(1 + \epsilon^2)\bar{I}$
($\bar{I}$ lt)

III.163 $\bar{v}_A = -0.4\bar{i}$ ft/s
$\bar{v}_B = 6.8\bar{i} + 17.32\bar{j}$ ft/s
($\bar{i}$ rt, $\bar{j}$ up)

III.167 $d = (1 + \epsilon)v_1\left(\frac{2h_1}{g}\right)^{1/2}$
$h_2 = \epsilon^2 h_1$

III.169 $\bar{v}_{ball} = 9.15\bar{i} + 14.14\bar{j}$ ft/s
$\bar{v}_{block} = 1.456\,(-\bar{i} + \bar{j})$ ft/s
($\bar{i}$ 45° lt of vert, $\bar{k}$ in)

III.171 111 kN

III.173 $\ddot{\theta} + \frac{g}{l}\sin\theta = 0$

III.175 (a) 27.8 ft/s
(b) $13.11\bar{e}_R + 18.53\bar{e}_\phi$ ft/s

III.177 $-2.06\bar{e}_R + 1.50\bar{e}_\phi$ m/s

III.179 (a) 55.6 in./s
(b) 111.1 in./s
(c) $v_{circ} = 90.8$ in./s

III.181 15.71 rad/s (ccw)

III.183 $R\,\ddot{R} + 3\,\dot{R}\,\ddot{R} - \frac{2M}{mR^{1/2}} = 0$

III.185 $\dot{\omega} = -1.039$ rad/s²
$\dot{v}_C = 0.278$ ft/s² (dn)

III.187 $\ddot{\theta} + \frac{5}{9}\frac{g}{l}\sin\theta = 0$

IV.1 (a) $18.35i$ in./s²
(b) $-4.9\bar{i} - 355.2\bar{j}$ in./s²
(c) $-8.74\bar{i} - 307.5\bar{j}$ in./s²
($\bar{i}$ rt, $\bar{k}$ out)

IV.3 (a) $-0.707(\bar{i} + \bar{j})$ m/s
(b) $1.414\bar{k}$ rad/s
(c) $-4.83\bar{k}$ rad/s²
($\bar{i}$ rt, $\bar{k}$ out)

IV.5 $\bar{v}_B = -0.212(\bar{i} + \bar{j})$ m/s
$\bar{a}_B = -0.1016(\bar{i} + \bar{j})$ m/s²
($\bar{i}$ rt, $\bar{k}$ out)

IV.7 $\omega = 0$, $\alpha = 33.3$ rad/s² (ccw)

IV.9 $-17.38l\bar{i} - 12.57l\bar{j}$
($\bar{i}$ rt, $\bar{j}$ dn)

IV.11 $\bar{v}_A = \frac{vl}{a}\sin^2\theta\cos\theta\,\bar{I} + \left(v - \frac{vl}{a}\sin^3\theta\right)\bar{J}$
$\bar{a}_A = \frac{v^2l}{a^2}\sin^3\theta[(3\sin^2\theta - 2)\bar{I} + 3\sin\theta\cos\theta\bar{J}]$
($\bar{I}$ rt, $\bar{J}$ up)

IV.15 704 ft below the pivot
Independent of θ

IV.17 $\omega = 12$ rad/s (ccw)
$v_0 = 5$ ft/s (lt)

IV.19 (a) $-10.80\bar{k}$ rad/s
(b) $3.60\bar{i} - 1.80\bar{j}$ m/s
($\bar{i}$ rt, $\bar{k}$ out)

IV.21 25 rad/s (ccw)

IV.23 (a) $-20\bar{k}$ rad/s
(b) $-20\bar{k}$ rad/s²
(c) $40\bar{i}$ m/s
(d) $20\bar{i}$ m/s²
(e) $40\bar{i} - 400\bar{j}$ m/s²
($\bar{i}$ rt, $\bar{k}$ out)

IV.25 $\bar{v}_0 = \omega r\,\bar{e}_t$
$\bar{a}_0 = \alpha r\,\bar{e}_t + \frac{\omega r^2}{R-r}\,\bar{e}_n$

IV.27 (a) $\omega = 16$ rad/s (ccw)
$\alpha = 2$ rad/s² (cw)
(b) $v_C = 8$ m/s (up)
$a_C = 1.0$ m/s² (dn)

IV.29 $(5.40\omega^2 - 12.60\alpha)\bar{i} + (5.40\omega^2 - 5.40\alpha)\bar{j}$
($\bar{i}$ rt, $\bar{j}$ up)

IV.31 $\omega_{BC} = 1.778$ rad/s (ccw)
$v_C = 1067$ mm/s (up)

IV.33 (a) $\omega_{BC} = \omega\cos\left(\frac{l^2}{r^2} - \sin^2\theta\right)^{-1/2}$
(b) $v_C/v_B = \sin\theta + \sin\theta\cos\theta\,(4 - \sin^2\theta)^{-1/2}$

IV.35 54.6 rev/min

IV.37 $\omega_{BC} = \omega$ (cw)
$\omega_{CD} = 2\omega$ (cw)
$\alpha_{BC} = 6\omega^2$ (ccw)
$\alpha_{CD} = 14\omega^2$ (cw)

IV.39 $-128.20\bar{i} - 37.75\bar{j}$ m/s²
($\bar{i}$ rt, $\bar{j}$ up)

IV.41 $\omega_{BC} = 2.90$ rad/s (cw)
$\alpha_{BC} = 3.43$ rad/s² (cw)

IV.43 $\omega = 1.25/\sin\theta$ (cw)
$\alpha = 1.5625\cos\theta/\sin^3\theta$ (cw)

IV.45 (a) $0.8\bar{i} + 4.8\bar{j}$ m/s
(b) $0.9\bar{i} + 2.4\bar{j}$ m/s²
($\bar{i}$ lt, $\bar{k}$ out)

IV.47 (a) $48\bar{k}$ rad/s²
(b) $-32\bar{j}$ ft/s²
(c) $96\bar{i}$ ft/s²
($\bar{i}$ rt, $\bar{k}$ in)

IV.49 (a) 0.69 ft/s (rt)
(b) 8.07 ft/s² (lt)

IV.51 $-1.93\bar{i} + 3.35\bar{j}$ m/s
($\bar{i}$ rt, $\bar{k}$ out)

IV.55 $\alpha_{AB} = 374$ rad/s² (ccw)
$\alpha_{BC} = 872$ rad/s² (cw)

IV.57 14.05 m/s² (rt)

IV.59 $\bar{v}_E = \omega r \bar{i}$
$\bar{a}_E = -0.75 r\omega^2 \bar{j}$
($\bar{i}$ rt, $\bar{k}$ out)

IV.61 $\bar{v}_C = 10.47\bar{i} + 20.94\bar{j}$
($\bar{i}$ rt, $\bar{k}$ in)

IV.63 $\bar{v}_P = 12.57\bar{i}$ m/s
$\bar{a}_P = -65.8\bar{i} - 79.0\bar{j}$ m/s²
($\bar{i}$ rt, $\bar{k}$ in)

IV.65 $v_B = 2.78$ m/s (rt)
$a_B = 3.54$ m/s² (lt)

IV.67 (a) 0
(b) 1458 rad/s² (cw)

IV.69 $\alpha_r = 0$, $\alpha_{fr} = 0.343$ rad/s² (ccw)
$\alpha_{body} = 0.0080$ rad/s² (cw)

IV.71 2.17 rad/s² (cw)

V.1 $x_G = \dfrac{l}{2}$

V.3 $y_G = \dfrac{4r}{3\pi}$

V.5 $y_G = \dfrac{2r \sin \theta}{3\theta}$

V.7 $z_G = \dfrac{l}{2}$

V.9 $\dfrac{ml^2}{12}$

V.11 $0.3199 \, mr^2$

V.13 $mr^2\left(\dfrac{1}{2} - \dfrac{4}{9}\dfrac{\sin^2 \theta}{\theta^2}\right)$

V.15 $\frac{3}{2} mr^2$

V.17 $\bar{r}_G = -3.33\bar{i} - 2.0\bar{j}$ mm
$I_{\hat{z}\hat{z}} = 4.035(10^{-3})$ kg-m²

V.19 $\bar{r}_G = 129.52\bar{i} + 82.87\bar{j}$ mm
$I_{\hat{z}\hat{z}} = 4.947(10^{-3})$ kg-m²

V.21 $\bar{r}_G = 78.21\bar{k}$ mm
$I_{\hat{z}\hat{z}} = 4.837(10^{-3})$ kg-m²

V.23 $\bar{r}_G = 0.038\bar{i} - 7.537\bar{j}$ in.
$I_{\hat{z}\hat{z}} = 119.15$ lb-s²-ft

V.25 $T = 40.5$ N

V.27 $\mu_k = 0.1$
$T = 0.1 \, mg$ (tension)

V.29 (a) $N_{front} = 7.52$ kN
$N_{rear} = 6.96$ kN
(b) 1.70 m/s² (downhill)

V.31 $T = 5$ kips
$\bar{v} = 12.35$ ft/s² (uphill)

V.33 $0.457 < h < 1.143$ meters

V.35 $0.8 - \dfrac{1}{F}(343 + 784\mu) < h < 0.8$
$+ \dfrac{1}{F}(343 - 784\mu)$ meters
$h = 0.8$ m

V.37 (a) 194.4 ft
(b) 451.4 ft

V.39 $\bar{a}_G = g \cos \theta \, \bar{i}$
$T_{DC} = \dfrac{10g \sin \theta \, (2\sin \theta + \cos \theta)}{2\cos \theta + 3\sin \theta}$
$T_{BA} = \dfrac{10g \sin \theta \, (\cos \theta + \sin \theta)}{2\cos \theta + 3\sin \theta}$
($\bar{j} = \bar{e}_{A/B}$, $\bar{k}$ out)

V.41 1.81 lb-in.

V.43 (a) 7.55 kN
(b) 9.61 kN

V.45 (a) 16 rad/s² (cw)
(b) 8.81 rad/s² (cw)

V.47 4.99 in.

V.49 3.20 seconds

V.51 $\alpha = 24.1$ rad/s (cw)
$F_A = 3.75$ lb (up)

V.53 (a) $5.14 \, \dfrac{P}{mL}$ (ccw)
(b) $0.285 \, P \, \bar{i} + mg \, \bar{j}$
($\bar{i}$ rt, $\bar{j}$ up)

V.55 (a) $\ddot{\theta} + 0.947 \, \dfrac{g}{b} \sin \theta = 0$
(b) $1.056 \, b$

V.57 $f_A = 2.49$ lb (rt)
$f_B = 0$

V.59 (a) 1.810 kN
(b) 16.6 m (wound on)

V.61 (a) $a_A = 0.8915$ m/s² (up)
(b) $v_C = 2.67$ m/s

V.63 (a) $a_{axle} = 1.365$ m/s² (up)
$a_{CD} = 4.095$ m/s² (dn)
(b) 8.19 m/s (dn)

V.65 (a) $\frac{2}{3}g \sin \theta$ (dn parallel to incline)
(b) $\frac{1}{3} \tan \theta$

V.67 $a_G = 5.23\bar{i}$ m/s²
$\bar{R} = -6.54\bar{i} + 64.71\bar{j}$ N
($\bar{j} = \bar{e}_n$, $\bar{k}$ in)

V.69 $\alpha = 6.84$ rad/s² (ccw)
$a_C = 54.7$ in./s² (lt)

V.71 (a) $8\bar{i}$ m/s
(b) $-66.67\bar{i}$ N
($\bar{i}$ rt)

V.73 $a_G = 1.657$ m/s² (lt)
$\alpha = 8.29$ rad/s² (ccw)

V.75 (a) $\dfrac{\omega_1 R}{2}$
(b) $\dfrac{\omega_1 R}{2\mu_k g}$

V.77 $\mu = 0.40$

V.79 (a) 0.092
(b) $\dfrac{0.261g}{R}$ (cw)

V.81 0.558 m/s² (lt)

V.83 $T_A = 1245$ lb
$T_B = 1204$ lb

V.85 (a) 3491 kN (up)
(b) 55.3 m/s² (up)

V.87 4.90 m/s² (dn)

V.89 $T_A = 41.34$ lb
$T_B = 38.29$ lb

V.91 (a) $T_A = 0$
$T_B = 28.3$ lb (up)
(b) $-6.10\bar{i} - 5.24\bar{j}$ ft/s²
($\bar{i}$ lt, $\bar{j}$ dn)

V.93 (a) 0.818 rad/s² (ccw)
(b) $\bar{R}_B = 851\bar{i} + 4255\bar{j}$ N
$\bar{R}_A = 158\bar{i} + 32\bar{j}$ N
($\bar{i}$ rt, $\bar{j}$ up)

V.95 (a) 0.688 kN-m (ccw)
(b) $\bar{R}_A = 12.17\bar{i} + 9.39\bar{j}$ kN
$\bar{R}_B = -19.77\bar{i} + 13.77\bar{j}$ kN
($\bar{i}$ lt, $\bar{j}$ up)

V.97 47.72 N

V.99 0

V.101 $a_A = 3.82$ m/s² (lt)
$a_B = 0.98$ m/s² (rt)

V.103 $\ddot{\theta}_1 + \dfrac{9g}{8l} \sin \theta_1 + \dfrac{3}{8} \ddot{\theta}_2 \cos(\theta_1 - \theta_2)$
$+ \dfrac{3}{8} \dot{\theta}_2^2 \sin(\theta_1 - \theta_2) = 0$
$\ddot{\theta}_2 + \dfrac{3g}{2l} \sin \theta_2 + \dfrac{3}{2} \ddot{\theta}_1 \cos(\theta_1 - \theta_2)$
$- \dfrac{3}{2} \dot{\theta}_1^2 \sin(\theta_1 - \theta_2) = 0$

V.105 $T_{disk} = 2m \, r^2\omega^2$
$T_{BC} = \dfrac{m \, \omega^2 r^2}{2}$
$T_{CD} = \dfrac{m \, \omega^2 r^2}{6}$

V.107 525.8 J

V.109 1347 ft-lb

V.111 246.7 N-m

V.113 104.54 rev

V.115 12.81 (dn)

V.117 5.46 m

V.119 (a) 6.69 ft/s (rt)
(b) 4.65 ft/s (rt)

V.121 (a) 8.31 ft/s (tang to circle, dn)
(b) 25.71 lb

V.123 55.25°

V.125 (a) 27.8 ft/s (up)
(b) 23.5 ft/s (up)

V.127 $\omega_{AB} = 3.25$ rad/s (cw)
$\omega_{BC} = 3.25$ rad/s (ccw)

V.129 (a) 3.09 ft/s
(b) 1 ft

V.131 (a) $145.77\bar{i}$ kg-m/s
(b) 163.13 kg-m²/s
($\bar{i}$ lt, 28° below horiz)

V.133 6.51 seconds

V.135 $v_1 = \frac{2}{5}g \sin \theta t$
$v_{sph} = \frac{5}{7}g \sin \theta t$

V.137 5.11 seconds

V.139 110.95 m/s (uphill)

V.141 (a) 1.732 rad/s
(b) $0.75\bar{e}_R + 1.73\bar{e}_\theta$ ft/s

V.143 $h = 0.450$ m
$\omega = 40$ (cw)

V.145 (a) $8.07\bar{i}$ ft/s
$F_{lt} = 540$ lb, $F_{rt} = 188$ lb
($\bar{i}$ rt, 30° above horiz)

V.147 (a) 28 mm
(b) 10 m/s (rt)

V.149 $v_{top} = 0$
$v_{bot} = 54.45$ ft/s (rt)

V.151 0.99 ft/s

V.153 $v_A = 7.03$ ft/s (up)
$v_B = 25.49$ ft/s (dn)

V.155 $\bar{v}_G = 5.4\bar{i} + 8\bar{j}$
$\bar{\omega} = 100\bar{k}$ rad/s
($\bar{i}$ up, $\bar{j}$ rt)

V.157 $\bar{v} = 13.86\bar{i} + 10\bar{j}$
($\bar{i}$ 60° lt of vert, $\bar{j}$ 30° rt of vert)

VI.1 (a) 1.0 ($\bar{i} + \bar{j}$) m/s
(b) $2.75\bar{i} - 0.75\bar{j}$ m/s
($\bar{i} = \bar{e}_{B/A}$, $\bar{k}$ out)

VI.3 $\bar{v}_A = -0.10\bar{i} - 1.766\bar{j}$ m/s
$\bar{v}_B = 1.020\bar{i} - 0.006\bar{j}$ m/s
($\bar{i}$ rt, $\bar{j}$ up)

VI.5 (a) 17.25 rad/s
(b) 28.12°

VI.7 (a) $4\bar{i}$ m/s
(b) $-4\bar{j} - 10\bar{k}$ m/s
(c) $2\bar{i} - 3.46\bar{j} - 8.66\bar{k}$ m/s
($\bar{i}$ rt, $\bar{j}$ up)

VI.9 (a) & (b) $- 6.98\bar{j} - 0.60\bar{k}$ m/s
($\bar{i} = \bar{e}_{P/O}$, $\bar{k}$ out)

VI.11 $\bar{a}_A = -4.0\bar{i} + 2.4\bar{j}$ m/s²
$\bar{a}_B = 3.68\bar{i} - 1.76\bar{j}$ m/s²
($\bar{i}$ rt, $\bar{j}$ up)

VI.13 $\bar{a}_A = -7.46\bar{i} + 0.40\bar{j}$ m/s²
$\bar{a}_B = 0.22\bar{i} - 3.76\bar{j}$ m/s²
($\bar{i}$ rt, $\bar{j}$ up)

VI.15 (a) $-800\bar{i} - 8\bar{j} - 170\bar{k}$ m/s²
(b) $-1608\bar{i} - 20\bar{k}$ m/s²
(c) $-1500\bar{i} - 4\bar{j} - 99\bar{k}$ m/s²
($\bar{i}$ rt, $\bar{j}$ up)

VI.17 $-16.35\bar{i} - 21.88\bar{j} - 0.88\bar{k}$ m/s²
($\bar{i} = \bar{e}_{P/O}$, $\bar{k}$ out)

VI.19 (a) $\bar{\alpha} = 150\pi\bar{j} - 10\bar{k}$ rad/s²
(b) $(10R - 25L)\bar{i} - [(900\pi^2 + 25)R + 10L]\bar{j}$
($\bar{i} = \bar{e}_{B/A}$, $\bar{k}$ out)

VI.21 $-3.87\bar{i} + 1.4709(10^4)\bar{k}$ N
($\bar{i}$ east, $\bar{j}$ north)

VI.23 $a = 2.53(10^{-3})$ ft/s² (south)
$\rho = 1.584(10^4)$ ft

VI.25 $-2m\,\omega_e\,gt\cos\lambda\,\bar{i}$
($\bar{i}$ east)

VI.27 (a) $\bar{v}_{B/A} = -161.3\bar{i}$ ft/s
$\bar{a}_{B/A} = -0.723\bar{j}$ ft/s²

(b) $(\bar{v}_{B/A})_{rel} = -152.3\bar{i}$ ft/s
$(\bar{a}_{B/A})_{rel} = 27.4\bar{j}$ ft/s²
$[\bar{i} = (\bar{e}_t)_A, \bar{j} = (\bar{e}_n)_A]$

VI.29 $(\bar{v}_{A/B})_{rel} = 625\bar{i} - 625\bar{j}$ m/s
$(\bar{a}_{A/B})_{rel} = -260\bar{i} - 304\bar{j}$ m/s²
($\bar{i}$ south, $\bar{j}$ east)

VI.31 (a) $\bar{a}_P = -21\bar{i} + 519\bar{j}$ m/s²
(b) $v_P = 13.22$ m/s
$\theta = 10.89°$ (cw from $\bar{i}$)
($\bar{i} = \bar{e}_R, \bar{j} = \bar{e}_\phi$)

VI.33 $\omega_{AB} = 57.73$ rad/s (ccw)
$\dot{\omega}_{AB} = 15795$ rad/s² (cw)

VI.35 $\bar{a}_E = -100\bar{i} - 100\bar{j}$ m/s²
$\rho = 0.354$ ft
($\bar{i}$ lt, $\bar{j}$ up)

VI.37 $\alpha_{AB} = 0.1519$ rad/s² (ccw)
$\alpha_{CD} = 0.0969$ rad/s² (cw)

VI.39 $\bar{a}_t = -\sin\theta\cos\theta\,\bar{j} + 4\cos\theta\,\bar{k}$ in./s²
$\bar{a}_n = -(4 + \sin^2\theta)\bar{i}$ in./s²
$|\bar{a}_t|_{max} = 4$ in./s², $|\bar{a}_n|_{max} = 5$ in./s²
($\bar{i} = \bar{e}_n$, $\bar{k}$ out)

VI.41 $\bar{\omega} = 3\bar{j} + 2\bar{k}$ rad/s
$\bar{\alpha} = 6\bar{i} - 2.67\bar{k}$ rad/s²
($\bar{i}$ rt, $\bar{j}$ up)

VI.43 0.3 m/s (dn)

VI.45 0

VI.47 400 ft/s² (in)

VI.49 $\bar{\omega} = -104.7\bar{i} - 10\bar{k}$ rad/s
$\bar{\alpha} = 1047\bar{j} - 2\bar{k}$ rad/s²
($\bar{i} = \bar{e}_{B/A}, \bar{j} = \bar{e}_{C/B}$)

VI.51 $\bar{\omega} = -125.66\bar{i} - 0.264\bar{k}$ rad/s
$\bar{\alpha} = 33.2\bar{j}$ rad/s²
($\bar{i}$ rt, $\bar{k}$ up)

VI.53 (a) $\bar{\omega} = -50\bar{i} + 10\bar{j}$ rad/s
$\bar{\alpha} = -100\bar{i} + 20\bar{j} + 500\bar{k}$ rad/s²
(b) $\bar{v}_C = -5\bar{k}$ m/s
$\bar{a}_C = -75\bar{i} - 125\bar{j} - 10\bar{k}$ m/s²
($\bar{i} = \bar{e}_{B/O}, \bar{j}$ up)

VI.55 $\bar{\omega} = -35.2\bar{j} + 0.40\bar{k}$ rad/s
$\bar{\alpha} = 14.01\bar{i}$ rad/s²
($\bar{i} = -\bar{e}_t$, $\bar{k}$ 60° above $\bar{e}_n$)

VI.57 $\bar{\omega} = -0.0667\bar{i} - 0.125\bar{k}$ rad/s
$\bar{\alpha} = 0.5\bar{i} + 0.0083\bar{j}$ rad/s²
($\bar{j}$ rt, $\bar{k}$ up)

VI.59 $\bar{\omega} = (0.5\sin\frac{\pi t}{5} - 30\pi)\bar{i} + 0.866\sin\frac{\pi t}{5}\bar{j}$ rad/s

$\bar{\alpha} = 0.1\pi\cos\frac{\pi t}{5}\bar{i} + 0.1732\pi\cos\frac{\pi t}{5}\bar{j} + 25.98\pi\sin\frac{\pi t}{5}\bar{k}$ rad/s²

($\bar{i}$ 60° lt of vert, $\bar{j}$ 30° rt of vert)

VI.63 $\bar{\omega} = -0.5\bar{\imath} + 0.3\bar{\jmath}$ rad/s
$\bar{\alpha} = -0.6\bar{\jmath}$ rad/s^2

VI.65 $\bar{v} = -1.80\bar{\imath} + 4.80\bar{\jmath} + 9.60\bar{k}$ m/s
$\bar{a} = 0.408\bar{\imath} - 0.804\bar{\jmath} - 0.024\bar{k}$ m/s^2

VI.67 (a) $\bar{v}_C = 5\bar{\imath} + 0.20\bar{\jmath} - 1.0\bar{k}$ m/s
$\bar{a}_C = -4\bar{\imath} + 148\bar{\jmath} - 1000\bar{k}$ m/s^2
(b) $\bar{v}_D = -5\bar{\imath} + \bar{\jmath} - 11\bar{k}$ m/s
$\bar{a}_D = -1010\bar{\imath} - 92\bar{\jmath}$ m/s^2
($\bar{\jmath}$ rt, $\bar{k}$ up)

VI.69 $\bar{\omega} = -5\bar{\imath} + 63.3\bar{\jmath} + 0.5\bar{k}$ rad/s
$\bar{\alpha} = -64.4\bar{\imath} - 4.3\bar{\jmath} - 311.7\bar{k}$ rad/s^2
($\bar{\jmath}$ 60° rt of vert, $\bar{k}$ 30° lt of vert)

VII.1 $-\frac{1}{4} mr^2\omega\bar{\imath}$

VII.3 $-0.5\omega_2\bar{\imath} - \omega_1\bar{k}$

VII.5 $0.10053\bar{\imath} + 0.16056\bar{\jmath}$ kg-m^2/s

VII.7 $\frac{1}{4}mr^2[1.732(\omega_1 - \omega_2)\bar{\imath} + 0.5(\omega_1 + \omega_2)\bar{\jmath}]$

VII.13 $I_{xx} = \frac{1}{8}m\,(b^2 + c^2)$
$I_{yy} = \frac{1}{8}m\,(2a^2 + c^2)$
$I_{zz} = \frac{1}{8}m\,(2a^2 + b^2)$

VII.15 $I_{xx} = I_{zz} = \frac{1}{8}m\,(3h^2 + a^2)$
$I_{yy} = \frac{1}{3}ma^2$

VII.17 $I_{xy} = \frac{1}{20}mab, \; I_{xz} = \frac{1}{20}\,mac$
$I_{yz} = \frac{1}{20}\,mbc$

VII.19 (a) $1.1052(10^5)$ kg/m^5
(b) $z_G = 0.2063$ m
(c) $I_{\hat{x}\hat{x}} = I_{\hat{y}\hat{y}} = 0.4244$ kg-m^2
$I_{\hat{z}\hat{z}} = 0.4419$ kg-m^2

VII.21 $\dfrac{3m}{20}\,(r^2 + 4h^2)$

VII.23 (a) $I_{xx} = I_{yy} = I_{zz}$
$= \frac{2}{5}m\left[\dfrac{5r^4h + 10r^3h^2 + 10r^2h^3 + 5rh^4 + h^5}{3r^2h + 3rh^2 + h^3}\right]$
(b) $I_{xx} = I_{yy} = I_{zz} = \frac{2}{3}\,mr^2$

VII.25 $I_{\hat{x}\hat{x}} = I_{\hat{y}\hat{y}} = 2.002$ kg-m^2, $I_{\hat{z}\hat{z}} = 3.497$ kg-m^2
($\hat{z}$ is axis of symmetry)

VII.27 (a) 2095 lb-s^2-ft
(b) 2038 lb-s^2-ft

VII.29 $z_G = 0.00641$ (lt of centroid of cylinder)
$I_{xx} = I_{yy} = 24.96$ kg-m^2
$I_{zz} = 10.98$ kg-m^2
(z is axis of symmetry)

VII.31 $I_{xx} = \frac{10}{3}\,ml^2$
$I_{yy} = \frac{22}{3}\,ml^2$
$I_{zz} = \frac{32}{3}\,ml^2$
$I_{xy} = -2ml^2$
$I_{xz} = I_{yz} = 0$

VII.33 $\frac{7}{16}\,mr^2$

VII.35 $\dfrac{-\sqrt{3}}{16}\,mr^2$

VII.37 0.11426 kg-m^2

VII.39 0

VII.41 $-\dfrac{100\pi}{3}\,\bar{k}$ rad/s^2

VII.43 (a) $\bar{F}_s = \frac{1}{6}\,ml\omega^2\,(\sin\theta\,\bar{\imath} - \cos\theta\,\bar{\jmath}) \pm mg\,\bar{k}$
$\bar{M}_s = \frac{1}{9}\,ml^2\omega^2 \sin\theta \cos\theta\,\bar{k} \pm \frac{1}{6}\,mgl\,\bar{\jmath}$
(b) $\bar{F}_s = (\frac{1}{6}\,ml\omega^2 \pm mg)(\sin\theta\,\bar{\imath} - \cos\theta\,\bar{\jmath})$
$\bar{M}_s = (\frac{1}{9}\,ml^2\omega^2 \sin\theta \cos\theta \pm \frac{1}{6}mgl \cos\theta)\bar{k}$
($\bar{\imath}$ along short leg of bar, $\bar{k}$ out)

VII.45 (a) $\bar{A} = -\bar{B} = 22.11\bar{I} + 7.29\bar{K}$ N
(b) $\bar{A} = -\bar{B} = 5.53\bar{I} + 7.29\bar{K}$ N
(a) & (b) $\dot{\omega} = -208$ rad/s^2
($\bar{J} = \bar{e}_{B/A}$, $\bar{K}$ up)

VII.47 (a) $\bar{H}_A = -93.76\bar{\imath} + 611.8\bar{k}$ lb-s-in.
$\dot{\bar{H}}_A = -468.8\bar{\jmath} - 1224\bar{k}$ lb-in.
(b) $\bar{H}_A = 77.72\bar{k}$ lb-s-in.
$\dot{\bar{H}}_A = -155.43\bar{k}$ lb-in.
($\bar{\imath} = \bar{e}_{B/A}$, $\bar{k}$ in)

VII.49 $(0.433\,\omega_1\omega_2 - 0.1083\omega_1^2)mr^2\bar{k}$

VII.51 $\bar{H}_A = -48.8\,mr^2\bar{\imath} + 278.2\,mr^2\bar{\jmath}$

VII.53 (a) $(\bar{H}_C)_A = -0.4\,mr^2\omega_1\bar{\imath} + (0.4mr^2 + ml^2)\Omega\bar{\jmath}$
$(\dot{\bar{H}}_C)_A = 0.4mr^2\omega_1\Omega\,\bar{k}$
$(\bar{H}_C)_B = 0.4\,mr^2\omega_2\bar{\imath} + (0.4mr^2 + ml^2)\Omega\,\bar{\jmath}$
$(\dot{\bar{H}}_C)_B = -0.4mr^2\omega_2\Omega\bar{k}$
(b) $0.4r^2(\omega_1 - \omega_2)\Omega\bar{k}$
($\bar{\imath} = \bar{e}_{B/C}$, $\bar{\jmath}$ up)

VII.55 $\dot{\omega} = -9376$ rad/s^2
$\bar{A} = 167\bar{J} - 3563\bar{K}$ N
$\bar{B} = 9\bar{J} - 188\bar{K}$ N
($\bar{I} = \bar{e}_{B/A}$, $\bar{J}$ up)

VII.57 $\dot{\omega} = -8.220$ rad/s^2
$\bar{A} = -3654\bar{I} - 66\bar{K}$ N
$\bar{B} = -5115\bar{I} - 96\bar{K}$ N
($\bar{J} = \bar{e}_{B/A}$, $\bar{K}$ rt)

VII.59 $N_A = N_B = 68.6$ newtons
$f_B = -f_A = -v^2$ newtons

VII.61 (a) $-1.302\bar{\jmath}$ lb-ft
(b) $1.302\bar{\imath}$ lb-ft

VII.63 $\Sigma\bar{F} = m_1\bar{a}_{G_1} + m_2\bar{a}_{G_2}$
$\Sigma\bar{M}_P = \dot{\bar{H}}_{G_1} + \dot{\bar{H}}_{G_2} + \bar{r}_{G_1/P} \times m_1\bar{a}_{G_1} + \bar{r}_{G_2/P} \times m_2\bar{a}_{G_2}$

VII.65 (a) 1.190 N-m (opposite $\bar{v}_{car}$)
(b) $\Delta N_{outer} = -\Delta N_{inner} = 1.190$ N

VII.67 (a) $-m\,[k_1^2\omega_1\omega_2 \sin\theta + (k_1^2 - k_2^2)\omega_2^2 \sin\theta \cos\theta]\bar{k}$
[$\bar{\imath}$ at θ rt of vert, $\bar{\jmath}$ at $(90° - \theta)$ lt of vert]

VII.69 $\bar{F}_A = 2.49(10^3)\bar{i} + 9.04(10^3)\bar{j} - 1.12(10^3)\bar{k}$ lb
$\bar{M}_A = 3.58(10^4)\bar{j} + 2.19(10^5)\bar{k}$ lb-ft
$(\bar{i} = \bar{e}_{G/A}, \bar{k}$ out)

VII.71 394 N

VII.73 $\bar{M}_A = 18.70\bar{k}$ lb-in.
$\bar{F}_A = 0.526\bar{i} + 1.973\bar{j}$ lb
$(\bar{i} = \bar{e}_{G/A}, \bar{k}$ out)

VII.75 $N = m(g + \frac{1}{2}r\omega_2^2) > 0$ always

VII.77 $(\omega_1)_{min} = 3.08$ rad/s

VII.79 (a) $\cos\theta = \dfrac{(3r^2 - 4l^2)\omega_1\omega_2 - 4gl}{(r^2 - 4l^2)\omega_2^2}$

(b) $\omega_2 = \dfrac{4gl}{3r^2 - 4l^2}\,\omega_1$

VII.81 (a) $(0.4r^2 + L^2)\ddot{\theta} + gL\sin\theta - L^2\omega_1^2\cos\theta\sin\theta = 0$

(b) $\cos\theta = \dfrac{g}{L\omega_1^2}$

(c) $\omega_1 < \left(\dfrac{g}{l}\right)^{1/2}$

VII.83 $\omega_1^2 L(\frac{1}{2} + \frac{7}{12}\sin\theta)\cos\theta = mg\sin\theta$

VII.85 $\bar{M}_G = -174.0\bar{k}$ N-m
$\bar{F}_G = 5767\bar{k}$ N
$(\bar{i} = \bar{e}_t, \bar{k}$ up)

VII.87 $\bar{M}_G = -3.15(10^8)\bar{k}$ N-m
$\bar{F}_G = 2.45(10^8)\bar{k}$ N

VII.89 (a) 41.09°
(b) 7.88 rad/s
(c) 8.43 rad/s

VII.91 $\dot{\omega}_2 = 45.5$ rad/s²
$\ddot{\theta} = 27.5$ rad/s²

VII.93 $\dot{\omega}_2 = 1355$ rad/s²
$\ddot{\theta} = 25.5$ rad/s²

VII.95 24.34 J

VII.97 7.32 rad/s

VII.99 0.285 lb-ft

VII.101 4.86 m/s

VII.103 (a) 16.16 m/s
(b) $0.80\bar{j}$ rad/s

VII.105 $x_P + y_P = \frac{2}{3}l$

VII.107 $\Delta\bar{v}_G = 5\bar{i}$ m/s
$\bar{\omega}_f = 27.5\bar{j} + 7.5\bar{k}$ rad/s

VII.109 $\bar{v}_G = -3.680\bar{k}$ m/s
$\bar{\omega} = 33.0\bar{i} + 50.1\bar{j}$ rad/s
$(\bar{k}$ up, $\bar{j}$ lt along flat edge)

VII.111 $\dot{\theta} = 8.705$ rad/s
$\omega_1 = 2.038$ rad/s

VIII.1 $k_2/k_1 = 1/4$

VIII.3 $m = 2.56$ kg
$k = 650$ kN/m

VIII.5 (a) 0.10 m
(b) $\ddot{s} + 4(10^3)s = 0.4(10^3)$
(c) $\ddot{x} + 4(10^3)x = 0$
(d) 63.2 rad/s
(e) parts (a) and (b)

VIII.7 (a) $2\pi\left(\dfrac{\delta_{st}}{g}\right)^{1/2}$

(b) δ_{st}

VIII.9 $\ddot{R} + \left(\dfrac{k}{m} - \Omega^2\right)R = \dfrac{k}{m}R_0$

$\omega_n = \left(\dfrac{k}{m} - \Omega^2\right)^{1/2}$

$\Omega < \left(\dfrac{k}{m}\right)^{1/2}$

VIII.11 $m\ddot{x} + k\left[1 + \dfrac{(T - kl)}{k(x^2 + L^2)^{1/2}}\right]x = 0$

$m\ddot{x} + \dfrac{T}{L}x = 0$, for $x << L$

VIII.13 (a) 0.355 seconds
(b) For $\theta = -2.25°$, $t = 0.2364$ seconds
$\bar{v} = 0.151\bar{e}_\theta$ m/s
$\bar{a} = -0.045\bar{e}_R + 0.385\bar{e}_\theta$ m/s²
For $\theta = 2.25°$, $t = 0.4729$ seconds
$\bar{v} = 0.151\bar{e}_\theta$ m/s
$\bar{a} = -0.045\bar{e}_R + 0.385\bar{e}_\theta$ m/s²

VIII.15 (a) 1961 N/m
(b) $x = -0.020\cos 22.14t$

VIII.17 (a) 5.0 ft
(b) $x = C\sin(143.5\,t + \phi)$
(c) $C = 0.1251$ ft
$\omega_n = 143.5$ rad/s
$\phi = 0.050$ rad

VIII.19 (a) 0.0349 seconds
(b) 3.90 m/s
(c) $C = 54.9$ mm
$\omega_n = 60$ rad/s
$\phi = 0.932$ rad

VIII.21 $\omega_d = 209.4$ rad/s
$\beta = 0.036$
$x_1 = 4.48$ mm

VIII.23 (a) $100\{1 - \exp[-4\pi\beta/(1 - \beta^2)^{1/2}]\}\%$
(b) T + V decreases by a constant % in each cycle

VIII.27 489 N-s/m

VIII.29 (a) $3.065\sin(9.787t)\exp(-7.363t)$
(b) $t = 0.0946$ seconds, $x = 1.2205$ m
(c) 2.45 m

VIII.31 $x = -129.3 \sin 31.62t + 3.21 \sin 8t$
 $\quad + 4.03 \sin 16t$ mm

VIII.33 (a) $x = 0.417(1 - \cos 13.89t)$ ft
 (b) $x = 0.513 \sin (13.89t - 0.663)$ ft

VIII.35 (a) 44.29 rad/s
 (b) 1.067 mm

VIII.37 (a) $\Omega > (2g/3R)^{1/2}$
 (b) not possible

VII.39 (a) $\ddot{y} + 160\, y = 6.40 \sin \dfrac{\pi vt}{2}$

 (b) 8.05 m/s
 (c) 73.8 mm

VIII.41 (a) $\Omega = \omega_n(1 - 2\beta^2)^{1/2}$
 (b) $D = F/[2k\beta(1 - \beta)]$
 $\quad \psi = \tan^{-1}[(1 - 2\beta^2)^{1/2}/\beta]$
 $\quad \beta > \sqrt{2}/2$

VIII.43 (a) $x = D \sin (\Omega t + \alpha + \psi)$ where:
 $\quad D = 0.005[(1 - \Omega^2/2000)^2 + \Omega^2/2500]^{-1/2}$
 $\quad \alpha = \tan^{-1}(0.02\,\Omega)$
 $\quad \psi = \tan^{-1}[0.02\,\Omega\,(1 - \Omega^2/2000)^{-1}]$
 (b) $\Omega = 6.5$Hz: $a_{max} = 10.0$ m/s²
 $\quad \Omega = 7.5$Hz: $a_{max} = 11.7$ m/s²

VIII.45 $x = x_0 \cos(\Omega t + \psi)$ where
 $\dfrac{x_0}{\delta_0} = \dfrac{\Omega^2}{\omega_n{}^2}\left[\left(1 - \dfrac{\Omega^2}{\omega_n{}^2}\right)^2 + 4\beta^2\,\dfrac{\Omega^2}{\omega_n{}^2}\right]^{-1/2}$
 $\psi = \tan^{-1}\left[2\beta\,\dfrac{\Omega}{\omega_n}\left(\dfrac{\Omega^2}{\omega_n{}^2} - 1\right)^{-1}\right]$

VIII.47 $\ddot{\theta} + \dfrac{8g}{3\pi R}\,\theta = 0$

VIII.49 $\ddot{\theta} + 7.5(10^3)\theta = 0$

VIII.51 $ml^2\ddot{\theta} + c\,a^2\dot{\theta} + k\,a^2\theta = 0$

VIII.53 $\ddot{\theta} + \dfrac{2g}{3mr}\,\theta = 0$

VIII.55 19.65 rad/s

VIII.57 9.90 rad/s

VIII.59 $I_{zz} = \dfrac{\tau^2}{4\pi^2}\,k$

VIII.61 $(\omega_n)_{el} = \left[\dfrac{6}{5}\left(\dfrac{k}{m} + \dfrac{g}{l}\right)\right]^{1/2}$
 □stable for all k
 $(\omega_n)_{rt}\left[\dfrac{6}{5}\left(\dfrac{k}{m} - \dfrac{g}{l}\right)\right]^{1/2}$
 □unstable when $k < mg/l$

VIII.63 $R^{-1} = 3.29(10^{-8})[0.280 \cos \phi + 1]$
 $v_A = 1.548(10)^4$ ft/s
 $v_P = 2.75(10)^4$ ft/s

VIII.65 $R^{-1} = 7.32(10^{-5}) \cos (\phi - 89.04°) + 1.333(10^{-4})$
 $\quad$ 1/km
 $e = 0.549$, closed orbit

VIII.67 18.75°

VIII.69 (a) 1.695
 (b) 2.49(10⁴) km
 (c) 31.69°

VIII.71 51.46°

VIII.73 (a) 2.398(10⁴) mile²/s
 (b) −6.790 mile²/s
 (c) 0.380
 (d) 2.469 mile/s
 (e) 9712 miles

VIII.75 $\bar{v} = -11.34\bar{e}_R + 8.63\bar{e}_\phi$ km/s

VIII.77 (a) $3.82\bar{e}_\phi$ km/s
 (b) $\pm 2.55\bar{e}_R + 6.37\bar{e}_\phi$ km/s

VIII.79 $\dot{v} = -5.35(10^3)$ mile/hr²
 $\rho = 5.18(10^4)$ miles

VIII.81 (a) $\dot{v} = -2.85(10^{-3})$ km/s²
 $\quad \rho = 2.06(10^4)$ km
 (b) $\dot{v} = -1.273(10^{-3})$ km/s²
 $\quad \rho = 9.90(10^4)$ km

VIII.83 $\bar{v} = 3.67\bar{e}_R + 3.65\bar{e}_\phi$ km/s

VIII.85 53.90 hrs

VIII.87 178.75 days

VIII.89 3.452 hours

VIII.91 85.08°

VIII.93 (a) 1000 N
 (b) 2500 N

VIII.95 (a) $(20 + 0.12434\,t)\dot{v} + 0.12434\,v = 2.984$
 (b) at $t = 0.5$ s, $v = 0.0744$ ft/s
 $\quad$ at $t = 1.0$ s, $v = 0.1483$ ft/s

VIII.97 50.98 lb more than tang comp of weight

VIII.99 (a) $\sigma[g(l - s) + v^2]$
 (b) σgs

VIII.101 $R = 4516$ N (rt)
 $M = 0$

VIII.103 12.51 m

VIII.105 (a) 9.09 lb
 (b) 11.57 lb

VIII.107 $|\bar{R}| = 345.1$ N, midpoint of vane along
 outward normal

VIII.109 $\bar{A} = 621.3\bar{I} + 3750\bar{J}$ N
 $\bar{B} = 8177\bar{J}$ N
 ($\bar{I}$ rt, $\bar{J}$ up)

VIII.111 $\bar{A} = 6903\bar{I}$ N

$\bar{C} = -3170\bar{I} + 12,492\bar{J}$ N

($\bar{I}$ rt, $\bar{J}$ up)

VIII.113 $3.14(10^4)$ N

VIII.115 $0.2131v^3$ ft-lb/s

increases monotonically

VIII.117 40.47 rad/s

VIII.119 (a) for $\dot{m} = 30/g$, $T = 9325$ lb

(b) for $\dot{m} = 60/g$, $T = 18,650$ lb

VIII.121 $\dot{m}_P = \dfrac{m_0\,(\dot{v} + g)}{(\dot{v} + g)t + u}$

VIII.123 single stage: $h_{burnout} = 1.1205(10^5)$ m

two stage: $h_{burnout} = 2.497(10^5)$ m

VIII.125 $T_{engine} = 4555$ lb (opp $\bar{v}$)

VIII.127 $\bar{a} = 3.37\bar{I} - 0.92\bar{J}$ m/s^2

($\bar{I}$ rt, $\bar{J}$ up)

INDEX

CONVERSION FACTORS

ALPHABETICAL LIST OF UNITS
U.S.–BRITISH UNITS TO SI UNITS

To convert from	to	Multiply by
atmosphere	pascal (Pa)	$1.013\ 25\ (10^5)$
bar	pascal (Pa)	$1.000\ 000\ (10^5)$*
British thermal unit	joule (J)	$1.055\ 056\ (10^3)$
Btu/hour	watt (W)	$2.930\ 711\ (10^{-1})$
centimeter of mercury (0°C)	pascal (Pa)	$1.333\ 22\ (10^3)$
dyne	newton (N)	$1.000\ 000\ (10^{-5})$*
dyne-centimeter	newton-meter (N-m)	$1.000\ 000\ (10^{-7})$*
dyne/centimeter^2	pascal (Pa)	$1.000\ 000\ (10^{-1})$*
erg	joule (J)	$1.000\ 000\ (10^{-7})$*
erg/second	watt (W)	$1.000\ 000\ (10^{-7})$*
foot	meter (m)	$3.048\ 000\ (10^{-1})$*
foot^2	meter^2 (m^2)	$9.290\ 304\ (10^{-2})$*
foot^3	meter^3 (m^3)	$2.831\ 685\ (10^{-2})$
foot/second	meter/second (m/s)	$3.048\ 000\ (10^{-1})$*
foot-pound	joule (J)	$1.355\ 818$
foot-pound/second	watt (W)	$1.355\ 818$
foot/second^2	meter/second^2 (m/s^2)	$3.048\ 000\ (10^{-1})$*
gallon (U.S. liquid)	meter^3 (m^3)	$3.785\ 412\ (10^{-3})$
gram/centimeter^3	kilogram/meter^3 (kg/m^3)	$1.000\ 000\ (10^3)$*
horsepower (550 ft-lb/s)	watt (W)	$7.456\ 999\ (10^2)$
inch	meter (m)	$2.540\ 000\ (10^{-2})$*
inch^2	meter^2 (m^2)	$6.451\ 600\ (10^{-4})$*

*Denotes an exact quantity.